Euclidean and Non-Euclidean Geometries

Euclidean and Non-Euclidean Geometries

DEVELOPMENT AND HISTORY

FOURTH EDITION

MARVIN JAY GREENBERG
EMERITUS, UNIVERSITY OF CALIFORNIA

W. H. Freeman and Company
NEW YORK

Senior Publisher: Craig Bleyer
Senior Acquisitions Editor: Terri Ward
Associate Editor: Brendan Cady
Assistant Editor: Laura Capuano
Senior Executive Media Editor: Roland Cheyney
Executive Marketing Manager: Robin O'Brien
Project Editor: Vivien Weiss
Design Manager: Vicki Tomaselli
Text Designer: Marsha Cohen
Illustrations: Fine Line
Senior Illustration Coordinator: Bill Page
Production Manager: Julia DeRosa
Composition: Matrix Publishing Services
Printing and Binding: RR Donnelley

Library of Congress Control Number: 2007928758
ISBN-13: 978-0-7167-9948-0
ISBN-10: 0-7167-9948-0

Printed in the United States of America

First printing

W. H. Freeman and Company
41 Madison Avenue
New York, NY 10010
Houndmills, Basingstoke RG21 6XS, England
www.whfreeman.com

*To the memory of my main mathematical friends
and teachers, with deep gratitude:
Emil Artin, Errett Bishop, Shiing-shen Chern, Claude Chevalley,
Paul Cohen, Karel De Leeuw, Samuel Eilenberg, Azriel Evyatar,
Daniel Gorenstein, Alexandre Grothendieck, Serge Lang,
Max Rosenlicht, Gideon Schwarz, Edwin Spanier,
and Oscar Zariski.*

The moral of this book is: Check your premises.

Contents

CHAPTER 6 *The Discovery of Non-Euclidean Geometry* 239

CHAPTER 7 *Independence of the Parallel Postulate* 289

This book presents the discovery of non-Euclidean geometry and the subsequent reformulation of the foundations of Euclidean geometry as a suspense story. The mystery of why Euclid's parallel postulate could not be proved remained unsolved for more than two thousand years, until the discovery of non-Euclidean geometry and its Euclidean models revealed the impossibility of any such proof. This discovery shattered the traditional conception of geometry as the true description of physical space. Mainly through the influence of David Hilbert's *Grundlagen der Geometrie,* a new conception emerged in which the existence of many equally consistent geometries was acknowledged, each being a purely formal logical discipline that may or may not be useful for modeling physical reality. Albert Einstein stated that without this new conception of geometry, he would not have been able to develop the theory of relativity (see Einstein, 1921, Chapter I). The philosopher Hilary Putnam stated that "the overthrow of Euclidean geometry is the most important event in the history of science for the epistemologist." Chapter 8 of this book reveals the philosophical dilemma that persists to this day.

This text is useful for several kinds of students. Prospective high school and college geometry teachers are presented with a rigorous treatment of the foundations of Euclidean geometry and an introduction to hyperbolic geometry (with emphasis on its Euclidean models). General education and liberal arts students are introduced to the history and philosophical implications of the discovery of non-Euclidean geometry (for example, the book was used very successfully as part of a course on scientific revolutions at Colgate University). Mathematics majors are given, in addition, detailed instruction in transformation geometry and hyperbolic trigonometry, challenging exercises, and a historical perspective that, sadly, is lacking in most mathematics texts.

A unique feature of this book is that some new results are developed *in the exercises* and then built upon in subsequent chapters. My experience teaching from earlier versions of this text convinced me that this method is very valuable for deepening students' understanding (students not only learn by doing, they enjoy developing new results on their own). *If students do not do a good number of exercises, they will have difficulty following subsequent chapters.* There are two sets of exercises for the first six chapters; the "major" exercises are the more challenging ones, which all students should attempt, but which mathematics majors are more likely to solve. This distinction is dropped in the last four chapters; most of the exercises for Chapters 7, 9, and 10 are "major," whereas the exercises for Chapter 8 are unusual for a mathematics text, consisting of historical and philosophical essay topics. Hints are given for most of the exercises. A solutions manual is available for instructors. All chapters also have projects at the end for further research in the library and/or on the Web.

I have used the development of non-Euclidean geometry to revive interest in the study of Euclidean geometry. I believe that this approach makes a traditional college course in Euclidean geometry more interesting: To identify the flaws in various attempted proofs of the Euclidean parallel postulate, we carefully examine the axiomatic foundations of Euclidean geometry; to prove the relative consistency of hyperbolic geometry, the properties of inversion in Euclidean circles are studied; to justify János Bolyai's construction of the limiting parallel rays, some ideas from projective geometry (cross-ratios, harmonic tetrads, perspectivities) are introduced.

I have used modified versions of Hilbert's axioms for Euclidean geometry, instead of the ruler-and-protractor postulates customary in current high school texts. In a rigorous, historically motivated presentation of the foundations of geometry, it is vital to separate the purely geometric ideas of Euclid from the numerical methods that came much later, in the seventeenth century, with Descartes and Fermat; even then, Descartes defined his numbers and five algebraic operations upon them geometrically, in terms of constructions on segments. We explain what he did in a section of Chapter 1 new to this edition.

Emphasis on Elementary Geometry
Our emphasis throughout most of this book is on *elementary geometry*, the geometry of lines and circles—i.e., the geometry of *straightedge-and-compass constructions*. For elementary geometry, it is not necessary to use the enormous power of the real number system. In fact, a much

smaller number system for doing elementary analytic geometry is the field **K** of *constructible numbers*[1]—all those numbers obtained from rational numbers by repeatedly applying Descartes' five algebraic operations. It is the constructible number system that enables proofs of the impossibility of solving the classical straightedge-and-compass construction problems in a Euclidean plane—trisecting arbitrary angles, duplicating the side of arbitrary cubes, and squaring all circles (described in another new section of Chapter 1).

There are a few places in elementary geometry where a continuity argument is needed, such as in the proof of Euclid's very first proposition. It turns out that only "quadratic continuity" is needed—analytically, that means the existence of positive solutions to certain quadratic equations, solutions which exist in the field of constructible numbers. Geometrically, it means assuming elementary principles such as the line-circle principle (a line that passes through a point inside a circle intersects the circle in two points) or the circle-circle principle (a circle that passes through points inside and outside another circle intersects that other circle in two points). We discuss those principles and their relationship in Chapter 3.

Another assumption that appears surreptitiously in Euclid's *Elements* and that was generally accepted by later geometers is *Archimedes' axiom*. We will discuss very carefully how that axiom is used and how it can either be dispensed with or replaced with the weaker, purely geometric *axiom of Aristotle*. Hilbert called Archimedes' axiom "the axiom of measurement" because it is used to measure segments and angles by real numbers. We will indicate how that is done in the discussion after Theorem 4.3, Chapter 4, but we emphasize that for developing elementary geometry it is not necessary to introduce Archimedes' axiom. The only reason we bring it up in Chapter 4 is to simplify our language and to skip the technicalities involved in circumventing it. An important advantage in avoiding the use of Archimedes' axiom is that geometrical models can be exhibited in which *infinitesimal* elements exist, and they can be used decisively to prove that certain geometric statements are *unprovable* from the given axioms (see the note for advanced students in Chapter 4, as well as Appendix B).

For readers interested in the technicalities omitted here, please refer to Robin Hartshorne's superb treatise *Geometry: Euclid and Beyond* (Springer, 2000). In fact, we will repeatedly refer to Hartshorne's treatise

---

[1] Called the *surd field* in Moise (1990). His text will be referred to often.

for technical details and results that are at a more advanced level than this text, so all citations of his work that do not specify a date are references to his treatise. I am very grateful to Robin for all our interactions about geometry and mathematics history. Hartshorne has the same philosophy as I have of doing elementary geometry without real numbers. It was also Hilbert's philosophy in his later writings (see the quotations from him in Appendix B); in fact, the title of this book might well be "Elementary Euclidean and Non-Euclidean Geometries According to Hilbert."

Our main topic is Euclid's fifth postulate (an equivalent postulate is often referred to as "the Euclidean parallel postulate")—what can be learned from some of the many failed attempts over two thousand years to prove it from his other postulates, and the non-Euclidean geometry which results from replacing it with Hilbert's hyperbolic parallel postulate. See the Introduction for a detailed description of our development.

▓▓ Terminology

Terminology and notation throughout the book are reasonably standard. I have followed W. Prenowitz and M. Jordan in using the term "neutral geometry" for the part of Euclidean geometry that is independent of the parallel postulate (the traditional name "absolute geometry" misleadingly implies that all other geometries depend on it). I have introduced the names "asymptotic" and "divergent" for the two types of parallels in hyperbolic geometry; I consider these a definite improvement over the welter of names in the literature. The theorems, propositions, and figures are numbered by chapter; for example, Theorem 4.1 is the first theorem in Chapter 4. Such directives as "see Coxeter (1968)" refer to the Bibliography at the back of the book (the Bibliography is arranged topically rather than strictly alphabetically).

▓▓ Suggested Curricula for Different Courses

Here are some suggested curricula for different courses:

1. A one-term course for prospective geometry teachers and/or mathematics majors, *with students of average ability.* Cover Chapters 1–7, skipping the notes for advanced students, skimming over the complicated axiomatic development in Chapter 6 to emphasize more the visualization possible via the Euclidean models in Chapter 7, then adding Chapter 8 if there is time. In assigning exercises, omit the Major Exercises (except possibly for Chapter 1); omit most of the

Exercises on Betweenness from Chapter 3; omit Exercises 22–30 and
33 from Chapter 4; omit Exercises 10–18 from Chapter 5; and assign
only the Review Exercise, Exercises K-1, K-2, K-3, K-5, K-11, K-12,
K-17, K-18, and K-20, and selected P-exercises from Chapter 7.

2. A one-term course for prospective geometry teachers and/or mathematics majors, *with better than average students*. Add to the cur-
   riculum of (1) the remainder of Chapter 7 and many of the exercises omitted in (1). These students should at least browse through
   Chapter 10 and study as much of it as time permits, for it is quite
   concrete.

3. A one-term course for general education and/or liberal arts students.
   The core of this course would be Chapters 1, 2, and 5, the first three
   sections of Chapters 6 and 7, and all of Chapter 8. In addition, the
   instructor should selectively discuss material from Chapters 3–6
   (such as Hilbert's axioms, the Saccheri–Legendre theorem, and some
   of the theorems in hyperbolic geometry), but should not impose too
   many proofs on these students. The essay topics of Chapter 8 are
   particularly appropriate for such a course.

4. A two-term course for mathematics majors. Cover as much of the
   book as time permits. If only one term is available, assign the rest
   of the book to interested mathematics majors for independent study.

5. Mathematics majors and other advanced students may feel that they
   know Euclidean geometry, know how to do proofs, and therefore
   want mainly to learn some non-Euclidean geometry. I recommend
   learning the axioms for a Hilbert plane in Chapter 3, carefully studying the section Axioms of Continuity, and doing the exercises related to that section; they can skim the rest of the first three chapters. Chapter 4 should be completely worked through if they wish
   to approach the subject axiomatically without bringing in real numbers; otherwise, this chapter can also be skimmed once they have
   mastered the material on Saccheri and Lambert quadrilaterals, as
   can Chapter 5 on the history of attempts to prove Euclid V. These
   students can then get to work on Chapters 6, 7, and 10 about hyperbolic geometry, noting the side remarks about elliptic geometry.
   If there is time they can go back to Chapter 9 to study the motions
   in a hyperbolic plane and their algebraic description in the Poincaré
   upper half-plane model.

Thus this book is a resource for a wide variety of students, from
the naive to the sophisticated, from the nonmathematical-but-educated
to the mathematical wizards.

The late Errett Bishop once taught a liberal arts course in logic during which he realized the questionable nature of classical logic and wrote a book about doing mathematical analysis constructively. My own book has evolved from a liberal arts course in geometry that I taught. I am very pleased by the warm reception accorded earlier editions of this book for its unusual combination of rigor and history. It indicates that there is a real need to "humanize" mathematics texts and courses. For example, when I taught calculus to a large class, I was astonished at how much livelier the students (mainly nonmathematicians) became after they researched and then wrote essays about the history of calculus (many were fascinated by the strange personality of Isaac Newton), about the relevance of calculus to their own fields, and about their fear of this awesome subject. Also, such essays provide good practice in improving writing skills, which many students need. Instructors can assign essays from the Projects at the end of Chapters 1–6 and the topics in Chapter 8.

New to This Edition
The extensive improvements in this fourth edition are as follows: In **Chapter 1,** the introductory sections have been rewritten with more ancient history. In the section The Power of Diagrams, where simple diagrams indicating proofs of the Pythagorean theorem are shown, the distinction between "content" and "area" is briefly explained in terms of dissection. Three new sections on (1) constructions (including "neusis" constructions), (2) Descartes' analytic geometry, and (3) the number π have been added. Almost all the chapters now end with summaries entitled Conclusion. A new major exercise and six new projects have been added to Chapter 1.

In **Chapter 2,** the sections on logic have been improved. The rules of generalization and specification are introduced, as are the rules for equality (Logic Rule 12). There is a brief new section on the historical development of mathematical logic. For beginners, I show how one might find a proof for Proposition 2.2 of incidence geometry; teaching students how to construct proofs is a major pedagogical issue! In the section on models, I mention how Kepler did not believe that the regular heptagon existed because it has no straightedge-and-compass construction; I added a project to report on Viète's neusis construction of it. In the section on consistency, I added a note for advanced students about that, including mention of Gödel's theorems. Another new section for advanced students discusses the problem of consistency; that discussion is continued in the comments for Chapter 2 in the revised

Instructors' Manual, along with references on the controversy about potential versus completed infinity. The section on affine and projective planes has been expanded. It includes some material previously left as exercises, such as a proof of the principle of duality (the first meta-mathematical theorem in history). A brief section on the history of projective geometry has been added. There are also many new exercises and projects about projective planes, including the finite ones, which are admittedly a digression from the main topic but which are extremely interesting.

The content of **Chapter 3** remains mostly the same, although I have rewritten parts of it. The proof of the important crossbar theorem is now in the text instead of being left as an exercise. I have accepted Hartshorne's terminology of a *Hilbert plane* as a model of our thirteen axioms of incidence, betweenness, and congruence, and his definition of a *Euclidean plane* as a Hilbert plane also satisfying Hilbert's Euclidean axiom of parallelism and the circle-circle continuity principle. Dedekind's axiom is *not* assumed in this edition, so a Euclidean plane could be coordinatized by any Euclidean field, not just the real numbers. This permits the new example of the constructible Euclidean plane coordinatized by the field **K** of constructible numbers. I added notes for advanced students on the relative consistency of plane Euclidean geometry and on the existence of certain geometric sets. There are new major exercises showing the equivalence of the circle-circle continuity principle with the converse to the triangle inequality. I also added a major exercise about *taxicab geometry*. The old projects have been scrapped and six new ones added, including some about Pythagorean ordered fields.

There have been major revisions to **Chapter 4,** Neutral Geometry. The first four and sixth (now fifth) sections are substantially the same, except that the proof of the existence of midpoints has been moved out of the exercises into the text, and the discussion of Theorem 4.3 on measurement explains why measurement is convenient but not mathematically necessary. The old fifth section, which used Archimedes' axiom to prove the Saccheri–Legendre theorem, has been replaced with the completely new sixth section entitled "Saccheri and Lambert Quadrilaterals," in which the theory of those quadrilaterals is developed without using Archimedes' axiom. Some of this material previously was in the exercises of old Chapter 5 and in the text of old Chapter 6, but I've added important results which were not in the previous editions. The main one is the *uniformity theorem* (whose proof is left for the new major exercises), which states that there are three distinct types of

Hilbert planes: those which satisfy the acute angle hypothesis, the obtuse angle hypothesis, or the right angle hypothesis (called *semi-Euclidean*), respectively. The section Angle Sum of a Triangle is also much improved. It includes my new non-obtuse-angle theorem, which states that the obtuse angle hypothesis cannot hold in a Hilbert plane satisfying Aristotle's axiom—the angle sum of every triangle must be ≦180°. This is a stronger result than the Saccheri–Legendre theorem, which is mentioned but whose proof is now left for the exercises. In the note for advanced students about non-Archimedean geometries, Dehn's examples using infinitesimals are given to show that a semi-Euclidean plane need not be Euclidean and that Hilbert planes satisfying the obtuse angle hypothesis do exist. The exercises have been augmented to further the understanding of the new material. The old projects have been scrapped, replaced with six new ones intended for advanced students.

In **Chapter 5,** a new section has been added about Clavius' axiom that equidistant curves are lines. This axiom is shown to be equivalent to the plane being semi-Euclidean. In the section on Clairaut's axiom that rectangles exist, that axiom is also shown to be equivalent to the plane being semi-Euclidean; my new theorem named after Proclus is proved, asserting that a semi-Euclidean plane satisfies Hilbert's Euclidean axiom of parallelism if and only if it satisfies Aristotle's angle unboundedness axiom—thus Aristotle's axiom is a *missing link*. In the section on Wallis, his long dispute with the philosopher Hobbes is mentioned. In both this chapter and the adjacent ones, much more attention and credit are given to Saccheri for his remarkable development of elementary hyperbolic geometry so far ahead of his time. There are fewer exercises in Chapter 5 now because many of the old exercises have been covered in the text. Three of the old projects have been scrapped, replaced by two new projects, one of which asks for a report on the theory of similar triangles developed in Hartshorne's treatise without using real numbers; the approach with real numbers is still given in our exercises.

Chapter 6 has been thoroughly revised. The early sections on the history of the discovery of hyperbolic geometry have small but significant improvements (e.g., the revelation that János Bolyai left 14,000 pages of mathematical notes!). I call your attention to the new historical references mentioned in the footnotes. The sections that develop hyperbolic geometry are very different now because we no longer assume Dedekind's axiom and the axiomatic development is more subtle. Most important is the new discussion of Hilbert's hyperbolic axiom of parallelism—the existence of two limiting parallel non-opposite rays emanating from a point P not on a line l and situated symmetrically about the perpendicular from P to l. We do show (as in previous editions) how that existence can be proved from the negation of Hilbert's Euclidean axiom of parallelism when Dedekind's axiom is assumed. I also mention without proof my "advanced theorem" showing that instead of Dedekind's axiom, one need only assume the line-circle continuity principle and Aristotle's axiom in order to prove existence of limiting parallel rays (and those conditions are necessary, unlike Dedekind's axiom). The lengthy proofs of all the basic theorems in elementary hyperbolic geometry from Hilbert's axioms are not provided in the body of the chapter—that work is only indicated in the exercises and major exercises, with a few references to Hartshorne. For a student learning hyperbolic geometry for the first time, it is discouraging to struggle through the purely axiomatic development (in my experience), so I recommend devoting more time and effort to studying the Euclidean models in Chapter 7. At the end of Chapter 6, four of the old projects have been scrapped and replaced with three new advanced ones.

In **Chapter 7,** I have rewritten the first section discussing the relative consistency of hyperbolic geometry and the impossibility of proving the Euclidean parallel postulate in neutral geometry. Then I've added a new section about the contributions of Eugenio Beltrami, whose work in getting hyperbolic geometry accepted by the mathematical community, and incidentally proving consistency, has been underestimated in the past. The next four sections remain basically the same, though I have improved the biographical information about Klein and Poincaré. There are significant improvements in the section now entitled "Inversion in Circles, Poincaré Congruence." I introduce Hartshorne's *multiplicative length* for segments in the Poincaré models, which will be shown in Appendix B to be the key to important new results in elementary hyperbolic geometry. I give a second verification of Axiom C-1 for the P-model which does not appeal to continuity. I added a new Proposition 7.12 showing that a P-circle is a Euclidean circle, and I show how to construct its P-center and its E-center. In the final section, I added a proof of the Bolyai–Lobachevsky formula for the Klein model, simpler than the proof given earlier for the Poincaré model. The exercises for Chapter 7 provide a mini-course in advanced Euclidean geometry. Some of the exercises have been expanded or replaced: For example, K-19(c) exhibits a *new special point* of a triangle in the hyperbolic plane; K-20 is a new exercise about *symmetric parallelograms*;

H-11 indicates how to prove Pappus' theorem, and the new H-12 shows how to prove Pascal's theorem for a circle. Three new projects have been added. I call particular attention to Project 3, inviting readers skilled in computer graphics to draw accurate diagrams in the models for perpendicular bisectors of a hyperbolic triangle that meet (a) in an ultra-ideal point, instead of the distorted Figure 6.22, p. 274, and (b) in an ideal point.

Chapter 8 has a new title: Philosophical Implications, Fruitful Applications. It concludes with a new section describing the latter. Since previous editions discussed only philosophical disagreements and the impasse about them, which may be a bit discouraging, I added a joyous new section describing the amazing applications of hyperbolic geometry to other branches of mathematics, to cosmology, and even to art. Most of those applications are quite advanced, so the presentation is intended only to give readers a taste of them and to suggest further reading about them. Probably the most impressive application is to William Thurston's *geometrization conjecture*, which implies the famous *Poincaré conjecture* (which has apparently been proved by Grisha Perelman using the work of Richard Hamilton).

Chapter 9 is the least changed of all the chapters. One improvement is the recognition that Archimedes' axiom is needed in order that automorphisms be familiar geometric transformations. There are some interesting new exercises and a new advanced project inviting readers to do further study of the implications of Klein's ideas about groups and geometry.

Chapter 10, in which real numbers are finally exploited, has some very significant additions: Dini's flowering surface, another example of a surface of constant negative curvature, is illustrated; curvature explains the constant that puzzled J. Bolyai; the "reality" of the hyperbolic and elliptic planes, which cannot be embedded in $\mathbb{R}^3$, as abstract Riemannian manifolds of dimension 2; four specific and useful examples of circumference and circular area calculations; the right and equilateral triangle construction theorems; and a new model of the Euclidean plane within the hyperbolic plane. A new concluding section is about Bolyai's constructions in a hyperbolic plane, exhibiting his great theorem (whose proof was completed in 1995 by Will Jagy) determining the pairs of circles and regular 4-gons having the same area which can *both* be constructed (with straightedge and compass); the answer is in terms of Gauss' classical determination of those numbers n for which the angle $2\pi/n$ can be constructed (they are certain products of

Fermat numbers). I have replaced old Exercises 30–33 with new ones and added ten new projects.

Appendix A has been greatly expanded. It now presents a survey of basic concepts and results in differential and Riemannian geometry, with emphasis on dimension 2. I thank Robert Osserman for the helpful comments he made about my earlier draft.

Appendix B has also been enlarged and deepened. In addition to a more extensive discussion of elementary geometry without real numbers, including Pejas' classification of all Hilbert planes, new material is presented explaining Hilbert's field of ends used to coordinatize a hyperbolic plane and applying it to Hartshorne's proof of my conjecture that a segment in a hyperbolic plane is constructible if and only if its multiplicative length (in the field of ends) is a constructible number.

The **Bibliography** has been considerably enlarged, with about 60 new listings, the updating of several old listings, and some deletions. Instead of the "Suggested Further Reading" I wrote in previous editions, I now suggest that interested readers browse in their libraries through all the listings that catch their attention, many of which are not referenced in the text.

Acknowledgments

The history of the discovery of non-Euclidean geometry provides a valuable and accessible case study in the enormous difficulty we humans have in letting go of entrenched assumptions and opening ourselves to a new paradigm. It is delightfully instructive to observe the errors made by very capable people as they struggled with strange new possibilities they or their culture could not accept—Saccheri, working out the new geometry but rejecting it because it was "repugnant"; Legendre, giving one clever but incorrect proof after another of Euclid's parallel postulate; Lambert, speculating about a possible geometry on a "sphere of imaginary radius"; Farkas Bolyai, publishing a false proof of Euclid's parallel postulate after his son János had already published a non-Euclidean geometry; Gauss, afraid to publish his discoveries and not recognizing that his surfaces of constant negative curvature provided the tool for a proof that non-Euclidean geometry is consistent; or Charles Dodgson (alias Lewis Carroll), defending Euclid against his "modern rivals." It is inspiring to witness the courage it took János Bolyai and

Lobachevsky to put forth the new idea before the surrounding culture could grasp it, and sad to see how little they were appreciated during their lifetimes.

Werner Erhard, who founded the *est* training taken by about a million people, understood the nontechnical message of this book. He read the Bolyai correspondence in Chapters 5 and 6 to thousands of people at an *est* gathering in San Francisco. I am happy to express my appreciation to him and to my students, whose enthusiasm for "having their minds blown" by this course has boosted my morale (especially Robert Curtis, Steven Krantz, and Thomas Shea). Suggestions from readers over the years have been helpful in improving the book, and I do welcome them.

I'm very grateful for the excellent three-dimensional illustrations drawn for Appendix A and throughout the book by Thomas Banchoff and his associates. Tom's editorial comments on the early chapters were also helpful. Thanks also to Moses Ma, who provided computer graphics and important technical support. I thank Will Jagy for valuable interactions about constructions in a hyperbolic plane, following up on his great *Intelligencer* article (1995). I also thank Fu Yu for her sweet inspiration. Finally, I feel enormous gratitude to Reece Thomas Harris for our many email conversations, for quotes such as this: "The presence of beauty beyond the usual borders of our comprehension breaks those borders and allows new creative forces to emerge."

My additional thanks to all the friendly people at W. H. Freeman and Company who helped produce this book, including Terri Ward, Brendan Cady, and Laura Capuano, as well as the late John Staples, without whose openness to innovation this book might not have appeared. Thanks also to Craig Bleyer for inviting me to write a fourth edition and to Vivien Weiss for her excellent editorial assistance.

Introduction

Most people are unaware that in the early nineteenth century a revolution took place in the field of geometry that was as scientifically profound as the Copernican revolution in astronomy and, in its impact, as philosophically important as the Darwinian theory of evolution. "The effect of the discovery of hyperbolic geometry on our ideas of truth and reality has been so profound," wrote the great Canadian geometer H. S. M. Coxeter, "that we can hardly imagine how shocking the possibility of a geometry different from Euclid's must have seemed in 1820." Today, however, we have all heard of the space-time geometry in Einstein's theory of relativity. "In fact, the geometry of the space-time continuum is so closely related to the non-Euclidean geometries that some knowledge of [these geometries] is an essential prerequisite for a proper understanding of relativistic cosmology."

Euclidean geometry is the kind of geometry you learned in high school, the geometry most of us use to visualize the physical universe. It comes from the text by the Greek mathematician Euclid, the *Elements*, written around 300 B.C. Our picture of the physical universe based on this geometry was painted largely by Isaac Newton in the late seventeenth century.

Geometries that differ from Euclid's own arose out of a deeper study of *parallelism*. Consider this diagram of two rays perpendicular to segment PQ:

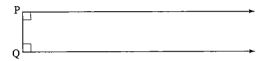

In Euclidean geometry, the perpendicular distance between the rays remains equal to the distance from P to Q as we move to the right. However, in the nineteenth century two alternative geometries were proposed. In hyperbolic geometry (from the Greek *hyperballein*, "to exceed"), the distance between the rays increases without bound. In elliptic geometry (from the Greek *elleipein*, "to fall short"), the distance decreases and the rays eventually meet. These non-Euclidean geometries were later incorporated in a much more general geometry developed by G. F. B. Riemann (it is this more general geometry that is used in Einstein's general theory of relativity).[1]

We will concentrate on Euclidean and hyperbolic geometries in this book. Hyperbolic geometry requires a change in only one of Euclid's axioms, and can be as easily grasped as high school geometry. Elliptic geometry, on the other hand, involves the new topological notion of "non-orientability," since all the points of the elliptic plane not on a given line lie on the same side of that line. This geometry cannot easily be approached in the spirit of Euclid. I have therefore made only brief comments about elliptic geometry in the body of the text, with further indications in Appendix A. (Do not be misled by this, however; elliptic geometry is no less important than hyperbolic.) Riemannian geometry requires a thorough understanding of differential and integral calculus and is therefore beyond the scope of this book (it is discussed in Appendix A).

---

[1] Einstein's special theory of relativity, which is needed to study subatomic particles, is based on a simpler geometry of space-time due to H. Minkowski. The names "hyperbolic geometry" and "elliptic geometry" were coined by F. Klein; some authors misleadingly call these geometries "Lobachevskian" and "Riemannian," respectively.

   Chapter 1 begins with a brief history of geometry in ancient times and emphasizes the development of the axiomatic method by the Greeks. It presents Euclid's five postulates and includes one of Legendre's attempted proofs of the fifth postulate. In order to detect the flaw in Legendre's argument (and in other arguments), it will be necessary to carefully reexamine the foundations of geometry. However, before we can do any geometry at all, we must be clear about some fundamental principles of logic. These are reviewed informally in Chapter 2. In this chapter, we consider what constitutes a rigorous proof, giving special attention to the method of indirect proof, or *reductio ad absurdum*. Chapter 2 introduces the very important notion of a *model* for an axiom system, illustrated by finite models for the axioms of incidence as well as projective and affine models.

   Chapter 3 begins with a discussion of some flaws in Euclid's presentation of geometry. These are then repaired in a thorough presentation of David Hilbert's axioms (slightly modified) and their elementary consequences. You may become restless over the task of proving results that appear self-evident. Nevertheless, this work is essential if you are to steer safely through non-Euclidean space.

   Our study of the consequences of Hilbert's axioms, with the exception of the Euclidean parallel postulate, is continued in Chapter 4; this study is called *neutral geometry*. We will prove some familiar Euclidean theorems (such as the exterior angle theorem) by methods different from those used by Euclid, a change necessitated by gaps in Euclid's proofs. We will also prove some theorems that Euclid would not recognize (such as the theorems about Saccheri and Lambert quadrilaterals, the uniformity theorem, and the non-obtuse-angle theorem).

   Supported by the solid foundation of the preceding chapters, we will be prepared to analyze in Chapter 5 several important attempts to prove the parallel postulate (in the exercises you will have the opportunity to find flaws in still other attempts). Following that, your Euclidean conditioning should be shaken enough so that in Chapter 6 we can explore "a strange new universe," one in which triangles have the "wrong" angle sums, rectangles do not exist, and parallel lines may diverge or converge asymptotically. In doing so, we will see unfolding the historical drama of the almost simultaneous discovery of hyperbolic geometry by Gauss, J. Bolyai, and Lobachevsky in the early nineteenth century.

   This geometry, however unfamiliar, is just as consistent as Euclid's. This is demonstrated in Chapter 7 by studying three Euclidean models that also aid in visualizing hyperbolic geometry. The Poincaré models have the advantage that angles are measured in the Euclidean way; the Beltrami–Klein model has the advantage that lines are represented by segments of Euclidean lines. In Chapter 7, we will also discuss topics in Euclidean geometry not usually covered in high school, such as inversions in circles, the pole of a chord, cross-ratios, the Poincaré center of a Euclidean circle, perspectivities, and harmonic homologies.

Chapter 8 takes up in a general way some of the philosophical implications of non-Euclidean geometries. The presentation is deliberately controversial, and the essay topics are intended to stimulate further thought and reading. The last section of Chapter 8 provides many examples of the fruitfulness of hyperbolic geometry, with suggestions for further study of them.

Chapter 9 introduces the new insights gained for geometry by the transformation approach (Felix Klein's *Erlanger Programme*). We classify all the motions of Euclidean and hyperbolic planes, use them to solve geometric problems, describe them analytically in the Cartesian and Poincaré models, characterize groups of transformations that are compatible with our congruence axioms, and introduce the fascinating topic of symmetry, determining all finite symmetry groups (essentially known by Leonardo da Vinci).

Chapter 10 is mainly devoted to the trigonometry of the hyperbolic plane, touching also upon area theory and surfaces of constant negative curvature. Among other results, we prove the hyperbolic analogue of the Pythagorean theorem and derive formulas for the circumference and area of a circle, for the relationships between right triangles and Lambert quadrilaterals, and for the circumscribed cycle of a triangle. We define various coordinate systems used to do analytic geometry in the hyperbolic plane. The polar coordinate system is used to exhibit a model of the Euclidean plane within the hyperbolic plane, a recent discovery. Finally, there is a detailed discussion of János Bolyai's remarkable results on straightedge-and-compass constructions in the hyperbolic plane, with recent improvements.

Appendix A tells more about elliptic geometry, which is mentioned throughout the book. We then introduce differential geometry, sketching the magnificent insights of Gauss and Riemann.

Appendix B is about elementary geometry without real numbers, about W. Pejas' classification of the planes Hilbert axiomatized for neutral geometry, about Hilbert's construction of a *field of ends* used to coordinatize hyperbolic planes, and about the application of that field to characterize constructible segments and angles in hyperbolic planes.

It is very important that you do as many exercises as possible, since new results are developed in the exercises and then built on in subsequent chapters. By working all the exercises, you may come to enjoy geometry as much as I do.[2]

Hyperbolic geometry used to be considered a historical curiosity. Some practical-minded students always ask me what it is good for. Following Euclid's example, I may give them a coin (not having a servant to hand it to them) and tell them that I earn a living from it. Sometimes I ask them what great music and art are good for, or I refer them to essay topics 5 and 8 in Chapter 8. If they persist, I refer them to Luneburg's research on binocular vision, to classical mechanics, and to current research in topology, ergodic theory, arithmetic algebraic geometry, and automorphic function theory (see the last section of Chapter 8). This book and the course using it provide practical-minded people an opportunity to stretch their minds. As the great French mathematician Jacques Hadamard said, "Practical application is found by not looking for it, and one can say that the whole progress of civilization rests on that principle." Only impractical dreamers spent two thousand years wondering about proving Euclid's parallel postulate, and if they hadn't done so, there would be no spaceships exploring the galaxy today.

---

[2] The mathematical physicist Freeman Dyson wrote: "The difference between a text without problems and a text with problems is like the difference between learning to read a language and learning to speak it. I intended to speak the language of Einstein, and so I worked my way through the problems." (See his *Disturbing the Universe*, New York: Basic Books, 1979, p. 13.)

1

Euclid's Geometry

> If Euclid failed to kindle your youthful enthusiasm, then you were
> not born to be a scientific thinker.
>
> Albert Einstein

Very Brief Survey of the Beginnings of Geometry

The word "geometry" comes from the Greek *geometrein* (*geo-*, "earth,"
and *metrein*, "to measure"); geometry was originally the craft of measuring land. The Greek historian Herodotus (fifth century B.C.) credits
Egyptian surveyors ("rope stretchers") with having originated the subject of geometry. The Greek philosopher Aristotle credits the Egyptian
priestly leisure class with the further development of their mathematics, which they kept secret from the public. They found the correct formula for the volume of a truncated square pyramid—a remarkable accomplishment—and of course the Egyptians built (around 2500 B.C.)
those magnificent pyramids, their greatest achievement. But basically
Egyptian geometry was a miscellaneous collection of rules for calculation—some correct, some not—without any justification provided. For
example, according to the Rhind papyrus, written before 1700 B.C. by
the Egyptian priest Ahmes, they thought that the area of a circular disk
was equal to the area of the square on eight-ninths of the diameter.
Ahmes called his writing *Directions for knowing all dark things*!

Babylonian mathematics was more advanced than Egyptian. The term "Babylonian" refers not just to the inhabitants of the lost city of Babylon, located just south of Baghdad, but more generally to peoples who lived in a region then called Mesopotamia, which is now part of Iraq. The surviving clay tablets from which historians learned about their mathematics date primarily from two eras: first, around 2000 B.C., and second, from 600 B.C. forward for around 900 years. The Babylonians had a highly developed arithmetic that used positional notation resembling our decimal system, but they used the base 60 (hexagesimal system), not our base 10. Their positional notation included fractions as well as whole numbers. They could solve some quadratic and cubic equations.

Geometry played a lesser role for them. Some of their calculations of areas and volumes were correct, some were not. They did know the Pythagorean theorem at least a thousand years before Pythagoras was born, and they found many Pythagorean triples, integers satisfying $a^2 + b^2 = c^2$, such as (3456, 3367, 4825). They knew that corresponding sides of similar triangles are proportional. The division of a circle into 360° originated with Babylonian astronomy.

The Hindu civilization of ancient India developed geometric information related to the shapes and sizes of altars and temples. Historians have not been able to accurately date the beginning of Indian verbal empirical rules for areas and volumes. Their *Sulbasutra*, the oldest mathematics texts currently known, are compilations of oral teachings that may go back to around 2000 B.C. In Sutra 50 of Baudhayana's *Sulbasutram* is found a version of the Pythagorean theorem, which he uses to show how to construct a square having the same area as a given rectangle. It was the Indians who much later made one of the greatest mathematical inventions of all time: the number *zero*.

The ancient Chinese were mainly concerned with practical matters; their classic *Jiuzhang suanshu* (*Nine Chapters on the Mathematical Art*) included hundreds of problems on surveying, agriculture, engineering, taxation, etc. Its Chapter 9, devoted to right triangle problems, displays familiarity with the Pythagorean theorem and exhibits Pythagorean triples such as (48, 55, 73). A Chinese diagram indicating why the Pythagorean theorem is valid is the oldest such known.

All these civilizations knew how to calculate the areas of simple rectilinear shapes. They guessed that the ratio of circumference to diameter in circles is constant, and they obtained rough approximations to that constant (William Jones called it π in 1706). The Babylonians and Chinese knew that the area of a circle is half the circumference times half the diameter.

Mathematics in these four ancient civilizations evolved in an intuitive and experimental manner. It was developed mainly to solve practical problems and referred to the physical world. The authors of works that have come down to us state problems in numbers and solve them by recipes for which they do not provide justification.

It was the Greeks, beginning with the legendary Thales of Miletus in the sixth century B.C., who came to insist that geometric statements be established by careful deductive reasoning rather than by trial and error. Furthermore, those statements did not refer to physical objects. They were about idealizations such as a line segment that had length but no breadth. The *orderly development of theorems with proofs about abstract entities* became characteristic of Greek mathematics and was entirely new.[1] This was the first major revolution in the history of mathematics.

How this revolution came about is not well understood by historians. Among Greek philosophers, *dialectics*, the art of arguing well, which originated in Parmenides' Eleatic school of philosophy, played an important role. And undoubtedly proofs were an outgrowth of the need to convince others in a debate.

The first serious historian of mathematics in ancient Greece was Eudemus of Rhodes. His works have been lost, but we know about them from Proclus in the fifth century, who quotes from the *Eudemian summary*. Much of what Greek mathematical history we know derives from that source.

The Pythagoreans

The systematization begun by Thales was continued over the next two centuries by Pythagoras and his disciples. Pythagoras was a spiritual teacher. He taught the immortality of the soul. He organized a brotherhood of spiritual seekers that had its own purification and initiation rites, had a meditation practice, followed a vegetarian diet, and shared all property (including credit for intellectual discoveries) communally. The Pythagoreans differed from other religious sects in their belief that the pursuit of philosophical, musical, and mathematical studies provided a moral basis for the conduct of life. Pythagorean philosophy was directed to the goal of sane, civilized living.

---

[1] J. L. Heilbron wrote: "Students should not become impatient if they do not immediately understand the point of geometrical proofs. Entire civilizations missed the point altogether!"

In music, which was absolutely central to their philosophy, the Pythagoreans observed that when the lengths of vibrating strings are expressible as ratios of small numbers, the tones will be harmonious. If a given string sounds the note C when plucked, then a similar string twice as long will sound the note C an octave below. Tones between these two notes are emitted by strings whose lengths have intermediate ratios: 16:9 for D, 8:5 for E, 3:2 for F, 4:3 for G, 6:5 for A, and 16:15 for B. Thus the Pythagoreans discovered what is possibly the oldest of all quantitative physical laws.

In mathematics, the Pythagoreans taught the mysterious and wonderful properties of numbers. By "number" the Pythagoreans meant what we call a "whole or natural number" or "positive integer." Their motto was *"All is number."*[2] Philolaus said: *"All things which can be known have number; for it is impossible that without number anything can be conceived or known."*

They discovered some basic results in what we now call *number theory*, but they also viewed each number as having a specific quality—belief in numerology was common among ancient civilizations. For example, 10 was considered the number of "perfection." They believed that there must be a Central Fire hidden from us on the other side of the sun in order that there would be 10 major heavenly bodies, not just the 6 planets then known plus the earth, sun, and moon.

A fraction was considered by them to be a relation (ratio or proportion) between two whole numbers, not in itself a number. To avoid unnecessary circumlocutions, we will say simply that they accepted what we call *positive rational numbers*. We will say that once a unit of measurement was arbitrarily chosen, the Pythagoreans originally believed that all geometric magnitudes (length, area, volume) were measured by rational numbers.

So the Pythagoreans were greatly shocked when they discovered (around 430 B.C.) irrational lengths, such as $\sqrt{2}$; we will give Aristotle's proof of that irrationality in Chapter 2 when we discuss *reductio ad absurdum* reasoning. In their geometric language, they said that *the diagonal of a square is incommensurable with the side*, meaning that there was no unit of measure for which the diagonal and the side both have lengths that are whole numbers (the same applies to the diagonal

---

[2] Kurt Gödel showed in 1931 that so far as formally axiomatized mathematics is concerned, this Pythagorean doctrine is correct. He showed, by his scheme for numbering all the formulas and sentences in any given formal theory, how the statements of that theory can all be translated into statements about numbers. He used that numbering to prove his famous incompleteness theorems (see Chapter 8).

and side of a regular pentagon). Proclus wrote: "It is well known that the man who first made public the theory of incommensurables perished in a shipwreck, in order that the inexpressible and unimaginable should ever remain veiled." Historians consider that a myth, but this discovery precipitated the first major crisis in the foundations of mathematics.[3] Since the Pythagoreans certainly did not consider $\sqrt{2}$ to be a number, they transmuted their algebra into geometric form in order to represent $\sqrt{2}$ and other irrational lengths by line segments. Euclid followed that path later.

The Pythagoreans were unable to develop a theory of proportions that was also valid for irrational lengths. This was later achieved brilliantly by Plato's pupil Eudoxus, whose very modern theory was incorporated into Book V of Euclid's *Elements*.

The development of plane geometry by the Pythagorean school was brought to a conclusion around 400 B.C. in the work *Elements* by the mathematician Hippocrates of Chios (not to be confused with the famous physician of the same name). Although this treatise has been lost, historians believe that it covered most of Books I–IV of Euclid's *Elements,* which appeared about a century later. Hippocrates is also known for his proof that the area of a certain *lune* (a region bounded by two circular arcs) is equal to the area of a certain triangle, a result that gave hope for "squaring a circle."

With the Pythagoreans, mathematics became more closely related to a love of knowledge for its own sake than to the needs of practical life. Yet we owe a great debt to the Pythagoreans for also recognizing that Nature can be understood through abstract mathematics.

Plato

The fourth century B.C. saw the flourishing of Plato's Academy of science and philosophy in Athens, which attracted the leading scholars of that era (such as Aristotle, who later founded his own Lyceum). In the *Republic,* Plato wrote: "The study of mathematics develops and sets into operation a mental organism more valuable than a thousand eyes, because through it alone can truth be apprehended." Above the gate

---

[3] Subsequent major crises were caused by the nonrigorous use of infinitesimals in the calculus, by the discovery of non-Euclidean geometries, by the Dedekind–Cantor introduction of infinite sets into algebra and analysis, by Cantor's theory of their cardinal and ordinal numbers, and by paradoxes in the early development of set theory (see Chapter 8).

to the Academy was the proclamation: "Let no one ignorant of geometry enter here." Plato claimed that reasoning about geometric objects trains the mind for the more difficult task of ascending to knowledge of what he called "The Good." Plato taught that the world of Ideas is more important than the material world of the senses. The errors of the senses must be corrected by concentrated thought, which is best learned by studying mathematics. Certainly we are able to imagine perfect geometric figures—perfectly straight lines with no breadth, etc. Plato maintained that these ideal figures not only exist in our imaginations but also exist in a world of perfect Ideas, of universal eternal truths. Human minds are not eternal, but he believed that our minds have the ability to perceive aspects of the eternal world of Ideas. Many prominent mathematicians over the centuries have subscribed to Plato's view that the truths of mathematics reside in an objective reality outside of our individual minds; others consider this viewpoint a psychologically useful myth, while still others reject it entirely.[4]

Plato cited the proof for the irrationality of the length of a diagonal of the unit square as a dramatic illustration of the power of the method of indirect proof (reductio ad absurdum—see Chapter 2). Aristotle considered this method Zeno's invention—a type of argument that begins by assuming some statement accepted by an opponent and then seeking to extract an unacceptable consequence from it, forcing the opponent to retract his commitment. Plato emphasized that the irrationality of length could never have been discovered empirically by physical measurements. A practical civilization such as the Egyptian was perfectly content to treat $\sqrt{2}$ as 7/5 or some other rational approximation. Greek civilization had moved to a new level of abstract thinking that emphasized *exactness*, not approximations, and had made new conceptual discoveries as a result.

Plato was a philosopher, not a mathematician, but Plato knew Archytas, the last great Pythagorean mathematician; and at Plato's Academy were the most important Greek mathematicians of that age to whom, before Euclid, the axiomatic-deductive method has been ascribed: Theodorus, Eudoxus, Theaetetus. In Plato's dialogue about Theaetetus, Socrates asks him what an irrational is. Theaetetus replies that he is very confused about it and does not know, but he has concerns about it. Euclid later incorporated Theaetetus' work on irrationals in his Book X.

---

[4] Eric Temple Bell considered it "fantastic nonsense of no possible value to anyone." You see that the philosophy of mathematics—unlike most of mathematics itself—is replete with controversies (see Chapter 8).

Plato may have been largely responsible for the restriction of geometric constructions to those effected with circles and lines only, because he considered them ideal geometric figures. Plutarch wrote of Plato's indignation at the use of a new mechanism invented by Eudoxus and Archytas for solving geometric problems, considering that they had shamefully turned their backs upon the nonphysical objects of pure intelligence, corrupting the major benefit of geometry—training the mind in abstract thinking.

Eudoxus was certainly the greatest mathematician of the era before Archimedes. He invented the *method of exhaustion,* an unintentionally humorous name for what we now call the *limiting process* used to determine curved lengths, areas, and (curved or rectilinear) volumes, a process that is the essential basis of the integral calculus. By that method he demonstrated that the areas of two circles are to each other as the squares on their diameters. He eventually left Plato's Academy to found his own school. He was primarily responsible for turning astronomy into a mathematical science, using a complicated model of several spheres to account for the motions around the earth of the sun, moon, and six planets then known. His model placed the stars on an outermost sphere of a universe he considered to be finite in extent.

Euclid of Alexandria

The beautiful city of Alexandria was founded in 331 B.C., at the point where the river Nile meets the Mediterranean Sea, by the conqueror Alexander the Great. It developed into a center for science, art, and culture and became the capital of Egypt. After Alexander died, the first King Ptolemy, who was an enlightened ruler, established in Alexandria a school and institute known as The Museum. He recruited the top scholars of that time to teach and work there. One of them was Euclid.

Very little is known personally about Euclid of Alexandria. From the material in the books he wrote, it is presumed that he studied either at Plato's Academy or with students of that Academy. Later he started his own school in Alexandria, where his most famous student was Apollonius of Perga, who developed the advanced theory of conic sections, building upon a treatise (since lost) by Euclid on that subject.

Euclid authored about a dozen treatises on various subjects, including optics, astronomy, music, mechanics, and spherical geometry. Unfortunately, all but five of them have been lost. His most famous one, the *Elements*, written around 300 B.C., has survived, though not as an original manuscript written by Euclid himself. The version we use today has been reconstructed from a tenth-century Greek copy found around 1800 in the Vatican Library and from Arabic translations of other lost Greek copies and revisions. We are greatly indebted to the medieval Arab scholars for preserving much of classical Greek mathematics. The first printed version of the *Elements* appeared in Venice in 1482 (Campanus' translation from the Arabic), and since then hundreds of editions have been published. A new Greek text was compiled in the 1880s by Heiberg, and that was translated into English in 1908 by Sir Thomas Heath; it is the version to which English speakers mainly refer.

The *Elements* is a definitive treatment in 13 volumes of Greek plane and solid geometry and number theory. We do not know which of its material is original with Euclid, but we do know that in compiling this masterpiece Euclid built on the achievements of his predecessors: the Pythagoreans, Hippocrates, Archytas, Eudoxus, and Theaetetus.

- Books I–IV and VI are about plane geometry.

- Books XI–XIII are about solid geometry.

- Book V gives Eudoxus' theory of proportions.

- Books VII–IX treat the theory of whole numbers. The last proposition of Book IX (Proposition 36) provides a method of constructing a *perfect* number—a number that is equal to the sum of its proper divisors, such as 6, 28, or 496. To this day no other method has been found.

- Book X presents Theaetetus' classification of certain types of irrationals; curiously, Euclid did not include a proof that the diagonal of a square is incommensurable with its side, though the Italian translation by Commandino in 1575 does add a proof of that. Book II provides a geometric method for solving certain quadratic equations (without algebraic notation, which came many centuries later). Also, in Euclid's treatment of whole numbers, stemming from the Pythagoreans, it is a peculiarity that 1 was not considered a number! It was the unit or "the monad."

In this text we will redo much of the plane geometry in the *Elements*. We will use notation such as I.47 to refer to the 47th proposition in Book I of the *Elements* (it's the Pythagorean theorem).

Euclid's *Elements* is not just about geometry and number theory; it is about how to think logically, how to build and organize a complicated theory, step by logical step. Euclid's approach to geometry dominated the teaching of the subject for over two thousand years. The axiomatic method used by Euclid is the prototype for all of what we now call *pure mathematics*. It is pure in the sense of "pure thought": No physical experiments could be performed to verify that the statements about ideal objects are correct—only the reasoning in the demonstrations can be checked.

Euclid's *Elements* is pure also in that the work includes no practical applications. Of course, Euclid's geometry has had an enormous number of applications to practical problems in engineering, architecture, astronomy, physics, etc., but none are mentioned in the *Elements*. According to legend, a beginning student of geometry asked Euclid, "What shall I get by learning these things?" Euclid called his servant, saying, "Give him a coin, since he must make gain out of what he learns."

Later Greek mathematicians did concern themselves with applications and other sciences—notably Archimedes with his mechanics and hydrostatics, Eratosthenes with his remarkable estimate of the circumference of the earth, Hipparchus and Claudius Ptolemy with their astronomy, and Heron with his optics and mechanics.

Aristotle and the Greek astronomers did not consider that the mathematical abstraction "Euclidean space" described all of actual physical space because they believed the universe was finite in extent (bounded). Thus the "truth" of Euclidean geometry for them is puzzling to us. It was the work of Isaac Newton many centuries later that led to the identification of those two "spaces" in people's minds, which lasted until Einstein and other cosmologists proposed other possible geometric models for vast physical space.

The Axiomatic Method

Mathematicians can make use of trial and error, computation of special cases, informed guessing, flashes of insight, drawing diagrams, or any other method to discover their results. *The axiomatic method is a method of proving that the results are correct and organizing them into a logical structure.* Some of the most important results in mathematics were originally given only incomplete proofs (we shall see that even Euclid was guilty of this). No matter—correct, complete proofs would be supplied later (sometimes very much later), and mathematicians would be satisfied.

So proofs give us assurance that results are correct. In many cases, they also give us more *general* results. For example, the Egyptians, Babylonians, and Indians inferred by experiment that if a triangle has sides of lengths 3, 4, and 5, it is a right triangle. But later mathematicians proved that if a triangle has sides of lengths a, b, and c and if $a^2 + b^2 = c^2$, then the triangle is a right triangle. It would take an infinite number of experiments to check this result, and, anyhow, experiments measure things only approximately. Finally, proofs give us tremendous insight into relationships among different things we are studying, forcing us to organize our ideas in a coherent way. You will appreciate this by the end of Chapter 6 (if not sooner). Gauss gave many proofs of the fundamental theorem of algebra and of the quadratic reciprocity theorem in number theory. In so doing, he was not trying to convince himself and others of the correctness of those statements; he was seeking deeper insights, different relationships to help understand why those statements were valid.

Other important scientific works besides Euclid's proceeded axiomatically: Archimedes' Book 1 on theoretical mechanics proved 15 propositions from 7 postulates. Newton's *Principia* deduced the laws of motion from his well-known laws assumed at the start. In the twentieth century, theoretical physicists Mach, Einstein, and Dirac used the axiomatic method in some of their works.

What is the axiomatic method? If I wish to persuade you by *pure deductive reasoning* to believe some statement S_1, I could show you how this statement follows logically from some other statement S_2 that you may already accept. However, if you don't believe S_2, I would have to show you how S_2 follows logically from some other statement S_3. I might have to repeat this procedure several times until I reach some statement that you already accept, one that I do not need to justify. That statement plays the role of an *axiom* or *postulate*. If I cannot reach a statement that you will accept as the basis of my argument, I will be caught in an "infinite regress," giving one demonstration after another without end.

So there are two requirements that must be met for us to agree that a proof is correct:

REQUIREMENT 1. Acceptance of certain statements called *axioms* or *postulates* without further justification.

REQUIREMENT 2. Agreement on how and when one statement "follows logically" from another, i.e., agreement on certain rules of logic.

Euclid's monumental achievement was to single out a few simple postulates, statements that were acceptable to his peers without further justification, and then to deduce from them all the conclusions known at that time in elementary geometry—many of the results not at all obvious—without there being any vicious circles in his reasoning and with most of his proofs being correct. One reason the *Elements* is such a beautiful work is that so much has been deduced from so little!

However, such a marvelous organization of results did not spring fully developed from Euclid's head the way the goddess Athena in Greek mythology sprang fully grown from the head of the god Zeus. Geometric results had been accumulated over many years by the Greeks, and unfortunately all those earlier works have been lost to us. We know that they existed from reports by later commentators such as Proclus. Euclid singled out (most of) the basic assumptions needed to prove all the other results. Such an axiomatization and organization can only be done successfully for a mature subject that has already been considerably developed in a perhaps disorganized way (e.g., the axioms for the real numbers came very late in their history).

Undefined Terms

We have been discussing what is required for us to agree that a proof is correct. Here is an additional requirement that we took for granted:

REQUIREMENT 0. Mutual understanding of the meaning of the words and symbols used in the discourse.

There should be no problem in reaching mutual understanding so long as we use terms familiar to both of us and use them consistently. If I use an unfamiliar term, you have the right to demand a *definition* of this term. Definitions cannot be given arbitrarily; they are subject to rules of reasoning also. For example, if I defined a right angle to be a 90° angle and then defined a 90° angle to be a right angle, I would violate the rule against *circular reasoning*. Sometimes a proof must first be given in order for a definition to be acceptable—e.g., if I define the specific number π to be the ratio of the circumference of any circle to the length of its diameter, I am tacitly assuming that that ratio is *constant*; that definition will not be valid until a proof of constancy is supplied (incredibly, very few books supply the proof, even the books specifically devoted to the amazing history of this number).

Also, we cannot define every term that we use. In order to define one term we must use other terms, and to define these terms we must use still other terms, and so on. If I were not allowed to leave some terms *undefined,* we would get involved in infinite regress (that's why dictionaries are circular). The undefined terms are also called *primitive terms.*

Euclid did attempt to define all his geometric terms, which was a surprising mistake, since Aristotle had already explained the necessity for undefined terms. Euclid defined a "straight line" to be "that which lies evenly with the points on itself." This definition is not very useful; so it is better to take "line" as an undefined term. Similarly, Euclid defined a "point" as "that which has no part"—again, not a very informative definition. So we will also accept "point" as an undefined term. Fortunately, nowhere in the *Elements* does Euclid use in his proofs those of his "definitions" that are vague. They are more like guides to visualizing the geometry.

Here are the five primitive geometric terms that we will use as our basis for defining all other geometric terms in plane geometry:

point
line
lie on (as in "two points *lie on* a unique line")
between (as in "point C is *between* points A and B")
congruent

For solid geometry, we would have to introduce a further undefined geometric term, "plane," and extend the relation "lie on" to allow points and lines to lie on planes. *In this book we will restrict our formal development to plane geometry*—to one single plane, if you like. We will not use this term in our formal development, though we will mention it informally.

There are expressions that are often used synonymously with "lie on." Instead of saying "point P *lies on* line *l*," we sometimes say "*l passes through* P" or "P is *incident with l*," denoted by P I *l*. If point P lies on both line *l* and line *m*, we say that "*l* and *m have* point P *in common*" or that "*l* and *m intersect* (or *meet*) in the point P."

Our undefined term, "line," replaces what is usually called "a straight line." The adjective "straight" is problematic when it modifies the noun "line," so we won't use it.[5] Nor will we talk about "curved lines." Although the word "line" will not be defined, its use will be

---

[5] Euclid did use the expression "straight line" and allowed the word "line" to also be used for what we call "curves"; e.g., he defined a "circle" as a certain kind of line. It is much simpler to avoid using "straight" in our formal discussions, though we will have to use that word occasionally informally. See Chapter 2 and Appendix A for more on straightness.

Figure 1.1

restricted by the axioms for our geometry. For instance, the first axiom states that two given points lie on only one line. Thus, in Figure 1.1, l and m could not both represent lines in our geometry since they both pass through the distinct points P and Q.

It is natural to ask how to understand our five undefined terms. The traditional method is something like this: You know how to draw a "segment" with a straightedge. You can repeatedly extend the segment in both directions with your straightedge. So *imagine* the drawn segment already extended indefinitely longer in both directions with no ends; at the same time, *imagine* such a drawing becoming thinner and thinner until it has no breadth yet has not vanished—or if you can't imagine that, picture it as having a tiny breadth but then ignore that breadth, as we do when we look at a geometric diagram. Similarly, you know what a dot drawn on paper looks like—it occupies a tiny area; *imagine* that area shrinking to zero without the dot disappearing to give an idealized "point" that is pure position. You know what it means for a dot you draw to lie on your drawn segment, though you could quibble about the dot lying "partly on" it because your drawing has breadth—just idealize the drawing in your imagination. The relation "between" for dots will only refer to three dots lying on a drawn segment; in that case, you know what it means for one dot to lie between the other two.

We will discuss visualizing congruence below. In studying these imaginary objects, we are dispensing with the features of physical objects that are irrelevant to what we are trying to accomplish. We are simplifying the subject matter. All of science depends on idealized, simplified ideas like this.

Alternatively, you could do what a blind person (who does not use the sense of touch) or a computer must do: Having no image for our undefined terms, just reason carefully about those terms using only the properties we will assume about them in our axioms. While psychologically more difficult, that would be preferable because in later chapters we will provide alternative interpretations of some of the undefined terms that may startle you. The visualizations are purely heuristic, not part of the formal mathematics, and the flexibility to interpret the undefined terms in a manner not originally intended often leads to some very important new mathematics. That is the modern point of view.

There are other mathematical terms we will use that could be added to our list of undefined terms since we won't define them; they have been omitted from the list because they are not specifically geometric in nature. Nevertheless, since there may be some uncertainty about these terms, a few remarks are in order.

The word "set" is fundamental in all of mathematics today; it is now used in elementary schools, so undoubtedly you are familiar with its use. Think of it as a "collection of objects." A related notion is "belonging to" a set or "being an element (or member) of" a set. If every element of a set S is also an element of a set T, we say that S is "contained in" or "part of" or "a subset of" T. We will define "segment," "ray," "circle," and other geometric terms to be certain sets of points. A "line," however, is not a set of points in our treatment.[6] When we need to refer to the set of all points lying on a line l, we will denote that set by $\{l\}$.

For us, the word "equal" will mean "identical." Euclid used the word "equal" in different undefined senses, as in his assertion that "base angles of an isosceles triangle are equal." We understand him to be asserting that base angles of an isosceles triangle have an equal number of degrees, not that they are identical angles. So to avoid confusion we will not use the word "equal" in Euclid's sense. Instead, we will use the undefined term "congruent" and say that "base angles of an isosceles triangle are *congruent*. Similarly, we don't say that "if AB *equals* AC, then △ABC is isosceles." (If AB *equals* AC, then following our use of the word "equals," △ABC is not a triangle at all, only a segment.) Instead, we say that "if AB is *congruent to* AC, then △ABC is isosceles." This use of the undefined term "congruent" is more general than the one to which you may be accustomed; it applies not only to triangles but to angles and segments as well, but it only applies to objects of the same kind (e.g., it would be nonsensical to say that some angle is congruent to a segment). To understand the use of this word, picture congruent objects as "having the same size and shape." Alternatively, *imagine* that you could move one object, without changing its size and shape, and superimpose it to fit exactly on the other object. This is a heuristic, informal, visual image of congruence, which is not to be used in proofs.

Of course, we must specify (as Euclid did for "equals" in his "common notions") that "a thing is congruent to itself" and that "things congruent to the same thing are congruent to each other." Statements

---

[6] For reasons of duality in projective planes in Chapter 2. Also, the Greeks denied that a line was made up of points.

like these will later be included among our axioms of congruence (Chapter 3).

Our list of undefined geometric terms is due to David Hilbert (1862–1943). His treatise *Grundlagen der Geometrie* (*Foundations of Geometry*), first edition 1899 (later editions have important supplements by Hilbert and Paul Bernays), clarified Euclid's definitions, filled the gaps in some of Euclid's proofs, added more axioms that Euclid tacitly assumed, and provided brand new important insights into the foundations of geometry. We will elaborate on that in Chapters 3–4.

Hilbert built on earlier work by Moritz Pasch, who in 1882 published the first treatise on geometry that met the new standards of rigor of his time; Pasch made explicit Euclid's unstated assumptions about *betweenness* (the axioms of betweenness will be studied in Chapter 3). Some other mathematicians who have worked to establish rigorous foundations for Euclidean geometry are G. Peano, M. Pieri, G. Veronese, O. Veblen, G. de B. Robinson, E. V. Huntington, H. G. Forder, and G. Birkhoff. These mathematicians used lists of undefined terms different from the one used by Hilbert. Pieri used only the two undefined terms "point" and "motion"; as a result, however, his axioms were more complicated. *The selection of undefined terms and axioms is arbitrary and a matter of convenience and aesthetics.* Hilbert's selection is popular because it leads to an elegant development of geometry quite similar to Euclid's presentation.

Euclid's First Four Postulates

Euclid based his geometry on five fundamental *axioms* or *postulates*. (Aristotle made a distinction between those two words that is no longer accepted.) We will slightly rephrase Euclid's postulates for greater clarity and precision.

EUCLID'S POSTULATE I. For every point P and for every point Q not equal to P there exists a unique line that passes through P and Q.

This postulate is sometimes expressed informally by saying that "two points determine a unique line." We will denote the unique line that passes through P and Q by $\overleftrightarrow{PQ}$. Actually, Euclid forgot to assume that the line is unique, and since he tacitly used uniqueness in his proofs (e.g., his proof of I.4), his first postulate was amended by subsequent commentators.

To state the second postulate, we must present our first definition.

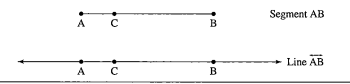

Figure 1.2

DEFINITION. Given distinct points A and B. The *segment* AB is the set whose members are the points A and B and all points C that lie on the line $\overleftrightarrow{AB}$ and are between A and B (Figure 1.2). The two given points A and B are called the *endpoints* of the segment AB.[7]

EUCLID'S POSTULATE II. For every segment AB and for every segment CD there exists a unique point E on line $\overleftrightarrow{AB}$ such that B is between A and E and segment CD is congruent to segment BE (Figure 1.3).

This postulate is expressed informally by saying that "any segment AB can be *extended* (or *produced*) by a segment BE congruent to a given segment CD." Notice that in this postulate we have used the undefined term "congruent" in the new way, and we use the usual notation CD $\cong$ BE to express the fact that CD is congruent to BE.

Euclid did not think of his lines as being infinitely long in both directions as we do, but rather as being segments extendable arbitrarily in both directions. The ancient Greeks did not accept the existence of infinite entities. Aristotle taught that the universe is finite in extent, so *the infinite should only be thought of as potential, not actual.* Thus, Euclid's lines are potentially infinite insofar as we can keep extending them as much as we like, by Postulate 2. The Greek expression *to apeiron* means not only infinitely large but also undefinable, hopelessly complex, that which cannot be handled. Proclus wrote: *"Just as sight recognizes darkness by the experience of not seeing, so imagination recognizes the infinite by not understanding it."*

Aristotle's philosophical view of the infinite became a dogma that slowed the advance of mathematics for thousands of years. It was finally overthrown in the late nineteenth century by Richard Dedekind and Georg Cantor. It is difficult for some of us today to comprehend why Aristotle and his successors (including the great mathematician Gauss) were so afraid of abstract infinite things; after all, if we can

---

[7] Warning on notation: In many high school geometry texts, the notation $\overline{AB}$ is used for "segment AB."

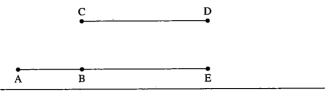

Figure 1.3 CD ≅ BE.

imagine an abstract line without breadth and an abstract point that has no part, neither of which exists in the physical world, why are we forbidden to imagine an abstract infinitely long line or an infinite set?

In order to state the third postulate, we must introduce another definition.

DEFINITION. Given distinct points O and A. The set of all points P such that segment OP is congruent to segment OA is called *the circle with O as center and OA as radius*. For each point P in that set, we say that P *lies on* the circle and OP is called a *radius* of the circle.

It follows from our version of Euclid's previously mentioned common notion that "a thing is congruent to itself" that OA ≅ OA, so point A lies on the circle. Also, if P lies on the circle and OP ≅ OQ, then Q also lies on the circle because of Euclid's common notion that "things congruent to the same thing are congruent to each other." (In Chapter 3, we will state these common notions as additional axioms.) The term "radius" does not appear in Euclid's work; he only spoke of a *diameter* of a circle, defined as a segment whose endpoints lie on the circle (i.e., a *chord*) and which passes through the center of the circle.

EUCLID'S POSTULATE III. For every point O and every point A not equal to O, there exists a circle with center O and radius OA (Figure 1.4).

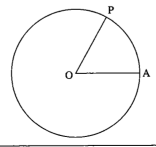

Figure 1.4 Circle with center O and radius OA.

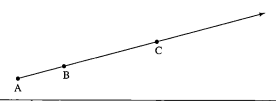

Figure 1.5 Ray $\overrightarrow{AB}$.

Actually, because we are using the language of sets rather than
that of Euclid, it is not really necessary to assume this postulate; it is
a consequence of a set theory axiom that the subset of all points P such
that OP $\cong$ OA exists. Of course, set theory did not yet exist in 300 B.C.
Euclid talked of *drawing* the circle with center O and radius OA. Our
formal treatment purifies[8] Euclid by eliminating references to drawing.
(Notice that when we illustrate in Figure 1.4 what a circle looks like,
we are tacitly working in one plane, as we stated. If instead we were
working in three dimensions, the set of all points P such that OP $\cong$ OA
would be the sphere with center O and radius OA.)

DEFINITION. The *ray* $\overrightarrow{AB}$ is the following set of points lying on the
line $\overleftrightarrow{AB}$: those points that belong to the segment AB and all points C
on $\overleftrightarrow{AB}$ such that B is between A and C. The ray $\overrightarrow{AB}$ is said to *emanate
from the vertex* A and to be *part* of line $\overleftrightarrow{AB}$ (see Figure 1.5).

DEFINITION. Rays $\overrightarrow{AB}$ and $\overrightarrow{AC}$ are *opposite* if they are distinct, if they
emanate from the same point A, and if they are part of the same line
$\overleftrightarrow{AB} = \overleftrightarrow{AC}$ (Figure 1.6.).

DEFINITION. An "*angle* with *vertex* A" is a point A together with two
distinct non-opposite rays $\overrightarrow{AB}$ and $\overrightarrow{AC}$ (called the *sides* of the angle)
emanating from A (see Figure 1.7).[9]

Figure 1.6 $\overrightarrow{AB}$ and $\overrightarrow{AC}$.

---

[8] However, by bringing in set theory, as Hilbert did, we are sullying Euclid. To avoid
that, many of the terms we define as sets would have to be left undefined and new
axioms would have to be added to characterize them. The Greeks believed that lines
and circles were *not* made up of points.

[9] According to this definition, there is no such thing in our treatment as a "straight angle," nor is there such a thing as a "zero angle." We eliminated those expressions because most of the assertions we will make about angles do not apply to them.

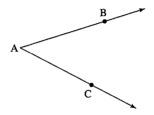

Figure 1.7 Angle with vertex A.

We use the notation ∢A, ∢BAC, or ∢CAB for this angle. If $r = \overrightarrow{AB}$ and $s = \overrightarrow{AC}$, then rays r, s are said to be *coterminal* (meaning they emanate from the same vertex), and the angle is also denoted ∢(r, s).

DEFINITION. If two angles ∢DAB and ∢CAD have a common side $\overrightarrow{AD}$ and the other two sides $\overrightarrow{AB}$ and $\overrightarrow{AC}$ form opposite rays, the angles are *supplements* of each other, or *supplementary angles* (Figure 1.8).

DEFINITION. An angle ∢BAD is a *right angle* if it has a supplementary angle to which it is congruent (Figure 1.9).

We have thus succeeded in defining a right angle without referring to "degrees," by using the primitive notion of congruence of angles. Degrees will not be introduced formally until Chapter 4, although we will occasionally refer to them in informal discussions. We can now state Euclid's fourth postulate.

EUCLID'S POSTULATE IV. All right angles are congruent to one another.

This postulate expresses a homogeneity of the plane; two right angles "have the same size and shape" no matter where they are located

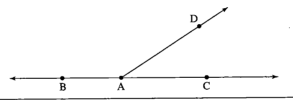

Figure 1.8 ∢BAD and ∢DAC are supplementary angles.

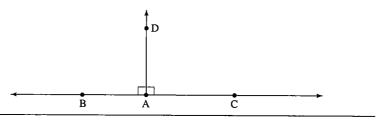

Figure 1.9 Right angles ∢BAD ≅ ∢CAD.

in the plane. The fourth postulate provides an "intrinsic" standard of measurement for angles since right angles have been geometrically defined and other angles can be compared with them.[10]

The Parallel Postulate

Euclid's first four postulates have always been readily accepted by mathematicians. The fifth postulate—the "parallel postulate"—however, became highly controversial. As we shall see later, consideration of alternatives to Euclid's parallel postulate resulted in the development of non-Euclidean geometries. At this time we are not going to state the fifth postulate in its original form as it appeared in the *Elements*. Instead, we will present a simpler postulate, which we will show (in Chapter 4) is logically equivalent to Euclid's original. This version is sometimes called *Playfair's postulate* because it appeared in John Playfair's formulation of Euclidean geometry published in 1795—though it was first presented by Proclus in the fifth century. We will call it the *Euclidean parallel postulate* because it distinguishes Euclidean geometry from other geometries based on parallel postulates. The most important definition in this book is the following:

DEFINITION. Two lines l and m are *parallel* if they do not intersect, i.e., if no point lies on both of them. We denote this by $l \parallel m$.

Notice first that in making this definition we assume the lines lie in the same plane (because of our convention that all points and lines lie in one plane unless stated otherwise); in solid geometry, there are

---

[10] On the contrary, there is no intrinsic standard of measurement for segments in Euclidean geometry (this will be proved in Chapter 9). Units of length (1 foot, 1 meter, etc.) must be chosen arbitrarily. The remarkable fact about elliptic and hyperbolic geometries, on the other hand, is that they do admit an intrinsic standard of length (see Chapters 6 and 9).

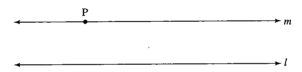

Figure 1.10 *m* is the unique line through P parallel to *l*.

non-coplanar lines that fail to intersect, and they are called *skew lines*, not "parallel" lines. Speaking informally, notice second what the definition does *not* say: It does not say that the lines are "equidistant," i.e., it does not say that the "distance" between the two lines is everywhere the same. Don't be misled by drawings of parallel lines in which the lines appear to be equidistant, like railroad tracks. To be rigorous we must not introduce assumptions that have not been stated explicitly. At the same time, don't jump to the conclusion that parallel lines are *not* equidistant. We are not committing ourselves either way and shall reserve judgment until we study the matter further. At this point, the only thing we know for sure about parallel lines is that they do not meet.[11]

THE EUCLIDEAN PARALLEL POSTULATE. For every line *l* and for every point P that does not lie on *l*, there exists a unique line *m* through P that is parallel to *l* (see Figure 1.10).

Once again, this is an axiom for plane geometry; in solid geometry, there are infinitely many lines through P that do not intersect *l*.

Why was this postulate so controversial? It may seem "obvious" to you, perhaps because you have been conditioned to think in Euclidean terms. However, if we consider the axioms of geometry as abstractions from experience, we can see a difference between this postulate and the other four. The first two postulates are abstractions from our experiences drawing with a straightedge; the third postulate derives from our experience drawing with a compass. The fourth postulate is less obvious as an abstraction. One could argue that it derives from our experience measuring angles with a protractor, where the sum of supplementary angles is always 180°, so that if supplementary angles are congruent to each other, they must each measure 90°; if we think of congruence for angles in terms of having the same number of degrees when measured by a protractor, then indeed all right angles are

---

[11] I have found two books about mathematics for educated lay readers, written by well-known, respected authors, which claim that the Euclidean parallel postulate asserts that "parallel lines never meet." That is a definition, not a postulate!

congruent. (Don't interpret what was just said as any kind of proof of
the fourth postulate; it is just a heuristic argument to make that assumption plausible from our experience.)

The parallel postulate is different in that we cannot verify empirically whether two drawn lines meet since we can draw only segments,
not complete lines. We can extend the segments further and further to
see if the lines containing them meet, but we cannot go on extending
them forever. Our only recourse is to verify parallelism indirectly by
using criteria other than the definition.

What is another criterion to test whether l is parallel to m? Euclid
suggested drawing a *transversal* (i.e., a line t that intersects both l and
m in distinct points) and considering the interior angles α and β on
one side of t. He predicted that if the "sum" of angles α and β turns
out to be less than two right angles, the line segments, if produced sufficiently far, would meet on the same side of t as angles α and β (see
Figure 1.11). This, in fact, is the content of Euclid's fifth postulate
(which we will refer to as Euclid V). It is a criterion for l and m to *not*
be parallel, and it tells on which side of the transversal they meet.

We have stated this criterion unofficially because it involves terms
that we will only be able to define precisely later (interior angles, same
side of the transversal, sum of angles). We are appealing to your previous experience with geometry and to the diagram so that you will
understand the content of Euclid V.

Reece Thomas Harris pointed out that what Euclid V in fact does
is grant the power to construct triangles by extending segments until
the lines meet (and it doesn't mention parallels). Indeed, we will later
use that power to construct a triangle that is similar to a given one on
a given segment (see Wallis' postulate, Chapter 5). However, the difficulty with this construction is that it does not provide any bound for
how far we have to extend the line segments to find the third vertex of
the triangle. We have the same difficulty as before in accepting it.

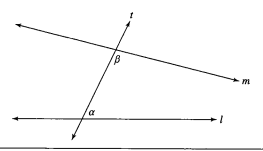

Figure 1.11

Euclid himself must have recognized the controversial nature of his fifth postulate, for he postponed using it for as long as he could—until the proof of I.29, which is the converse of the alternate interior angle theorem I.27 for parallel lines; then he used it for his results on parallelograms. That use of Euclid V may be why it has been incorrectly called "the parallel postulate."

We know from Aristotle that in his time the theory of parallels had not yet been put on a rigorous basis. Undoubtedly the formulation of a postulate which does provide a rigorous foundation for that theory is Euclid's original contribution.

Attempts to Prove the Parallel Postulate

Remember that an axiom was originally supposed to be so simple and obvious that no educated person could doubt its validity. From the very beginning, however, the parallel postulate was attacked as insufficiently plausible to qualify as an unproved assumption. For about two thousand years, mathematicians tried to derive it from the other four postulates or to replace it with another postulate, one more self-evident. All attempts to derive it from the first four postulates turned out to be unsuccessful because the so-called proofs always entailed a hidden assumption that was unjustifiable. The substitute postulates, purportedly more self-evident, turned out to be logically equivalent to the parallel postulate, so that nothing was gained logically by the substitution. We will examine these attempts in detail in Chapter 5, for they are very instructive. For the moment, let us consider one such effort.

Adrien-Marie Legendre (1752–1833) was one of the best mathematicians of his time, contributing important discoveries to many different branches of mathematics. Yet he was so obsessed with proving the parallel postulate that over a period of 29 years, he published one attempt after another in 20 different editions of his *Éléments de Géometrie*. Here is one attempt (see Figure 1.12).

Given P not on line l. Drop perpendicular PQ from P to l at Q. Let m be the line through P perpendicular to $\overleftrightarrow{PQ}$. Then m is parallel to l since l and m have the common perpendicular $\overleftrightarrow{PQ}$. Let n be any line through P distinct from m and $\overleftrightarrow{PQ}$. We must show that n meets l. Let $\overrightarrow{PR}$ be a ray of n between $\overrightarrow{PQ}$ and a ray of m emenating from P. There is a point R' on the opposite side of $\overrightarrow{PQ}$ from R such that $\sphericalangle QPR' \cong \sphericalangle QPR$. Then Q lies in the interior of $\sphericalangle RPR'$. Since line l passes through the point Q interior to $\sphericalangle RPR'$, l must intersect one of the sides of this

Adrien-Marie Legendre

angle. If l meets side $\overrightarrow{PR}$, then certainly l meets n. Suppose l meets side $\overrightarrow{PR'}$ at a point A. Let B be the unique point on side $\overrightarrow{PR}$ such that PA $\cong$ PB. Then $\triangle$PQA $\cong$ $\triangle$PQB (SAS); hence $\sphericalangle$PQB is a right angle, so that B lies on l (and n).

You may feel that this argument is plausible enough. Yet how could you tell whether it is correct? You would have to justify each step, first defining each term carefully. For instance, you would have to define

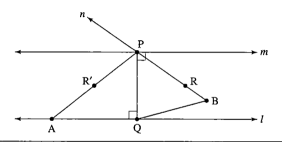

Figure 1.12

what is meant by two lines being "perpendicular"—otherwise, how could you justify the assertion that lines l and m are parallel simply because they have a common perpendicular? (You would first have to prove that as a separate theorem, if you could.) You would have to justify the side-angle-side (SAS) criterion of congruence in the last statement. You would have to define the "interior" of an angle and prove that a line through the interior of an angle must intersect one of the sides. In proving all of these things, you would have to be sure to use only the first four postulates and not any statement equivalent to the fifth; otherwise the argument would be circular.

Thus, there is a lot of work that must be done before we can detect the flaw. In the next few chapters, we will do this preparatory work so that we can confidently decide whether or not Legendre's proposed proof is valid. (Legendre's argument contains several statements that cannot be proved from the first four postulates.) As a result of this work, we will be better able to understand the foundations of Euclidean geometry. We will discover that a large part of this geometry is independent of the theory of parallels and is equally valid in hyperbolic geometry.

The Danger in Diagrams

Diagrams have always been helpful in understanding geometry—they are included in Euclid's *Elements*, and they are included in this book. But there is a danger that a diagram may suggest a fallacious argument. A diagram may be slightly inaccurate or it may represent only a special case. If we are to recognize the flaws in arguments such as Legendre's, we must not be misled by diagrams that *look* plausible.

What follows is a well-known and rather involved argument that pretends to prove that all triangles are isosceles. Place yourself in the context of what you know from high school geometry. (After this chapter you will have to put that knowledge on hold.) Find the flaw in the argument.

Given $\triangle ABC$. Construct the bisector of $\angle A$ and the perpendicular bisector of side BC opposite to $\angle A$. Consider the various cases (Figure 1.13).

▓▓▓ **CASE 1.** The bisector of $\angle A$ and the perpendicular bisector of segment BC are either parallel or identical. In either case, the bisector of $\angle A$ is perpendicular to BC and hence, by definition, is an altitude.

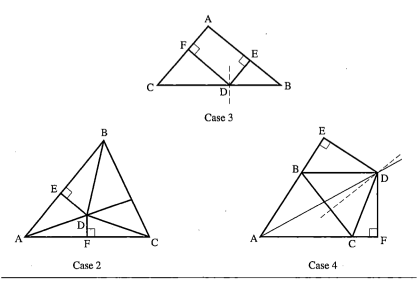

Figure 1.13

Therefore, the triangle is isosceles. (The conclusion follows from the Euclidean theorem: If an angle bisector and altitude from the same vertex of a triangle coincide, the triangle is isosceles.)

Suppose now that the bisector of ∢A and the perpendicular bisector of the side opposite are not parallel and do not coincide. Then they intersect in exactly one point, D, and there are three cases to consider:

▓▓▓ **CASE 2.** The point D is inside the triangle.

▓▓▓ **CASE 3.** The point D is on the triangle.

▓▓▓ **CASE 4.** The point D is outside the triangle.

For each case, construct DE perpendicular to AB and DF perpendicular to AC, and for cases 2 and 4 join D to B and D to C. In each case, the following proof now holds (see Figure 1.13).

DE ≅ DF because all points on an angle bisector are equidistant from the sides of the angle; DA ≅ DA, and ∢DEA and ∢DFA are right angles; hence △ADE is congruent to △ADF by the hypotenuse-leg theorem of Euclidean geometry. (We could also have used the SAA theorem with DA ≅ DA, and the bisected angle and right angles.) Therefore, we have AE ≅ AF. Now, DB ≅ DC because all points on the perpendicular bisector of a segment are equidistant from the ends of the segment. Also, DE ≅ DF, and ∢DEB and ∢DFC are right angles. Hence, △DEB is congruent to △DFC by the hypotenuse-leg theorem, and hence FC ≅ BE. It follows that AB ≅ AC—in cases 2 and 3 by addition and in case 4 by subtraction. The triangle is therefore isosceles.

Henri Poincaré said: "Geometry is the art of reasoning well from badly drawn diagrams." J. L. Lagrange, the great master of dynamics after Newton, prided himself that his *Analytic Mechanics* (published in 1788) contained not a single diagram. Jean Dieudonné, in his *Linear Algebra and Geometry* (first published in 1969), also omitted all diagrams, contending that they are "unnecessary." But Hilbert did include diagrams in his *Grundlagen der Geometrie.*

The Power of Diagrams

Geometry, for human beings, is a visual subject, and many people think visually more than symbolically. Correct diagrams can be extremely helpful in understanding proofs and in discovering new results. For example, the great physicist Richard Feynman invented a new type of diagram (now named after him) to understand and do research in quantum electrodynamics.

One of the best illustrations of the power of diagrams is Figure 1.14, which reveals immediately the validity of the Pythagorean theorem in Euclidean geometry.

Figure 1.15 is a simpler diagram suggesting a proof by dissection. (Euclid's argument was much more complicated—see his proof of I.47.)

Algebra did not blossom with more-or-less its current symbolism until the eighteenth century. It was developed by the Arabs and Hindus, with earlier work by the Babylonians and the Alexandrian Greek number theorist Diophantus. It took a while for the novel idea of performing arithmetic operations with letters, instead of numbers, to become commonplace—François Viète in sixteenth century France originated that.

So our idea that the Pythagorean theorem asserts that $a^2 + b^2 = c^2$, where a, b are the lengths of the legs and c is the length of the hypotenuse of a right triangle, is a relatively modern idea. If you read I.47, Euclid's statement of that theorem, it does not display such an equation. It states: "In a right triangle, the square on the side subtending the right angle is equal to the squares on the sides containing

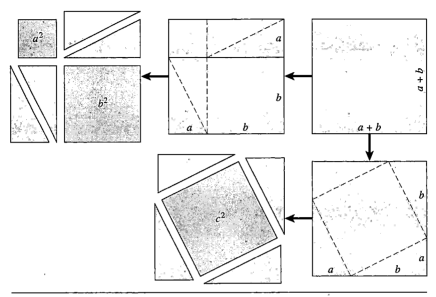

Figure 1.14

the right angle." We interpret this to mean that the *area* of the square having the hypotenuse of the right triangle as a side is equal to the *sum* of the areas of the squares having the legs of the right triangle as their sides. If we think of area as a number, then by the definition of the area of a geometric square as the numerical square of the length of its side, this statement is equivalent to the equation above, which we may call the *Pythagorean equation.*

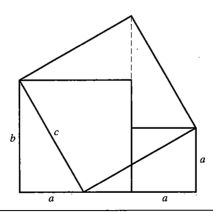

Figure 1.15

However, just as Euclid did not have numbers as lengths of segments, he did not have numbers as areas of plane figures. For example, the square on a segment of length $1 + \sqrt{2}$ would have irrational area if he attempted that. Instead, Euclid considered area to be another kind of *magnitude,* a term he did not define (*volume* is a third kind of magnitude for solid geometry). To distinguish that concept from numerical area, modern mathematicians define what it means for two plane polygonal figures to *have equal content*—informally, it means that you can dissect one figure into polygonal pieces and then reassemble those pieces to construct the other figure. That's exactly what we illustrated with the second diagrammatic "proof" of the Pythagorean theorem in Figure 1.15.

Figure 1.14 illustrates a possibly weaker result. If we adjoin to each of the figures another figure consisting of four copies of the original right triangle, then the resulting figures will have equal content.

We will not develop these ideas in this text. For more details on the interesting theory of equal content, see Hartshorne (2000), Chapter 5.

Straightedge-and-Compass Constructions, Briefly

In our heuristic discussion of Euclid's postulates, we mentioned drawings with straightedge and compass as the experiential basis for accepting the first three postulates. We rephrased those postulates to be compatible with today's rigorous style of expressing abstract mathematics. However, we can follow Euclid's style and informally talk about drawing, provided that you understand how to translate such figures of speech into our precise language. Here is Euclid's version of the first three postulates in Heath's translation:

1. To draw a straight line from any point to any point.
2. To extend a finite straight line continuously in a straight line.
3. To describe a circle with any center and any distance.

This language shows that Euclid thought about geometric existence *constructively,* in the sense of idealized straightedge-and-compass constructions. Such idealized constructions became very important in the history of elementary geometry. They are just a figure of speech for obtaining the existence of certain points by intersecting lines and/or circles with lines and/or circles, as well as for the existence of lines and circles as guaranteed by Postulates I–III in the form we have stated them. Euclid never mentions straightedge or compass, though he does use words like "draw," "describe," and "extend" in Heath's translation.

Here is a list of those propositions in Books I–IV of the *Elements* which are constructions:

I.1.   To construct an equilateral triangle on a given segment.

I.2.   To draw a segment congruent to a given segment at a given point.

I.3.   To cut off a smaller segment from a larger segment.

I.9.   To bisect an angle.

I.10. To bisect a segment.

I.11. To erect a perpendicular to a line at a given point on the line.

I.22. To construct a triangle, given three sides, provided any two are greater than the third.

I.23. To reproduce a given angle at a given point and side.

I.31. To draw a line parallel to a given line through a given point not on that line.

I.42. To construct a parallelogram with a given angle equal in content to a given triangle.

I.44. To construct a parallelogram with given side and angle equal in content to a given triangle.

I.45. To construct a parallelogram with a given angle equal in content to a given figure.

I.46. To construct a square on a given segment.

II.14. To construct a square equal in content to a given figure.

III.1. To find the center of a circle.

III.17. To draw a tangent to a circle from a point outside the circle.

IV.1. To inscribe a given segment in a circle.

IV.2. To inscribe a triangle, equiangular to a given triangle, in a circle.

IV.3. To circumscribe a triangle, equiangular to a given triangle, around a circle.

IV.4. To inscribe a circle in a triangle.

IV.5. To circumscribe a circle around a triangle.

IV.10. To construct an isosceles triangle whose base angles are twice the vertex angle.

IV.11. To inscribe a regular pentagon in a circle.

IV.12. To circumscribe a regular pentagon around a circle.

IV.15. To inscribe a regular hexagon in a circle.

IV.16. To inscribe a regular 15-sided polygon in a circle.

The remaining propositions in Books I–IV are about relationships and non-relationships among geometric figures. Here are several notable examples of such propositions:

I.5.   Base angles of an isosceles triangle are congruent.

I.15.   Vertical angles are congruent.

I.16.   An exterior angle of a triangle is greater than either opposite interior angle.

I.17.   Any two angles of a triangle together are less than two right angles.

I.20.   Any two sides of a triangle together are greater than the third.

I.27.   Congruence of alternate interior angles implies the lines are parallel.

I.29.   If two lines are parallel, then alternate interior angles cut by any transversal are congruent.

I.32.   The angle sum of a triangle is two right angles, and an exterior angle equals the sum of opposite interior angles.

I.34.   Opposite sides and angles of a parallelogram are congruent, respectively.

I.47.   Theorem of Pythagoras.

I.48.   Converse of the theorem of Pythagoras.

III.5.   If two circles intersect, they do not have the same center.

III.10. Two circles can intersect in at most two points.

III.20. The angle at the center is twice the angle at a point of the circumference subtending a given arc of a circle.

III.21. Two angles from points of a circle subtending the same arc are congruent.

III.22. Opposite angles of a quadrilateral inscribed in a circle add up to two right angles.

III.31. An angle with vertex on a circle and subtending a semicircle of that circle is a right angle.

Many of these propositions should be recognizable to you from your previous course in Euclidean geometry. Proposition III.31 is attributed to Thales.

    After constructing the beautifully symmetric equilateral triangle on a given segment in his very first proposition, why did Euclid wait until his 46th proposition to construct the beautifully symmetric square on a given segment? Because that construction depends on using the fifth postulate.

There were three famous straightedge-and-compass construction problems in ancient Greek geometry and a fourth that was less famous but equally important:

1. Trisect any angle.
2. Square any circle (i.e., construct a square having the same area as the given circle).
3. Duplicate any cube (i.e., construct a segment such that the cube on that segment has twice the volume of the given cube).
4. For any $n > 6$, construct a regular n-gon (i.e., an n-sided convex polygon in which all sides and all angles are congruent to one another, respectively).

The construction of a regular n-gon for $n = 3, 4, 5, 6$ was carried out in Euclid's *Elements*. The first unsolved case is $n = 7$.

The critical difficulty in these problems is the restriction to using straightedge and compass alone (or, more precisely, to using only lines and circles and no other curves). From the point of view of a design engineer, say, that restriction can be circumvented by using other instruments. However, from the point of view of a pure mathematician, that restriction poses an interesting theoretical problem that eventually led to some extremely interesting mathematics.

It turned out that *none* of these constructions could be carried out *in general*. For certain special cases, the construction could be done— e.g., a right angle can be trisected with straightedge and compass alone (just bisect the angle of an equilateral triangle). A regular octagon can easily be constructed by circumscribing a square with a circle, perpendicularly bisecting the sides of the square and joining the four points where those perpendicular bisectors hit the circle to the adjacent vertices of the square. Thanks to the work of C. F. Gauss, we know exactly for which n Problem 4 is solvable, and the answer very surprisingly depends on certain prime numbers that were first investigated by Fermat and have been named after him. There are only three Fermat primes $n > 6$ for which the regular n-gon is currently known to be constructible: 17, 257, and 65,537; that is because it's a currently unsolved problem as to whether there are any other Fermat primes.

The impossibility of these constructions in general could only be proved after analytic geometry was invented, and these geometric problems were successfully translated into purely algebraic ones in the early nineteenth century. That was a great triumph for the use of algebra in geometry (a vindication of Descartes and Fermat, who pioneered such use).

That impossibility has been thoroughly explained in several other texts, so we won't go into it here. See Hartshorne or Moise, for example. Very briefly and sketchily, using Cartesian coordinates of points, the algebraic analogue of any straightedge-and-compass construction involves the determination of certain numbers obtained from rational numbers by repeatedly using the four arithmetic operations and the operation of taking the square root of a positive number. The general-case analogue of Problems 1 and 3 involves solving cubic equations, and it can be proved that roots of irreducible cubic equations with rational number coefficients cannot be obtained using only those five operations. Problem 2 was shown unsolvable when Lindemann proved the much stronger result that π is transcendental—it is not a root of any polynomial with integer coefficients.

Descartes (and long before him, Pappus in the third century) conjectured the impossibility in general of those first three constructions. Kepler argued for the impossibility of constructing the regular heptagon (seven-sided) and asserted, as a result, that it was simply "unknowable." He also claimed that the regular p-gon for a prime $p > 5$ could not be constructed; he did not know about the exceptions $p = 17$, 257, and 65,537 found by Gauss.

Certain Greek mathematicians of antiquity invented interesting tools and methods other than straightedge and compass to construct those desired geometric objects. As the simplest example, Archimedes showed how to trisect any angle using a *marked* straightedge (see Exercise 16). Many centuries later, Viète proposed to add a new axiom to geometry to permit such so-called *neusis* constructions, but Euclidean geometry was considered too sacrosanct by then for new axioms to be accepted. Here is *Viète's axiom*, in which segment AB plays the role of two marks on the straightedge:

VIÈTE'S AXIOM. Given a segment AB and point C. Let t, u be distinct lines or a line and a circle (Figure 1.16). Then there exists a point P on t and a point Q on u such that PQ $\cong$ AB and P, Q, C are collinear.

You are invited to explore some of these developments in the exercises. Regarding the impossibility results, mathematician Oscar Morgenstern said:

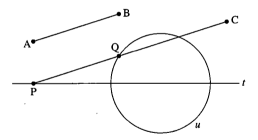

Figure 1.16 PQ ≅ AB.

Some of the profoundest insights the human mind has achieved are
stated in negative form. . . . Such insights are that there can be no per-
petuum mobile, that the speed of light cannot be exceeded, that the circle cannot be squared using ruler and compass only, that similarly an
angle cannot be trisected, and so on. Each of these statements is the
culmination of great intellectual effort. All are based on centuries of
work. . . . Though stated negatively, these and other discoveries are
positive achievements and great contributions to human knowledge.

Descartes' Analytic Geometry and
Broader Idea of Constructions

Although coordinates had been used long before their work (e.g., in
astronomy and geography), historians give René Descartes and Pierre
Fermat equal credit for the invention of analytic geometry, in which
numerical coordinates and algebraic equations in those coordinates are
used to obtain geometric results. Descartes was the first to publish in
1637, as an appendix (*La Géométrie*, in three parts) to his very influential *Discourse on Method*, his philosophical method for finding and
recognizing correct knowledge. Fermat never did publish his work; instead, he communicated his results in private letters to a few colleagues,
and his work was made public only in 1679, fourteen years after he
died. Curiously, although both these men were outstanding mathematicians, mathematics was not their profession. Fermat was a jurist
who did mathematics as a hobby. He is best known for his work in
number theory; his famous "last theorem" was finally proved in 1995,
as a corollary to Andrew Wiles' proof of the main part of the profound

René Descartes

Shimura–Taniyama conjecture. Fermat also discovered the basic idea of the differential calculus before Newton and Leibniz. Descartes contributed to other sciences besides mathematics, but he was primarily a philosopher whose writings had a great impact on the way educated people viewed the world.

Both men initially introduced their algebraic methods in order to solve problems from classical Greek geometry, recognizing that the new methods had great potential to solve other problems. Their successors over many decades realized that potential. Descartes' stated goal was to provide general methods, using algebra, to "solve any problem in geometry." He did not see geometry as an axiomatic deductive science that derives theorems about geometric objects.

In the time of Descartes, the tradition was that algebra was a completely separate subject from geometry. That tradition was breaking down with the work of Viète in the sixteenth century, and both Descartes and Fermat built on Viète's ideas.

Descartes defined the five algebraic operations of addition, subtraction, multiplication, division, and extraction of square roots as

geometric constructions on line segments and showed how those operations could be performed in the Euclidean plane by straightedge-and-
compass constructions. Thus, those algebraic operations were a legitimate part of classical Euclidean plane geometry; they were operations
on geometric objects, not operations on numbers.

Particularly innovative was his simple definition of multiplication of
segments in terms of similar triangles once a unit segment had been arbitrarily chosen. Viète thought of the product of two segments as representing the area of a rectangle having those segments as its sides (in
solid geometry, the product of three segments was thought to represent
a volume). An algebraic expression such as $a^2 + b$ made no sense to
Viète, for how could one add an area to a segment? With Descartes'
definition, it made perfectly good sense as the sum of two segments.
Moreover, with Descartes' definition, expressions involving products of
four or more terms now made geometric sense as segments, whereas
previously they had been rejected as meaningless because space had
only three dimensions. Thus, Descartes could carry out geometrically
all algebra involving those five operations. For example, Descartes
showed how to solve geometrically all quadratic equations in one unknown having positive roots; he did not deal with negative roots because at that time they were considered "false." He stated his general
method as follows:

> If we wish to solve any geometric problem, we first suppose the solu-
> tion already effected, and give names to all the segments needed for its
> construction—to those that are unknown as well as to those that are
> known. Then, making no distinction between known and unknown seg-
> ments, we must unravel the difficulty in any way that shows most nat-
> urally the relations between these segments, until . . . we obtain an
> equation in a single unknown.

He then developed geometric techniques for solving polynomial equations in a single unknown, at least for equations of degree at most 6.
To geometrically solve equations of degree 3 or 4, he had to intersect
conics—parabolas or ellipses (including circles) or hyperbolas—with
each other or with lines, for he recognized that their solutions could
not generally be constructed by straightedge and compass alone (that
was not proved rigorously until the nineteenth century). To solve equations of degree 5 or 6, he had to introduce cubic curves. The study of
conics and higher-degree curves belongs to what used to be called
higher geometry; this text is primarily about elementary geometry, so

we won't delve into that important subject. Descartes was certainly not the first to use constructions other than straightedge and compass ones (see Project 8); his new idea was to study them *algebraically*.

When Descartes gave a name or letter to the solution sought and then reasoned from there, he was using the method that had classically been called "analysis"—reasoning from the conclusion until one arrives at propositions previously established or at an axiom. By reversing the order of the steps—*if possible*—one obtains a demonstration of the result. Analysis is a systematic method of discovering *necessary* conditions for the result to hold; synthesis would then hopefully show that those conditions are *sufficient*. It is because of this method that Descartes' geometry is called *analytic*. (Later in the history of mathematics, "analysis" came to have a completely different meaning: It was the branch of mathematics dealing with limiting processes—the calculus and its more advanced developments. So it would be more appropriate to call it "coordinate geometry" rather than "analytic geometry," and some authors do call it that.)

Most of the proofs in this book are *synthetic,* as in Euclid. Only in the much later chapters will we use some analytic geometry.

It took many years before analytic geometry was well understood and accepted into mainstream mathematics. Blaise Pascal objected to the use of algebra in geometry because it had no axiomatic foundation at that time. What also slowed its acceptance was Descartes' style of writing, which was deliberately difficult to understand. Descartes warned his readers that *"I shall not stop to explain this in more detail, because I should deprive you of the pleasure of mastering it yourself."*

Isaac Newton was ambivalent about the proper role of analytic geometry. In an appendix to his *Opticks* (published in 1704, composed in 1676), he used analytic geometry to exhibit 72 species of curves given by third-degree polynomial equations (cubics) in two unknowns and plotted them. Newton thereby opened an entirely new field of geometry for study: higher-degree plane algebraic curves (later, transcendental curves—not given by a polynomial equation but given by transcendental functions such as the logarithm or trigonometric functions—were studied). Before the invention of analytic geometry, only a dozen or so curves were known to the Greeks.

But in his *Arithmetica universalis* (published in 1707 but written a quarter-century earlier), Newton said:

Equations are Expressions of Arithmetical Computation, and properly have no place in Geometry . . . these two sciences ought not to be

> confounded. The Ancients did so industriously distinguish them from
> one another that they never introduced Arithmetical Terms into Geom-
> etry. And the Moderns, by confounding both, have lost the Simplicity
> in which all the Elegancy of Geometry consists.

It is clear from most of Newton's writings that he fully realized and utilized the value and power of coordinate numerical algebraic meth-ods. Undoubtedly what Newton intended by this declaration was that coordinate methods should not be used when dealing with *elementary geometry*, i.e., Euclid's geometry of lines and circles, but they are ac-ceptable in higher geometry.

Nevertheless, Newton wrote his monumental *Principia* in the syn-thetic style of Euclid because that was the style of mathematics that was considered rigorous in his time. Newton later admitted that he orig-inally discovered and elaborated his results by analytic methods.

Descartes and Fermat brought algebraic techniques into geometry in a convincing manner that eventually revolutionized the subject. Their analytic geometry was more limited than ours—e.g., they usually did not allow negative coordinates. John Wallis, in his *Arithmetica Infini-torum* in 1655, was the first to do that systematically (we shall en-counter his work again in Chapter 5). Hence all the loci of Descartes and Fermat were restricted to the first quadrant.

Briefly on the Number π

All the ancient civilizations guessed that the ratio of circumference C to diameter d of a circle was constant. For example, by marking a start-ing point and an ending point for a circular wheel rolling on a flat surface, it could be seen that the wheel advanced forward a bit over three diameters when it went through one revolution. The same ap-proximate result was obtained no matter what the size of the wheel, indicating that the ratio was independent of the size of the wheel.

In 1706, William Jones denoted that constant as π, and Leonhard Euler subsequently popularized this symbol in his voluminous writings. The ancient Egyptians had various estimates of π, one such being $22/7 = 3.\overline{142857}$.

The ancients also guessed from experience that the ratio of the area A of a circular disk to the square r^2 of its radius was constant. The Babylonians and ancient Chinese recognized that constant to be the same π because they knew, in our notation, the formula $A = Cr/2$; we don't know how they arrived at this result. Archimedes, in the century following Euclid, proved that formula, expressing it geometrically by saying, "*The circle equals in area the right triangle with base equal to its circumference and altitude equal to its radius.*" He proved his result using Eudoxus' method of exhaustion—a limiting argument. Archimedes also approximated a circle by inscribed and circumscribed regular polygons. Using 96-sided polygons, he obtained after a very lengthy calculation the estimate 3.1416 for π; he also obtained a crude bound for how much his estimate might be off. We know now that it is correct to four decimal places. Some early mathematicians thought of a circle as being a "regular polygon with infinitely many sides."

To treat these ideas rigorously yet on a relatively elementary level, if C is defined as the limit of the perimeters of the inscribed and circumscribed regular polygons, then after first proving that those limits exist and are the same, the constancy of C/d can be proved by applying theorems about similar triangles to those regular polygons. (See Moise, 1990, Section 21.2.)

However, we will learn, in Chapter 6, that in non-Euclidean geometry, similar triangles do not exist (except for congruent triangles, which are trivially similar). So that proof breaks down in non-Euclidean geometry, and in fact we will show in Chapter 10 that *C/d is not constant in real non-Euclidean geometry!* The reason C/d appears to be constant in our local physical world is that real Euclidean geometry provides a very good approximate model for that local world, as we all know. In the vast global world of the universe as a whole, Euclidean geometry may not be the best model, as we will discuss in Chapter 8.

Now the number π occurs in many formulas in many branches of mathematics, branches such as probability and statistics and complex analysis that have nothing to do with Euclidean geometry. Yet the definition of π we have indicated above depends on a Euclidean result. While it is correct that the number π was discovered historically via real Euclidean geometry, it is not logically correct to define π that way if we used the integral calculus to prove that $C = \pi d$; that would be circular reasoning.

(For those readers who know calculus, determination of the arc length of a quarter of a circle of radius R comes down to multiplying R by the integral $\int_{0}^{1} dt/\sqrt{1 - t^2}$. To obtain the answer $\pi/2$ for this integral, one must have already defined and studied the arcsin function, determined its derivative, and know that arcsin(0) = 0 and arcsin(1) = $\pi/2$, where

π has been previously defined analytically or is simply defined as 2 arcsin(1).)

The correct method is to define π as the limit of a certain sequence[12] of rational numbers, which can be done in many ways. Similarly, although the trigonometric functions were discovered historically via real Euclidean geometry and defined in terms of ratios of sides of right triangles, definitions that made sense because of the Euclidean theorem that corresponding sides of similar triangles are proportional, one logically correct definition of the trigonometric functions independent of Euclidean geometry is in terms of certain absolutely convergent infinite series.[13] Then those functions can be used in real non-Euclidean geometries as well, where the Euclidean theory of similar triangles is inoperative. See Chapter 10 and any rigorous treatise on analysis.

The number π has fascinated mathematicians (amateur as well as professional) throughout the ages. Several attractive books devoted entirely to this number have been published in recent years (see the bibliography at the back of this book). Incredibly, none of those books provide a proof that π is well defined, i.e., that C/d is constant in Euclidean geometry!

Conclusion

We have briefly discussed many historical facts and ideas in this chapter to provide the background for what will follow. You have the opportunity to explore them further in the exercises and projects for this chapter.

In subsequent chapters, we will hone in on a rigorous presentation of plane Euclidean geometry, placing special emphasis on the role played by the parallel postulate. We will then be able to analyze other attempts to prove that postulate besides the attempt by Legendre discussed in this chapter. After that we will see the dramatic story unfold of the discovery and ultimate validation of non-Euclidean geometry.

---

[12] For example,

$$\frac{\pi}{4} = 1 - \frac{1}{3} + \frac{1}{5} - \frac{1}{7} + \cdots.$$

[13] For example,

$$\sin x = x - \frac{x^3}{3!} + \frac{x^5}{5!} - \frac{x^7}{7!} + \cdots.$$

Review Exercise

Which of the following statements are correct?

(1) The Euclidean parallel postulate states that for every line l and for every point P not lying on l there exists a unique line m through P that is parallel to l.

(2) An "angle" is defined as the space between two rays that emanate from a common point.

(3) Most of the results in Euclid's *Elements* were discovered by Euclid himself.

(4) By definition, a line m is "parallel" to a line l if for any two points P, Q on m, the perpendicular distance from P to l is the same as the perpendicular distance from Q to l.

(5) It was unnecessary for Euclid to assume the parallel postulate because the French mathematician Legendre proved it.

(6) A "transversal" to two lines is another line that intersects both of them in distinct points.

(7) By definition, a "right angle" is a 90° angle.

(8) "Axioms" or "postulates" are statements that are assumed, without further justification, whereas "theorems" or "propositions" are proved using the axioms.

(9) We call $\sqrt{2}$ an "irrational number" because it cannot be expressed as a quotient of two whole numbers.

(10) The ancient Greeks were the first to insist on proofs for mathematical statements to make sure they were correct.

(11) Archimedes was the first to develop a theory of proportions valid for irrational lengths.

(12) The precise technology of measurement available to us today confirms the Pythagoreans' claim that $\sqrt{2}$ is irrational.

(13) The ancient Greek astronomers did not believe that three-dimensional Euclidean geometry was an idealized model of the entire space in which we live because they believed the universe is finite in extent, whereas Euclidean lines can be extended indefinitely.

(14) Descartes brought algebra into the study of geometry and showed he could solve every geometric problem with his method.

(15) The meaning of the Greek word "geometry" is "the art of reasoning well from badly drawn diagrams."

(16) A great many of Euclid's propositions can be interpreted as constructions with straightedge and compass, although he never mentions those instruments explicitly.
(17) Euclid provided constructions for bisecting and trisecting any angle.
(18) Although π is a Greek letter, in Euclid's *Elements* it did not denote the number we understand it to denote today.

Exercises

In Exercises 1–4, you are asked to define some familiar geometric terms. The exercises provide a review of these terms as well as practice in formulating definitions with precision. In making a definition, you may use the five undefined geometric terms and all other geometric terms that have been defined in the text so far or in any preceding exercises.

Making a definition sometimes requires a bit of thought. For example, how would you define *perpendicularity* for two lines l and m? A first attempt might be to say that "l and m intersect and at their point of intersection these lines form right angles." It would be legitimate to use the terms "intersect" and "right angle" because they have been previously defined. But what is meant by the statement that *lines* form right angles? Surely, we can all draw a picture to show what we mean, but the problem is to express the idea verbally using only terms introduced previously. According to the definition on page 18, an angle is formed by two nonopposite *rays* emanating from the same vertex. We may therefore define l and m as *perpendicular* if they intersect at a point A and if there is a ray $\overrightarrow{AB}$ that is part of l and a ray $\overrightarrow{AC}$ that is part of m such that ⦜BAC is a right angle (Figure 1.17). We denote this by $l \perp m$.

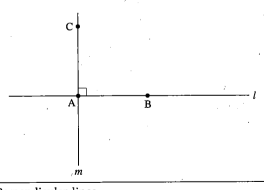

Figure 1.17 Perpendicular lines.

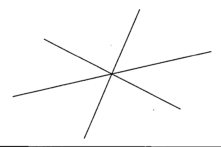

Figure 1.18 Concurrent lines.

1. Define the following terms:
   (a) *Midpoint* M of a segment AB.
   (b) *Perpendicular bisector* of a segment AB (you may use the term "midpoint" since you have just defined it).
   (c) Ray $\overrightarrow{BD}$ *bisects* angle ⊀ABC (given that point D is between A and C).
   (d) Points A, B, and C are *collinear*.
   (e) Lines *l*, *m*, and *n* are *concurrent* (see Figure 1.18).
2. Define the following terms:
   (a) The *triangle* △ABC formed by three noncollinear points A, B, and C.
   (b) The *vertices, sides*, and *angles* of △ABC. (The "sides" are segments, not lines.)
   (c) The sides *opposite to* and *adjacent to* a given vertex A of △ABC.
   (d) *Medians* of a triangle (see Figure 1.19).
   (e) *Altitudes* of a triangle (see Figure 1.20).
   (f) *Isosceles* triangle, its *base*, and its *base angles*.
   (g) *Equilateral* triangle.
   (h) *Right* triangle.

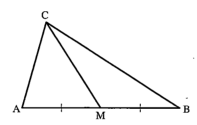

Figure 1.19 Median.

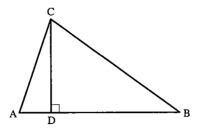

Figure 1.20   Altitude.

3. Given four points, A, B, C, and D, no three of which are collinear and such that any pair of the segments AB, BC, CD, and DA either have no point in common or have only an endpoint in common. We can then define the *quadrilateral* □ABCD to consist of the four segments mentioned, which are called its *sides*, the four points being called its *vertices* (see Figure 1.21). (Note that the order in which the letters are written is essential. For example, □ABCD may not denote a quadrilateral because, for example, AB might cross CD. If □ABCD did denote a quadrilateral, it would not denote the same one as □ACDB. Which permutations of the four letters A, B, C, and D do denote the same quadrilateral as □ABCD?) Using this definition, define the following notions:
   (a) The *angles* of □ABCD.
   (b) *Adjacent* sides of □ABCD.
   (c) *Opposite* sides of □ABCD.
   (d) The *diagonals* of □ABCD.
   (e) A *parallelogram*. (Use the word "parallel.")
4. Define *vertical angles* (Figure 1.22). How would you attempt to prove that vertical angles are congruent to each other? (Just sketch a plan for a proof—don't carry it out in detail.)
5. Use a common notion to prove the following result: If P and Q are any points on a circle with center O and radius OA, then OP ≅ OQ.

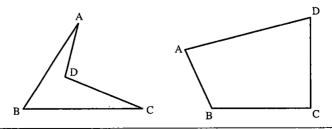

Figure 1.21   Quadrilaterals.

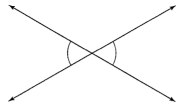

Figure 1.22 Vertical angles.

6. (a) Given two points A and B and a third point C between them. (Recall that "between" is an *undefined* term.) Can you think of any way to prove from the postulates that C lies on line $\overleftrightarrow{AB}$?

   (b) Assuming that you succeeded in proving C lies on $\overleftrightarrow{AB}$, can you prove from the definition of "ray" and the postulates that $\overrightarrow{AB} = \overrightarrow{AC}$?

7. If S and T are any sets, their *union* ($S \cup T$) and *intersection* ($S \cap T$) are defined as follows:

   (i) Something belongs to $S \cup T$ if and only if it belongs either to S or to T (or to both of them).

   (ii) Something belongs to $S \cap T$ if and only if it belongs both to S and to T.

   Given two points A and B, consider the two rays $\overrightarrow{AB}$ and $\overrightarrow{BA}$. Draw diagrams to show that $\overrightarrow{AB} \cup \overrightarrow{BA} = \overleftrightarrow{AB}$ and $\overrightarrow{AB} \cap \overrightarrow{BA} = AB$. What additional axioms about the undefined term "between" must we assume in order to be able to *prove* these equalities?

8. To further illustrate the need for careful definition, consider the following possible definitions of *rectangle*:

   (i) A quadrilateral with four right angles.

   (ii) A quadrilateral with all angles congruent to one another.

   (iii) A parallelogram with at least one right angle.

   In this book we will take (i) as our definition. Your experience with Euclidean geometry may lead you to believe that these three definitions are equivalent; sketch informally how you might prove that and notice carefully which theorems you are tacitly assuming. In hyperbolic geometry, these definitions give rise to three different sets of quadrilaterals (see Chapter 6).

9. Can you think of any way to prove from the postulates that for every line l

   (a) There exists a point lying on l?

   (b) There exists a point not lying on l?

10. Can you think of any way to prove from the postulates that the plane is nonempty, i.e., that points and lines exist? (Discuss with

your instructor what it means to say that mathematical objects, such as points and lines, "exist.")

11. Do you think that the Euclidean parallel postulate is "obvious"? Write a brief essay explaining your answer.

12. What is the flaw in the "proof" that all triangles are isosceles? (All the theorems from Euclidean geometry used in the argument are correct.)

13. If the number π is defined as the ratio of the circumference of any circle to its diameter, what theorem must first be proved to legitimize this definition? For example, if I "define" a new number φ to be the ratio of the area of any circle to its diameter, that would not be legitimate. Explain why not.

14. In this exercise, we will review several basic Euclidean constructions with a straightedge and compass. Such constructions fascinated mathematicians from ancient Greece until the nineteenth century, when all classical construction problems were finally solved.

   (a) Given a segment AB. Construct the perpendicular bisector of AB. (Hint: Make AB a diagonal of a rhombus, as in Figure 1.23.)

   (b) Given a line l and a point P lying on l. Construct the line through P perpendicular to l. (Hint: Make P the midpoint of a segment of l.)

   (c) Given a line l and a point P *not* lying on l. Construct the line through P perpendicular to l. (Hint: Construct isosceles triangle △ABP with base AB on l and use (a).)

   (d) Given a line l and a point P not lying on l. Construct a line through P parallel to l. (Hint: Use (b) and (c).)

   (e) Construct the bisecting ray of an angle. (Hint: Use the Euclidean theorem that the perpendicular bisector of the base on an

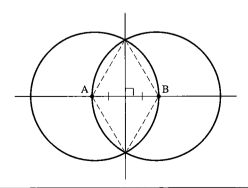

Figure 1.23

isosceles triangle is also the angle bisector of the angle opposite the base.)

(f) Given △ABC and segment DE ≅ AB. Construct a point F on a
given side of line $\overleftrightarrow{DE}$ such that △DEF ≅ △ABC.

(g) Given angle ∢ABC and ray $\overrightarrow{DE}$. Construct F on a given side of
line $\overleftrightarrow{DE}$ such that ∢ABC ≅ ∢FDE.

15. Euclid assumed the compass to be *collapsible*. That is, given two
points P and Q, the compass can draw a circle with center P passing through Q (Postulate III); however, the spike cannot be moved
to another center O to draw a circle of the same radius. Once the
spike is moved, the compass collapses. Check through your constructions in Exercise 14 to see whether they are possible with a
collapsible compass. (For purposes of this exercise, being "given"
a line means being given two or more points on it.)

(a) Given three points P, Q, and R. Construct with a straightedge
and collapsible compass a rectangle □PQST with PQ as a side
and such that PT ≅ PR (see Figure 1.24).

(b) Given a segment PQ and a ray $\overrightarrow{AB}$. Construct the point C on
$\overrightarrow{AB}$ such that PQ ≅ AC. (Hint: Using part (a), construct rectangle □PAST with PT ≅ PQ and then draw the circle centered
at A and passing through S.)

Part (b) shows that you can transfer segments with a collapsible
compass and a straightedge, so you can carry out all constructions
as if your compass did not collapse.

16. The straightedge you used in the previous exercises was supposed
to be *unruled* (if it did have marks on it, you weren't supposed to
use them). Now, however, let us mark two points on the straightedge so as to mark off a certain distance d. Archimedes showed
how we can then trisect an arbitrary angle.

For any angle, draw a circle γ of radius d centered at the vertex O of the angle. This circle cuts the sides of the angle at points
A and B. Place the marked straightedge so that one mark gives a

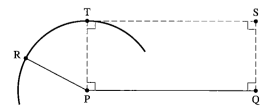

Figure 1.24

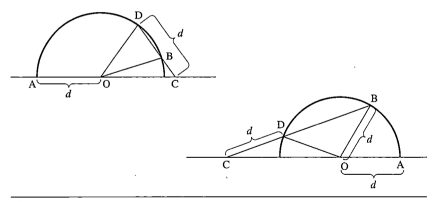

Figure 1.25

point C on line $\overleftrightarrow{OA}$ such that O is between C and A, the other mark gives a point D on circle γ, and the straightedge must simultaneously rest on the point B, so that B, C, and D are collinear (Figure 1.25). Prove that $\angle COD$ so constructed is one-third of $\angle AOB$. (Hint: Use Euclidean theorems on exterior angles and isosceles triangles.)

Major Exercises

1. The number $\rho = (1 + \sqrt{5})/2$ was called the *golden ratio* by the Greeks, and a rectangle whose sides are in this ratio is called a *golden rectangle*.[14] Prove that a golden rectangle can be constructed with straightedge and compass as follows:
   (a) Construct a square $\square$ABCD.
   (b) Construct midpoint M of AB.
   (c) Construct point E such that B is between A and E and MC $\cong$ ME (Figure 1.26).
   (d) Construct the foot F of the perpendicular from E to $\overleftrightarrow{DC}$.
   (e) Then $\square$AEFD is a golden rectangle (use the Pythagorean theorem for $\triangle$MBC).
   (f) Moreover, $\square$BEFC is another golden rectangle (first show that $1/\rho = \rho - 1$).

The next two exercises require a knowledge of trigonometry.

---

[14] For applications of the golden ratio to Fibonacci numbers and phyllotaxis, see Coxeter (2001), Chapter 11. Also see Livio (2005).

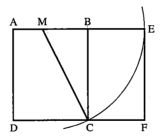

Figure 1.26

2. The Egyptians thought that if a quadrilateral had sides of lengths a, b, c, and d, then its area S was given by the formula $(a + c)(b + d)/4$. Prove that actually

$$4S \leq (a + c)(b + d)$$

with equality holding only for rectangles. (Hint: Twice the area of a triangle is $ab \sin \theta$, where θ is the angle between the sides of lengths a, b, and $\sin \theta \leq 1$, with equality holding only if θ is a right angle.)

3. Prove analogously that if a triangle has sides of lengths a, b, c, then its area S satisfies the inequality

$$4S\sqrt{3} \leq a^2 + b^2 + c^2$$

with equality holding only for equilateral triangles. (Hint: If θ is the angle between sides b and c, chosen so that it is at most $60°$, then use the formulas

$$2S = bc \sin \theta$$
$$2bc \cos \theta = b^2 + c^2 - a^2 \text{ (law of cosines)}$$
$$\cos (60° - \theta) = (\cos \theta + \sqrt{3} \sin \theta)/2$$

4. Let $\triangle ABC$ be such that AB is not congruent to AC. Let D be the point of intersection of the bisector of $\sphericalangle A$ and the perpendicular bisector of side BC. Let E, F, and G be the feet of the perpendiculars dropped from D to $\overleftrightarrow{AB}$, $\overleftrightarrow{AC}$, $\overleftrightarrow{BC}$, respectively. Prove that:
   (a) D lies outside the triangle on the circle through ABC.
   (b) One of E or F lies inside the triangle and the other outside.
   (c) E, F, and G are collinear.
   (Use anything you know, including coordinates if necessary.)

5. Figure out an algebraic proof that if a natural number n is not the square of some other natural number, then $\sqrt{n}$ is irrational. (If you

are stymied, see Barry Mazur's essay "How did Theaetetus prove his theorem?" at www.math.harvard.edu/~mazur/preprints/Eva. Nov.20.pdf. In this essay, Pappus is quoted as saying "Ignorance of the fact that incommensurables exist is a brutish and inhuman state." Do you agree or disagree? Explain.)

Projects

1. (a) Report on at least three other proofs. of the Pythagorean theorem besides the ones illustrated in this chapter. (Suggestion: See Maor, 2007.) If you find further interesting historical information about this great theorem, report on that too (e.g., President Garfield's proof).

   (b) A *Pythagorean triple* is a triple (a, b, c) of positive integers satisfying the Pythagorean equation. The triple is *primitive* if the integers have no common factor. A general Pythagorean triple is a positive integer multiple of a primitive one (cancel the gcd). Find polynomials p, q, r of degree 2 in two integer variables such that every primitive Pythagorean triple is given by $a = p(m, n)$, $b = q(m, n)$, and $c = r(m, n)$ and conversely these equations provide a primitive Pythagorean triple for every pair of unequal relatively prime positive integers (m, n). (Hint: Show that this problem is equivalent to finding all points on the unit circle with rational coordinates and solve that using the pencil of lines through $(-1, 0)$.) Search the web for further results on Pythagorean triples and report on the results you find most interesting.

2. From the long list of propositions in Euclid's *Elements* that were described in this chapter as straightedge-and-compass constructions, choose five that have not been discussed in Exercise 14 and report in detail on how Euclid's proofs of those propositions can be interpreted as such constructions.

3. Write a paper explaining in detail why it is impossible to trisect an arbitrary angle or square a circle using only a compass and unmarked straightedge (see Jones, Morris, and Pearson, 1991; Eves, 1972; or Moise, 1990). Explain how arbitrary angles can be trisected if in addition we are allowed to draw a parabola or a hyperbola or a conchoid or a limaçon (see Peressini and Sherbert, 1971).

4. Here are two other famous results in the theory of constructions:

(a) Mathematicians G. Mohr of Denmark and L. Mascheroni of Italy discovered independently that all Euclidean constructions of points can be made with a compass alone. A line, of course, cannot be drawn with a compass, but it can be determined with a compass by constructing two points lying on it. In this sense, Mohr and Mascheroni showed that the straightedge is unnecessary.

(b) On the other hand, German mathematician J. Steiner and Frenchman J. V. Poncelet showed that all Euclidean constructions can be carried out with a straightedge alone if we are first given a single circle and its center.

Report on these remarkable discoveries (see Eves, 1972).

5. Given any △ABC. Draw the two rays that trisect each of its angles and let P, Q, and R be the three points of intersection of adjacent trisectors. Prove Morley's theorem[15] that △PQR is an equilateral triangle (see Figure 1.27 and Coxeter, 2001, Section 1.9).

6. An n-sided polygon is called *regular* if all its sides (respectively, angles) are congruent to one another. Construct a regular pentagon and a regular hexagon with straightedge and compass. The regular septagon cannot be so constructed; in fact, Gauss proved the remarkable theorem that the regular n-gon is constructible if and only if all odd prime factors of n occur to the first power and have the form $2^{2^m} + 1$ (e.g., 3, 5, 17, 257, 65,537). Report on this result, using Klein (2007). Primes of that form are called *Fermat primes*. The five listed are the only ones known at this time. Gauss did not actually construct the regular 257-gon or 65,537-gon; he showed only that the minimal polynomial equation satisfied by $\cos(2\pi/n)$ for such n could be solved in the surd field (see Moise, 1990). Other

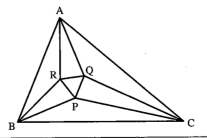

Figure 1.27 Morley's theorem.

[15] For a converse and generalization of Morley's theorem, see Kleven (1978).

devoted (obsessive?) mathematicians carried out the constructions. The constructor for $n = 65,537$ labored for 10 years and was rewarded with a Ph.D. degree; what is the reward for checking his work?

7. Research and report on *neusis* constructions, mentioned in this chapter and illustrated in Exercise 16. (Search on the web or use Bos (2001) as a reference.) Give and explain your opinion on whether Viète's axiom for neusis constructions should be accepted in elementary geometry. Discuss in your report Viète's construction of the regular heptagon using compass and marked ruler as another example of how useful this axiom is (see Hartshorne, Problem 30.4). Describe the primes p for which the regular p-gon can be constructed with compass and marked straightedge using the marks between lines only (Hartshorne, Corollary 31.9). Is $p = 11$ one of them?

8. Report on solutions in antiquity to the three classical construction problems using curves other than lines and circles (e.g., the quadratrix, the conchoid, the cissoid, etc.). Use Bos (2001) as a great reference to report on the historical issue of what constitutes an "exact" construction in geometry and for a thorough analysis of what Descartes did.

9. Write a report on the invention/discovery of *analytic geometry*. Your report should explain the differences and similarities between the works of Descartes and Fermat.

10. In chronological order of birth, Eudoxus, Archimedes, and Apollonius were the greatest mathematicians of ancient Greece. Choose one of them and report on his work.

11. Report on episodes that interest you in the history of irrational numbers (use the web or a good history text such as Katz, 1998).

12. Report on Descartes' *La Géométrie* (1954, in its English translation, if necessary). Do you agree with his statement that explaining the subject in too much detail deprives the reader of the pleasure of mastering it himself?

13. Comment on the following quotes:
    (a) The axiomatic method has many advantages over honest work—Bertrand Russell.
    (b) Our difficulty is not in the proofs, but in learning what to prove—Emil Artin.

2

Logic and Incidence Geometry

Elementary Logic

In the previous chapter, we introduced the postulates and basic definitions of Euclid's plane geometry, slightly rephrased for greater precision. We would like to begin proving some theorems or propositions that are logical consequences of the postulates. However, certain exercises in the previous chapter may have alerted you to expect some difficulties that we must first clear up. For example, there is nothing in the postulates that guarantees that a line has any points lying on it (or off it)! You may feel this is ridiculous—it wouldn't be a line if it didn't have any points lying on it. Your protest is legitimate, for if my concept of a line were so different from yours, then we would not understand each other, and Requirement 0—that there be mutual under-standing of words and symbols used—would be violated.

So let's be clear: We must play this game according to the rules, the rules mentioned in Requirement 2 but not spelled out. Unfortunately, to discuss them thoroughly would require changing the content of this text from geometry to mathematical logic. Instead, I will simply remind you of some basic rules of reasoning that you, as a rational being, already know and have used in your previous work in mathematics. Some ideas and notation from mathematical logic will be introduced. If you have a good deal of experience in mathematics, I recommend that you quickly skim this material on logic and move ahead to the section on models.

LOGIC RULE 0. No unstated assumptions may be used in a proof.

The reason for taking the trouble to list all our axioms is to be explicit about our basic assumptions, including the most obvious. Although it may be "obvious" that two points lie on a unique line, Euclid stated this as his first postulate. So if in some proof we want to say that every line has points lying on it, we should list this as another postulate (or prove it, but we can't). In other words, all our cards must be out on the table, and we will have to add two other axioms in the section on incidence geometry to guarantee that existence.

Perhaps you have realized by now that there is a vital relation between axioms and undefined terms. As we have seen, we must have undefined terms in order to avoid infinite regress. But this does not mean we can use these terms in any way we choose. The axioms tell

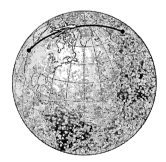

Figure 2.1 The shortest path between two points on a sphere is an arc of a great circle (a circle whose center is the center of the sphere and whose radius is the radius of the sphere, e.g., the equator).

us exactly what properties of undefined terms we are allowed to use in our arguments. You may have some other properties in your mind when you think about these terms, but you're not allowed to use them in a proof (Rule 0). For example, when you think of the unique line segment determined by two points, you probably think of it as being "straight," or as "the shortest path between the two points." Euclid's postulates alone do not allow us to assume these properties. Besides, from one viewpoint, these properties could be considered contradictory. If you were traveling over the surface of the earth, idealized as a sphere, say from San Francisco to Moscow, the shortest path would be an arc of a great circle (a straight path would bore through the earth). Indeed, a pilot making that trip nonstop normally takes the great circle route (see Figure 2.1).

Theorems and Proofs

All mathematical theorems are conditional statements, statements of the form

> If [hypothesis] *then* [conclusion]

In some cases, a theorem may state only a conclusion; the axioms of the particular mathematical system are then implicitly assumed as a hypothesis. If a theorem is not written in the conditional form, it can nevertheless be translated into that form. For example,

> *Base angles of an isosceles triangle are congruent*

can be translated as

> *If a triangle has two congruent sides, then the*
> *angles opposite those sides are congruent.*

Put another way, a conditional statement says that one condition (the hypothesis) implies another (the conclusion). If we denote the hypothesis by H, the conclusion by C, and the word "implies" by a double arrow $\Rightarrow$, then every theorem has the form $H \Rightarrow C$. (In the example above, H is "two sides of a triangle are congruent" and C is "the angles opposite those sides are congruent.")

Not every conditional statement is a theorem. For example, the statement

> *If $\triangle ABC$ is any triangle, then it is isosceles*

is not a theorem. Why not? You might say that this statement is "false," whereas theorems are "true." Let's avoid the loaded words "true" and "false" as much as we can, for they beg the question and lead us into much more complicated philosophical issues.

In a given mathematical system, the only statements we call *theorems* are those statements for which a *correct proof* has been supplied. (We also call them *propositions, corollaries,* or *lemmas.* "Theorem" and "proposition" are interchangeable words, though usually the word "theorem" is reserved for a particularly important proposition. A "corollary" is an immediate consequence of a theorem, and a "lemma" is a "helping or subsidiary theorem." Logically, they all mean the same; the title is just an indicator of the author's emphasis.) The statement that every triangle is isosceles has not been given a correct proof (I hope you found the flaw in the pretended proof in Chapter 1). You will later refute that statement in Euclidean geometry by proving there exists a triangle that is not isosceles.

The crux of the matter, then, is the notion of *proof.* By definition, a proof is a list of statements, together with a justification for each statement, ending up with the conclusion desired. Usually, each statement in a formal proof will be numbered in this book, and the justification for it will follow in parentheses.

LOGIC RULE 1. The following are the six types of justifications allowed for statements in proofs:
   (1) "By hypothesis . . ."
   (2) "By axiom . . ."
   (3) "By theorem . . ." (previously proved)
   (4) "By definition . . ."
   (5) "By step . . ." (a previous step in the argument)
   (6) "By rule . . . of logic."

Later in this text our proofs will be less formal, and justifications may be omitted when they are clear. (Be forewarned, however, that these omissions can lead to incorrect results.) Also, *a justification may include several of the above types.*

Having described proofs, it would be nice to be able to tell you how to find or construct them. Yet that is the artistry, the creativity, of doing mathematics. Certain techniques for proving theorems are learned by experience, by imitating what others have done. If the problem is not too complicated, you can figure out a proof using your natural reasoning ability. But there is no mechanical method for proving or dis-

" I THINK YOU SHOULD BE
MORE EXPLICIT HERE IN STEP TWO."

proving *every* statement in mathematics. (The nonexistence of such a mechanical method is, when stated precisely, a deep theorem in mathematical logic and is the reason why computers as we know them today will never put mathematicians out of business—see any advanced text on mathematical logic. Of course, there has been progress in automatic theorem proving for small portions of mathematics.) There is a mechanical method for *verifying* that a proof, presented formally, is correct—just check the justification for each step. In the discussion of Proposition 2.2 ahead, indication is given of how its proof might have been discovered.

Some suggestions may help you construct proofs. First, make sure you clearly understand the meaning of each term in the statement of the proposed theorem. If necessary, review their definitions. Second, keep reminding yourself of what it is you are trying to prove. If it

involves parallel lines, for example, look up previously proved propositions that give you information about parallel lines. If you find another proposition that seems to apply to the problem at hand, check
carefully to see whether it really does apply. Draw diagrams to help
you visualize the problem.

RAA Proofs

The most common type of proof in this book is proof by reductio ad
absurdum, abbreviated RAA. In this type of proof, you want to prove
a conditional statement, $H \Rightarrow C$, and you begin by assuming the contrary of the conclusion you seek. We call this contrary assumption the
RAA hypothesis to distinguish it from the hypothesis H. The RAA hypothesis is a temporary assumption from which we derive, by reasoning, an *absurd statement* ("absurd" in the sense that it denies something known to be valid). Such a statement might deny the hypothesis
of the theorem or the RAA hypothesis; it might deny a previously proved
theorem or an axiom. Once it is shown that the negation of C leads to
an absurdity, it follows that C must be valid. This is called the *RAA
conclusion*. To summarize:

LOGIC RULE 2. To prove a statement $H \Rightarrow C$, assume the negation of
statement C (RAA hypothesis) and deduce an absurd statement, using
the hypothesis H if needed in your deduction.

Let us illustrate this rule by proving the following proposition
(Proposition 2.1): If l and m are distinct lines that are not parallel, then
l and m have a unique point in common.

PROOF:

(1) Because l and m are not parallel, they have a point in common
    (by definition of "parallel").
(2) Since we want to prove uniqueness for the point in common,
    we will assume the contrary, that l and m have two distinct
    points A and B in common (RAA hypothesis).
(3) Then there is more than one line on which A and B both lie
    (step 2 and the hypothesis of the theorem, $l \neq m$).
(4) A and B lie on a unique line (Euclid's Postulate I).
(5) Intersection of l and m is unique (step 3 contradicts step 4, RAA
    conclusion). ◄

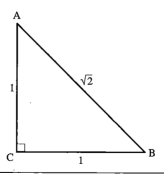

Figure 2.2

Notice that in steps 2 and 5, instead of writing "Logic Rule 2" as justification, we wrote the more suggestive "RAA hypothesis" and "RAA conclusion," respectively.

As another illustration, consider one of the earliest RAA proofs, given by Aristotle and presumably discovered by the Pythagoreans (to their great dismay). In giving this proof, we will use some facts that you know about Euclidean geometry, algebra, and numbers, and we will be informal.

Suppose $\triangle ABC$ is a right isosceles triangle with right angle at C. We can choose our unit of length so that the legs have length 1. The theorem then says that the length of the hypotenuse is irrational (Figure 2.2).

By the Pythagorean equation, the length of the hypotenuse is $\sqrt{2}$, so we must prove that $\sqrt{2}$ is an irrational number, i.e., that it is not a rational number.

What is a rational number? It is a number that can be expressed as a quotient p/q of two integers p and q. For example, 2/3 and 5 = 5/1 are rational numbers. We want to prove that $\sqrt{2}$ is not one of these numbers.

We begin by assuming the contrary, that $\sqrt{2}$ is a rational number (RAA hypothesis). In other words, $\sqrt{2} = p/q$ for certain unspecified whole numbers p and q. We may assume, from our knowledge of numbers and fractions, that after canceling out any common 2's, p and q are not both even numbers.

Next, we clear denominators

$$\sqrt{2}q = p$$

and square both sides:

$$2q^2 = p^2.$$

This equation says that p^2 is an even number (since p^2 is twice another whole number, namely, q^2). If p^2 is even, p must be even, for the square of an odd number is odd, as you know. Thus,

$$p = 2r$$

for some whole number r (by definition of "even"). Substituting $2r$ for p in the previous equation gives

$$2q^2 = (2r)^2 = 4r^2.$$

We then cancel 2 from both sides to get

$$q^2 = 2r^2.$$

This equation says that q^2 is an even number; hence, as before, q must be even.

We have shown that numerator p and denominator q are both even. Now this is absurd because all common 2 factors were canceled. Thus $\sqrt{2}$ is irrational (RAA conclusion). ◄

Negation

In an RAA proof, we begin by "assuming the contrary." Sometimes the contrary or negation of a statement is not obvious, so you should know the rules for negation.

First, some remarks on notation. If S is any statement, we will denote the negation or contrary of S by $\sim S$. For example, if S is the statement "p is even," then $\sim S$ is the statement "p is not even" or "p is odd."

The rule below applies to those cases where S is already a negative statement. The rule states that two negatives make a positive.

Logic Rule 3. The statement "$\sim(\sim S)$" means the same as "S."

We followed this rule when we negated the statement "$\sqrt{2}$ is irrational" by writing the contrary as "$\sqrt{2}$ is rational" instead of "$\sqrt{2}$ is not irrational."

Another rule we have already followed in our RAA method is the rule for negating an implication. We wish to prove $H \Rightarrow C$, and we assume, on the contrary, H does not imply C, i.e., that H holds and at the same time $\sim C$ holds. We write this symbolically as $H \ \& \ \sim C$, where & is the abbreviation for "and." A statement involving the connective "and" is called a *conjunction*. Thus:

LOGIC RULE 4. The statement "$\sim[H \Rightarrow C]$" means the same thing as "H & $\sim C$."

Let us consider, for example, the conditional statement "If 3 is an odd number, then 3^2 is even." According to Rule 4, the negation of this is the declarative statement "3 is an odd number and 3^2 is odd."

How do we negate a conjunction? A conjunction S_1 & S_2 means that statements S_1 and S_2 both hold. Negating this would mean asserting that one of them does not hold, i.e., asserting the negation of one or the other. Thus:

LOGIC RULE 5. The statement "$\sim[S_1$ & $S_2]$" means the same as "$[\sim S_1 \bigvee \sim S_2]$."

Here we have introduced the logic symbol "$\bigvee$" to abbreviate *or*. A statement involving "$\bigvee$" is called a *disjunction*. The mathematical "or" is not exclusive like the "or" in everyday usage. When a mathematician writes "$S_1 \bigvee S_2$," what is meant is "either S_1 holds or S_2 holds *or they both hold.*"

Now let us clarify what is meant by "an absurd statement" in Rule 2 (RAA): It is a *contradiction*, a statement of the form "S & $\sim S$." Usually in an RAA argument, statement S will occur in one line of the proof and statement $\sim S$ will occur in another line. By the meaning of "and" we can then infer S & $\sim S$, but we will usually not bother with that and will just point out that the line with $\sim S$ contradicts the line with S.

Quantifiers

Most mathematical statements involve *variables*. For instance, the Pythagorean theorem states that for any right triangle, if a and b are the lengths of the legs and c the length of the hypotenuse, then $c^2 = a^2 + b^2$. Here a, b, and c are variable numbers, and the triangle whose sides they measure is a variable triangle.

Variables can be quantified in two different ways. First, in a *universal* way, as in the expressions:

"For any x, . . ."
"For every x, . . ."
"For all x, . . ."
"Given any x, . . ."
"If x is any . . ."

Second, in an *existential* way, as in the expressions:

"For some x, . . ."
"There exists an x . . ."
"There is an x . . ."
"There are x . . ."

Consider Euclid's first postulate, which states informally that two points P and Q determine a unique line l. Here P and Q may be any two points, so they are quantified universally, whereas l is quantified existentially since it is asserted to exist once P and Q are given.

It must be emphasized that a statement beginning with "For every . . ." does not imply the existence of anything. The statement "Every unicorn has a horn on its head" does not imply that unicorns exist.

If a variable x is quantified universally, this is usually denoted as $\forall x$ (read as "for all x"). If x is quantified existentially, this is usually denoted as $\exists x$ (read as "there exists an x . . ."). After a variable x is quantified, some statement is made about x, which we can write as $S(x)$ (read as "statement S about x"). Thus, a universally quantified statement about a variable x has the form $\forall x S(x)$.

We wish to have rules for negating quantified statements. How do we deny that statement $S(x)$ holds for all x? We can do so clearly by asserting that for some x, $S(x)$ does not hold.

LOGIC RULE 6. The statement "$\sim[\forall x S(x)]$" means the same as "$\exists x \sim S(x)$."

For example, to deny "All triangles are isosceles" is to assert "There is a triangle that is not isosceles."

Similarly, to deny that there exists an x having property $S(x)$ is to assert that all x fail to have property $S(x)$.

LOGIC RULE 7. The statement "$\sim[\exists x S(x)]$" means the same as "$\forall x \sim S(x)$."

For example, to deny "There is an equilateral right triangle" is to assert "Every right triangle is nonequilateral" or, equivalently, to assert "No right triangle is equilateral."

Since in practice quantified statements involve several variables, the above rules will have to be applied several times. Usually, common sense will quickly give you the negation. If not, follow the above rules.

Let's work out the denial of Euclid's first postulate. This postulate is a statement about all pairs of points P and Q; negating it would mean, according to Rule 6, asserting the existence of points P and Q that do not satisfy the postulate. Postulate I involves a conjunction, asserting that P and Q lie on some line *l and* that *l* is unique. In order to deny this conjunction, we follow Rule 5. The assertion becomes either "P and Q do not lie on any line" *or* "they lie on more than one line." Thus, the negation of Postulate I asserts: "There are two points P and Q that either do not lie on any line or lie on more than one line."

If we return to the example of the surface of the earth, thinking of a "line" as a great circle, we see that there do exist such points P and Q—namely, take P to be the north pole and Q the south pole. Infinitely many great circles pass through both poles (see Figure 2.3).

Mathematical statements are sometimes made informally, and you may sometimes have to rephrase them before you will be able to negate them. For example, consider the following statement:

> *If a line intersects one of two parallel lines,*
> *it also intersects the other.*

This appears to be a conditional statement, of the form "if . . . then . . ."; its negation, according to Rule 4, would appear to be:

> *A line intersects one of two parallel lines*
> *and does not intersect the other.*

If this, seems awkward, it is because the original statement contained *hidden quantifiers* that have been ignored. The original statement refers to *any* line that intersects one of two parallel lines, and these are *any* parallel lines. There are universal quantifiers implicit in

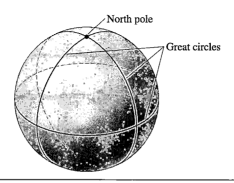

Figure 2.3

the original statement. So we have to follow Rule 6 as well as Rule 4 in forming the correct negation, which is:

> *There exist two parallel lines and a line that intersects*
> *one of them and does not intersect the other.*

Here are two other ways we will work with quantifiers in proofs. Suppose it has been previously proved or an axiom states that there exists some object with a certain property. We are then permitted to say "Let . . . be an object with that property." This amounts to naming an exemplar of what has been proved to exist. For example, Euclid's third postulate asserts the existence of a circle with a given radius. So we can say "Let γ be a circle with radius . . ." and refer to Euclid III for justification. This naming method is called *specification*. Going in the other direction, suppose we wish to prove that all objects of a certain type have a certain property. We begin by naming an arbitrary object of that type. Then we prove that it has the property we seek. Since the object was arbitrary, we are allowed to conclude that all objects of that type have the desired property. That method is called *generalization*. For example, suppose we want to prove that the square of every odd number is odd. We start by saying "Let n be an odd number" and justify this as our hypothesis. Then we prove that the square of n is odd. That will usually be the end of our proof, it being understood that since n was arbitrary, we have proved the assertion for all n.

Implication

Another rule, called the *rule of detachment*, or *modus ponens*, is the following:

LOGIC RULE 8. If $P \Rightarrow Q$ and P are steps in a proof, then Q is a justifiable step.

This rule is almost a definition of what we mean by implication. For example, we have an axiom stating that if $\angle A$ and $\angle B$ are right angles, then $\angle A \cong \angle B$ (Postulate IV). Now in the course of a proof, we may come across two right angles. Rule 8 allows us to assert their congruence as a step in the proof.

You should beware of confusing a conditional statement $P \Rightarrow Q$ with its *converse* $Q \Rightarrow P$. For example, the converse of Postulate IV states that if $\angle A \cong \angle B$, then $\angle A$ and $\angle B$ are right angles, which is not valid.

However, it may sometimes happen that both a conditional statement and its converse are valid. In the case where $P \Rightarrow Q$ and $Q \Rightarrow P$ both hold, we write simply $P \Leftrightarrow Q$ (read as *"P if and only if Q"* or *"P is logically equivalent* to Q"). All definitions are of this form. For example, three points are collinear if and only if they lie on a line. Some theorems are also of this form, such as the theorem "a triangle is isosceles if and only if two of its angles are congruent to each other." An abbreviation for *"P if and only if Q"* is *"P iff Q."*

The next rule gives a few more ways that "implication" is often used in proofs.

LOGIC RULE 9.

(a) $[[P \Rightarrow Q] \ \& \ [Q \Rightarrow R]] \Rightarrow [P \Rightarrow R]$

(b) $[P \ \& \ Q] \Rightarrow P, \ [P \ \& \ Q] \Rightarrow Q$

(c) $[\sim Q \Rightarrow \sim P] \Leftrightarrow [P \Rightarrow Q]$

Part (c) states that every implication $P \Rightarrow Q$ is logically equivalent to its *contrapositive* $\sim Q \Rightarrow \sim P$. For example, the statement "If two sides of a triangle are congruent, then the angles opposite those sides are congruent" is logically equivalent to the statement "If the angles opposite two sides of a triangle are *not* congruent, then the two sides are *not* congruent." You can verify this logical equivalence by using the RAA rule and Rule 3. Part (a) expresses the *transitivity* of implication. Part (b) gives the connection between conjunction and implication.

All parts of Rule 9 are called *tautologies* because they are valid just by their form, not because of what P, Q, and R mean; by contrast, the validity of a formula such as $P \Rightarrow Q$ does depend on the meaning of its constituents P and Q, as we have seen. There are infinitely many tautologies, and the next rule gives the most controversial one (see the historical discussion below).

Law of Excluded Middle and Proof by Cases

LOGIC RULE 10. For every statement P, "$P \vee \sim P$" is a valid step in a proof (law of excluded middle).

For example, given point A and line l, we may assert that either A lies on l or it does not. If this is a step in a proof, we will usually then break the rest of the proof into cases—giving an argument under the case assumption that A lies on l and giving another argument under the case assumption that A does not. Both arguments must be given, or else the proof is incomplete. A proof of this type is given in Chapter 3 for the proposition that there exists a line through A perpendicular to l.

Sometimes there are more than two cases. For example, it is a theorem that either an angle is acute or it is right or it is obtuse—three cases. We will have to give three arguments—one for each case assumption. You will give such arguments when you prove the SSS (side-side-side) criterion for congruence of triangles in Exercise 32 of Chapter 3. This method of *proof by cases* was used (correctly) in the incorrect attempt in Chapter 1 to prove that all triangles are isosceles.

LOGIC RULE 11.   Suppose the disjunction of statements S_1 or S_2 or . . . or S_n is already a valid step in a proof. Suppose that proofs of C are carried out from each of the *case assumptions* $S_1, S_2, \ldots, S_n$. Then C can be concluded as a valid step in the proof (proof by cases).

Finally, we will state Euclid's "common notions" for equality as a rule of logic.

LOGIC RULE 12.
    (1) $\forall X\ (X = X)$
    (2) $\forall X\ \forall Y\ (X = Y \Leftrightarrow Y = X)$
    (3) $\forall X\ \forall Y\ \forall Z\ ((X = Y\ \&\ Y = Z) \Rightarrow X = Z)$
    (4) If $X = Y$ and $S(X)$ is a statement about X, then $S(X) \Leftrightarrow S(Y)$.

Statement (1) says equality is *reflexive*; (2) says equality is *symmetric*; and (3) says equality is *transitive*. The conjunction of (2) and (3) gives us Euclid's common notion that "things equal to the same thing are equal to each other." Later on we will encounter other binary relations having these three properties—congruence, for example. Such relations are called *equivalence relations*. They play an extremely important role in modern mathematics. Statement (4) says that "equals can be *substituted* for equals" in any statement. This informal assertion must be qualified *when quantifiers are part of the statement*, for in that case *you are only allowed to substitute for* "free" *occurrences of the variable* X. See any logic textbook for the details.

Brief Historical Remarks

This concludes our list of rules for elementary logic. No claim is made that all the basic rules of logic have been listed, just that those listed suffice for our purpose of developing elementary geometry (we have skipped many technical details, including the *careful development of a formal language*—all "statements" we discuss must be expressed in that language). Euclid took the rules of reasoning for granted, but if we are committed to making all our assumptions explicit, we should do so not only for our geometric assumptions but also for our assumptions about logic.

Aristotle was the first to formulate basic principles of logic in his system of *syllogisms*. However, mathematicians in ancient Greece did not use Aristotle's syllogistic forms. Instead, they basically followed the forms of argument delineated in the third century B.C. by the Stoic (Megarian) philosophers—most prominently Chrysippus, considered a greater logician than Aristotle, but his works mostly have been lost.

It was Gottfried Wilhelm von Leibniz in his 1666 publication *De Arte Combinatoria* who first proposed the idea of an algebra of logic. He wished to develop a symbolic language for reasoning with a simple set of basic rules to do logic algebraically. It was not until the middle of the nineteenth century that George Boole and Augustus de Morgan began to carry out his idea. *Boolean algebra* is now the foundation for computer arithmetic and is very important in pure mathematics.

In 1879 Gottlob Frege brought quantifiers into logic, introducing what is now known as *the predicate calculus*, but with terrible notation. Most of the currently used notation and methods of mathematical logic stem from the society of logicians founded in the 1880s by Giuseppe Peano along with Mario Pieri. They emphasized the importance of a formal symbolic language for mathematics to remove the ambiguities of natural languages, to make mathematics utterly precise, and to permit the mathematical study of entire mathematical theories. Many years later, this formalization also enabled the programming of computers to do mathematics.

The discovery and validation of non-Euclidean geometries, together with Georg Cantor's invention of set theory and Karl Weierstrass' rigorous presentation of analysis, caused mathematicians to study axiomatics seriously for the first time. It was not until 1889 that axioms for the arithmetic of natural numbers were satisfactorily formulated— by Peano, based on Richard Dedekind's set-theoretic development using the *successor* function (and influenced by earlier algebraic work of Herman Grassmann). Peano's 1899 "first-order" axioms did not refer to sets. They included the basic algebraic laws of addition and multiplication and, most importantly, the principle of *mathematical induction*, which mathematicians had been using informally since at least the time of Fermat and Pascal. The formal system based on those axioms is called *Peano arithmetic*, denoted PA. Hilbert's set-theoretic axiomatization of elementary geometry also appeared in 1899.

In the twentieth century, mathematical logic came into its own as a very important branch of mathematics. The most influential foundational works in logic in the early twentieth century were the *Principia Mathematica* of Bertrand Russell and A. N. Whitehead; the work of David Hilbert with his associates Wilhelm Ackermann, Paul Bernays, and John von Neumann; and the contributions of Thoralf Skolem. By formalizing all rules of reasoning and axioms in a purely symbolic language, mathematicians were able to study entire branches of their subject, such as Peano arithmetic and elementary geometry and Zermelo–Fraenkel set theory. They were then able to prove theorems about those branches—theorems that are called *metamathematical* because they are about mathematical theories, not about numbers or geometric figures or sets. The most important metamathematical theorems are the completeness and incompleteness theorems of Kurt Gödel from the early 1930s, which revolutionized our thinking about the nature of mathematics. Also vitally important in the 1930s were the equivalent determinations of the class of *effectively computable number-theoretic functions* by Alan Turing, Alonzo Church, Emil Post, and Gödel.

The rules of logic we have listed come from what is known as *classical two-valued* logic. Just as there are non-Euclidean geometries, in which certain axioms of Euclidean geometry are changed, there are also non-classical logics in which certain rules are changed or dropped. For example, constructivist mathematicians such as L. E. J. Brouwer and Errett Bishop reject the use of the law of excluded middle when applied to infinite sets; Arend Heyting developed the so-called *intuitionist formal logic* for reasoning without that law. Constructivists believe that it is meaningless to assert that a statement either holds or does not hold when we have no method of deciding which one is the case (so for them statements have three values: true, false, and presently indeterminate). They also reject Logic Rule 6 when applied to infinite sets because they insist that in order to meaningfully assert that a mathematical object *exists*, one must supply an "effective" method for constructing it; they consider it inadequate merely to assume that the object does not exist (RAA hypothesis) and then derive a contradiction. Such a derivation for them merely proves $\sim \sim Q$, where Q is the existence assertion; for them, $\sim \sim Q$ does not automatically imply Q (they deny Logic Rule 3).

The constructivists do not challenge the use of classical logic in elementary geometry.

Incidence Geometry

Let us apply the logic we have developed to a very basic part of geometry, incidence geometry. This is a geometry of straightedge drawing alone, if you like—no circles are given, only lines and points. We will see that there are many different examples of such a geometry. We assume only the undefined terms *point* and *line* and the undefined relation *incidence* between a point and a line, expressed as "P lies on *l*" or "*l* passes through P," as before. We will also use the abbreviation P ɪ *l* in formulas. We don't discuss "betweenness" or "congruence" or distance in this restricted geometry. We are now beginning the new axiomatic development of geometry that fills the gaps in Euclidean geometry and applies to other geometries as well; that development will continue in later chapters.

These undefined terms will be subjected to three axioms, the first of which is the same as Euclid's first postulate.

INCIDENCE AXIOM 1. For every point P and for every point Q not equal to P, there exists a unique line *l* incident with P and Q.

We say that "*l* is the line joining P to Q," and we denote it, as before, by $\overleftrightarrow{PQ}$.

INCIDENCE AXIOM 2. For every line *l*, there exist at least two distinct points incident with *l*.

INCIDENCE AXIOM 3. There exist three distinct points with the property that no line is incident with all three of them.

The last two axioms fill the gap mentioned in the Exercises of Chapter 1. We can now assert that every line has points lying on it—at least two, possibly more—and that all the points do not lie on one single line. Moreover, we know that the geometry must have at least three distinct points in it, by the third axiom and Rule 9(b) of logic. Namely, Incidence Axiom 3 is a conjunction of two statements:

1. There exist three distinct points.
2. For every line, at least one of these points does not lie on that line.

Rule 9(b) tells us that a conjunction of two statements implies each statement separately, so we can conclude that three distinct points exist (applying Rule 8, modus ponens). Applying Incidence Axiom 1 to any pair of those three points, we deduce that the geometry must also have at least three distinct lines.

When we refer to these axioms in our justifications, we will denote them as I-1, I-2, and I-3.

Incidence geometry has some defined terms, such as "collinear," "concurrent," and "parallel," defined exactly as they were in Chapter 1. To repeat:

DEFINITION. Three or more points A, B, C, . . . are *collinear* if there exists a line incident with all of them.

Axiom I-3 can be rewritten as "There exist three distinct noncollinear points."

DEFINITION. Three or more lines l, m, n, . . . are *concurrent* if there exists a point incident with all of them.

As before, if point P lies on both l and m, we say that "l and m intersect or meet at P" or "l and m have point P in common." Notice that "concurrent" is the *dual* notion to "collinear" in the sense that it is defined the same way except that the roles of point and line are interchanged.

DEFINITION. Lines l and m are *parallel* if they are distinct lines and no point is incident with both of them.

We use the notation $l \parallel m$ for "l and m are parallel." Notice that according to Axiom I-1, the dual notion for points to the notion of parallel lines is vacuous—there are no such pairs of points.

For the fun of it, let us write our three axioms in symbolic logic notation, with the understanding that capital letters denote points and italic lowercase letters denote lines. We will use the abbreviation $\exists!$ to mean "There exists a *unique* . . . (having a certain property)." We also abbreviate $\sim(P = Q)$ by $P \neq Q$.

AXIOM I-1. $\forall P \forall Q ((P \neq Q) \Rightarrow \exists! \, l \, (P \text{ I } l \, \& \, Q \text{ I } l))$

AXIOM I-2. $\forall l \, \exists P \exists Q \, (P \neq Q \, \& \, (P \text{ I } l \, \& \, Q \text{ I } l))$

AXIOM I-3. $\exists A \exists B \exists C \, ((A \neq B \, \& \, A \neq C \, \& \, B \neq C) \, \& \sim \exists l \, (A \text{ I } l \, \& \, B \text{ I } l \, \& \, C \text{ I } l))$

What sort of results can we prove using this meager collection of axioms? None that are very exciting, but here are five easy ones. We proved the first one previously.

PROPOSITION 2.1. If l and m are distinct lines that are not parallel, then l and m have a unique point in common.

PROPOSITION 2.2. There exist three distinct lines that are not concurrent.

For students new to doing proofs, permit me to think "out loud" slowly just to illustrate how one might discover a proof of this. If I were not familiar with the notion of concurrence, I would reread the definition to make sure I understood it. I might draw one or more diagrams to help me visualize three nonconcurrent lines. Then I might get annoyed at having to prove something so obvious but would remind myself that we're learning to be rigorous, which will turn out to be a useful skill. I look at the axioms to see which ones tell me that lines exist. Not I-3, because the only line mentioned there is said *not* to exist. Not I-2, because although it says that every line has a certain property, I remember that that doesn't guarantee existence ("Every unicorn . . ."). So I have to use I-1, which does assert the existence of a line, but it is a conditional existence—first I have to be given two points. Where will I find them? Aha! I-3 gives me three distinct points A, B, C, and they're not collinear. Then I can apply I-1 and join those points in pairs to obtain three lines that are distinct because the points are not collinear. Are those lines concurrent? Certainly not, but to *prove* it I could first prove a lemma that if three lines are concurrent, the point at which they meet is unique. This follows from Proposition 2.1 already proved. So I can finish the argument using RAA: If those joins were concurrent, then A = B = C, contradicting the way we obtained those points. Done!

I leave it as an exercise to rewrite that argument as a formal proof and to find proofs for the following three propositions. Remember that you can use results previously proved.

PROPOSITION 2.3. For every line, there is at least one point not lying on it.

PROPOSITION 2.4. For every point, there is at least one line not passing through it.

PROPOSITION 2.5. For every point P, there exist at least two distinct lines through P.

Models

In reading over the axioms of incidence in the previous section, you may have imagined drawing dots for points and, with a straightedge, long dashes to illustrate lines. With this representation in mind, the axioms appear to be correct statements (ignoring as usual the breadth of the drawn dots and dashes). We will take the point of view that these idealized dots and dashes are a model for incidence geometry.

More generally, if we have any formal system, suppose we interpret the undefined terms in some way—i.e., give the undefined terms a particular meaning—and then interpret statements about those undefined terms by substituting the interpreted meanings. We call this an *interpretation* of the system. We can then ask whether the axioms, so interpreted, are correct statements. If they are, we call the interpretation a *model* of the axioms. When we take this point of view, interpretations of the undefined terms "point," "line," and "incident" other than the usual dot-and-dash drawings become possible. That is something Euclid never imagined. Moritz Pasch said in 1882:

If geometry is to be deductive, the deduction must everywhere be independent of the *meaning* of geometrical concepts, just as it must be independent of the diagrams; only the *relations* specified in the postulates and definitions employed may legitimately be taken into account.

■ **EXAMPLE 1.** Consider a set {A, B, C} of three distinct letters. We interpret "point" to be any one of those letters. "Lines" will be those subsets that contain exactly two letters—{A, B}, {A, C}, and {B, C}. A "point" will be interpreted as "incident" with a "line" if it is a member of that subset. Thus, under this interpretation, A lies on {A, B} and {A, C} but does not lie on {B, C}. In order to determine whether this interpretation is a model, we must check whether the interpretations of the axioms are correct statements. For Incidence Axiom 1, if P and Q are any two of the letters A, B, and C, then {P, Q} is the unique "line" on which they both lie. For Axiom I-2, if {P, Q} is any "line," P and Q are two distinct "points" lying on it. For Axiom I-3, we see that A, B, and C are three distinct "points" that are not collinear.

What is the use of models? The main property of any model of an axiom system is that all theorems of the system are correct statements in the model. This is because logical consequences of correct statements are themselves correct. (By the definition of "model," axioms are correct statements when interpreted in models; theorems are logical consequences of axioms. We are assuming that the rules of logic we have listed apply to our models.) Thus, we immediately know that the five propositions in the previous section hold when interpreted in the three-point model of Example 1. Check them if you are not convinced.

Suppose we have a statement in the formal system but don't yet know whether it can be proved. We can look at our models and see whether the statement is correct in the models. If we can find one model where the interpreted statement fails to hold, we can be sure that no proof is possible. You are undoubtedly familiar with testing for the correctness of geometric statements by drawing diagrams. Of course, the converse does not work; just because a drawing makes a statement look right does not guarantee that you can prove it. This was illustrated in Chapter 1.

The advantage of having several models is that a statement may hold in one model but not in another. Models are "laboratories" for experimenting with the formal system.

Let us experiment with the Euclidean parallel postulate. This is a statement in incidence geometry: "For every line l and every point P not lying on l, there exists a unique line through P that is parallel to l." This statement appears to be correct if we imagine our drawings are on an infinite flat sheet of paper (can you see that on a finite sheet there would be many parallels through P?). But what about our three-point model? It is immediately apparent that *no parallel lines exist* in this model: {A, B} meets {B, C} in the point B and meets {A, C} in the point A; {B, C} meets {A, C} in the point C. (We say that this model has the *elliptic parallel property,* as shown in Figure 2.4.)

Thus, we can conclude that *no proof of the Euclidean parallel postulate from the axioms of incidence alone is possible; in fact, from the*

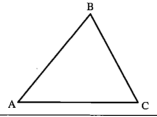

Figure 2.4 Elliptic parallel property (no parallel lines). A three-point incidence geometry.

axioms of incidence geometry alone, it is impossible to prove that parallel lines exist. Similarly, the statement "any two lines have a point in common" (the elliptic parallel property) cannot be proved from the axioms of incidence geometry, for if you could prove it, it would hold in the idealized drawn model and in the models that will be described in Examples 3 and 4.

The technical description for this situation is that the statement "Parallel lines exist" is "independent" of the axioms of incidence. We call a statement *independent of* or *undecidable from* given axioms if it is impossible to either prove or disprove the statement from those axioms. Independence may be demonstrated by constructing two models for the axioms: one in which the statement holds and one in which it does not hold. This method will be used very decisively in Chapter 7 to settle once and for all the question of whether the Euclidean parallel postulate can be proved using all the other axioms we will later introduce. For now, we know that the incidence axioms alone are too weak to prove it.

▨▨▨ **EXAMPLE 2.** Suppose we interpret "points" as points on a sphere, "lines" as great circles on the sphere, and "incidence" in the usual sense, as a point lying on a great circle. In this interpretation there are again no parallel lines because any pair of great circles on a sphere intersect in two points that are *antipodal* (meaning the straight line in three-space joining them passes through the center of the sphere—like the north and south poles). However, this interpretation is *not a model* for incidence geometry, for the uniqueness part of the interpretation of Axiom I-1 fails to hold—e.g., there are infinitely many great circles passing through the north and south poles on the sphere, all the "circles of longitude" (see Figure 2.3, p. 63).

▨▨▨ **EXAMPLE 3.** Let the "points" be the four letters A, B, C, and D. Let the "lines" be all six sets containing exactly two of these letters: {A, B}, {A,C}, {A, D}, {B, C}, {B, D}, and {C, D}. Let "incidence" be set membership, as in Example 1. As an exercise, you can verify that this is a model for incidence geometry and that in this model the Euclidean parallel postulate does hold (see Figure 2.5). By Examples 1 and 3, the Euclidean parallel postulate is independent of the axioms of incidence geometry.

▨▨▨ **EXAMPLE 4.** Let the "points" be the five letters A, B, C, D, and E. Let the "lines" be all 10 sets containing exactly two of these letters.

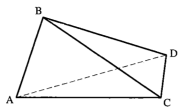

Figure 2.5 Euclidean parallel property. A four-point incidence geometry.

Let "incidence" be set membership, as in Examples 1 and 3. You can verify that in this model the following statement about parallel lines, called *the hyperbolic parallel property*, holds: "*For every line l and every point P not on l, there exist at least two lines through P parallel to l*" (see Figure 2.6).

(The figures illustrating Examples 1, 3, and 4 are only meant to be suggestive. They have features not included in the definition of those models in terms of letters. For example, in Figure 2.5, the dash illustrating the "line" {A, D} appears to intersect the dash illustrating the "line" {B, C} when those "lines" are actually parallel, so it's better to view Figures 2.5 and 2.6 as three-dimensional drawings.)

Let us summarize the significance of models. *Models can be used to demonstrate the impossibility of proving or disproving a statement from the axioms.* We just showed the undecidability of the Euclidean, elliptic, and hyperbolic statements about parallel lines in incidence geometry. Moreover, if an axiom system has many models that are essentially different from one another, as are the models in Examples 1, 3, and 4, then that system has a wide range of applicability. Propositions proved from the axioms of such a system are automatically

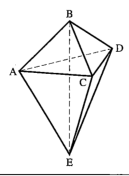

Figure 2.6 Hyperbolic parallel property. A five-point incidence geometry.

correct statements within any of the models. Mathematicians often discover that an axiom system they constructed with one particular model in mind has applications to completely different models they never dreamed of, as we will see.

As we mentioned in Chapter 1, Johannes Kepler believed that the regular heptagon (seven-sided) was "unknowable" because he argued that there is no way to construct it using straightedge and compass alone. For Kepler, knowledge in geometry meant constructibility by straightedge and compass. After Kepler died and by the time *real* analytic geometry became generally accepted, no mathematician of any importance denied knowledge of the regular heptagon or, more generally, the regular n-gon, because they accepted the possibly nonconstructible existence of the angle whose radian measure is $2\pi/n$. By successively laying off this angle with vertex at the origin n times, the points where those rays intersect a fixed circle centered at the origin will form the vertices of a regular n-gon. We will see from our study of models of our axioms of geometry that existence may depend on which model you're looking at.

In Chapter 3, we will exhibit a model (coordinatized by the field of constructible numbers) of the elementary Euclidean axioms in which regular heptagons do not exist. So from our current point of view, Kepler was merely restricting his attention to such a model. You can see that "existence" is a tricky notion! (See Project 7, Chapter 1 for a reference to Viète's neusis construction of the regular heptagon.)

Consistency

An axiomatized theory is called *consistent* if no contradiction can be proved from the axioms. Notice that *in an inconsistent theory, every statement is provable* because of the RAA rule: Given any statement S, assume $\sim S$ (RAA hypothesis). Since the theory is inconsistent, it has proved some contradiction (we don't care which). Hence, by RAA conclusion, S is proved in that theory. This is a three-step proof of S in the inconsistent theory. Review the RAA rule if you don't follow this. Obviously, an inconsistent theory is worthless.

Models provide evidence for the consistency of the axiom system. For example, if incidence geometry were inconsistent, there would exist a proof of the statement $\forall P \forall Q \ (P = Q)$ (since, as we just showed, any statement in the language of an inconsistent theory could be proved). Translating that alleged proof into the language of the three-point model of Example 1, we would have a proof that A = B, for instance. But we chose our model so as to have three distinct letters A, B, and C. Hence, *we know that incidence geometry is consistent.*

We will discuss the question of consistency of Euclidean and non-Euclidean geometries in Chapter 7. For now, note that the consistency of Euclidean geometry was never doubted because it was believed to describe, in an idealized fashion, the space in which we all live. In that sense, it was believed that its axioms are "true." We will see that the discovery of non-Euclidean geometry shattered the belief in the "truth" of Euclidean geometry. However, all classical mathematicians believe that Euclidean geometry is consistent, especially since no contradiction has popped up in over 2400 years. No one has any idea how to prove that in an absolute manner similar to the proof that incidence geometry is consistent. We will later discuss Hilbert's *relative* proof of the consistency of real Euclidean geometry—relative to the consistency of the theory of the real number system.

NOTE FOR ADVANCED STUDENTS. Mathematicians first became seriously concerned about consistency after it was discovered that Georg Cantor's set theory contained contradictory statements about the set of all sets or the set of all ordinal numbers. Bertrand Russell's famous paradox (see Exercise 19) showed that Gottlob Frege's system of logic and classes was inconsistent.

It is generally very difficult if not impossible to convincingly prove that complicated mathematical theories are consistent. The simplest such proof of any importance is the one that *propositional logic*—logic without quantifiers—*is consistent.* The key to that proof is to introduce suitable "truth tables" for statements in propositional logic. A *tautology* is a statement whose truth table has only "true" in all its entries, no matter what the "truth values" of its constituents are (e.g., $P \Rightarrow P$ is "true" no matter what P is). After stating suitable axioms, the key is to prove that *all theorems in propositional logic are tautologies* (and conversely). Since P & $\sim P$ is not a tautology (it is "false" no matter what P is), it cannot be proved. Hence propositional logic is consistent. For details, see any good mathematical logic text. Notice, however, that although we have used loaded words like "true" and "false" here, because of the historical and psychological origin of these ideas, we could just as easily have used any two distinct signs, such as 1 and 0.

You will find in many reputable books and articles the *claim* that "If a formal axiomatic theory $\mathcal{T}$ has a model $\mathcal{M}$, then $\mathcal{T}$ is consistent"—some books even describe this claim as a "theorem" in metamathematics. The idea behind such a claim is that the model "exists in reality," so it is meaningful to assert that statements in the model are either "true" or "false." Now by definition of a model, the axioms of $\mathcal{T}$ are true when interpreted in $\mathcal{M}$, and since our logic is designed to be truth-preserving, all the statements in $\mathcal{T}$ proved from those axioms must also be true when interpreted in $\mathcal{M}$. Hence if a contradiction were provable in $\mathcal{T}$, the interpretation in $\mathcal{M}$ of that contradiction, which is also a contradiction, would be true in $\mathcal{M}$. But contradictions are false, not true. Therefore, $\mathcal{T}$ must be consistent if it has a model $\mathcal{M}$.

We used that strategy to prove that incidence geometry is consistent, arguing that if it was inconsistent, then we could prove A = B in the model of Example 1, a statement we know is false. The point is that for such a trivial three-point model, the notions of "truth" and "falsity" for statements in the set-theoretic language of that model are straightforward and can be rigorously defined (e.g., using a method of Alfred Tarski).

However, when we are dealing with *infinite models*, the notions of "truth" and "falsity" are not so clear, and there is even disagreement among reputable mathematicians and philosophers as to whether such models "really exist." Therefore, that *claim* would not be a theorem in metamathematics until further hypotheses are added.

For example, PA (Peano arithmetic) has as its "standard" model the infinite system $\mathbb{N}$ of natural numbers. Reputable mathematicians like Gauss did not accept that $\mathbb{N}$ exists because it is an infinite set (Gauss, as we mentioned, accepted Aristotle's doctrine that infinity is only potential—one cannot collect all the natural numbers in a set). In formal set theory, an axiom is required to obtain the existence of $\mathbb{N}$—it is simply *assumed* to exist. And even for those mathematicians who do accept its existence, the concept of truth in $\mathbb{N}$ is not generally clear to all—constructivists don't accept it and philosophers are still arguing about it.

If the model $\mathbb{N}$ guaranteed, as claimed, that PA is consistent, then why did Hilbert and his associates work so hard trying to prove consistency by "finitary" methods? Although they obtained finitary consistency proofs for some simpler arithmetical theories, they couldn't prove by finitary methods that the full PA was consistent and complete. Then Gödel, in 1931, proved that it is *impossible* to prove the consistency of PA by methods considered to be "finitary"—that was his famous second incompleteness theorem (his first incompleteness theorem proved that PA is incomplete by constructing a formal statement that was undecidable from the axioms of PA).

Virtually all mathematicians believe that PA is consistent, but after Gödel's result, we have no hope of proving consistency by methods considered to be finitary. The attitude of most mathematicians is that the belief in the consistency of PA and Zermelo–Fraenkel (ZF) set theory is based on experience: No contradiction has been found in all the many years we have been working with those systems. We have confidence that if a contradiction is ever found, mathematicians will adjust their axioms a bit to get rid of it, as was done when Cantor's informal infinite set theory was formalized by Ernest Zermelo and Abraham Fraenkel in the early twentieth century. Nicholas Bourbaki wrote: "Historically speaking, it is of course untrue that mathematics is free from contradiction; non-contradiction appears as a goal to be achieved, not as a God-given quality that has been granted us once for all."

Model theory is a very important branch of mathematical logic. It was via infinite model theory that Abraham Robinson, in 1960, discovered an extension of the real number system—now called *the hyperreal numbers*—in which infinitesimals and infinitely large numbers exist. His *nonstandard analysis* showed how to use them to justify the use of infinitesimal and infinite methods in the differential and integral calculus—methods that were freely used without justification by Newton, Leibniz, Euler, *et al.*

The three-, four-, and five-point models we have exhibited are trivial examples of *finite incidence geometries*. Finite geometries have turned out to be surprisingly important (see Project 7).

Isomorphism of Models

We now make precise the important notion of two models being "essentially the same," or *isomorphic*. For incidence geometries, this will mean that there exists a one-to-one correspondence $P \leftrightarrow P'$ between the points of the models and a one-to-one correspondence $l \leftrightarrow l'$ between the lines of the models such that $P \text{ I } l$ if and only if $P' \text{ I } l'$; such a correspondence is called an *isomorphism* from one model onto the other.

▨▨▨ **EXAMPLE 5.** Consider a set $\{a, b, c\}$ of three letters, which we will call "lines" now. "Points" will be those subsets that contain exactly two letters—$\{a, b\}$, $\{a, c\}$, and $\{b, c\}$. Let incidence be set membership; for example, "point" $\{a, b\}$ is "incident" with "lines" a and b but not with c. This model is the *dual* of the three-point model in Example 1—all we've done is interchange the interpretations of "point" and "line." It certainly seems to be structurally the same. (However, the duals of Examples 3 and 4 are not structurally the same as the originals—in fact, they're not even models. Can you see why not?) An explicit isomorphism of Example 5 with Example 1 is given by the following correspondences:

$$A \leftrightarrow \{a, b\} \qquad \{A, B\} \leftrightarrow b$$
$$B \leftrightarrow \{b, c\} \qquad \{B, C\} \leftrightarrow c$$
$$C \leftrightarrow \{a, c\} \qquad \{A, C\} \leftrightarrow a.$$

Note that A lies on $\{A, B\}$ and $\{A, C\}$ only; its corresponding "point" $\{a, b\}$ lies on the corresponding "lines" b and a only. Similar checking with B and C shows that incidence is preserved by our correspondence. On the other hand, if we had used a correspondence such as

$$\{A, B\} \leftrightarrow a$$
$$\{B, C\} \leftrightarrow b$$
$$\{A, C\} \leftrightarrow c$$

for the "lines," keeping the same correspondence for the "points," we would not have an isomorphism because, for example, A lies on $\{A, C\}$ but the corresponding "point" $\{a, b\}$ does not lie on the corresponding "line" c.

To further illustrate the idea that isomorphic models are "essentially the same," consider two models with different parallelism properties, such as one with the elliptic property and one with the Euclidean. We claim that these models are not isomorphic: Suppose, on the contrary, that an isomorphism could be set up. Given line l and point P not on it, then every line through P meets l, by the elliptic property. Hence every line through the corresponding point P′ meets the corresponding line l', but that contradicts the Euclidean property of the second model.

Later on, we will need to use the concept of "isomorphism" for models of a geometry more complicated than incidence geometry—neutral geometry. In neutral geometry, we will have betweenness and congruence relations, in addition to the incidence relation, and we will require an "isomorphism" to preserve those relations as well.

The general idea is that *an isomorphism of two models of an axiom system is a one-to-one correspondence between the basic objects of the system that preserves all the basic relations of the system.*

Another example (to be discussed in Chapter 9) is the axiom system for a "group." Roughly speaking, a group is a set with a multiplication for its elements satisfying a few familiar axioms of algebra. An "isomorphism" of groups will then be a one-to-one mapping $x \rightarrow x'$ of one set onto the other, which preserves the multiplication, i.e., for which $(xy)' = x'y'$.

Projective and Affine Planes

We now briefly discuss two types of models of incidence geometry that are particularly significant. During the Renaissance, in the fifteenth century, artists developed a theory of perspective in order to realistically paint two-dimensional representations of three-dimensional scenes. Their theory described the *projection* of points in the scene onto the artist's canvas by lines from those points to a fixed viewing point in one of the artist's eyes; the intersection of those lines with the plane of the canvas was used to construct the painting. The mathematical formulation of this theory was called *projective geometry*. In this technique of projection, parallel lines that lie in a plane cutting the plane of the canvas are painted as meeting (visually, they appear to meet at a point on the faraway horizon, as shown in Figure 2.7).

This suggested an extension of Euclidean geometry in which parallel lines "meet at infinity," so that the Euclidean parallel property is replaced by the elliptic parallel property in the extended plane. We will carry out this extension rigorously. First, some definitions.

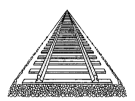

Figure 2.7 Parallel railroad tracks appear to converge as they recede into the distance.

DEFINITION. A *projective plane* is a model of incidence geometry having the elliptic parallel property (any two lines meet) and such that every line has at least three distinct points lying on it (strengthened Incidence Axiom 2).

Our proposed extension of the Euclidean plane uses only its incidence properties (not its betweenness and congruence properties); the purely incidence part of Euclidean geometry is called *affine geometry*, which leads to the next definition.

DEFINITION. An *affine plane* is a model of incidence geometry having the following Euclidean parallel property:

$$\forall l \ \forall P \ (\sim(P \ \text{I} \ l) \Rightarrow \exists! m \ (P \ \text{I} \ m \ \& \ l \parallel m)).$$

So the idea in extending an affine plane to a projective plane is to add enough new "points at infinity" so that all lines parallel to any given line will now meet at one such point. Moreover, in order to satisfy Axiom I-1, we need to join those "points at infinity" by inventing a new "line at infinity" that intuitively corresponds to the horizon in the example above. Here we see mathematical imagination at its best! The technicality in our construction is that we will be working within set theory, and we have to define those new objects as certain sets. It may be awkward psychologically at first for you to think of those sets as "points" and a "line," but remember that we are free to interpret those undefined terms any way we choose so long as we can prove that the axioms are satisfied in that interpretation. That's what we'll do.

Example 3 in this chapter illustrated the smallest affine plane (four points, six lines).

Let $\mathcal{A}$ be any affine plane. We introduce a relation $l \sim m$ on the lines of $\mathcal{A}$ to mean "$l = m$ or $l \parallel m$." This relation is obviously *reflexive* ($l \sim l$) and *symmetric* ($l \sim m \Rightarrow m \sim l$). Let us prove that it is *transitive* ($l \sim m$ and $m \sim n \Rightarrow l \sim n$): If any pair of these lines are equal, the conclusion is immediate, so assume that we have three distinct lines such that $l \parallel m$ and $m \parallel n$. Suppose, on the contrary, that l meets n at point P. P does not lie on m because $l \parallel m$. Hence we have two distinct parallels n and l to m through P, which contradicts the Euclidean parallel property of $\mathcal{A}$.

A relation that is reflexive, symmetric, and transitive is called an *equivalence relation*. Such relations occur frequently in mathematics and are very important. Whenever they occur, we consider the equivalence classes determined by the relation: For example, the *equivalence class* [*l*] of *l* is defined to be the set consisting of all lines equivalent to *l*—i.e., of *l* and all the lines in $\mathscr{A}$ parallel to *l*. In the familiar Cartesian model of the Euclidean plane, the set of all horizontal lines is one equivalence class, the set of verticals is another, the set of lines with slope 1 is a third, etc. Equivalence classes take us from equivalence to equality: $l \sim m \Leftrightarrow [l] = [m]$.

For historical and visual reasons, we call these equivalence classes *points at infinity*; we have made this vague idea precise within modern set theory. We now enlarge the model $\mathscr{A}$ to a new model $\mathscr{A}^*$ by adding these points, calling the points of $\mathscr{A}$ "ordinary" points for emphasis. We further enlarge the incidence relation by specifying that each of these equivalence classes lies on every one of the lines in that class: [*l*] lies on *l* and on every line *m* such that $l \parallel m$. Thus, in the enlarged plane $\mathscr{A}^*$, *l* and *m* are no longer parallel, but they meet at [*l*].

We want $\mathscr{A}^*$ to be a model of incidence geometry also, which requires one more step. To satisfy Euclid's Postulate I, we need to add one new line on which all (and only) the points at infinity lie: Define the *line at infinity* l_∞ to be the set of all points at infinity. Let us now check that $\mathscr{A}^*$ is a projective plane, called the *projective completion* of $\mathscr{A}$.

VERIFICATION OF I-1. If P and Q are ordinary points, they lie on a unique line of $\mathscr{A}$ (since I-1 holds in $\mathscr{A}$) and they do not lie on l_∞. If P is ordinary and Q is a point at infinity [*m*], then either P lies on *m* and $\overleftrightarrow{PQ} = m$, or, by the Euclidean parallel property, P lies on a unique parallel *n* to *m* and Q also lies on *n* (by the definition of incidence for points at infinity), so $\overleftrightarrow{PQ} = n$. If both P and Q are points at infinity, then $\overleftrightarrow{PQ} = l_\infty$.

VERIFICATION OF STRENGTHENED I-2. Each line *m* of $\mathscr{A}$ has at least two points on it (by I-2 in $\mathscr{A}$), and now we've added a third point [*m*] at infinity. That l_∞ has at least three points on it follows from the existence in $\mathscr{A}$ of three lines that intersect in pairs (such as the lines joining the three noncollinear points furnished by Axiom I-3); the equivalence classes of those three lines do the job.

VERIFICATION OF I-3. It holds already in $\mathscr{A}$.

VERIFICATION OF THE ELLIPTIC PARALLEL PROPERTY. If two ordinary lines do not meet in $\mathscr{A}$, then they belong to the same equivalence class and meet at that point at infinity. An ordinary line *m* meets l_∞ at [*m*]. ◄

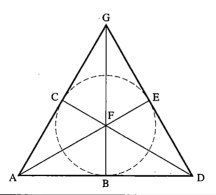

Figure 2.8 The smallest projective plane (seven points).

▓▓▓ **EXAMPLE 6.** Figure 2.8 illustrates the smallest projective plane, projective completion of the smallest affine plane; it has seven points and seven lines. The dashed line could represent the line at infinity, for removing it and the three points C, B, and E that lie on it leaves us with a four-point, six-line affine plane isomorphic to the one in Example 3, Figure 2.5.

Informally, the usual Euclidean plane, regarded just as a model of incidence geometry (ignoring its betweenness and congruence structures), is referred to as the *real affine plane,* and its projective completion is called the *real projective plane* (see Example 8 for a formal definition).

Notice what happens to a line in the real affine plane after it has been extended with a point at infinity: It becomes a closed curve in the real projective plane. Namely, imagine two horizontal parallel lines in the real affine plane. They have to meet at a point at infinity on the right and also at a point at infinity on the left. But those points at infinity must be the same because of Proposition 2.1: The point of intersection of two lines is unique. So when you travel along one line out to infinity to the right, after you "reach infinity," if you keep going in the same direction you will be returning from the left to where you started. (This is loose talk, of course; there is no notion of distance in incidence geometry and "infinity" is just a figure of speech suggested by perspective drawing.)

▓▓▓ **EXAMPLE 7.** To visualize the projective completion $\mathscr{A}^*$ of the real affine plane $\mathscr{A}$, picture $\mathscr{A}$ as the plane T tangent to a sphere S in Euclidean three-space at its north pole N (Figure 2.9). If O is the cen-

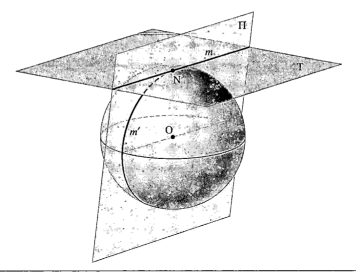

Figure 2.9

ter of sphere S, we can join each point P of T to O by a Euclidean line that will intersect the northern hemisphere of S in a unique point P′; this gives a one-to-one correspondence between the points P of T and the points P′ of the northern hemisphere of S (N corresponds to itself). Similarly, given any line m of T, we join m to O by a plane Π through O that cuts out a great circle on the sphere and a great semicircle $m′$ on the northern hemisphere; this gives a one-to-one correspondence between the lines m of T and the great semicircles $m′$ of the northern hemisphere, a correspondence that clearly preserves incidence.

Now if $l \parallel m$ in T, the planes through O determined by these parallel lines will meet in a line lying in the plane of the equator, a line that (since it goes through O) cuts out a pair of antipodal points on the equator. Thus, the line at infinity of $\mathscr{A}^*$ can be visualized under our isomorphism as the equator of S with antipodal points identified (they must be identified, or else Axiom I-1 will fail). In other words, $\mathscr{A}^*$ can be described as the northern hemisphere with antipodal points on the equator pasted to each other; however, we can't visualize this pasting very well because it can be proved that the pasting cannot be done in Euclidean three-space without tearing the hemisphere.

Projective planes are the most important models of pure incidence geometry. We will see in Chapter 9 that Euclidean, hyperbolic, and, of course, elliptic geometry can all be considered "subgeometries" of projective geometry. This discovery by Cayley led him to exclaim that "projective geometry is all of geometry," which turned out to be an oversimplification.

▓▓ EXAMPLE 8. ALGEBRAIC MODELS OF AFFINE AND PROJEC- TIVE PLANES. If you've taken a course in abstract algebra, you know what an abstract *field* F is. If not, think of the following specific fields that are familiar:

$\mathbb{Q}$ = the field of all rational numbers
$\mathbb{R}$ = the field of all real numbers
$\mathbb{C}$ = the field of all complex numbers.

Let F be any field. Let F^2 be the set of all ordered pairs (x, y) of ele- ments of F. We give F^2 the structure of an affine plane by taking its elements as our "points." A "line" will be the set of all solutions to a linear equation

$$ax + by + c = 0,$$

where at least one of the coefficients a, b is nonzero. Point (x, y) will be interpreted as "incident" with that line if it satisfies "the" equation (notice that multiplying the coefficients a, b, c by a nonzero constant yields the same "line"). With these interpretations, we claim that F^2 becomes an affine plane called the *affine plane over* F. By the defini- tion of "affine plane," we must verify the interpretations of the three incidence axioms and we must verify the Euclidean parallel property. If you've taken a course in analytic geometry, you know how to ver- ify those. We sketch a few of the ideas:

1. To verify I-3, show that the points $(0, 0)$, $(0, 1)$, and $(1, 0)$ are not collinear by showing that any linear equation they all satisfy must have all three coefficients equal to 0.
2. To verify I-2, say coefficient $a \neq 0$. Then $(-c/a, 0)$ is one point on the line. Find another depending on whether b is 0 or not.
3. To verify I-1, let (u, v) and (s, t) be distinct points. Use your knowl- edge of analytic geometry to write a linear equation satisfied by those points. To show uniqueness, use Cramer's rule to find the unique solution to a pair of linearly independent linear equations.
4. To verify the Euclidean parallel property, first establish the result that two lines are parallel iff they have the same slope (handle the case of vertical lines separately). Then use the point-slope formula to determine the unique line parallel to a given line through a given point not on that line.

Next we briefly describe the *projective plane over* F, denoted $P^2(F)$. Here both "points" and "lines" are equivalence classes of triples (x, y, z) of elements of F that are not all zero, where two such triples are considered *equivalent* if one is a nonzero constant multiple of the other. You can easily verify that this is an equivalence relation. Each such triple is referred to as *homogeneous coordinates,* and its equivalence class will be denoted $[x, y, z]$. We interpret incidence by the linear homogeneous equation

$$ax + by + cz = 0$$

when $[x, y, z]$ is a "point" and $[a, b, c]$ is a "line."

We show that $P^2(F)$ is isomorphic to the projective completion of F^2 as follows: Map each "point" (x, y) of F^2 to the "point" $[x, y, 1]$ of $P^2(F)$. Map each "line"

$$\{(x, y) \mid ax + by + c = 0\}$$

of F^2 to the "line" $[a, b, c]$ of $P^2(F)$. Verify easily that these mappings are one-to-one and preserve "incidence" for the affine plane. Next map the line at infinity in the projective completion to the "line" $[0, 0, 1]$, i.e., to the "line" whose equation is $z = 0$; it is the only "line" in $P^2(F)$ that is not the image under our mapping of an affine line. A "point" on this line has homogeneous coordinates of the form $[a, b, 0]$, where at least one of a, b is nonzero. We let this point correspond to the point at infinity common to all the lines parallel to the affine line $ax + by = 0$. It is straightforward to verify that these mappings provide the desired isomorphism.

It follows from this isomorphism that $P^2(F)$ is a projective plane since an interpretation isomorphic to a projective plane is easily seen to satisfy all the requirements to be a projective plane. Let us check, for example, that every line has at least three points on it. If it is the image of an affine line, we know by I-2 that the affine line has at least two points on it, and the projective line also has the point at infinity of that affine line. If it is the image of the line at infinity, it has the three distinct points $[1, 0, 0]$, $[0, 1, 0]$, and $[1, 1, 0]$ lying on it.

Hopefully, with this model you see that there is nothing mysterious about the "line at infinity," for under our isomorphism it is just given by the equation $z = 0$. Nor is there any mystery about the "point at infinity" common to all the parallel affine lines $ax + by = t$, where a and b are fixed (not both zero) and t varies through all the elements of F; under our isomorphism, it is the "point" $[b, -a, 0]$.

A projective plane isomorphic to $P^2(F)$ for some field F is said to be *coordinatized* by F.

▨▨▨ **EXAMPLE 9. DUALITY IN PROJECTIVE GEOMETRY.** Let $\mathscr{P}$ be a projective plane. Define the *dual* interpretation $\mathscr{P}^*$ of $\mathscr{P}$ to have as its points the lines of $\mathscr{P}$, as its lines the points of $\mathscr{P}$, and as its incidence the same incidence relation. Let us verify that $\mathscr{P}^*$ is also a projective plane:

1. To verify the interpretation of I-1 in $\mathscr{P}^*$, we must show that any two lines of $\mathscr{P}$ meet in a unique point. That they meet is the elliptic parallel property, which holds in $\mathscr{P}$ by the definition of a projective plane. That the point of intersection is unique was proved in Proposition 2.1.
2. To verify the interpretation of I-2 in $\mathscr{P}^*$, refer to Proposition 2.5 for $\mathscr{P}$, which you will prove as an exercise.
3. To verify the interpretation of I-3 in $\mathscr{P}^*$, refer to Proposition 2.2 proved for $\mathscr{P}$.
4. To verify the elliptic parallel property for $\mathscr{P}^*$, observe that it is just the interpretation of I-1 for $\mathscr{P}$.
5. Finally, we must show that every line of $\mathscr{P}^*$ has at least three points on it, which means showing every point A of $\mathscr{P}$ has at least three lines through it. By Proposition 2.4, there exists a line l that does not pass through A. By the definition of projective plane, l has at least three points lying on it. Joining three points of l to A then provides three lines that we seek.

The fact that $\mathscr{P}^*$ is also a projective plane explains the *principle of duality in plane projective geometry*: If a statement has been proved to hold in all projective planes, then the dual statement obtained by interchanging "point" and "line" automatically holds as well—no further proof is required. Caveat: If the statement involves defined notions (such as "collinear"), you must replace those notions by their duals ("concurrent" in this case). *This was probably the first metamathematical theorem in history.*

You can see duality very clearly in the algebraic model $P^2(F)$. A "point" in that model is an equivalence class $[x, y, z]$ of triples of not-all-zero elements of F under the equivalence relation that the triple (x, y, z) is equivalent to (x', y', z') iff there is a nonzero k in F such that $x' = kx$, $y' = ky$, and $z' = kz$. But a "line" in that model is exactly the same thing, except that we have been using letters from the beginning of the alphabet for "lines." And incidence is given by the same linear homogeneous equation.

Brief History of Real Projective Geometry

An important 1822 text on synthetic real projective geometry was composed by Frenchman J.-V. Poncelet while he was incarcerated in a Russian prison after being captured from Napoleon's invading army. He introduced the points at infinity officially into geometry, though the idea had already appeared in a piece by Johannes Kepler in 1604 and in the neglected treatise by Girard Desargues in 1639. Desargues and Kepler thought that the points at infinity formed a "circle of infinite radius" (which they seem to do when viewed affinely), but Poncelet correctly recognized that they formed a line (no different from any other line when considered projectively). Blaise Pascal was another earlier contributor to projective geometry with his *mystic hexagram theorem* of 1639, discovered when he was only 16 (see Project 4).

The principle of duality was first expounded in 1825–1827 by J.-D. Gergonne. Poncelet knew about duality but thought it resulted from Apollonius' idea of the poles and polars determined by a conic; in the case of a circle, that *polarity* (see Project 2) will play an important role in our work in Chapter 7. The most famous dual theorems are those of Pascal and C.-L. Brianchon about a hexagon inscribed in (respectively, circumscribed about) a conic.

The algebraic approach to projective geometry via homogeneous coordinates and homogeneous equations was introduced by A. F. Möbius in 1827 and then vastly developed into higher dimensions by J. Plücker in the 1830s. There was an acrimonious dispute during the nineteenth century between the projective geometers who worked algebraically and those who worked synthetically—over which was the proper approach. Poncelet was a strident synthesist, declaring publicly that algebraic methods were inferior, yet it was discovered from his private notes long after he died that he (like Newton) secretly used algebraic methods to discover some of his results. Another leading synthesist was K. C. G. von Staudt, who, in the 1850s, eliminated any references to number and distance from projective geometry. Plücker's work was not appreciated until decades later, so he became a physicist and made important contributions to that science.

Projective geometry is the best setting for the study of algebraic geometry. For a simple example, the theorem of Bézout states that a plane algebraic curve of degree m intersects another plane algebraic curve of degree n in mn points if the intersections are counted with multiplicities. This theorem is generally valid only in the projective plane, not in the affine plane, because the intersections at infinity must be counted, and only when the plane is coordinatized by numbers from an algebraically closed field such as $\mathbb{C}$ (otherwise, the two curves might not intersect at all—consider a line and a circle in the real Euclidean plane; in case the line is tangent to the circle, the point of tangency must be counted with multiplicity 2 for Bézout's theorem to work).

Algebraic curves of degree 2 are *conics*. The nondegenerate affine ones are the ellipse, parabola, and hyperbola. Desargues recognized that they can be distinguished in the real affine plane by the number of points each has at infinity, namely, 0, 1, 2, respectively. In the projective plane, they cannot be distinguished—they all look like ellipses (more precisely, they are all projectively equivalent).

We will not develop projective geometry very deeply in this text, using it mainly in Chapters 7 and 10 to facilitate our understanding of non-Euclidean geometries. See the projects in this chapter for further interesting theorems.

Conclusion

This chapter has two main themes: The first is logic, and the second is incidence geometry. Experienced students of mathematics probably were able to quickly review the classical principles of logic presented in the first few sections, but even they need to study the sections on models and should take note of the RAA proof that $\sqrt{2}$ is irrational. Mathematical logic, insofar as it is the study of correct reasoning (it also studies other important topics such as computability), traditionally has two aspects: *syntax* and *semantics*. Generally speaking, syntax studies the *form* of reasoning and is a purely formal study of the connectives $\Rightarrow$, &, $\vee$, $\sim$; the quantifiers $\forall$, $\exists$; predicates such as = and $\in$; variables, etc. Semantics, on the other hand, *interprets* the formal symbols and gives them various meanings, and we are only concerned with mathematical interpretations.

A formal mathematical theory starts with undefined terms and axioms about those terms, which can be written in a symbolic language (as we did "for the fun of it" with Axioms I-1, I-2, and I-3) or which can be written in a natural language such as English for easier comprehension. Using the rules of logic, propositions were then proved from the axioms, and we described precisely what proofs are. When axioms have been given, what we are interested in is interpretations that satisfy those axioms. Those are called *models*.

Our main application of these ideas so far is to *incidence geometry*. We gave its three basic axioms and stated five propositions that can easily be proved from those axioms. That was purely formal, although we used undefined terms *point, line, incidence,* which are suggestive of familiar geometric notions. However, since the terms are undefined, we took the liberty of interpreting them in unfamiliar ways, such as in our three-, four-, and five-point models, which have three different parallel properties. The notion of *parallel lines* is the main topic studied in this text. What we accomplished with those models was to show that incidence geometry is a *consistent* theory and to show the *impossibility* of proving various different statements about parallel lines if we only assume the axioms of incidence geometry. The demonstrations of those impossibilities belong to a subject that may be new to you: *metamathematics.*

Then we returned to mathematics itself and gave the two most important examples of incidence geometries (i.e., models of the axioms of incidence geometry): *affine planes,* which are models in which the Euclidean parallel property holds, and *projective planes,* which are models in which parallel lines do not exist (and in which every line has at least three points lying on it). We proved the main result that every affine plane can be naturally *completed* to a projective plane by adjoining points "at infinity" and a "line at infinity" on which all those points lie. We then presented the main example of affine and projective planes coordinatized by a field (such as the field of real numbers or the field of complex numbers).

Finally, we proved another metamathematical theorem, the *principle of duality* for projective planes.

Affine geometry is Euclidean geometry without betweenness and congruence. In the next chapter, we will add betweenness and congruence to our structure.

Exercises

1. (a) What is the negation of $P \lor Q$?
   (b) What is the negation of $P \ \& \ {\sim}Q$?
   (c) Using Logic Rules 3, 4, and 5, show that $P \Rightarrow Q$ means the same as $[{\sim}P \lor Q]$.
2. State the negation of Euclid's fourth postulate.
3. State the negation of the Euclidean parallel postulate. (This will be very important later.)

4. State the converse of each of the following statements:
   (a) If lines l and m are parallel, then a transversal t to lines l and m cuts out congruent alternate interior angles.
   (b) If the sum of the degree measures of the interior angles on one side of transversal t is less than 180°, then lines l and m meet on that side of transversal t.
5. Rewrite the informal argument given in the text to prove Proposition 2.2 as a formal proof, i.e., as a list of steps with each step numbered and with a justification for each step given. The justification must be one of the six types allowed by Logic Rule 1. Use your own argument if you have a better one.
6. Give formal proofs of Propositions 2.3, 2.4, and 2.5.
7. For each pair of axioms of incidence geometry, invent an interpretation in which those two axioms are satisfied but the third axiom is not. (This will show that the three axioms are *independent* in the sense that it is impossible to prove any one of them from the other two. It is more economical and elegant to have axioms that are independent, but it is not essential for developing an interesting theory.)
8. Show that the interpretations in Examples 3 and 4 of this chapter are models of incidence geometry and that the Euclidean and hyperbolic parallel properties, respectively, hold for them.
9. In each of the following interpretations of the undefined terms, which of the axioms of incidence geometry are satisfied and which are not? Tell whether each interpretation has the elliptic, Euclidean, or hyperbolic parallel property.
   (a) "Points" are lines in Euclidean three-dimensional space, "lines" are planes in Euclidean three-space, "incidence" is the usual relation of a line lying in a plane.
   (b) Same as in part (a), except that we restrict ourselves to lines and planes that pass through a fixed point O.
   (c) Fix a circle in the Euclidean plane. Interpret "point" to mean a Euclidean point inside the circle, interpret "line" to mean a chord of the circle, and let "incidence" mean that the point lies on the chord. (A *chord* of a circle is a segment whose endpoints lie on the circle.)
   (d) Fix a sphere in Euclidean three-space. Two points on the sphere are called *antipodal* if they lie on a diameter of the sphere; e.g., the north and south poles are antipodal. Interpret a "point" to be a set {P, P'} consisting of two points on the sphere that are antipodal. Interpret a "line" to be a great circle on the sphere.

Interpret a "point" $\{P, P'\}$ to "lie on" a "line" C if both P and
P' lie on C (actually, if one lies on C, then so does the other,
by the definition of "great circle").

10. (a) Show that when each of two models of incidence geometry has
exactly three "points" in it, the models are isomorphic.

   (b) Must two models having exactly four "points" be isomorphic? If
you think so, show this; if you think not, give a counterexample.

   (c) Show that the models in Exercises 9(b) and 9(d) are isomorphic. (Hint: Take the point O of Exercise 9(b) to be the center
of the sphere in Exercise 9(d) and cut the sphere with lines
and planes through point O to get the isomorphism.)

11. Invent a model of incidence geometry that has neither the elliptic,
hyperbolic, nor Euclidean parallel properties. These properties refer
to any line l and any point P not on l. Invent a model that has different parallelism properties for different choices of l and P. (Hint:
Five points suffice for a finite example, or you could find a suitable
piece of the Euclidean plane for an infinite example, or you could
refer to a previous exercise. Or invent a fourth example.)

12. (a) Show that in any affine plane, $\forall l \; \forall m \; \forall n \; (l \parallel m \; \& \; m \parallel n \; \&$
$l \neq n \Rightarrow l \parallel n)$. This property is called *transitivity of parallelism*.

   (b) Why must we assume $l \neq n$ in defining this property?

   (c) Show that, conversely, a model with this property must be an
affine plane.

   (d) Exhibit a model of incidence geometry in which parallel lines
exist but parallelism is not transitive.

13. Suppose that in a given model for incidence geometry, every "line"
has at least three distinct "points" lying on it. What is the least number of "points" and the least number of "lines" such a model can
have? Suppose further that the model has the Euclidean parallel property, i.e., is an affine plane. Show that 9 is now the least number of
"points" and 12 the least number of "lines" such a model can have.

14. (a) Let S be the following statement in the language of incidence
geometry: If l and m are any two distinct lines, then there exists a point P that does not lie on either l or m. Show that S is
not a theorem in incidence geometry, i.e., cannot be proved
from the axioms of incidence geometry.

   (b) Show, however, that statement S holds in every projective
plane. Hence $\sim S$ cannot be proved from the axioms of incidence geometry either, so S is independent of those axioms.

   (c) Use statement S to prove that in a finite projective plane, all
the lines have the same number of points lying on them. (Hint:

Map the points on l onto points on m by projecting from the point P. This mapping is called a *perspectivity with center* P.)

(d) Prove that in a finite affine plane, all the lines have the same number of points lying on them. (Hint: Apply part (c) to the projective completion or find a direct affine proof.)

15. (a) Four distinct points, no three of which are collinear, are said to form a *quadrangle*. Let $\mathscr{P}$ be a model of incidence geometry for which every line has at least three distinct points lying on it. Show that a quadrangle exists in $\mathscr{P}$.

(b) Now suppose $\mathscr{P}$ is a projective plane. Four distinct lines, no three of which are concurrent, are said to form a *quadrilateral*. Use the principle of duality to prove that a quadrilateral exists in $\mathscr{P}$.

(c) Give an example of a statement that holds in all affine planes but whose dual never holds. Thus the principle of duality is not valid for affine planes.

16. (a) Fill in the missing details in Example 8.

(b) Generalize the definition of $P^2(F)$ to construct $P^3(F)$, *projective three-space coordinatized by the field* F. Interpret "points," "lines," and "planes" in $P^3(F)$. If you have some experience with analytic geometry in three dimensions, show that any two planes in $P^3(F)$ have a line in common. Show that three non-collinear points lie in a unique plane; what is the three-dimensional dual to this statement?

(c) Propose undefined terms and axioms for three-dimensional projective geometry.

17. The following whimsical syllogisms are by Lewis Carroll. They are intended to illustrate that logical syntax depends only on the form of the argument, not on the meaning or truth of the statements. Which of them are correct arguments?

(a) No frogs are poetical; some ducks are unpoetical. Hence, some ducks are not frogs.

(b) Gold is heavy; nothing but gold will silence him. Hence, nothing light will silence him.

(c) All lions are fierce; some lions do not drink coffee. Hence, some creatures that drink coffee are not fierce.

(d) Some pillows are soft; no pokers are soft. Hence, some pokers are not pillows.

18. Here is a whimsical question: We think of the lines in the real affine plane as "straight." When we completed that plane to the real projective plane, we added just one point at infinity to each affine line.

As we indicated, this extended line is now a closed curve. How did the line lose its "straightness" just by adding one point at infinity? Or, could a closed curve be "straight"? Can you picture the real projective plane as some smooth surface in Euclidean three-space? Discuss this question informally.

19. (a) Let *S* be the following self-referential statement: "Statement *S* is false." Show that *S* is true iff *S* is false. This is the liar paradox. Does it imply that some statements are neither true nor false? (Kurt Gödel used the variant "This statement is unprovable" as the starting point for his famous incompleteness theorem in mathematical logic.)

(b) A set is intuitively any collection of things, and those things are the elements of that set. Suppose we collect all the sets *S* with the property that $S \notin S$ and only those sets. Call that set *C*. By the law of excluded middle, either $C \in C$ or $C \notin C$. Show that in either case, a contradiction can be deduced. This is Bertrand Russell's paradox. Does it imply that set theory is inconsistent? Discuss this question with your instructor.

Major Exercises

1. Consider the following interpretation of incidence geometry. Begin with a punctured sphere in Euclidean three-space, i.e., a sphere with one point N removed. Interpret "points" as points on the punctured sphere. For each circle on the sphere passing through N, interpret the punctured circle obtained by removing N as a "line." Interpret "incidence" in the usual sense of a point lying on a punctured circle. Is this interpretation a model? If so, what parallel property does it have? Is it isomorphic to any other model you know? (Hint: If N is the north pole, project the punctured sphere stereographically from N onto the plane Π tangent to the sphere at the south pole, as shown in Figure 2.10. Use the fact that planes through N other than the tangent plane cut out circles on the sphere and lines in Π. For an amusing discussion of this interpretation, refer to Chapter 3 of Sved, 1991.)

2. Show that every projective plane $\mathscr{P}$ is isomorphic to the projective completion of some affine plane $\mathscr{A}$. (Hint: Pick any line *m* in $\mathscr{P}$, pretend that *m* is "the line at infinity," remove *m* and all the points lying on it, and then show that what remains is an affine plane $\mathscr{A}$ and that $\mathscr{P}$ is isomorphic to the completion $\mathscr{A}^*$ of $\mathscr{A}$.)

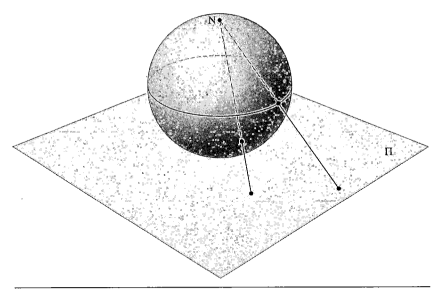

Figure 2.10 Stereographic projection.

3. Let $\mathscr{P}$ be a finite projective plane so that, according to Exercise 14(c), all lines in $\mathscr{P}$ have the same number of points lying on them; call this number $n + 1$, with $n \geq 2$. Show the following:
   (a) Each point in $\mathscr{P}$ has $n + 1$ lines passing through it.
   (b) The total number of points in $\mathscr{P}$ is $n^2 + n + 1$.
   (c) The total number of lines in $\mathscr{P}$ is $n^2 + n + 1$.
       The number n is called the *order* of the finite projective plane.
4. Let $\mathscr{A}$ be a finite affine plane so that, according to Exercise 14(d), all lines in $\mathscr{A}$ have the same number of points lying on them; let n be this number, with $n \geq 2$. Show the following:
   (a) Each point in $\mathscr{A}$ has $n + 1$ lines passing through it.
   (b) The total number of points in $\mathscr{A}$ is n^2.
   (c) The total number of lines in $\mathscr{A}$ is $n(n + 1)$.
       The number n is called the *order* of the finite affine plane.
5. Let F be the field with two elements $\{0, 1\}$ whose multiplication and addition have the usual tables except that $1 + 1 = 0$. Show that F^2 is isomorphic to the smallest affine plane, described in Example 3 of the text. Show that $P^2(F)$ is isomorphic to the projective plane described in Example 6 of the text. This is the smallest projective plane; it has order 2 and is called the *Fano plane* in honor of Gino Fano, who worked with finite geometries in 1892 (K. G. C. von Staudt was the first to consider them).

6. Recall from Exercise 15 that four points, no three of which are collinear, form a *quadrangle*. The four points are called the *vertices*, and the six lines obtained by joining pairs of vertices are called the *sides* of the quadrangle. (Note that sides are lines, not segments, because segments are defined by betweenness and we have no betweenness in pure incidence geometry.) Suppose we are working in a projective plane, so that every pair of sides will intersect. Pairs of sides that do not intersect at a vertex are called *opposite sides*, and there are three of those pairs; the points at which those pairs intersect are called the *diagonal points* of the quadrangle. *Fano's axiom* for projective planes asserts that *the diagonal points of any quadrangle are not collinear*. Show that Fano's axiom fails for the Fano plane.

   In $P^2(F)$, where F is any field, show that the four points at $[1, 0, 0]$, $[0, 1, 0]$, $[0, 0, 1]$, and $[1, 1, 1]$ are vertices of a quadrangle. Determine the equations for the six sides, tell which pairs are opposite sides, find the coordinates of the diagonal points, and tell whether or not those points are collinear.

7. Some authors characterize projective planes by three axioms: Axiom I-1, the elliptic parallel property, and the existence of a quadrangle. Show that a model of those axioms is a projective plane under our definition, and conversely.

8. Figure 2.11 is a symmetric depiction of the projective plane of order 3. The outer circle represents the line at infinity, and the black dots on it represent the points at infinity except that pairs of antipodal points on that circle are considered to be the same.

   Let F be the field with three elements $\{0, 1, -1\}$, whose multiplication and addition have the usual tables except that $1 + 1 = -1$ and $1 = (-1) + (-1)$ (addition mod 3). Label the 13 points in the

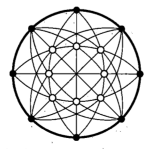

Figure 2.11

diagram with their homogeneous coordinates from F to illustrate the fact that this plane is isomorphic to $P^2(F)$.

Advanced Projects on Projective Planes

1. The following statement is by Desargues: "If the vertices of two triangles correspond in such a way that the lines joining corresponding vertices are concurrent, then the intersections of corresponding sides are collinear." This statement is independent of the axioms for projective planes. It holds only in those projective planes that can be embedded in a projective three-space. For example, if you regard Figure 2.12 as a three-dimensional picture in which the shaded triangles are in different planes, the line that Desargues asserts to exist is just the intersection of those two planes (the two triangles are in perspective from the point of concurrence P outside those planes). Report on this independence result and give an example of a non-Desarguesian projective plane (the best known example is due to Frederick Moulton in 1902; it is described in the English translation of Hilbert's *Grundlagen*). State the dual to Desargues' statement and compare that to its converse: What do you observe about them? (Note: A *triangle* in incidence geometry is defined to be a set of three distinct noncollinear points. The *sides* of the triangle are the three lines joining pairs of vertices. We cannot consider the sides as being segments because we do not have a notion of betweenness in pure incidence geometry.)

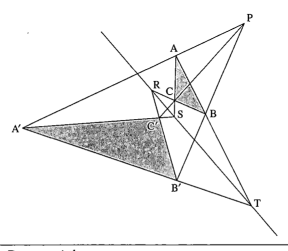

Figure 2.12 Desargues' theorem.

2. An isomorphism of a projective plane $\mathscr{P}$ onto its dual plane $\mathscr{P}^*$ is called a *polarity* of $\mathscr{P}$. It assigns to each point A of $\mathscr{P}$ a line $p(A)$ of $\mathscr{P}$ called the *polar* of A, and to each line m of $\mathscr{P}$ a point $P(m)$ of $\mathscr{P}$ called the *pole* of m, in such a way that A lies on m if and only if $P(m)$ lies on $p(A)$, and the correspondences are one-to-one onto. The set of all points A such that A lies on its polar is called the *conic* determined by this polarity, and for A on the conic, the polar $p(A)$ is called the *tangent* to the conic at A.

   This very abstract definition of "conic" (which does not refer to distances) can be reconciled with more familiar descriptions, such as the solution set to a homogeneous quadratic equation in three variables, when the plane can be coordinatized by a field. The theory of conics is one of the most important topics in plane projective geometry. Report on this theory. (The German poet Goethe said: "Mathematicians are like Frenchmen: Whatever you say to them, they translate it into their own language and forthwith it is something entirely different.")

3. Pappus of Alexandria (fourth century) was the last great Greek mathematician. His *Collection*, in eight volumes, is an invaluable compilation of the mathematical achievements of the ancient Greek world. He also contributed much original mathematics of his own. The theorem of Pappus in geometry states: "If A, B, and C are three distinct points on one line and if A', B', and C' are three other distinct points on a second line, then the intersections of lines AC' and CA', AB' and BA', and BC' and CB' are collinear." (See Figure 2.13.) Pappus' theorem can be proved for a projective plane $P^2(F)$ coordinatized by a field—in particular, for the real projective plane. G. Hessenberg proved, conversely, that if Pappus' statement holds in a projective plane, then it can be coordinatized by a field; his proof is based on ideas originating with von Staudt and later work by Hilbert.

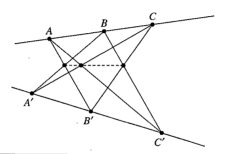

Figure 2.13 Pappus' theorem.

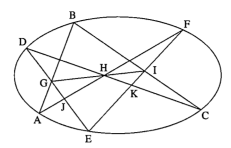

Figure 2.14 Pascal's mystic hexagram theorem.

Since $P^2(F)$ can be embedded in $P^3(F)$, it follows that Pappus' statement implies Desargues' (this was also proved directly in the plane by G. Hessenberg). The converse does not hold (see Project 5). Report on these results.

4. A pair of lines is a degenerate form of a conic. Pascal, at the age of 16, generalized Pappus' theorem to all conics in the real projective plane (as a result, some authors such as Hilbert refer to Pappus' theorem as Pascal's theorem). See Figure 2.14 and state the theorem. Brianchon's theorem was discovered 167 years afterward. Geometers subsequently noticed that it follows immediately from duality. See Figure 2.15 and state the theorem (Note: A tangent to a conic is the dual to a point on a conic.)

5. A *division ring* or a *skew field* has the same algebraic structure as a field except that multiplication is not necessarily commutative—i.e., $ab = ba$ may not hold for all a, b. An example is the skew field of *quaternions,* denoted **H** in honor of William Rowan Hamilton, who discovered them in 1853. (His close friend John Graves discovered the *octonions,* but Arthur Cayley published information about them first, so they are sometimes called the *Cayley numbers;* they do not form a division ring because the associative law $a(bc) = (ab)c$ does not hold for all octonions.)

If F is any division ring, we can construct the projective plane $P^2(F)$ coordinatized by F the same way as before, just being careful about the commutative law. A beautiful theorem relating algebra to geometry states that a projective plane can be coordinatized by some division ring if and only if Desargues' theorem holds in that plane. Furthermore, that division ring is a field—i.e., multiplication is commutative—if and only if Pappus' theorem holds in that plane.

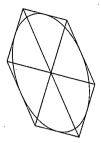

Figure 2.15 Brianchon's theorem.

A non-Desarguesian projective plane can be coordinatized only by an algebraic structure called a *ternary ring*. The octonions provide an example. Report on all these results.

6. The principle of duality is that once a statement S has been proved for *all* projective planes, its dual statement S^* is automatically also a theorem because S^* is just S applied to the dual plane. But as was pointed out, the statements of Desargues, Pappus, and Fano do not hold in all projective planes. Nevertheless, it is the case that if one of these three statements holds for a particular projective plane, then so does its dual, and that requires proof in each case—you cannot just invoke the principle of duality. Find or report on proofs that each of these statements implies its dual.

   However, suppose some statement S has been proved for all projective planes coordinatized by a field, or at least for all fields F of a certain type. In that case, S^* does hold automatically for those planes because the dual plane is also coordinatized by that same field, as we have seen. For example, Fano's statement holds for all planes coordinatized by a field or division ring of *characteristic* different from 2, i.e., one in which $1 + 1 \neq 0$. Report on this and the converse, that if the plane is coordinatized by a division ring and Fano's statement holds, then the division ring has characteristic $\neq 2$. Fano's and Pappus' statements are taken as axioms in those treatments of projective plane geometry which focus on generalizing classical results that hold in the real projective plane (Coxeter, 2003).

7. If F is a finite field, it is an elementary result in abstract algebra that the number of elements in F is a prime power p^k. Conversely, for every prime power p^k, there exists a finite field (unique up to isomorphism) with p^k elements. Since the order of the projective plane $P^2(F)$ is equal to the number of elements of F, it follows that

there exist projective planes of every prime power order. So there exist projective planes of orders 2, 3, 4, 5, 7, 8, 9, 11, ... It is known, however, that not every finite projective plane is coordinatized by a field (e.g., there are four different projective planes of order 9, up to isomorphism). The first example of a finite non-Desarguesian plane was published by O. Veblen and J. H. M. Wedderburn in 1907. A *finite* Desarguesian projective plane automatically satisfies Pappus' theorem; no geometric proof of this is known, but it follows from another famous *theorem of Wedderburn* that a finite division ring must be commutative (compare Project 5).

It is conjectured that the order of a finite projective plane must be a prime power. Orders 6, 14, 21, 22, and infinitely many others were shown to be impossible by the *Bruck–Ryser theorem*: Suppose that n is not a prime power and $n \equiv 1$ or 2 (mod 4). If n is not the sum of two squares, then no projective plane of order n exists.

Now $10 \equiv 2$ (mod 4), but 10 is the sum of two squares, so the Bruck–Ryser theorem does not apply. It was shown in 1989 by C. Lam and associates, after several years of computer searching, that there is no projective plane of order 10. They used results from 1970 by F. J. MacWilliams, N. J. A. Sloane, and J. G. Thompson to narrow the search to a few big computations. The next three unknown cases are $n = 12$, 15, and 18. Report on all these results.

As often happens in pure mathematics, the abstract subject of finite geometries turns out to have important connections to other subjects, e.g., to finite groups, cryptography, combinatorics, design theory, and quantum information theory. If he were alive today, Signor Fano would be very happy to see that his idea of finite geometries was so useful!

3

> *The value of Euclid's work as a masterpiece of logic has been very grossly exaggerated.*
>
> *Bertrand Russell*

Flaws in Euclid

Having specified our rules of reasoning in Chapter 2, let us return to Euclid. In the exercises of Chapter 1, we saw that Euclid neglected to state his assumptions that points and lines exist, that not all points are collinear, and that every line has at least two points lying on it. We made these assumptions explicit in Chapter 2 by adding two more ax-ioms of incidence, I-2 and I-3, to Euclid's first postulate, I-1. We proved a few consequences of those three axioms, we showed that those ax-ioms alone do not lead to any contradictions, and we briefly studied two main types of models of those axioms: affine planes, in which the Euclidean parallel postulate holds but which can be somewhat differ-ent from our usual Euclidean plane (e.g., they can be finite, and they have only an incidence structure), and projective planes, which are very different in that parallel lines do not exist in them. We showed the intimate connection between these two models: Each affine plane can be completed to a projective plane by adding a point at infinity to each line and the line at infinity upon which all those points lie;

David Hilbert

inversely, by removing one line and all the points on it from a projective plane, an affine plane is obtained.

In other exercises of Chapter 1, we saw that some assumptions about betweenness are needed. Euclid never mentioned this notion explicitly but tacitly assumed certain facts about it that seem obvious in diagrams. Gauss pointed out this omission in an 1831 letter to Farkas Bolyai, but he did not carry out the task of stating the required new axioms and deducing theorems from them. That was eventually done in 1882 by Moritz Pasch, and David Hilbert later incorporated Pasch's work as part of his *Grundlagen der Geometrie* (1899). Pasch has been called "the father of rigor in geometry" by the mathematician and historian Hans Freudenthal.

Several of Euclid's proofs are based on reasoning from diagrams. To make these proofs rigorous, a much larger system of explicit axioms is needed. We will present a modified version of David Hilbert's system of axioms, which are perhaps the most intuitive and are certainly the closest in spirit to Euclid's.[1] Hilbert's axioms are divided into five groups: incidence, betweenness, congruence, continuity, and parallelism. In the following sections, we will introduce the remaining four groups.

During the first quarter of the twentieth century, David Hilbert was considered the leading mathematician of the world (only Henri Poincaré could be considered his rival in that era). He made outstanding, original contributions to a wide range of mathematical fields as well as to theoretical physics (the infinite-dimensional spaces used in quantum mechanics are named after him). In addition to his work in geometry, he is perhaps best known for his research in invariant theory, algebraic number theory, integral equations, functional analysis, the calculus of variations, and mathematical logic. At the International Congress of Mathematics in 1900, he challenged mathematicians with 23 problems that turned out to be some of the most important of the twentieth century (most of them have been solved, the best known unsolved one being to settle the Riemann hypothesis). He unwittingly started a new tradition: In 2000, a committee of top mathematicians chose what they considered to be the 7 most challenging problems for the new century. The Clay Mathematics Institute is offering a million-dollar prize to anyone who can solve one of them, and it appears that one of those problems, the Poincaré conjecture in three dimensions, may have been proved (the proof is being thoroughly checked). The Riemann hypothesis is one of the other 6 problems.

Hilbert made a famous proclamation in 1930 that exemplifies his courageous, optimistic attitude toward mathematical problems: *Wir müssen wissen, wir werden wissen.* (We must know, we shall know.)[2]

Axioms of Betweenness

So far we have considered the two undefined terms *point* and *line* and the undefined *incidence* relation of a point to a line. Our fourth undefined or primitive term is the relation of *betweenness* among three

[1] Let us not forget that no serious work toward constructing new axioms for Euclidean geometry had been done until the discovery of non-Euclidean geometry shocked mathematicians into reexamining the foundations of the former. We have the paradox of non-Euclidean geometry helping us to better understand Euclidean geometry!

[2] See the biography of Hilbert by Constance Reid (1970). It is nontechnical and conveys the excitement of the time when Göttingen was the capital of the mathematical world. And see Gray, J. J. 2000, *The Hilbert Challenge*, New York: Oxford University Press.

points. By introducing another relation to our system, we are adding more structure to our geometry, which will eliminate certain models of the previous structure (incidence geometry, in this case) that cannot support the new structure. For example, it will be shown as a consequence of the four betweenness axioms to be introduced shortly that every line must have infinitely many points lying on it; thus, all the nice finite geometries mentioned in the examples and exercises of Chapter 2 will no longer concern us. We will refer to the betweenness axioms briefly as B-1 through B-4.

The flaw in the argument from diagrams in Chapter 1 that all triangles are isosceles has to do with betweenness. As you were asked to show in Major Exercise 4 of that chapter, the intersection D of the perpendicular bisector of the base with the bisector of the opposite angle must lie *outside* the triangle if these lines are distinct, and only one of the two feet of the perpendiculars dropped from D to the other two sides lies *inside* the triangle. These notions of "inside" and "outside" will be defined in terms of betweenness.

The statement of Euclid's Postulate 5 refers to two lines meeting on one "side" of a transversal, but Euclid neither defines the notion of "side" nor gives axioms for an undefined notion of "side." We will define that notion using betweenness and study its properties. Also, when we come to the proof of the exterior angle theorem in Chapter 4, you will see that betweenness properties play a crucial role.

Here is another example to illustrate the need for betweenness. It is an attempt to prove that the base angles of an isosceles triangle are congruent. This attempt is not Euclid's somewhat complicated proof known as *pons asinorum*, which is flawed in other ways, but is rather a simple argument found in some high school geometry texts.

PROOF:

Given $\triangle ABC$ with $AC \cong BC$. To prove $\angle A \cong \angle B$ (see Figure 3.1):

(1) Let the bisector of $\angle C$ meet AB at D (every angle has a bisector).
(2) In triangles $\triangle ACD$ and $\triangle BCD$, $AC \cong BC$ (hypothesis).
(3) $\angle ACD \cong \angle BCD$ (definition of bisector of an angle).
(4) $CD \cong CD$ (things that are equal are congruent).
(5) $\triangle ACD \cong \triangle BCD$ (SAS).
(6) Therefore, $\angle A \cong \angle B$ (corresponding angles of congruent triangles). ◄

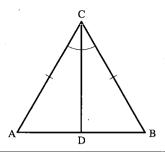

Figure 3.1

Consider the first step, whose justification is that every angle has a bisector. This is a correct statement and can be proved separately. But how do we know that the bisector of ⦟C meets $\overleftrightarrow{AB}$, or if it does, how do we know that the point of intersection D lies *between* A and B? This may seem obvious, but if we are to be rigorous, it requires proof. For all we know, the picture might look like Figure 3.2. If this were the case, steps 2–5 would still be correct, but we could conclude only that ⦟B is congruent to ⦟CAD, not to ⦟CAB, since ⦟CAD is the angle in △ACD that corresponds to ⦟B.

Once we state our four axioms of betweenness, it will be possible to prove (after a considerable amount of work) that the bisector of ⦟C does meet $\overleftrightarrow{AB}$ in a point D between A and B, so the above argument will be repaired (see the *crossbar theorem* later in this section). There is, however, an easier proof of the theorem (given in the next section). We will use the shorthand notation

$$A * B * C$$

to abbreviate the statement "point B is between point A and point C."

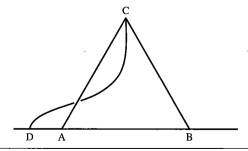

Figure 3.2

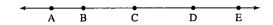

Figure 3.3

BETWEENNESS AXIOM 1. If A ∗ B ∗ C, then A, B, and C are three distinct points all lying on the same line, and C ∗ B ∗ A.

The first part of this axiom fills the gap mentioned in Exercise 6 of Chapter 1. The second part (C ∗ B ∗ A) makes the obvious remark that "between A and C" means the same as "between C and A"—it doesn't matter whether A or C is mentioned first.

BETWEENNESS AXIOM 2. Given any two distinct points B and D, there exist points A, C, and E lying on $\overleftrightarrow{BD}$ such that A ∗ B ∗ D, B ∗ C ∗ D, and B ∗ D ∗ E (Figure 3.3).

This axiom ensures that there are points between B and D and that the line $\overleftrightarrow{BD}$ does not end at either B or D. This axiom also shows that the points on a line do not form a *discrete* set like the natural numbers, where there are no natural numbers between n and $n + 1$ for any n.

BETWEENNESS AXIOM 3. If A, B, and C are three distinct points lying on the same line, then one and only one of the points is between the other two.

This axiom ensures that a line is not circular; if the points were on a simple closed curve like a circle, you would then have to say that each is between the other two or that none is between the other two—it would depend on which of the two arcs you look at (see Figure 3.4).
    Speaking intuitively, we have seen that when we complete the real affine plane to the real projective plane, a line becomes a closed curve.

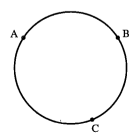

Figure 3.4

Thus, it is not possible to have a betweenness structure on the real
projective plane corresponding to our intuitive notion of betweenness
satisfying this axiom. In its place, a relation called *separation* among
four distinct points on a projective line can be introduced and studied—
see Appendix A.

Recall that the *segment* AB is defined as the set of all points between A and B together with the endpoints A and B. The *ray* $\overrightarrow{AB}$ is defined as the set of all points on the segment AB together with all points
C such that $A * B * C$. Axiom B-2 ensures that such points C exist,
B-3 ensures that C is not between A and B, and B-1 ensures that C is
not equal to either A or B; so the ray $\overrightarrow{AB}$ is larger than the segment AB.
Axiom B-1 also ensures that all points on ray $\overrightarrow{AB}$ lie on the line $\overleftrightarrow{AB}$.

PROPOSITION 3.1. For any two points A and B: (i) $\overrightarrow{AB} \cap \overrightarrow{BA} = AB$,
and (ii) $\overrightarrow{AB} \cup \overrightarrow{BA} = \{\overleftrightarrow{AB}\}$.

PROOF OF (i):

(1) By the definition of segment and ray, $AB \subset \overrightarrow{AB}$ and $AB \subset \overrightarrow{BA}$,
    so by the definition of intersection, $AB \subset \overrightarrow{AB} \cap \overrightarrow{BA}$.

(2) Conversely, let the point C belong to the intersection of $\overrightarrow{AB}$ and
    $\overrightarrow{BA}$; we wish to show that C belongs to AB.

(3) If $C = A$ or $C = B$, C is an endpoint of AB. Otherwise, A, B,
    and C are three collinear points (by the definition of ray and
    Axiom B-1), so exactly one of the relations $A * C * B$, $A * B *$
    C, or $C * A * B$ holds (Axiom B-3).

(4) If $A * B * C$ holds, then C is not on $\overrightarrow{BA}$; if $C * A * B$ holds, then
    C is not on $\overrightarrow{AB}$. In either case, C does not belong to both rays.

(5) Hence, the relation $A * C * B$ must hold, so C belongs to AB
    (definition of AB, proof by cases). ◄

The proof of (ii) is similar and is left as an exercise. (Recall that
$\{\overleftrightarrow{AB}\}$ is the set of points lying on the line $\overleftrightarrow{AB}$.)

Recall next that if $C * A * B$, then $\overrightarrow{AC}$ is said to be *opposite* to $\overrightarrow{AB}$
(see Figure 3.5). By Axiom B-1, points A, B, and C are collinear; by
Axiom 3, C does not belong to $\overrightarrow{AB}$, so rays $\overrightarrow{AB}$ and $\overrightarrow{AC}$ are distinct.
This definition is therefore in agreement with the definition given in
Chapter 1 (see Proposition 3.6). Axiom B-2 guarantees that every ray
$\overrightarrow{AB}$ has an opposite ray $\overrightarrow{AC}$.

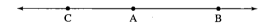

Figure 3.5

It seems clear from Figure 3.5 that every point P lying on the line
l through A, B, C must belong either to ray $\overrightarrow{AB}$ or to an opposite ray
$\overrightarrow{AC}$. This statement seems similar to the second assertion of Proposition 3.1, but it is actually more complicated; we are now discussing
four points, A, B, C, and P, whereas previously we had to deal with
only three points at a time. In fact, we encounter here another "picto-
rially obvious" assertion that cannot be proved without introducing another axiom (see Exercise 17).

Suppose we call the assertion "C * A * B and P collinear with A, B,
C $\Rightarrow$ P $\in \overrightarrow{AC} \cup \overrightarrow{AB}$" the *line separation property*. Some mathematicians
take this property as another axiom. However, it is considered inele-
gant in mathematics to assume more axioms than are necessary (although we pay for elegance by having to work harder to prove results).
So we will not assume the line separation property as an axiom; instead, we will prove it as a consequence of our previous axioms and
our last betweenness axiom, called the *plane separation axiom*.

DEFINITION. Let l be any line, and A and B any points that do not lie
on l. If A = B or if segment AB contains no point lying on l, we say A
and B are *on the same side of l*, whereas if A $\neq$ B and segment AB does
intersect l, we say that A and B are *on opposite sides* of l (see Figure
3.6). The law of the excluded middle (Logic Rule 10) tells us that A
and B are either on the same side or on opposite sides of l.

BETWEENNESS AXIOM 4 (PLANE SEPARATION). For every line l and
for any three points A, B, and C not lying on l:

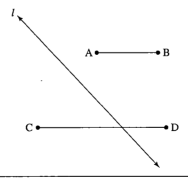

Figure 3.6 A and B are on the same side of l; C and D are on opposite
sides of l.

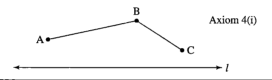

Figure 3.7

(i) If A and B are on the same side of l and if B and C are on the same side of l, then A and C are on the same side of l (see Figure 3.7).

(ii) If A and B are on opposite sides of l and if B and C are on opposite sides of l, then A and C are on the same side of l (see Figure 3.8).

COROLLARY. (iii) If A and B are on opposite sides of l and if B and C are on the same side of l, then A and C are on opposite sides of l.

Axiom 4(i) guarantees that our geometry is two-dimensional, since it does not hold in three-space. (Line l could be outside the plane of this page and cut through segment AC; this interpretation shows that if we assumed the line separation property as an axiom, we could not prove the plane separation property.) Betweenness Axiom 4 is also needed to make sense of Euclid's fifth postulate, which talks about two lines meeting on one "side" of a transversal. We can now define a *side* of a line l as the set of all ponts that are on the same side of l as some particular point A not lying on l. If we denote this side by H_A, notice that if C is on the same side of l as A, then by Axiom 4(i), $H_C = H_A$. (The definition of a *side* may seem circular because we use the word "side" twice, but it is not; we have already defined the compound expression "on the same side.") Another expression commonly used for a "side of l" is a *half-plane bounded by l*.

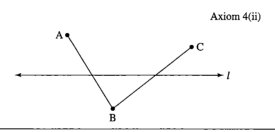

Figure 3.8

PROPOSITION 3.2. Every line bounds exactly two half-planes, and these half-planes have no point in common.

> *PROOF:*
> (1) There is a point A not lying on l (Proposition 2.3).
> (2) There is a point O lying on l (Incidence Axiom 2).
> (3) There is a point B such that B * O * A (Betweenness Axiom 2).
> (4) Then A and B are on opposite sides of l (by definition), so l has at least two sides.
> (5) Let C be any point distinct from A and B and not lying on l. If C and B are not on the same side of l, then C and A are on the same side of l (by the law of excluded middle and Betweenness Axiom 4(ii)). So the set of points not on l is the union of the side H_A of A and the side H_B of B.
> (6) If C were on both sides (RAA hypothesis), then A and B would be on the same side (Axiom 4(i)), contradicting step 4; hence the two sides are disjoint (RAA conclusion). ◄

We next apply the plane separation property to study betweenness relations among four points.

PROPOSITION 3.3. Given A * B * C and A * C * D. Then B * C * D and A * B * D (see Figure 3.9).

> *PROOF:*
> (1) A, B, C, and D are four distinct collinear points (see Exercise 1).
> (2) There exists a point E not on the line through A, B, C, D (Proposition 2.3).
> (3) Consider line $\overleftrightarrow{EC}$. Since (by hypothesis) AD meets this line in point C, points A and D are on opposite sides of $\overleftrightarrow{EC}$.
> (4) We claim A and B are on the same side of $\overleftrightarrow{EC}$. Assume on the contrary that A and B are on opposite sides of $\overleftrightarrow{EC}$ (RAA hypothesis).

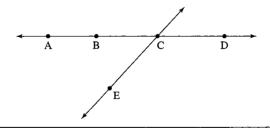

Figure 3.9

(5) Then $\overleftrightarrow{EC}$ meets $\overleftrightarrow{AB}$ in a point beween A and B (definition of "opposite sides").

(6) That point must be C (Proposition 2.1).

(7) Thus, A * B * C and A * C * B, which contradicts Betweenness Axiom 3.

(8) Hence, A and B are on the same side of $\overleftrightarrow{EC}$ (RAA conclusion).

(9) B and D are on opposite sides of $\overleftrightarrow{EC}$ (steps 3 and 8 and the corollary to Betweenness Axiom 4).

(10) Hence, the point C of intersection of lines $\overleftrightarrow{EC}$ and $\overleftrightarrow{BD}$ lies between B and D (definition of "opposite sides"; Proposition 2.1, i.e., that the point of intersection is unique).

A similar argument involving $\overleftrightarrow{EB}$ proves that A * B * D (Exercise 2(b)). ◀

COROLLARY.   Given A * B * C and B * C * D. Then A * B * D and A * C * D.

Finally we prove the *line separation property*.

PROPOSITION 3.4. If C * A * B and l is the line through A, B, and C (Betweenness Axiom 1), then for every point P lying on l, P lies either on ray $\overrightarrow{AB}$ or on the opposite ray $\overrightarrow{AC}$.

PROOF:

(1) Either P lies on $\overrightarrow{AB}$ or it does not (law of excluded middle).

(2) If P does lie on $\overrightarrow{AB}$, we are done, so assume it doesn't; then P * A * B (Betweenness Axiom 3).

(3) If P = C, then P lies on $\overrightarrow{AC}$ (by definition), so assume P ≠ C; then exactly one of the relations C * A * P, C * P * A, or P * C * A holds (Betweenness Axiom 3 again).

(4) Suppose the relation C * A * P holds (RAA hypothesis).

(5) We know (by Betweenness Axiom 3) that exactly one of the relations P * C * B, C * P * B, or C * B * P holds.

(6) If P * B * C, then combining this with P * A * B (step 2) gives A * B * C (Proposition 3.3), contradicting the hypothesis.

(7) If C * P * B, then combining this with C * A * P (step 4) gives A * P * B (Proposition 3.3), contradicting step 2.

(8) If B * C * P, then combining this with B * A * C (hypothesis and Betweenness Axiom 1) gives A * C * P (Proposition 3.3), contradicting step 4.

(9) Since we obtain a contradiction in all three cases, C * A * P does not hold (RAA conclusion).

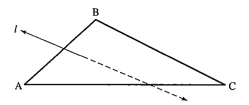

Figure 3.10

(10) Therefore, C * P * A or P * C * A (step 3), which means that P
  lies on the opposite ray $\overrightarrow{AC}$. ◄

The next theorem states a visually obvious property that Pasch discovered Euclid to be using without proof.

PASCH'S THEOREM. If A, B, C are distinct noncollinear points and l
is any line intersecting AB in a point between A and B, then l also intersects either AC or BC (see Figure 3.10). If C does not lie on l, then
l does not intersect both AC and BC.

   Intuitively, this theorem says that if a line "goes into" a triangle
through one side, it must "come out" through another side.

   PROOF:

   (1) Either C lies on l or it does not; if it does, the theorem holds
       (law of excluded middle).
   (2) A and B do not lie on l, and the segment AB does intersect l
       (hypothesis and Axiom B-1).
   (3) Hence, A and B lie on opposite sides of l (by definition).
   (4) From step 1 we may assume that C does not lie on l, in which
       case C is either on the same side of l as A or on the same side
       of l as B (separation axiom).
   (5) If C is on the same side of l as A, then C is on the opposite
       side from B, which means that l intersects BC and does not in-
       tersect AC; similarly, if C is on the same side of l as B, then l
       intersects AC and does not intersect BC (separation axiom).
   (6) The conclusion of Pasch's theorem holds (Logic Rule 11—proof
       by cases). ◄

   Here are some more results on betweenness and separation that you
will be asked to prove in the exercises.

PROPOSITION 3.5. Given A * B * C. Then AC = AB ∪ BC and B is the
only point common to segments AB and BC.

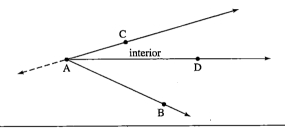

Figure 3.11

PROPOSITION 3.6. Given A * B * C. Then B is the only point common to rays $\overrightarrow{BA}$ and $\overrightarrow{BC}$, and $\overrightarrow{AB} = \overrightarrow{AC}$.

DEFINITION. Given an angle ⦨CAB, define a point D to be in the *interior* of ⦨CAB if D is on the same side of $\overleftrightarrow{AC}$ as B and if D is also on the same side of $\overleftrightarrow{AB}$ as C. (Thus, the interior of an angle is the intersection of two half-planes.) See Figure 3.11.

PROPOSITION 3.7. Given an angle ⦨CAB and point D lying on line $\overleftrightarrow{BC}$. Then D is in the interior of ⦨CAB if and only if B * D * C (see Figure 3.12).

WARNING Do not assume that every point in the interior of an angle lies on a segment joining a point on one side of the angle to a point on the other side. In fact, this assumption is false in hyperbolic geometry (see Exercise 19).

PROPOSITION 3.8. If D is in the interior of ⦨CAB, then (a) so is every other point on ray $\overrightarrow{AD}$ except A; (b) no point on the opposite ray to $\overrightarrow{AD}$ is in the interior of ⦨CAB; and (c) if C * A * E, then B is in the interior of ⦨DAE (see Figure 3.13).

DEFINITION. Ray $\overrightarrow{AD}$ is *between* rays $\overrightarrow{AC}$ and $\overrightarrow{AB}$ if $\overrightarrow{AB}$ and $\overrightarrow{AC}$ are not opposite rays and D is interior to ⦨CAB. (By Proposition 3.8(a), this definition does not depend on the choice of point D on $\overrightarrow{AD}$.)

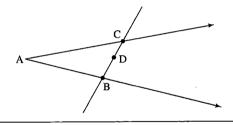

Figure 3.12

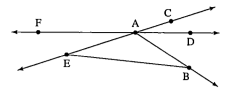

Figure 3.13

CROSSBAR THEOREM. If $\overrightarrow{AD}$ is between $\overrightarrow{AC}$ and $\overrightarrow{AB}$, then $\overrightarrow{AD}$ intersects segment BC (see Figure 3.14).

PROOF:

(1) D is in the interior of ⊰CAB (by hypothesis and definition of "betweenness" for rays).

(2) Let E be a point such that E ∗ A ∗ C (B-2; see Figure 3.13).

(3) Since line $\overleftrightarrow{AD}$ intersects segment EC in point A between E and C, E and C are on opposite sides of line $\overleftrightarrow{AD}$ (definition of "opposite sides").

(4) B is in the interior of ⊰DAE (step 1 and Proposition 3.8(c)).

(5) Hence B and E are on the same side of line $\overleftrightarrow{AD}$ (definition of "interior" of an angle).

(6) Therefore, B and C are on opposite sides of line $\overleftrightarrow{AD}$ (step 3 and corollary to B-4).

(7) Let G be the point between B and C that lies on line $\overleftrightarrow{AD}$ (step 6, definition of "opposite sides").

(8) G is in the interior of ⊰CAB (step 7 and Proposition 3.7).

(9) G lies either on ray $\overrightarrow{AD}$ or on its opposite ray (Proposition 3.4).

(10) Suppose G lies on the opposite ray (RAA hypothesis).

(11) Then G is not in the interior of ⊰CAB (Proposition 3.8(b)).

(12) Therefore, G lies on ray $\overrightarrow{AD}$ (step 11 contradicts step 8, RAA conclusion). ◄

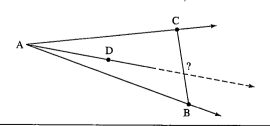

Figure 3.14

We call this result a *theorem* instead of a proposition to emphasize its importance (as was illustrated in the incomplete argument that base angles of an isosceles triangle are congruent).

DEFINITIONS. The *interior* of a triangle is the intersection of the interiors of its three angles. A point is *exterior* to the triangle if it is not in the interior and does not lie on any side of the triangle.

PROPOSITION 3.9. (a) If a ray r emanating from an exterior point of $\triangle ABC$ intersects side AB in a point between A and B, then r also intersects side AC or side BC. (b) If a ray emanates from an interior point of $\triangle ABC$, then it intersects one of the sides, and if it does not pass through a vertex, it intersects only one side.

You are asked to prove this also as an exercise.

▓▓▓ **EXAMPLE 1. AFFINE PLANES OVER ORDERED FIELDS.** We saw in Chapter 2 that if F is a field, then the set F^2 of ordered pairs (x, y) of elements of F can be given a natural structure of incidence plane, where lines are determined by linear equations and a point lies on a given line if and only if its coordinates satisfy the equation for that line. Moreover, the Euclidean parallel postulate holds in this plane, so it is (by definition) an affine plane.

Suppose now that F has the structure of an *ordered field*. This means that besides the algebraic operations of addition, subtraction, multiplication, and division for elements of F, there is a relation $a < b$ for elements of F that is compatible with the algebraic operations. (See p. 600 for the precise definition.) If you have not taken a course in abstract algebra, think of the familiar ordered fields of rational numbers $\mathbb{Q}$ or of real numbers $\mathbb{R}$ (later we will consider another important ordered field **K** called the *constructible field*—the closure of $\mathbb{Q}$ under the operation of taking square roots of positive numbers). Not every field can be given an order structure: One of the conditions for an ordered field is

For every a, b, c, if $a < b$ then $a + c < b + c$.
Another condition is that $0 < 1$.

Hence, $0 < 0 + 1 < 1 + 1 = 0 + 1 + 1 < 1 + 1 + 1 < \cdots$. Thus, by repeatedly adding 1's, we see that an ordered field must have infinitely many elements (in fact, it must contain an ordered subfield isomorphic to $\mathbb{Q}$). This eliminates all the finite fields we mentioned in the exercises for Chapter 2. Other conditions in an ordered field are that for every $a \neq 0$, we have $0 < a^2$ and that $-1 < 0$; hence -1 cannot have a square root in an ordered field. This eliminates the field $\mathbb{C}$ of complex numbers.

Given three distinct elements a, b, c in the ordered field F, we define b to be *between* a and c if either $a < b < c$ or $c < b < a$. For example, $\frac{1}{2}$ is between 1 and 0. Using this definition, we interpret betweenness for three distinct collinear points A, B, C in F^2 as follows:

CASE 1. The line they lie on has an equation of the form $y = mx + b$. Then A * B * C iff the first coordinate of B is between the first coordinates of A and C.

CASE 2. The line they lie on is vertical, i.e., has an equation of the form $x = k$, where k is constant. Then A * B * C iff the second coordinate of B is between the second coordinates of A and C.

We leave it as a major exercise for those readers familiar with ordered fields to verify that with this interpretation of betweenness, the interpretations of axioms B-1 through B-4 hold, so F^2 becomes a model of both our incidence axioms and our betweenness axioms. Let us illustrate Proposition 3.2: In Case 1, the two half-planes determined by that line are determined, respectively, by the inequalities $y < mx + b$ and $y > mx + b$; in Case 2, they are determined, respectively, by the inequalities $x < k$ and $x > k$. We call a model of both our incidence and betweenness axioms an *ordered incidence plane*.

NOTE. Since $\mathbb{Q}^2$ with the incidence and betweenness structures we have defined is an ordered incidence plane, we have shown that if the theory of the ordered field of rational numbers is consistent, then so is the theory of ordered incidence planes (because any proof of a contradiction in the latter theory could be translated via the above model into a contradiction in the former theory). This is a *relative consistency* demonstration, but it is important because we have more experience and confidence that the theory of the ordered field $\mathbb{Q}$ is consistent than we might have for this new theory of ordered incidence planes.

EXAMPLE 2. AN ORDERED INCIDENCE PLANE (THE DISK) WITH THE HYPERBOLIC PARALLEL PROPERTY. Let the open unit disk U in F^2, consisting of all points (x, y) in F^2 such that $x^2 + y^2 < 1$, be our new set of points. Interpret lines to be chords of the unit circle $x^2 + y^2 = 1$ and interpret incidence the same as before. You have already shown (at least informally) in Exercise 9(c) of Chapter 2 that this interpretation is an incidence plane satisfying the hyperbolic parallel property. If we restrict the relation of betweenness in F^2 to U, it is easy to see that the betweenness axioms so interpreted still hold. So U is another ordered incidence plane.

Axioms of Congruence

If we were more pedantic, *congruent*, the last of our undefined terms, would be replaced by two terms since it refers to either a relation between segments or a relation between angles. By "abuse of language" (as French mathematicians say—it is really a simplification of our language), we will not be so pedantic because the intuitive idea is the same for both types of congruence. We use the familiar symbol $\cong$ to denote congruence. The following definition provides further abuse because we will use the word "congruent" also as a defined term for a relation between triangles.

DEFINITION. Triangles $\triangle ABC$ and $\triangle DEF$ are *congruent* if there exists a one-to-one correspondence between their vertices such that corresponding sides are congruent and corresponding angles are congruent. We will use the notation $\triangle ABC \cong \triangle DEF$ to indicate not only that these triangles are congruent but that a correspondence demonstrating that congruence is such that A corresponds to D, B to E, and C to F (i.e., the order in which we write the vertices matters).

We will introduce six axioms for congruence, which will be referred to as C-1 through C-6.

CONGRUENCE AXIOM 1. If A and B are distinct points and if A' is any point, then for each ray r emanating from A' there is a *unique* point B' on r such that B' $\neq$ A' and AB $\cong$ A'B' (see Figure 3.15).

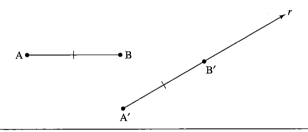

Figure 3.15

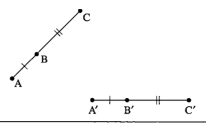

Figure 3.16

Intuitively speaking, this axiom says you can "move" the segment AB so that it lies on the ray r with A superimposed on A' and B superimposed on B'. (In Exercise 15(b), Chapter 1, you showed how to do this with a straightedge and a collapsible compass.)

CONGRUENCE AXIOM 2. If AB $\cong$ CD and AB $\cong$ EF, then CD $\cong$ EF. Moreover, every segment is congruent to itself.

This axiom replaces Euclid's first common notion since it says that segments congruent to the same segment are congruent to each other. It also replaces the fourth common notion since it says that segments that coincide are congruent.

CONGRUENCE AXIOM 3. If A * B * C, A' * B' * C', AB $\cong$ A'B', and BC $\cong$ B'C', then AC $\cong$ A'C' (see Figure 3.16).

This axiom replaces the second common notion since it says that if congruent segments are "added" to congruent segments, the sums are congruent. Here, "adding" means juxtaposing segments along the same line. For example, using Congruence Axioms 1 and 3, you can lay off a copy of a given segment AB two, three, . . . , n times, to get a new segment $n \cdot$ AB (see Figure 3.17).

CONGRUENCE AXIOM 4. Given any $\sphericalangle$BAC (where, by the definition of "angle," $\overrightarrow{AB}$ is not opposite to $\overrightarrow{AC}$) and given any ray $\overrightarrow{A'B'}$ emanating from a point A', then there is a *unique* ray $\overrightarrow{A'C'}$ on a given side of line $\overleftrightarrow{A'B'}$ such that $\sphericalangle$B'A'C' $\cong$ $\sphericalangle$BAC (see Figure 3.18).

This axiom can be paraphrased to state that a given angle can be "laid off" on a given side of a given ray in a unique way (see Exercise 14(g), Chapter 1).

Figure 3.17 AB″ = 3 · AB.

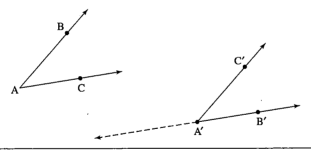

Figure 3.18 ∢ B'A'C ≅ ∢BAC.

CONGRUENCE AXIOM 5. If ∢A ≅ ∢B and ∢A ≅ ∢C, then ∢B ≅ ∢C. Moreover, every angle is congruent to itself.

This is the analogue for angles of Congruence Axiom 2 for segments; the first part asserts the transitivity and the second part the reflexivity of the congruence relation. Combining them, we can prove the symmetry of this relation: ∢A ≅ ∢B ⇒ ∢B ≅ ∢A.

PROOF:

∢A ≅ ∢B (hypothesis) and ∢A ≅ ∢A (reflexivity) imply (substituting A for C in Congruence Axiom 5) ∢B ≅ ∢A (transitivity). ◄

(By the same argument, congruence of segments is a symmetric relation.)

It would seem natural to assume next an "addition axiom" for congruence of *angles* analogous to Congruence Axiom 3 (the addition axiom for congruence of segments). We won't do this, however, because such a result can be proved using the next congruence axiom (see Proposition 3.19).

CONGRUENCE AXIOM 6 (SAS). If two sides and the included angle of one triangle are congruent, respectively, to two sides and the included angle of another triangle, then the two triangles are congruent (see Figure 3.19).

This side-angle-side criterion for congruence of triangles is a profound axiom. It provides the "glue" that binds the relation of congruence of segments to the relation of congruence of angles. It enables us to deduce all the basic results about triangle congruence with which you are presumably familiar. For example, here is one immediate consequence which states that we can "lay off" a given triangle on a given base and a given half-plane.

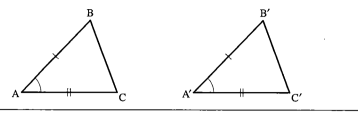

Figure 3.19 SAS.

COROLLARY TO SAS. Given △ABC and segment DE ≅ AB, there is a unique point F on a given side of line $\overleftrightarrow{DE}$ such that △ABC ≅ △DEF.

> *PROOF:*
>
> There is a unique ray $\overrightarrow{DF}$ on the given side such that ∢CAB ≅ ∢FDE, and F on that ray can be chosen to be the unique point such that AC ≅ DF (by Congruence Axioms 4 and 1). Then △ABC ≅ △DEF (SAS). ◄

As we said, Euclid did not take SAS as an axiom but tried to prove it as a theorem (Euclid I.4). His argument was essentially as follows. Move △A'B'C' so as to place point A' on point A and $\overrightarrow{A'B'}$ on $\overrightarrow{AB}$. Since AB ≅ A'B', by hypothesis, point B' must fall on point B. Since ∢A ≅ ∢A', $\overrightarrow{A'C'}$ must fall on $\overrightarrow{AC}$, and since AC ≅ A'C', point C' must coincide with point C. Hence, B'C' will coincide with BC and the remaining angles will coincide with the remaining angles, so the triangles will be congruent.

This argument is called *superposition*. It derives from the experience of drawing two triangles on paper, cutting out one, and placing it on top of the other. Although this argument is a good way to convince a novice in geometry to accept SAS, it is not a proof, and Euclid reluctantly used it in only one other proposition (I.8). It is not a proof because Euclid never stated an axiom that allows figures to be moved around without changing their size and shape.

Some modern writers introduce "motion" as an undefined term and lay down axioms for this term. (In fact, in Pieri's foundations of geometry, "point" and "motion" are the only undefined terms.) Or else, the geometry is first built up on a different basis, "distances" introduced, and a "motion" defined as a one-to-one transformation of the plane onto itself that preserves distance. Euclid can be vindicated by either approach. In fact, Felix Klein, in his 1872 *Erlanger Programme*, defined a geometry as the study of those properties of figures that remain invariant under a particular group of transformations. This idea will be developed in Chapter 9.

You will show in Exercise 35 that it is impossible to prove SAS or any of the other criteria for congruence of triangles (SSS, ASA, SAA) from the preceding axioms. As usual, the method for proving the impossibility of proving some statement S is to invent a model for the preceding axioms in which S is false.

As an application of SAS, the simple proof of Pappus for the theorem on base angles of an isosceles triangle follows.

PROPOSITION 3.10. If in $\triangle ABC$ we have $AB \cong AC$, then $\sphericalangle B \cong \sphericalangle C$ (see Figure 3.20).

PROOF:

(1) Consider the correspondence of vertices $A \leftrightarrow A$, $B \leftrightarrow C$, $C \leftrightarrow B$. Under this correspondence, two sides and the included angle of $\triangle ABC$ are congruent, respectively, to the corresponding sides and included angle of $\triangle ACB$ (by hypothesis and Congruence Axiom 5 that an angle is congruent to iself).

(2) Hence, $\triangle ABC \cong \triangle ACB$ (SAS), so the corresponding angles, $\sphericalangle B$ and $\sphericalangle C$, are congruent (by the definition of congruence of triangles). ◄

This proposition is Euclid I.5. Pappus' short proof was considered unacceptable by some because, if one thinks about triangle congruence as superposition, his proof seems to involve flipping the isosceles triangle through the third dimension; Pappus had the modern point of

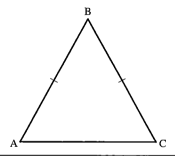

Figure 3.20

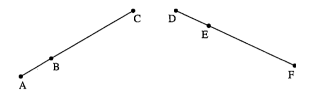

Figure 3.21

view of triangle congruence in terms of any one-to-one correspondence of vertices.[3]

Here are some more familiar results on congruence. We will prove some of them; if the proof is omitted, see the exercises.

PROPOSITION 3.11 (SEGMENT SUBTRACTION). If $A * B * C$, $D * E * F$, $AB \cong DE$, and $AC \cong DF$, then $BC \cong EF$ (see Figure 3.21).

PROPOSITION 3.12. Given $AC \cong DF$, then for any point B between A and C, there is a unique point E between D and F such that $AB \cong DE$.

  PROOF:

  (1) There is a unique point E on $\overrightarrow{DF}$ such that $AB \cong DE$ (Congruence Axiom 1).

  (2) Suppose E were not between D and F (RAA hypothesis; see Figure 3.22).

  (3) Then either $E = F$ or $D * F * E$ (definition of $\overrightarrow{DF}$).

  (4) If $E = F$, then B and C are two distinct points on $\overrightarrow{AC}$ such that $AC \cong DF \cong AB$ (hypothesis, step 1), contradicting the uniqueness part of Congruence Axiom 1.

  (5) If $D * F * E$, then there is a point G on the ray opposite to $\overrightarrow{CA}$ such that $FE \cong CG$ (Congruence Axiom 1).

  (6) Then $AG \cong DE$ (Congruence Axiom 3).

  (7) Thus, there are two distinct points B and G on $\overrightarrow{AC}$ such that $AG \cong DE \cong AB$ (steps 1, 5, and 6), contradicting the uniqueness part of Congruence Axiom 1.

  (8) $D * E * F$ (RAA conclusion). ◄

DEFINITION. $AB < CD$ (or $CD > AB$) means that there exists a point E between C and D such that $AB \cong CE$.

---

[3] In Appendix II of later editions of his *Grundlagen*, Hilbert (1988) did an advanced study of the role of the base angles of an isosceles triangle statement, constructing "non-Pythagorean" planes in which that statement and other familiar results fail. It also fails in the taxicab plane of Major Exercise 6.

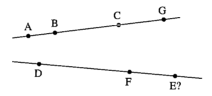

Figure 3.22

PROPOSITION 3.13 (SEGMENT ORDERING). (a) Exactly one of the following three conditions holds (*trichotomy*): AB < CD, AB ≅ CD, or AB > CD. (b) If AB < CD and CD ≅ EF, then AB < EF. (c) If AB > CD and CD ≅ EF, then AB > EF. (d) If AB < CD and CD < EF, then AB < EF (*transitivity*).

PROPOSITION 3.14. Supplements of congruent angles are congruent.

PROPOSITION 3.15. (a) Vertical angles are congruent to each other. (b) An angle congruent to a right angle is a right angle.

PROPOSITION 3.16. For every line *l* and every point P there exists a line through P perpendicular to *l*.

  PROOF:
  (1) Assume first that P does not lie on *l* and let A and B be any two points on *l* (Incidence Axiom 2). (See Figure 3.23.)
  (2) On the opposite side of *l* from P there exists a ray $\overrightarrow{AX}$ such that ⊀XAB ≅ ⊀PAB (Congruence Axiom 4).
  (3) There is a point P′ on $\overrightarrow{AX}$ such that AP′ ≅ AP (Congruence Axiom 1).

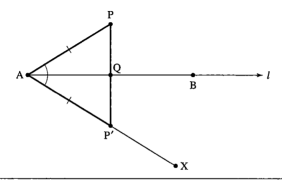

Figure 3.23

(4) PP' intersects l in a point Q (definition of opposite sides of l).

(5) If Q = A, then $\overleftrightarrow{PP'} \perp l$ (definition of $\perp$ and B-1).

(6) If Q ≠ A, then $\triangle PAQ \cong \triangle P'AQ$ (SAS).

(7) Hence, $\sphericalangle PQA \cong \sphericalangle P'QA$ (corresponding angles), so $\overleftrightarrow{PP'} \perp l$ (definition of $\perp$ and B-1).

(8) Assume now that P lies on l. Since there are points not lying on l (Proposition 2.3), we can drop a perpendicular from one of them to l (steps 5 and 7), thereby obtaining a right angle.

(9) We can lay off an angle congruent to this right angle with vertex at P and one side on l (Congruence Axiom 4); the other side of this angle is part of a line through P perpendicular to l (Proposition 3.15(b)). ◄

It is natural to ask whether the perpendicular to l through P constructed in Proposition 3.16 is unique. If P lies on l, Proposition 3.23 (later in this chapter) and the uniqueness part of Congruence Axiom 4 guarantee that the perpendicular is unique. If P does not lie on l, we will not be able to prove uniqueness for the perpendicular until the next chapter.

NOTE ON ELLIPTIC GEOMETRY. Informally, elliptic geometry may be thought of as the geometry on a Euclidean sphere with antipodal points identified (the model of incidence geometry first described in Exercise 9(d), Chapter 2). Its "lines" are the great circles on the sphere. Given such a "line" l, there is a point P called the "pole" of l such that every line through P is perpendicular to l! To visualize this, think of l as the equator on a sphere and P as the north pole; every great circle through the north pole is perpendicular to the equator (Figure 3.24).

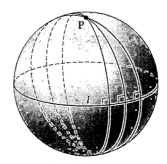

Figure 3.24

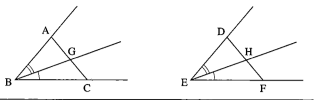

Figure 3.25

PROPOSITION 3.17 (ASA CRITERION FOR CONGRUENCE). Given △ABC and △DEF with ∡A ≅ ∡D, ∡C ≅ ∡F, and AC ≅ DF. Then △ABC ≅ △DEF.

PROPOSITION 3.18 (CONVERSE OF PROPOSITION 3.10). If in △ABC we have ∡B ≅ ∡C, then AB ≅ AC and △ABC is isosceles.

PROPOSITION 3.19 (ANGLE ADDITION). Given $\overrightarrow{BG}$ between $\overrightarrow{BA}$ and $\overrightarrow{BC}$, $\overrightarrow{EH}$ between $\overrightarrow{ED}$ and $\overrightarrow{EF}$, ∡CBG ≅ ∡FEH, and ∡GBA ≅ ∡HED. Then ∡ABC ≅ ∡DEF (see Figure 3.25).

PROOF:

(1) By the crossbar theorem,[4] we may assume G is chosen so that A * G * C.
(2) By Congruence Axiom 1, we may assume D, F, and H chosen so that AB ≅ ED, GB ≅ EH, and CB ≅ EF.
(3) Then △ABG ≅ △DEH and △GBC ≅ △HEF (SAS).
(4) ∡DHE ≅ ∡AGB, ∡FHE ≅ ∡CGB (step 3), and ∡AGB is supplementary to ∡CGB (step 1 and B-1).
(5) D, H, F are collinear, and ∡DHE is supplementary to ∡FHE (step 4, Proposition 3.14, and Congruence Axiom 4).
(6) D * H * F (Proposition 3.7, using the hypothesis on $\overrightarrow{EH}$).
(7) AC ≅ DF (steps 3 and 6, Congruence Axiom 3).
(8) ∡BAC ≅ ∡EDF (steps 3 and 6).
(9) △ABC ≅ △DEF (SAS; steps 2, 7, and 8).
(10) ∡ABC ≅ ∡DEF (corresponding angles). ◄

PROPOSITION 3.20 (ANGLE SUBTRACTION). Given $\overrightarrow{BG}$ between $\overrightarrow{BA}$ and $\overrightarrow{BC}$, $\overrightarrow{EH}$ between $\overrightarrow{ED}$ and $\overrightarrow{EF}$, ∡CBG ≅ ∡FEH, and ∡ABC ≅ ∡DEF. Then ∡GBA ≅ ∡HED.

---

[4] This renaming technique will be used frequently. G is just a label for any point ≠ B on the ray that intersects AC, so we may as well choose G to be the point of intersection rather than clutter the argument with a new label.

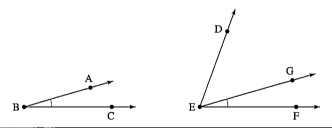

Figure 3.26

DEFINITION. ⊀ABC < ⊀DEF means there is a ray $\overrightarrow{EG}$ between $\overrightarrow{ED}$ and $\overrightarrow{EF}$ such that ⊀ABC ≅ ⊀GEF (see Figure 3.26).

PROPOSITION 3.21 (ORDERING OF ANGLES). (a) Exactly one of the following three conditions holds (*trichotomy*): ⊀P < ⊀Q, ⊀P ≅ ⊀Q, or ⊀Q < ⊀P. (b) If ⊀P < ⊀Q and ⊀Q ≅ ⊀R, then ⊀P < ⊀R. (c) If ⊀P > ⊀Q and ⊀Q ≅ ⊀R, then ⊀P > ⊀R. (d) If ⊀P < ⊀Q and ⊀Q < ⊀R, then ⊀P < ⊀R.

PROPOSITION 3.22 (SSS CRITERION FOR CONGRUENCE). If AB ≅ DE, BC ≅ EF, and AC ≅ DF, then △ABC ≅ △DEF.

The AAS criterion for congruence will be given in the next chapter because its proof is more difficult. The next proposition was assumed as an axiom by Euclid but can be proved from Hilbert's axioms.

PROPOSITION 3.23 (EUCLID'S FOURTH POSTULATE). All right angles are congruent to each other (see Figure 3.27).

PROOF:

(1) Given ⊀BAD ≅ ⊀CAD and ⊀FEH ≅ ⊀GEH (two pairs of right angles, by definition). Assume the contrary, that ⊀BAD is not congruent to ⊀FEH (RAA hypothesis).

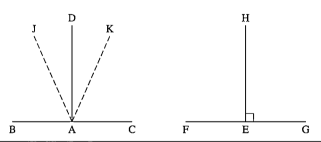

Figure 3.27

(2) Then one of these angles is smaller than the other—e.g., ∢FEH < ∢BAD (Proposition 3.21(a))—so that by definition there is a ray $\overrightarrow{AJ}$ between $\overrightarrow{AB}$ and $\overrightarrow{AD}$ such that ∢BAJ ≅ ∢FEH.

(3) ∢CAJ ≅ ∢GEH (Proposition 3.14).

(4) ∢CAJ ≅ ∢FEH (steps 1 and 3, Congruence Axiom 5).

(5) There is a ray $\overrightarrow{AK}$ between $\overrightarrow{AD}$ and $\overrightarrow{AC}$ such that ∢BAJ ≅ ∢CAK (step 1 and Proposition 3.21(b)).

(6) ∢BAJ ≅ ∢CAJ (steps 2 and 4, and Congruence Axiom 5).

(7) ∢CAJ ≅ ∢CAK (steps 5 and 6, and Congruence Axiom 5).

(8) Thus, we have ∢CAD greater than ∢CAK (by definition) and less than its congruent angle ∢CAJ (step 7 and Proposition 3.8(c)), which contradicts Proposition 3.21.

(9) ∢BAD ≅ ∢FEH (RAA conclusion). ◄

DEFINITIONS. An angle is *acute* if it is less than a right angle, *obtuse* if it is greater than a right angle.

According to Proposition 3.23 and Proposition 3.21(b) and (c), it doesn't matter which right angle is used for comparison in these definitions.

DEFINITION. A model of our incidence, betweenness, and congruence axioms is called a *Hilbert plane*.

Axioms of Continuity

There is a gap in the argument Euclid gives to justify his very first proposition. Here is his argument:

EUCLID'S PROPOSITION 1. Given any segment, there is an equilateral triangle having the given segment as one of its sides.

EUCLID'S PROOF:

(1) Let AB be the given segment. With center A and radius AB, let the circle BCD be described (Postulate III). (See Figure 3.28.)

(2) Again with center B and radius BA, let the circle ACE be described (Postulate III).

(3) From a point C in which the circles cut one another, draw the segments CA and CB (Postulate I).

(4) Since A is the center of the circle CDB, AC is congruent to AB (definition of circle).

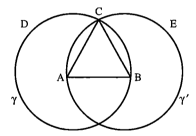

Figure 3.28

(5) Again, since B is the center of circle CAE, BC is congruent to
    BA (definition of circle).

(6) Since CA and CB are each congruent to AB (steps 4 and 5),
    they are congruent to each other (first common notion).

(7) Hence, △ABC is an equilateral triangle (by definition) having
    AB as one of its sides. ◄

Since every step has apparently been justified, you may not see the
gap in the proof. It occurs in the first three steps, especially in the third
step, which explicitly states that C is a point in which the circles cut
each other. (The second step states this implicitly by using the same
letter "C" to denote part of the circle, as in the first step.) The point
is: How do we know that such a point C exists?

If you believe it is obvious from the diagram that such a point C
exists, you are right—but you are not allowed to use the diagram to
justify this! We aren't saying that the circles constructed do not cut
each other; we're saying only that another axiom is needed to *prove*
that they do.

The gap can be filled using the following *circular* or *circle-circle continuity principle*:

CIRCLE-CIRCLE CONTINUITY PRINCIPLE. If a circle γ has one point
inside and one point outside another circle γ', then the two circles intersect in two points.

Here a point P is defined as *inside* a circle with center O and radius OR if OP < OR (*outside* if OP > OR). In Figure 3.28, point B is inside circle γ', and the point B' (not shown) such that A is the midpoint of BB' is outside γ'. This principle is also needed to prove Euclid
I.22, the converse to the triangle inequality (see Major Exercise 4).

Another gap occurs in Euclid's method of dropping a perpendicular to a line (Euclid I.12, our Proposition 3.16). His construction tacitly
assumes the *line-circle continuity principle*.

LINE-CIRCLE CONTINUITY PRINCIPLE. If a line passes through a point inside a circle, then the line intersects the circle in two points.

This follows from the circular continuity principle (see Major Exercise 1, Chapter 4); but our proof will use Proposition 3.16, so Euclid's argument must be discarded to avoid circular reasoning. Another useful consequence (see Major Exercise 2, Chapter 4) is the *elementary* or *segment-circle continuity principle*.

SEGMENT-CIRCLE CONTINUITY PRINCIPLE. If one endpoint of a segment is inside a circle and the other endpoint is outside, then the segment intersects the circle at a point in between.

Can you see why these are "continuity principles"? For example, in Figure 3.29, if you were drawing the segment with a pencil moving continuously from A to B, it would have to cross the circle (if it didn't, there would be a "hole" in the segment and the circle).

You may wonder why we have called these three statements "principles" instead of "theorems" or "axioms." The latter two would be theorems if we assumed the first one (as we will later show), but we do not wish to call the first one an axiom because we wish to illuminate exactly where it is needed, and then we will add it as a hypothesis. That will make the logical structure—which we emphasize in our treatment—clearer.

It is impossible to prove the circle-circle continuity principle from our incidence, betweenness, and congruence axioms alone. To demonstrate this independence result, one must exhibit a model of those axioms in which the circle-circle continuity principle is false. The construction of such a model is algebraic, requiring knowledge of Pythagorean ordered fields that are not Euclidean fields (see Hartshorne, Exercise 16.10). Also, Euclid I.1, the existence of equilateral triangles on any base, cannot be proved in arbitrary Hilbert planes without further assumption (see Hartshorne, Exercise 39.31).

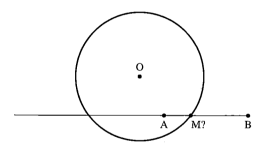

Figure 3.29

The next statement is not about continuity but rather about measurement. Archimedes was astute enough to recognize that a new axiom was needed. It is listed here because we will show that it is a consequence of Dedekind's continuity axiom, given later in this section. It is needed so that we can assign a positive real number as the *length* $\overline{AB}$ of an arbitrary segment AB, as will be explained in Chapter 4.

ARCHIMEDES' AXIOM. If CD is any segment, A any point, and r any ray with vertex A, then for every point $B \neq A$ on r there is a number n such that when CD is laid off n times on r starting at A, a point E is reached such that $n \cdot CD \cong AE$ and either $B = E$ or B is between A and E.

Here we use Congruence Axiom 1 to begin laying off CD on r starting at A, obtaining a unique point A_1 on r such that $AA_1 \cong CD$, and we define $1 \cdot CD$ to be AA_1. Let r_1 be the ray emanating from A_1 that is contained in r. By the same method, we obtain a unique point A_2 on r_1 such that $A_1A_2 \cong CD$, and we define $2 \cdot CD$ to be AA_2. Iterating this process, we can define, by induction on n, the segment $n \cdot CD$ to be AA_n.

For example, if AB were π units long and CD of 1 unit length, you would have to lay off CD at least four times to get to a point E beyond the point B (see Figure 3.30).

The intuitive content of Archimedes' axiom is that if you arbitrarily choose one segment CD as a unit of length, then every other segment has finite length with respect to this unit (in the notation of the axiom, the length of AB with respect to CD as unit is at most n units). Another way to look at it is to choose AB as unit of length. The axiom says that no other segment can be infinitesimally small with respect to this unit (the length of CD with respect to AB as unit is at least $1/n$ units).

The next statement is a consequence of Archimedes' axiom and the previous axioms (as you will show in Exercise 2, Chapter 5), but if one wants to do geometry with segments of infinitesimal length allowed, this statement can replace Archimedes' axiom (see my note "Aristotle's

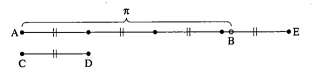

Figure 3.30

Axiom in the Foundations of Hyperbolic Geometry," *Journal of Geometry*, vol. 33, 1988). Besides, Archimedes' axiom is not a purely geometric axiom since it asserts the existence of a *number*.

ARISTOTLE'S ANGLE UNBOUNDEDNESS AXIOM. Given any side of an acute angle and any segment AB, there exists a point Y on the given side of the angle such that if X is the foot of the perpendicular from Y to the other side of the angle, XY > AB.

In other words, *the perpendicular segments from one side of an acute angle to the other are unbounded*—no segment AB can be a bound. In Chapter 5, where various attempts to prove Euclid V are analyzed, we will discuss how Proclus used this hypothesis in his attempt. Conversely, we will show in Chapter 4 that Euclid V implies Aristotle's axiom. Saccheri (whose work is discussed in Chapters 4–6) also recognized the importance of Aristotle's axiom and proved it using Archimedes' axiom.

IMPORTANT COROLLARY TO ARISTOTLE'S AXIOM. Let $\overrightarrow{AB}$ be any ray, P any point not collinear with A and B, and $\angle$XVY any acute angle. Then there exists a point R on ray $\overrightarrow{AB}$ such that $\angle$PRA < $\angle$XVY.

Informally, if we start with any point R on $\overrightarrow{AB}$, then as R "recedes endlessly" from the vertex A of the ray, $\angle$PRA decreases to zero (because it will eventually be smaller than any previously given angle $\angle$XVY). This result will be used in Chapter 6. Its proof uses Theorem 4.2 of Chapter 4 (the exterior angle theorem), and so it should be given after that theorem is proved, but we sketch the proof now for convenience of reference. You may skip it now and return when needed.

PROOF:

Let Q be the foot of the perpendicular from P to $\overleftrightarrow{AB}$. Since point B is just a label, we choose it so that Q ≠ B and Q lies on ray $\overrightarrow{BA}$. X and Y are arbitrary points on the rays r and s that are the sides of $\angle$XVY (see Figure 3.31). Let X' be the foot of the perpendicular from Y to the line containing r. By the hypothesis that the angle is acute and by the exterior angle theorem, we can show (by an RAA argument) that X' actually lies on r; so we can choose X to be X'.
  Aristotle's axiom guarantees that Y can be chosen such that XY > PQ. By Congruence Axiom 1, there is one point R on $\overrightarrow{QB}$ such that QR ≅ XV. We claim that $\angle$PRQ < $\angle$XVY. Assume the contrary. By trichotomy, there is a ray $\overrightarrow{RS}$ such that $\angle$QRS ≅ $\angle$XVY and $\overrightarrow{RS}$ either equals $\overrightarrow{RP}$ or is between $\overrightarrow{RP}$ and $\overrightarrow{RQ}$.

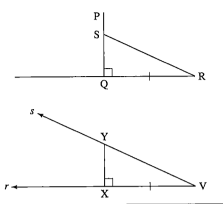

Figure 3.31

By the crossbar theorem, point S (which thus far is also merely a label) can be chosen to lie on segment PQ; then SQ is not greater than PQ. By the ASA congruence criterion, SQ ≅ XY. Hence XY is not greater than PQ, contradicting our choice of Y. Thus ∢PRQ < ∢XVY, as claimed. If R lies on ray $\overrightarrow{AB}$, then ∢PRQ = ∢PRA and we are done. If not, R and Q lie on the opposite ray. By the exterior angle theorem, if R' is any point such that Q * R * R', then ∢PR'Q < ∢PRQ < ∢XVY. We get ∢PBA = ∢PBQ < ∢XVY by taking R' = B.
◄

All four principles thus far stated are in the spirit of ancient Greek geometry. They are all consequences of the next axiom, which is utterly modern.

DEDEKIND'S AXIOM.[5] Suppose that the set $\{l\}$ of all points on a line l is the disjoint union $\Sigma_1 \cup \Sigma_2$ of two nonempty subsets such that no point of either subset is between two points of the other. Then there exists a unique point O on l such that one of the subsets is equal to a ray of l with vertex O and the other subset is equal to the complement.

Dedekind's axiom is a sort of converse to the line separation property stated in Proposition 3.4. That property says that any point O on

---

[5] This axiom was proposed by J. W. R. Dedekind in 1871; an analogue of it is used in analysis texts to express the completeness of the real number system. It implies that every Cauchy sequence converges, that continuous functions satisfy the intermediate value theorem, that the definite integral of a continuous function exists, and other important conclusions. Dedekind actually defined a "real number" as a Dedekind cut on the set of rational numbers, an idea Eudoxus had 2000 years earlier (see Moise, 1990, Chapter 20).

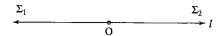

Figure 3.32

l separates all the other points on l into those to the left of O and those to the right (see Figure 3.32; more precisely, $\{l\}$ is the union of the two rays of l emanating from O). Dedekind's axiom says that, conversely, any separation of points on l into left and right is produced by a unique point O. A pair of subsets Σ_1 and Σ_2 with the properties in Dedekind's axiom is called a *Dedekind cut* of the line.

Loosely speaking, the purpose of Dedekind's axiom is to ensure that a line l has no "holes" in it, in the sense that for any point O on l and any positive real number x there exist unique points P_{-x} and P_x on l such that $P_{-x} * O * P_x$ and segments $P_{-x}O$ and OP_x both have length x (with respect to some unit segment of measurement). See Figure 3.33.

Without Dedekind's axiom there would be no guarantee, for example, of the existence of a segment of length π. With it, we can introduce a real number coordinate system into the plane and do geometry analytically. This coordinate system enables us to prove that our axioms for real Euclidean geometry are *categorical* in the sense that the system has a unique model (up to isomorphism—see the section Isomorphism of Models in Chapter 2), namely, the usual Cartesian coordinate plane of all ordered pairs of real numbers. (See Example 3 in the next section.)

The categorical nature of all the axioms is proved in Borsuk and Szmielew (1960, p. 276 ff.).

WARNING If you have never seen Dedekind's axiom before, arguments using it may be difficult to follow. Don't be discouraged. With the exception of Theorem 6.2 in hyperbolic geometry, it is not needed for studying the main theme of this book. I advise the beginning student to skip to the next section, Hilbert's Euclidean Axiom of Parallelism.

Let us sketch a proof that *Archimedes' axiom is a consequence of Dedekind's* (and the axioms preceding this section).

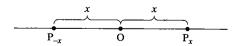

Figure 3.33

PROOF:

Given a segment CD and a point A on line l, with a ray r of l emanating from A. In the terminology of Archimedes' axiom, let Σ_1 consist of A and all points B on r reached by laying off copies of segment CD on r starting from A. Let Σ_2 be the complement of Σ_1 in r. We wish to prove that Σ_2 is empty, so assume the contrary.

In that case, let us show that we have defined a Dedekind cut of r (see Exercise 7(a)). Start with two points P, Q in Σ_2 and say A $*$ P $*$ Q. We must show that PQ $\subset \Sigma_2$. Let B be between P and Q. Suppose B could be reached, so that n and E are as in the statement of Archimedes' axiom; then, by Proposition 3.3, P is reached by the same n and E, contradicting P $\in \Sigma_2$. Thus PQ $\subset \Sigma_2$. Similarly, you can show that when P and Q are two points in Σ_1, PQ $\subset \Sigma_1$ (Exercise 7(b)). So we have a Dedekind cut. Let O be the point of r furnished by Dedekind's axiom.

CASE 1. O $\in \Sigma_1$. Then for some number n, O can be reached by laying off n copies of segment CD on r starting from A. By laying off one more copy of CD, we can reach a point in Σ_2, but by the definition of Σ_2, that is impossible.

CASE 2. O $\in \Sigma_2$. Lay off a copy of CD on the ray opposite to Σ_2 starting at O, obtaining a point P $\in \Sigma_1$. Then for some number n, P can be reached by laying off n copies of segment CD on r starting from A. By laying off one more copy of CD, we can reach O. That contradicts O $\in \Sigma_2$.

So in either case, we obtain a contradiction, and we can reject the RAA hypothesis that Σ_2 is nonempty. ◄

To further get an idea of how Dedekind's axiom gives us continuity results, we sketch a proof now of the segment-circle continuity principle from Dedekind's axiom (logically, this proof should be given later because it uses results from Chapter 4). Refer to Figure 3.29, p. 131.

PROOF:

By the definitions of "inside" and "outside" of a circle γ with center O and radius OR, we have OA < OR < OB. Let Σ_2 be the set of all points P on the ray $\overrightarrow{AB}$ that either lie on γ or are outside γ, and let Σ_1 be its complement in $\overrightarrow{AB}$. By trichotomy (Proposition 3.13(a)), Σ_1 consists of all points of the segment AB that lie inside γ. Applying Exercise 27 of Chapter 4, you can convince yourself that

(Σ_1, Σ_2) is a Dedekind cut. Let M be the point on $\overrightarrow{AB}$ furnished by Dedekind's axiom. Assume M does not lie on γ (RAA hypothesis).

CASE 1. OM < OR. Then $M \in \Sigma_1$. Let m and r be the lengths (defined in Chapter 4) of OM and OR, respectively. Since Σ_2 with M is a ray, there is a point $N \in \Sigma_2$ such that the length of MN is $\frac{1}{2}(r - m)$ (by laying off a segment whose length is $\frac{1}{2}(r - m)$). But by the *triangle inequality* (applied to $\triangle$OMN), the length of ON is less than $m + \frac{1}{2}(r - m) < m + (r - m) = r$, which contradicts $N \in \Sigma_2$.

CASE 2. OM > OR. The same argument applies, interchanging the roles of Σ_2 and Σ_1.

So in either case, we obtain a contradiction, and M must lie on γ. ◄

You will find a lovely proof of the circle-circle continuity principle from Dedekind's axiom on pp. 238–240 of Heath's translation and commentary on Euclid's *Elements* (1956). It assumes that Dedekind's axiom holds for semicircles, which you can easily prove, and also uses the triangle inequality and the fact that the hypotenuse is greater than the leg (proved in Chapter 4).

Euclid's tacit use of continuity principles can often be avoided. We did not use them in our proof of the existence of perpendiculars (Proposition 3.16). We did use the circular continuity principle to prove the existence of equilateral triangles on a given base, and Euclid used that to prove the existence of midpoints, as in your straightedge-and-compass solution to Major Exercise 1(a) of Chapter 1. But there is an

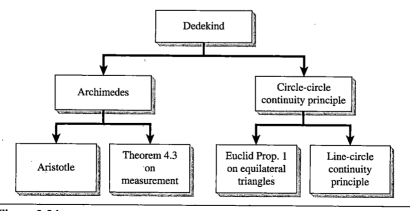

Figure 3.34

ingenious way to prove the existence of midpoints using only the very mild continuity given by Pasch's theorem (see Exercise 5, Chapter 4).

Figure 3.34 shows the implications discussed (assuming all the incidence, betweenness, and congruence axioms—especially SAS).

Hilbert's Euclidean Axiom of Parallelism

If we were to stop with the axioms we now have, we could do quite a bit of geometry, but not all of Euclidean geometry. We would be able to do what J. Bolyai called "absolute geometry." This name is misleading because it does not include elliptic geometry and other geometries (see Appendix B). Preferable is the name suggested by W. Prenowitz and M. Jordan, *neutral geometry*, so called because in doing this geometry we remain neutral about the one axiom from Hilbert's list left to be considered—historically the most controversial axiom of all.

HILBERT'S EUCLIDEAN AXIOM OF PARALLELISM. For every line l and every point P not lying on l there is at most one line m through P such that m is parallel to l (Figure 3.35).

Note that this axiom is weaker than the Euclidean parallel postulate introduced in Chapter 1. This axiom asserts only that *at most* one line through P is parallel to l, whereas the Euclidean parallel postulate asserts in addition that *at least* one line through P is parallel to l. The reason "at least" is omitted from Hilbert's axiom is that it can be proved from the other axioms (see Corollary 2 to Theorem 4.1 in Chapter 4); it is therefore unnecessary to assume this as part of an axiom. This observation is important because it implies that the elliptic parallel property (no parallel lines exist) is inconsistent with the axioms of neutral geometry. Thus, a different set of axioms is needed for the foundation of elliptic geometry (see Appendix A).

The axiom of parallelism completes our list of 15 axioms for *real* Euclidean geometry. A *real Euclidean plane* is a model of these axioms. In referring to these axioms, we will use the following shorthand: The

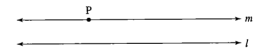

Figure 3.35

incidence axioms will be denoted by I-1, I-2, and I-3; the betweenness axioms by B-1, B-2, B-3, and B-4; the congruence axioms by C-1, C-2, C-3, C-4, C-5, and C-6 (or SAS). Dedekind's axiom and Hilbert's Euclidean parallelism axiom will be referred to by name.

The continuity axiom for a real Euclidean plane is Dedekind's axiom. This axiom is not needed to do elementary Euclidean geometry. Instead, the circle-circle continuity principle suffices to prove almost all the propositions in the first four volumes of Euclid's *Elements*.

DEFINITION. A *Euclidean plane* is a Hilbert plane in which Hilbert's Euclidean axiom of parallelism and the circle-circle continuity principle hold.

EXAMPLE 3. THE REAL EUCLIDEAN OR THE CARTESIAN PLANE. This is the model that most people have in mind when they talk about "the" Euclidean plane. In the treatise by Borsuk and Szmielew (1960) or in Hartshorne, it is proved that a real Euclidean plane is isomorphic to the model we are about to describe.

As we indicated, Dedekind's axiom provides a one-to-one correspondence between the points on a line and the ordered field $\mathbb{R}$ of real numbers. We have seen that $\mathbb{R}^2$ becomes a model of our incidence and betweenness axioms, as well as of Hilbert's Euclidean axiom of parallelism, with the interpretations discussed in Example 1 of this chapter. We now need to interpret the undefined term "congruence" to make $\mathbb{R}^2$ into a Euclidean plane. We do this via the familiar definition of *distance* or *segment length* in analytic geometry, based on the Pythagorean formula.

If $A = (a_1, a_2)$ and $B = (b_1, b_2)$ are two points in $\mathbb{R}^2$, define $d(A\ B)$ by

$$d(AB) = \sqrt{(a_1 - b_1)^2 + (a_2 - b_2)^2}.$$

Interpret $AB \cong CD$ to mean $d(A\ B) = d(C\ D)$; i.e., two segments are interpreted as congruent if they have the same length. To interpret congruence of angles, one could define a measure of angles by real numbers and interpret two angles to be congruent if they have the same angle measure; since that is not easy to do rigorously, we can use the following trick once we have verified the interpretation of C-1: Label the angle $\sphericalangle ABC$ with vertex B by letting A, C on the sides of the angle be the unique points such that $d(A\ B) = d(C\ B) = 1$. Label $\sphericalangle DEF$ similarly. Then interpret $\sphericalangle ABC \cong \sphericalangle DEF$ to mean $d(A\ C) = d(D\ F)$. (This is the SSS criterion in disguise.)

We leave it as Projects 1–4 to either verify that $\mathbb{R}^2$ with these interpretations of congruence satisfies our six congruence axioms and Dedekind's axiom, or to look up and report on the verification of those seven claims in other textbooks recommended. Hence $\mathbb{R}^2$ becomes a Euclidean plane with those interpretations—the *real Euclidean plane* (also referred to as the *Cartesian plane* in honor of Descartes' invention of analytic geometry, though Descartes had no precise notion of the real numbers and his coordinates were geometric segments).

EXAMPLE 4. THE CONSTRUCTIBLE EUCLIDEAN PLANE. In Example 3, we could try to use the same interpretations of congruence for the ordered rational affine plane $\mathbb{Q}^2$ instead of $\mathbb{R}^2$. Would that too become a model of our congruence axioms? The answer is NO! For instance, the interpretation of axiom C-1 fails. Consider the segment AB with A = (0, 0) and B = (1, 1). If one tries to lay off this segment on the ray r emanating from the origin A that passes through the point (1, 0) (i.e., the positive ray of the x-axis), we find that we cannot do that in $\mathbb{Q}^2$ because the point B′ on r which corresponds to B in $\mathbb{R}^2$ is B′ = $(\sqrt{2}, 0)$ since $d(AB) = \sqrt{2}$. Geometrically, the way we would construct the point B′ is to draw the circle γ centered at A of radius AB and then take B′ to be the point where that circle intersects the positive ray of the x-axis. When we restrict to points with rational coordinates, there is no such intersection point. We also see from this example that the segment-circle continuity principle fails in $\mathbb{Q}^2$: If C = (2, 0), segment AC has one endpoint A inside γ and the other endpoint C outside γ, yet there is no point in between in $\mathbb{Q}^2$ where γ intersects AC.

Joel Zeitlin informed me of another quirk in this interpretation. Consider point D = (1, 0). In $\mathbb{R}^2$, D is inside γ because $d(AD) = 1$ and γ has radius of length $\sqrt{2} > 1$. However, in $\mathbb{Q}^2$, D is not inside γ! The reason is that the point D′ in $\mathbb{R}^2$ between A and B for which $d(AD') = 1$ does not have rational coordinates (review the definition of "inside" and of < for segments). Similarly, D is neither outside nor on the circle γ. Trichotomy fails in this interpretation.

If you carry out or look up the verification of the interpretation of the congruence axioms and the circle-circle continuity principle in $\mathbb{R}^2$, you will see that the full power of the real number system is hardly used at all, only the fact that if a is a positive number, then $\sqrt{a}$ is in $\mathbb{R}$. The reason is that congruence is interpreted in terms of distance, and distance was defined as the square root of a positive number. As for the verification of the circle-circle continuity principle, it too comes

down to the existence of square roots of positive numbers because circles are represented in $\mathbb{R}^2$ by certain quadratic equations, and if the
hypothesis of the circle-circle principle is satisfied, then one can show
that the two quadratic equations for the two circles have two common
solutions obtained through use of the quadratic formula. This leads us
to the following definitions and theorem.

DEFINITION. A *Euclidean field* is an ordered field F with the property
that every positive element of F has a square root in F.

THEOREM. If F is a Euclidean field, then F^2, with congruence interpreted in the same way as in Example 3 above, is a Euclidean plane.

See the Projects for hints toward proving this.
Here is the most important example of a Euclidean field other
than $\mathbb{R}$.

DEFINITION. *The constructible field* **K** is the intersection of all Euclidean subfields of $\mathbb{R}$. (**K** is also called the *surd field* in Moise's 1990 text.)
An element of **K** is a real number that can be expressed in terms of rational numbers by finitely many applications of the five operations of
taking the square root of positive numbers, addition, subtraction, multiplication, and division. *The constructible Euclidean plane is* F^2, *where*
F = **K**.

For example, $(3 - \sqrt{2})^{1/2}$ is an element of **K**, but $\sqrt[3]{2}$ is not (that
requires proof). The latter result is the key to showing that duplication
of a cube is impossible using only straightedge and compass. In fact,
the Euclidean plane coordinatized by **K** is the key to proving the impossibility in general of the four classical straightedge-and-compass constructions discussed in Chapter 1. (See Hartshorne, Chapter 6, for all
the details.)

Note also that while the theory of *real* Euclidean planes is *categorical*—
all its models are isomorphic—the theory of Euclidean planes is not:
The plane coordinatized by $\mathbb{R}$ is not isomorphic to the plane coordinatized by **K**. For example, in $\mathbb{R}^2$ every angle has a trisector, but over
K the 60° angle does not have a trisector and the regular heptagon does
not exist (as Kepler observed).

**NOTE FOR ADVANCED STUDENTS ON THE RELATIVE CONSIS-
TENCY OF PLANE EUCLIDEAN GEOMETRY.** Hilbert used the result
that $\mathbb{R}^2$ is a model of his planar axioms to prove that if the theory of

the real numbers is consistent, then so is plane Euclidean geometry. Frankly, this result is of dubious value philosophically. Elementary plane Euclidean geometry is thousands of years older than the theory of the real numbers, and once the gaps in Euclid's presentation are filled by our—essentially Hilbert's—15 axioms for a Euclidean plane, we will have much more evidence to instill confidence that Euclidean geometry is consistent than we have for the consistency of the theory of $\mathbb{R}$. Or, if one seeks an algebraic proof of relative consistency, it is better to use the plane coordinatized by the field of constructible numbers $\mathbf{K}$ since $\mathbf{K}$ is a much more elementary field than $\mathbb{R}$ (e.g., $\mathbf{K}$ is a countably infinite field, an algebraic extension of $\mathbb{Q}$, whereas $\mathbb{R}$ is an uncountable transcendental extension of $\mathbb{Q}$, and its exact cardinality is a complete mystery to mathematicians because of the independence of the continuum hypothesis from the accepted axioms of set theory ZFC). $\mathbf{K}$ can be defined without referring to $\mathbb{R}$ by showing how to successively adjoin square roots of positive elements to fields built up that way starting from $\mathbb{Q}$ (see any good abstract algebra text).

Conclusion

The main purpose of this chapter is to fill in the gaps in Euclid's presentation of plane geometry. It is not claimed that we have filled in all of them—we have not, but almost all[6] the elementary synthetic Euclidean results you learned in high school can be proved from the 15 axioms for Euclidean planes.

The section on betweenness is probably new to you since Euclid did not consider that notion. The results on betweenness may seem obvious, yet they have profound significance. For one thing, they do not hold in elliptic geometry—the geometry of projective planes with the added structure of a four-point separation relation and a congruence relation (see Appendix A); in an elliptic plane, a line does not bound two half-planes (all the points not on the line are on the same side of the line). For another, they guarantee that we are working in two dimensions and that the plane is *orientable*—see Chapter 9, Exercise 23. Also review the warning in the betweenness section about one state-

---

[6] Euclid's theory of *content*—his version of area—requires Archimedes' axiom at certain points (see Hartshorne, Chapter 5).

ment you may consider "obvious" but which cannot be proved from our betweenness axioms (see Exercise 19).

The section on congruence contains results that should all be familiar. The main surprise, perhaps, is that the SAS triangle congruence criterion must be taken as an axiom—Euclid's superposition argument is good heuristics, but it is certainly not a proof in his system.

Euclid's fourth postulate (that all right angles are congruent to one another) is no longer an axiom in our system: It was proved as Proposition 3.23. Proclus, in his fifth-century commentary on Book I of the *Elements,* said Euclid IV should not be a postulate because it can be proved, and the idea for the proof we gave of Proposition 3.23 can be found in Proclus (1992, pp. 147–148). On the surface of a cone, right angles at the cone vertex are not congruent to right angles at other points of the cone (Henderson and Taimina, 2005, p. 58), so one can also argue that Euclid IV is not "obvious."

In the section on continuity, we showed how the circle-circle continuity principle fills the gap in Euclid's very first proposition, the construction of an equilateral triangle on any given base. We mentioned two other continuity principles that later will be shown to be consequences of circle-circle continuity and that fill other gaps in Euclid. We also introduced Aristotle's axiom, a very important elementary geometric axiom used by Proclus; Archimedes' axiom, which is not a purely geometric axiom but which is needed for measurement; and Dedekind's set-theoretic axiom, which turns out to be equivalent to coordinatizing our plane with real numbers.

Finally, we stated Hilbert's Euclidean axiom of parallelism, the last of our axioms for a Euclidean plane. In Chapter 4, we will show that it is equivalent to Euclid V. We have not derived any consequences of that axiom yet and will not do so for a while because we wish to remain neutral about it and see what can be proved without it. *None of the results in this chapter, including the results in the exercises, depend on Hilbert's Euclidean axiom of parallelism.* We provided (without proofs) two very important examples of Euclidean planes: the real Cartesian plane and the constructible Euclidean plane.

NOTE FOR ADVANCED STUDENTS ON THE EXISTENCE OF CERTAIN GEOMETRIC SETS. The astute reader may have noticed that while we have been very careful to add explicit axioms asserting the existence of certain points and lines, such as Axioms I-1, I-2, I-3, B-1, B-2, C-1, C-4, and the circle-circle continuity principle, and to carefully prove from those axioms other existence assertions (such as the existence of perpendiculars and parallels, the crossbar theorem, etc.), we have been rather casual about the existence of circles, segments, rays, half-planes, and so on. We either referred to "elementary set theory" as justification or just took their existence for granted. Let us be a little more precise here. Given distinct points O and A, the circle γ with center O and a radius OA is defined as

$$\gamma = \{P | OP \cong OA\}.$$

In words: Circle γ is the set of all points P satisfying the geometric condition that OP is congruent to the given segment OA. As another example, if A and B are distinct points,

$$AB = \{P | P = A \lor P = B \lor A * P * B\}.$$

In words: Segment AB is the set of all points P satisfying the geometric condition that either P is A, or P is B, or P is between A and B.

The general principle of set theory we are invoking is as follows: *For any geometric condition, the set of all points and lines satisfying that condition exists.* However, that set may be the empty set: As one example, the set of all triples of points A, B, C such that $A * B * C$ but A, B, C not collinear is empty, according to Axiom B-1. As another example, in a projective plane, the set of all lines parallel to a given line is empty.

What's missing here is a precise definition of "geometric condition." That would require a more systematic discussion of the mathematical logic underlying our theory. We would have to precisely define the language of our theory and what is a well-formed formula in that language. Then a geometric condition is just a well-formed formula in the language of elementary geometry with one or more free (i.e., unquantified) variables. We are not stating the above principle as another axiom in our system. Consider it rather as a background principle akin to Euclid's common notions.[7]

---

[7] To be totally faithful to the spirit of Euclid, one should not bring in set theory at all since it is a theory first presented rigorously in the twentieth century. In that case, one would have to replace everything we have done using sets with further undefined terms and further axioms about those terms (e.g., "circle" would become an undefined term). That is a complicated project. The interested reader is invited to learn about Tarski's different first-order primitive terms and axioms for elementary Euclidean geometry at http://en.wikipedia.org/wiki/Tarski's_axioms. Tarski's theory is *decidable* and *complete*— i.e., there is an algorithm for deciding whether any geometric statement in his language is provable or its negation is. One can question how "elementary" Tarski's axioms are since there are infinitely many continuity axioms (brought into one *axiom schema*).

Review Exercise

Which of the following statements are correct?

(1) Hilbert's axiom of parallelism is the same as the Euclidean parallel postulate given in Chapter 1.

(2) A * B * C is logically equivalent to C * B * A.

(3) In Axiom B-2, it is unnecessary to assume the existence of a point E such that B * D * E because this can be proved from the rest of the axiom and Axiom B-1, by interchanging the roles of B and D and taking E to be A.

(4) If A, B, and C are distinct collinear points, it is possible that *both* A * B * C *and* A * C * B.

(5) The "line separation property" asserts that a line has two sides.

(6) If points A and B are on opposite sides of a line l, then a point C not on l must be either on the same side of l as A or on the same side of l as B.

(7) If line m is parallel to line l, then all the points on m lie on the same side of l.

(8) If we were to take Pasch's theorem as an axiom instead of the separation axiom B-4, then B-4 could be proved as a theorem.

(9) The notion of "congruence" for two triangles is not defined in this chapter.

(10) It is an immediate consequence of Axiom C-2 that if AB $\cong$ CD, then CD $\cong$ AB.

(11) One of the congruence axioms asserts that if congruent segments are "subtracted" from congruent segments, the differences are congruent.

(12) In the statement of Axiom C-4, the variables A, B, C, A', and B' are quantified universally, and the variable C' is quantified existentially.

(13) One of the congruence axioms is the side-side-side (SSS) criterion for congruence of triangles.

(14) Euclid attempted unsuccessfully to prove the side-angle-side (SAS) criterion for congruence by a method called "superposition."

(15) We can use Pappus' method to prove the converse of the theorem on base angles of an isosceles triangle if we first prove the angle-side-angle (ASA) criterion for congruence.

(16) Archimedes' axiom is independent of the other 15 axioms for real Euclidean geometry given in this book.

(17) AB < CD means that there is a point E between C and D such that AB ≅ CE.

(18) All Euclidean planes are isomorphic to one another.

(19) $\sqrt[3]{2}$ is not a constructible number.

(20) A *Hilbert plane* is any model of the incidence, betweenness, and congruence axioms.

Exercises on Betweenness

1. Given A ∗ B ∗ C and A ∗ C ∗ D.
   (a) Prove that A, B, C, and D are four distinct points (the proof requires an axiom).
   (b) Prove that A, B, C, and D are collinear.
   (c) Prove the corollary to Axiom B-4.

2. (a) Finish the proof of Proposition 3.1 by showing that $\overrightarrow{AB} \cup \overrightarrow{BA} = \overleftrightarrow{AB}$.
   (b) Finish the proof of Proposition 3.3 by showing that A ∗ B ∗ D.
   (c) Prove the converse of Proposition 3.3 by applying Axiom B-1.
   (d) Prove the corollary to Proposition 3.3.

3. Given A ∗ B ∗ C.
   (a) Use Proposition 3.3 to prove that AB ⊂ AC. Interchanging A and C, deduce CD ⊂ CA; which axiom justifies this interchange?
   (b) Use Axiom B-4 to prove that AC ⊂ AB ∪ BC. (Hint: If P is a fourth point on AC, use another line through P to show P ∈ AB or P ∈ BC.)
   (c) Finish the proof of Proposition 3.5. (Hint: If P ≠ B and P ∈ AB ∩ BC, use another line through P to get a contradiction.)

4. Given A ∗ B ∗ C.
   (a) If P is a fourth point collinear with A, B, and C, use Proposition 3.3 and an axiom to prove that ~A ∗ B ∗ P ⟹ ~A ∗ C ∗ P.
   (b) Deduce that $\overrightarrow{BA} \subset \overrightarrow{CA}$ and, symmetrically, $\overrightarrow{BC} \subset \overrightarrow{AC}$.
   (c) Use this result, Proposition 3.1(a), Proposition 3.3, and Proposition 3.5 to prove that B is the only point that $\overrightarrow{BA}$ and $\overrightarrow{BC}$ have in common.

5. Given A ∗ B ∗ C. Prove that $\overrightarrow{AB} = \overrightarrow{AC}$, completing the proof of Proposition 3.6. Deduce that every ray has a *unique* opposite ray.

6. In Axiom B-2, we were given distinct points B and D, and we asserted the existence of points A, C, and E such that A ∗ B ∗ D, B ∗ C ∗ D, and B ∗ D ∗ E. We can now show that it was not necessary to assume the existence of a point C between B and D because

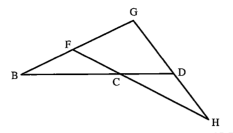

Figure 3.36

we can prove from our other axioms (including the rest of Axiom B-2) and from Pasch's theorem (which was proved without using Axiom B-2) that C exists.[8] Your job is to justify each step in the proof (some of the steps require a separate RAA argument).

PROOF (SEE FIGURE 3.36):

(1) There exists a line $\overrightarrow{BD}$ through B and D.
(2) There exists a point F not lying on $\overrightarrow{BD}$.
(3) There exists a line $\overrightarrow{BF}$ through B and F.
(4) There exists a point G such that B * F * G.
(5) Points B, F, and G are collinear.
(6) G and D are distinct points and D, B, and G are not collinear.
(7) There exists a point H such that G * D * H.
(8) There exists a line $\overrightarrow{GH}$.
(9) H and F are distinct points.
(10) There exists a line $\overleftrightarrow{FH}$.
(11) D does not lie on $\overleftrightarrow{FH}$.
(12) B does not lie on $\overleftrightarrow{FH}$.
(13) G does not lie on $\overleftrightarrow{FH}$.
(14) Points D, B, and G determine △DBG, and $\overleftrightarrow{FH}$ intersects side BG in a point between B and G.
(15) H is the only point lying on both $\overleftrightarrow{FH}$ and $\overleftrightarrow{GH}$.
(16) No point between G and D lies on $\overleftrightarrow{FH}$.
(17) Hence, $\overleftrightarrow{FH}$ intersects side BD in a point C between D and B.
(18) Thus, there exists a point C between D and B. ◄

7. (a) Define a Dedekind cut on a ray *r* the same way a Dedekind cut is defined for a line. Prove that the conclusion of Dedekind's

---

[8] Regarding superfluous hypotheses, there is a story that Napoleon, after examining a copy of Laplace's *Celestial Mechanics*, asked Laplace why there was no mention of God in the work. The author replied, "I have no need of this hypothesis."

axiom also holds for r. (Hint: One of the subsets, say, Σ_1, contains the vertex A of r; enlarge this set so as to include the ray
opposite to r and show that a Dedekind cut of the line l containing r is obtained.) Similarly, state and prove a version of
Dedekind's axiom for a cut on a segment.

(b) Supply the indicated arguments left out of the proof of
Archimedes' axiom from Dedekind's axiom.

8. From the three-point model (Example 1 in Chapter 2) we saw that
if we used only the axioms of incidence, we could not prove that
a line has more than two points lying on it. Using the betweenness
axioms as well, prove that every line has at least five points lying
on it. Give an informal argument to show that every segment (a fortiori, every line) has an infinite number of points lying on it (a formal proof requires the technique of mathematical induction).

9. Given a line l, a point A on l, and a point B not on l. Then every
point of the ray $\overrightarrow{AB}$ (except A) is on the same side of l as B. (Hint:
Use an RAA argument.)

10. Prove Proposition 3.7.

11. Prove Proposition 3.8. (Hint: For Proposition 3.8(c), prove in two
steps that E and B lie on the same side of $\overleftrightarrow{AD}$, first showing that
EB does not meet $\overrightarrow{AD}$ and then showing that EB does not meet the
opposite ray $\overrightarrow{AF}$. Use Exercise 9.)

12. Prove Proposition 3.9. (Hint: For Proposition 3.9(a), use Pasch's theorem and Proposition 3.7; see Figure 3.37. For Proposition 3.9(b),
let the ray emenate from point D in the interior of $\triangle ABC$. Use the
crossbar theorem and Proposition 3.7 to show that $\overrightarrow{AD}$ meets BC in
a point E such that A * D * E. Apply Pasch's theorem to $\triangle ABE$ and
$\triangle AEC$; see Figure 3.38.)

13. Prove that a line cannot be contained in the interior of a triangle.

14. If a, b, and c are rays, let us say that they are *coterminal* if they
emanate from the same point, and let us use the notation $a * b * c$

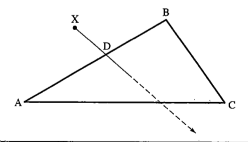

Figure 3.37

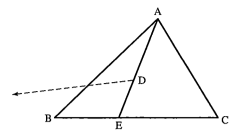

Figure 3.38

to mean that b is between a and c (as defined on p. 115). The analogue of Axiom B-1 states that if $a * b * c$, then a, b, c are distinct and coterminal and $c * b * a$; this analogue is obviously correct. State the analogues of Axioms B-2 and B-3 and Proposition 3.3 and tell which parts of these analogues are correct. (Beware of opposite rays!)

15. Find an interpretation in which the incidence axioms and the first two betweenness axioms hold but Axiom B-3 fails in the following way: There exist three collinear points, no one of which is between the other two. (Hint: In the usual Euclidean model, introduce a new betweenness relation A $*$ B $*$ C to mean that B is the midpoint of AC.)

16. Find an interpretation in which the incidence axioms and the first three betweenness axioms hold but the line separation property (Proposition 3.4) fails. (Hint: In the usual Euclidean model, pick a point P that is beween A and B in the usual Euclidean sense and specify that A will now be considered to be between P and B. Leave all other betweenness relations among points alone. Show that P lies neither on ray $\overrightarrow{AB}$ nor on its opposite ray $\overrightarrow{AC}$.)

17. A rational number of the form $a/2^n$ (with a, n integers) is called *dyadic*. In the interpretation of Example 1 (p. 117) for this chapter, restrict to those points which have dyadic coordinates and to those lines which pass through several dyadic points. The incidence axioms, the first three betweenness axioms, and the line separation property all hold in this dyadic rational plane; show that Pasch's theorem fails. (Hint: The lines $3x + y = 1$ and $y = 0$ do not meet in this plane.)

18. A set of points S is called *convex* if whenever two points A and B are in S, the entire segment AB is contained in S. Prove that a half-plane, the interior of an angle, and the interior of a triangle are all convex sets, whereas the exterior of a triangle is not convex. Is a triangle a convex set?

19. Fill in the details of Example 2 of this chapter to show informally that the open unit disk U in F^2 is an ordered incidence plane having the hyperbolic parallel property (existence of more than one parallel to a given line through a given point not on that line). Take F to be $\mathbb{R}$ if you are unfamiliar with ordered fields. Draw a diagram in this model to show that for any angle in this model, there exist points interior to that angle which do not lie on any line that intersects both sides of the angle. (Congruence in this model will be explained in Chapter 7.)

Exercises on Congruence

20. Justify each step in the following proof of Proposition 3.11:

   PROOF:
   (1) Assume on the contrary that BC is not congruent to EF.
   (2) Then there is a point G on $\overrightarrow{EF}$ such that BC $\cong$ EG.
   (3) G $\neq$ F.
   (4) Since AB $\cong$ DE, adding gives AC $\cong$ DG.
   (5) However, AC $\cong$ DF.
   (6) Hence, DF $\cong$ DG.
   (7) Therefore, F = G.
   (8) Our assumption has led to a contradiction; hence, BC $\cong$ EF. ◀

21. Prove Proposition 3.13(a). (Hint: In the case where AB and CD are not congruent, there is a unique point F $\neq$ D on $\overrightarrow{CD}$ such that AB $\cong$ CF (reason?). In the case where C * F * D, show that AB < CD. In the case where C * D * F, use Proposition 3.12 and some axioms to show that CD < AB.) Provide the details of the claim in Example 4 of this chapter that trichotomy sometimes fails in $\mathbb{Q}^2$.

22. Use Proposition 3.12 to prove Propositions 3.13(b) and (c).

23. Use the previous exercise and Proposition 3.3 to prove Proposition 3.13(d).

24. Justify each step in the following proof of Proposition 3.14 (see Figure 3.39).

   PROOF:
   Given $\sphericalangle$ABC $\cong$ $\sphericalangle$DEF. To prove $\sphericalangle$CBG $\cong$ $\sphericalangle$FEH:

   (1) The points A, C, and G being given arbitrarily on the sides of $\sphericalangle$ABC and the supplement $\sphericalangle$CBG of $\sphericalangle$ABC, we can choose the

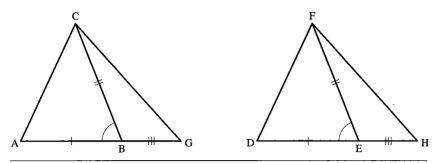

Figure 3.39

points D, F, and H on the sides of the other angle and its supplements so that AB ≅ DE, CB ≅ FE, and BG ≅ EH.

(2) Then, △ABC ≅ △DEF.
(3) Hence, AC ≅ DF and ∢A ≅ ∢D.
(4) Also, AG ≅ DH.
(5) Hence, △ACG ≅ △DFH.
(6) Therefore, CG ≅ FH and ∢G ≅ ∢H.
(7) Hence, △CBG ≅ △FEH.
(8) It follows that ∢CBG ≅ ∢FEH, as desired. ◄

25. Deduce Proposition 3.15 from Proposition 3.14.
26. Justify each step in the following proof of Proposition 3.17 (see Figure 3.40):

PROOF:

Given △ABC and △DEF with ∢A ≅ ∢D, ∢C ≅ ∢F, and AC ≅ DF.
To prove △ABC ≅ △DEF:

(1) There is a unique point B' on ray $\overrightarrow{DE}$ such that DB' ≅ AB.
(2) △ABC ≅ △DB'F.
(3) Hence, ∢DFB' ≅ ∢C.

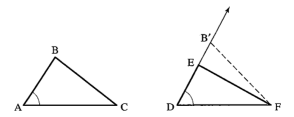

Figure 3.40

(4) This implies $\overrightarrow{FE} = \overrightarrow{FB'}$.

(5) In that case, B' = E.

(6) Hence, $\triangle ABC \cong \triangle DEF$. ◄

27. Prove Proposition 3.18.

28. Prove that an equiangular triangle (all angles congruent to one another) is equilateral.

29. Prove Proposition 3.20. (Hint: Use Axiom C-4 and Proposition 3.19.)

30. Given ∢ABC ≅ ∢DEF and $\overrightarrow{BG}$ between $\overrightarrow{BA}$ and $\overrightarrow{BC}$. Prove that there is a unique ray $\overrightarrow{EH}$ between $\overrightarrow{ED}$ and $\overrightarrow{EF}$ such that ∢ABG ≅ ∢DEH. (Hint: Show that D and F can be chosen so that AB ≅ DE and BC ≅ EF, and that G can be chosen so that A * G * C. Use Propositions 3.7 and 3.12 and SAS to get H; see Figure 3.25.)

31. Prove Proposition 3.21 (imitate Exercises 21–23).

32. Prove Proposition 3.22. (Hint: Use the corollary to SAS to reduce to the case where A = D, C = F, and the points B and E are on opposite sides of $\overleftrightarrow{AC}$.)

33. If AB < CD, prove that 2AB < 2CD.

34. (a) Prove Euclid's second postulate.

(b) Prove that the center of a circle is unique and its radius is unique up to congruence; that is, if points O, O' and radii OA, O'A', respectively, determine the same circle, then O = O' and OA ≅ O'A'.

35. In the real Euclidean plane of Example 3 in this chapter, we have defined the length of any segment by the Pythagorean formula. We will now distort that interpretation as follows: For segments on the x-axis only, redefine their length as twice what it was previously (e.g., the length of the segment from (1, 0) to (4, 0) is now 6 instead of 3). Reinterpret congruence of segments to mean that two segments in the plane have the same "length" in this perverse way of measuring (e.g., the segment from (0, 0) to (0, 6) on the y-axis is now congruent to the segment from (1, 0) to (4, 0) on the x-axis). Points, lines, incidence, and betweenness will have the same meaning as before and satisfy the same axioms as before. Congruence of angles will mean that the angles have the same number of degrees, i.e., the same meaning as in high school geometry (something we have not defined, but treat this example informally). Show informally that the first five congruence axioms and angle addition (Proposition 3.19) still hold in this interpretation but that SAS fails for certain pairs of triangles (see Figure 3.41). This shows that Axiom C-6 (SAS) is *independent* of the other 12 axioms for a Hilbert

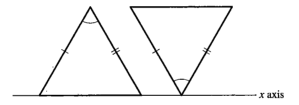

Figure 3.41

plane (it can neither be proved nor disproved from them). Draw diagrams to show that SSS and ASA also fail for certain pairs of triangles. Draw a diagram of a circle with center on the x-axis in this interpretation and use that diagram to show that the circle-circle continuity principle and the segment-circle continuity principle fail in this interpretation.

Major Exercises

1. In the real Euclidean plane, let γ be a circle with center A and radius of length r. Let γ' be another circle with center A' and radius of length r', and let d be the distance from A to A' (see Figure 3.42). There is a hypothesis about the numbers r, r', and d that ensures that the circles γ and γ' intersect in two distinct points. Figure out what this hypothesis is. (Hint: Its statement that certain numbers obtained from r, r', and d are less than certain others.)

    What hypothesis on r, r', and d ensures that γ and γ' intersect in only one point, i.e., that the circles are tangent to each other? (See Figure 3.43.)

2. Define the *reflection* in a line m to be the transformation R_m of the plane that leaves each point of m fixed and transforms a point A not on m as follows. Let M be the foot of the perpendicular from A to m. Then, by definition, $R_m(A)$ is the unique point A' such that

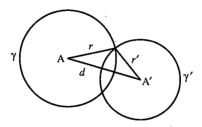

Figure 3.42

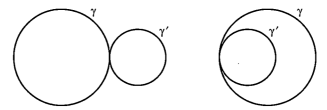

Figure 3.43

A′ * M * A and A′M ≅ MA (see Figure 3.44). This definition uses
the result from Chapter 4 that the perpendicular from A to m is
unique, so that the *foot* M is uniquely determined as the intersection with m. Prove that R_m is a *motion*, i.e., that AB ≅ A′B′ for any
segment AB. Prove also that AB ≅ CD ⇒ A′B′ ≅ C′D′ and that
∢A ≅ ∢B ⇒ ∢A′ ≅ ∢B′. (Chapter 9 will be devoted to a thorough
study of motions; the reflections generate the group of all such transformations.) (Hint: The proof breaks into the cases (i) A or B lies
on m, (ii) A and B lie on opposite sides of m, and (iii) A and B lie
on the same side of m. In (ii), let M, N be the midpoints of AA′,
BB′ and let C be the point at which AB meets m; prove that
A′ * C * B′ by showing that ∢A′CM ≅ B′CN and apply Axiom C-3.
In (iii), let C be the point at which AB′ meets m, and use B = (B′)′
and the first two cases to show that △ABC ≅ △A′B′C. Take care
not to use results that are valid only in Euclidean geometry.)

If F is an ordered field for which F^2 is a Hilbert plane, find the
explicit formula for the reflection across a line, treating separately
the cases where the line is vertical (given by an equation x = constant) and where it is not (hence given by an equation $y = mx + b$).
(Hint: In the latter case, a perpendicular to the line has slope $-1/m$.
Use that to find the coordinates of the foot M of the perpendicular
from A and then find the coordinates of A′ in terms of those of A.)

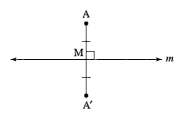

Figure 3.44

NOTE ON ELLIPTIC GEOMETRY. Consider the sphere with antipodal points identified, "lines" being great circles. The perpendicular from A to m is unique except for one point P called the *pole of m* (see Figure 3.24, p. 126, where m is the equator and P is the north pole); all perpendiculars to m pass through P. The definition of reflection is modified in this model so that $R_m(P) = P$, because the natural candidate for $R_m(P)$ is the point antipodal to P, but we have identified antipodal points. Show informally in this model that R_m is the same as the 180° rotation about the pole of m. When we study rotations in Hilbert planes in Chapter 9, we will prove that no rotation can be the same as a reflection, so this is another major difference between neutral geometry and elliptic geometry.

In the following exercises, we will assume that a segment AB has a *length* which has the familiar properties (they are spelled out in Theorem 4.3, Chapter 4). Here we denote that length by |AB|. You can think of it as a real number, as you did in high school, or you can read the more sophisticated treatment in Hartshorne's book, in which |AB| is the congruence class of segment AB and these classes can be added and ordered.

3. Let γ be a circle with center O. For any point P on γ, we have called segment OP a radius of γ. Let us call |OP| *the* radius of γ and denote it by r. Let γ' be another circle with center $O' \neq O$ and radius r' and let $d = |OO'|$. In the next chapter, we will prove the *triangle inequality* (Euclid I.20) for any Hilbert plane: If A, B, C are not collinear, then $|AB| + |BC| > |AC|$. Assume that result for now. Suppose that the hypothesis of the circle-circle continuity principle is satisfied—i.e., that there is a point of γ' inside γ and also another point of γ' outside γ. Show that the following three inequalities hold: $r + r' > d$, $r + d > r'$, and $r' + d > r$. (Hint: Use the triangles formed with O and O' by the point inside and by the point outside γ and apply the fact that if $a < b$, then for any c, $a + c < b + c$.)

4. *The converse to the triangle inequality* is the statement that if a, b, c are such that the sum of any two is greater than the third, then there exists a triangle whose sides have those lengths. Again, assuming the triangle inequality, which will be proved in Chapter 4, show that its converse implies the circle-circle continuity principle. (Hint: Apply the previous exercise to get one point of intersection and reflect across the line joining O and O' to get the other. Use

the uniqueness part of Axiom C-4 to prove that those two are the only points of intersection of the circles.)

5. Euclid I.22 is the converse to the triangle inequality. Here is Euclid's proof, which has a gap when he assumes without justification the existence of point K. Show that the gap can be filled by assuming the circle-circle continuity principle. Combining this with the previous two exercises, we obtain the following result: For all Hilbert planes,

Circle-circle continuity principle ⇔ converse to the triangle inequality.

Keep in mind that neither one of these has been proved by itself, only that they are logically equivalent given the 13 axioms for Hilbert planes (and the triangle inequality, which will be proved for all Hilbert planes in Chapter 4).

PROOF:
Choose notation for the three given lengths so that $a \geq b \geq c$. Take any point D and any ray $\overrightarrow{DE}$ emanating from D. Starting from D, lay off successively on that ray points F, G, H so that $a = |DF|$, $b = |FG|$, $c = |GH|$. Then the circle with center F and radius a meets the circle with center G and radius c at a point K, and $\triangle FGK$ is a triangle that has sides of "length" a, b, and c (see Figure 3.45). ◄

6. The *taxicab plane* is another example like Exercise 35 where distance is modified in $\mathbb{R}^2$ so that SAS and other familiar statements fail. Instead of using the Pythagorean formula to define the distance between two points $A = (a_1, a_2)$ and $B = (b_1, b_2)$, use *the taxicab formula:*

$$d^T(A, B) = |a_1 - b_1| + |a_2 - b_2|.$$

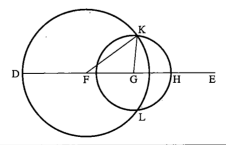

Figure 3.45

Diagramatically, if A and B are not on the same vertical or horizontal line, draw a horizontal line through A and drop a vertical perpendicular from B to that line with foot C. Then segment AB is the hypotenuse of a right triangle with right angle at C, and the ordinary distance $d(A, B)$ is the usual length of that hypotenuse. The taxicab distance $d^T(A, B)$ is the sum of the usual lengths of the legs of that triangle and is longer than the ordinary distance. If you were in a city with a rectangular grid of streets, it is the distance a taxicab would have to travel to get you around a corner at C from point A to point B. If, however, points A and B both lie on either a horizontal line, $y = $ constant, or on a vertical line, $x = $ constant, then $d^T(A, B) = d(A, B)$.

(a) With points, lines, incidence, betweenness, and congruence of angles interpreted as in Exercise 35 (the usual interpretation) but with congruence of segments interpreted via taxicab distance, exhibit a pair of triangles and a correspondence between their vertices for which SAS fails. Do the same for SSS and ASA. Show informally that the first five congruence axioms and angle addition (Proposition 3.19) still hold in this interpretation. (Hint: Verify C-1 using the formulas $x = r \cos \theta$ and $y = r \sin \theta$ relating rectangular to polar coordinates.)

(b) Exhibit an equilateral triangle in which one angle is a right angle and the other angles are acute. Since an equilateral triangle is an isosceles triangle, this is also an example in which the base angles of an isosceles triangle are not congruent.

(c) Exhibit a triangle in which two angles are congruent but the sides opposite those angles are not congruent.

(d) Show that a "circle" in taxicab geometry is a square in the real Euclidean plane but that not every Euclidean square is a taxicab circle. Give an example of two taxicab circles that satisfy the hypothesis of the circle-circle continuity principle but that intersect in infinitely many points.

(e) In the real Euclidean plane, the locus of points equidistant from two given points A, B is the perpendicular bisector of segment AB. What does that locus look like with respect to taxicab distance? (Hint: Work out some specific examples. The locus can have several shapes.)

(If you are stymied by this exercise, see the delightful little book *Taxicab Geometry: An Adventure in Non-Euclidean Geometry* (Dover, 1987), by Eugene F. Krause.)

Projects

1. Verify the claim in Example 3 that with the interpretation of congruence via the Pythagorean formula given there, the interpretations of the first five congruence axioms hold in the real Euclidean plane. (The nontrivial statements to verify are C-4, the laying off of an angle, and C-1, the laying off of a segment. If you are stymied, see Hartshorne.)

2. A *Pythagorean ordered field* is an ordered field F in which for every $c \in F$, $\sqrt{1 + c^2} \in F$. We see that $\mathbb{Q}$ is not Pythagorean by taking $c = 1$. Hilbert denoted the smallest Pythagorean subfield of $\mathbb{R}$ by the Greek letter Ω. An element of Ω is obtained from rational numbers by finitely many applications of the operations of addition, subtraction, multiplication, division, and taking the positive square root of a number of the form $1 + c^2$. Since $0 < 1 + c^2$, every Euclidean ordered field is Pythagorean, but the converse is false. If you have studied field theory, report on Exercise 16.10, p. 147, of Hartshorne where it is shown that Ω is strictly smaller than the constructible field $\mathbf{K}$ (e.g., $(1 + \sqrt{2})^{1/2} \notin \Omega$). If F is any ordered field, the interpretations of Axioms C-2 through C-5 hold in F^2, but the interpretation of C-1 will hold iff F is Pythagorean. Show this, referring to Hartshorne if you are stymied.

3. The standard method for verifying SAS (Axiom C-6) in F^2, when F is a Pythagorean ordered field, is to first establish the existence of enough motions in F^2 so that Euclid's idea of superposition can be made rigorous. Report on that method from Hartshorne, Chapter 3, Section 17. The difficulty of that verification shows that SAS is the deepest of the axioms. We will study motions of the plane in Chapter 9 (reflections have already been defined in Major Exercise 2 above).

4. Finally, to show that F^2 is a Euclidean plane when F is a Euclidean ordered field (in particular, when $F = \mathbb{R}$), one must verify the circle-circle continuity principle. In F^2, we interpreted segment congruence via the Pythagorean distance function $d(AB)$: $AB \cong CD$ iff $d(AB) = d(CD)$. Having verified the interpretation of C-1 in Project 2, we see also that $AB < CD$ iff $d(AB) < d(CD)$. If $A * B * C$, you can easily verify that $d(AC) = d(AB) + d(BC)$. You can directly verify the triangle inequality in F^2. Hence, Major Exercises 3–5 become applicable, and it suffices to verify the converse to the triangle inequality in order to verify the circle-circle continuity principle. If

you cannot verify that yourself, report on the verification found in Moise's *Elementary Geometry from an Advanced Standpoint*, 1990, 3rd ed., p. 239 ff. (where it is called *The Triangle Theorem*). In your report, highlight the step which uses the hypothesis that F is a Euclidean field (the step which uses $\sqrt{a} \in F$ for any $a > 0$).

5. If F is a Pythagorean ordered field that is not a Euclidean field (Project 2), then the interpretation of the circle-circle continuity principle fails in F^2—in fact, the line-circle continuity principle, which will be shown in Major Exercise 1 of Chapter 4 to be a consequence, fails. Here is an argument (due to Descartes!) to show that *the validity in F^2 of the line-circle continuity principle implies that F is Euclidean*: Given $a > 0$ in F. Let $r = \frac{1}{2}(a + 1)$ and let δ be the circle of center $(r, 0)$ and radius r. Let A = $(a, 0)$. Show that A lies inside δ. Hence the vertical line $x = a$ through A meets δ in two points, by the line-circle continuity principle. If B is the intersection point in the upper half-plane, show that B = $(a, \sqrt{a})$. Hence, $\sqrt{a} \in F$.

6. Combining the results of Projects 4 and 5, we see that for Hilbert planes of the form F^2, the line-circle continuity principle implies the circle-circle continuity principle. I do not know whether that implication holds for *all* Hilbert planes. If you can prove that it does, or provide a counterexample showing it doesn't, that would be a result worth publishing (provided that a search of the geometry literature verifies that someone else has not already published it).

4

Neutral Geometry

If only it could be proved . . . that "there is a Triangle whose angles are together not less than two right angles"! But alas, that is an ignis fatuus that has never yet been caught!

                    C. L. Dodgson (Lewis Carroll)

Geometry Without a Parallel Axiom

In the preceding chapters, we strengthened the foundations of Euclid's geometry by presenting 13 axioms plus continuity principles to replace his first 4 postulates. The 13 axioms (3 incidence, 4 betweenness, and 6 congruence axioms) are essentially those of David Hilbert, and in his honor a model of those axioms was called a *Hilbert plane.*

Euclid's fifth postulate will be discussed in this chapter, but it will not be assumed except when we explicitly announce it as a hypothesis. Instead, we will be studying some statements that we will show to be logically equivalent to it for Hilbert planes. One such statement is Hilbert's Euclidean axiom of parallelism introduced in Chapter 3. Our purpose is to develop as much elementary geometry as possible without assuming a parallel postulate, and that is what is meant by doing "neutral geometry"—adopting a neutral stance about a parallel postulate. All the elementary geometric results proved since we started assuming some of the Hilbert plane axioms are results in neutral geometry.[1] Euclid himself postponed invoking his fifth postulate for a proof until I.29, his 29th proposition of Book I. When we eventually bifurcate into studying Euclidean and hyperbolic geometries separately, all the results in neutral geometry will be valid in *both* geometries.

In all the propositions, theorems, corollaries, and lemmas of this chapter, the 13 axioms for a Hilbert plane will be assumed. Our proofs will be less formal henceforth.

What is the purpose of studying neutral geometry? We are not interested in studying it for its own sake. Rather, we are trying to clarify the role of the parallel postulate by seeing which theorems in the geometry do not depend on it, i.e., which theorems follow from the other axioms alone without ever using the parallel postulate in proofs. This will enable us to avoid many pitfalls and to see much more clearly the logical structure of our system. Certain questions that can be answered in Euclidean geometry (e.g., whether there is a unique parallel through a given point) may not be answerable in neutral geometry because its axioms do not give us enough information.

Alternate Interior Angle Theorem

The next theorem requires a definition: Let *t* be a transversal to lines *l* and *l'*, with *t* meeting *l* at B and *l'* at B'. Choose points A and C on *l* such that A ∗ B ∗ C; choose points A' and C' on *l'* such that A and A' are on the same side of *t* and such that A' ∗ B' ∗ C'. Then the following four angles are called *interior*: ∢A'B'B, ∢ABB', ∢C'B'B, ∢CBB'. The two pairs (∢ABB', ∢C'B'B) and (∢A'B'B, ∢CBB') are called pairs of *alternate interior angles* (see Figure 4.1).

ALTERNATE INTERIOR ANGLE (AIA) THEOREM 4.1. In any Hilbert plane, if two lines cut by a transversal have a pair of congruent alternate interior angles with respect to that transversal, then the two lines are parallel.

This is Euclid I.27. Our RAA proof will be less formal. The intuitive idea of the proof is that congruence of alternate interior angles implies that the lines are situated symmetrically about the transversal, so if by

---

[1] I am deliberately not defining "neutral geometry" precisely. In general, it will be the study of Hilbert planes, but occasionally a continuity axiom will also be explicitly assumed. The fundamental idea of neutral geometry is not to assume any parallel postulate or any statement equivalent to a parallel postulate.

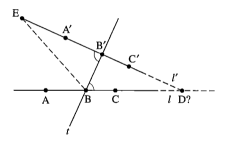

Figure 4.1

RAA hypothesis the lines met on one side of the transversal, we could reflect the triangle so formed over to the other side of the transversal to obtain a second meeting point, which violates Axiom I-1. Notice how crucial to this proof is Axiom B-4, which guarantees that a line has two disjoint sides.

PROOF:

Given ∢A′B′B ≅ ∢CBB′. Assume on the contrary that l and l' meet at a point D. Say D is on the same side of t as C and C′. There is a point E on $\overrightarrow{B'A'}$ such that B′E ≅ BD (Axiom C-1). Segment BB′ is congruent to itself, so that △B′BD ≅ △BB′E (SAS). In particular, ∢DB′B ≅ ∢EBB′. Since ∢DB′B is the supplement of ∢EB′B, ∢EBB′ must be the supplement of ∢DBB′ (Proposition 3.14 and Axiom C-4). This means that E lies on l, and hence l and l' have the two points E and D in common, which contradicts Proposition 2.1 of incidence geometry. Therefore, $l \parallel l'$. ◄

This theorem has two very important corollaries.

COROLLARY 1. Two lines perpendicular to the same line are parallel. Hence the perpendicular dropped from a point P not on line l to l is *unique* (and the point at which the perpendicular intersects l is called its *foot*).

PROOF:

If l and l' are both perpendicular to t, the alternate interior angles are right angles and hence are congruent (Proposition 3.23). ◄

COROLLARY 2 (EUCLID I.31). If l is any line and P is any point not on l, there exists at least one line m through P parallel to l (see Figure 4.2).

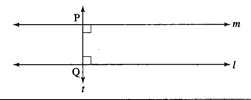

Figure 4.2

PROOF:

There is a line t through P perpendicular to l, and again there is a unique line m through P perpendicular to t (Proposition 3.16). Since l and m are both perpendicular to t, Corollary 1 tells us that $l \parallel m$. ◄

The construction of the parallel m to l through P given in the above proof will be used repeatedly. We will refer to it as the *standard construction*. Let Q be the foot of the perpendicular from P to l. For brevity, we will also call this the *standard configuration*, denoted PQlm.

WARNING You are accustomed in Euclidean geometry to use the *converse* of Theorem 4.1, which states, "If two lines are parallel, then alternate interior angles cut by a transversal are congruent." We haven't proved this converse, so don't use it!

Exterior Angle Theorem

An angle supplementary to an angle of a triangle is called an *exterior angle* of that triangle. The other two angles of the triangle are called *remote interior angles* relative to that exterior angle.

EXTERIOR ANGLE (EA) THEOREM 4.2. In any Hilbert plane, an exterior angle of a triangle is greater than either remote interior angle (see Figure 4.3).

To prove ∢ACD is greater than ∢B and ∢A:

PROOF:

Consider the remote interior angle ∢BAC. If ∢BAC ≅ ∢ACD, then $\overleftrightarrow{AB}$ is parallel to $\overleftrightarrow{CD}$ (Theorem 4.1), which contradicts the hypoth-

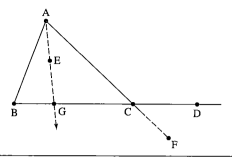

Figure 4.3

esis that these lines meet at B. Supose ∢BAC is greater than ∢ACD (RAA hypothesis). Then there is a ray $\overrightarrow{AE}$ between $\overrightarrow{AB}$ and $\overrightarrow{AC}$ such that ∢ACD ≅ ∢CAE (by definition). This ray $\overrightarrow{AE}$ intersects BC in a point G (crossbar theorem, Chapter 3). But according to Theorem 4.1, lines $\overleftrightarrow{AE}$ and $\overleftrightarrow{CD}$ are parallel. Thus ∢BAC cannot be greater than ∢ACD (RAA conclusion). Since ∢BAC is also not congruent to ∢ACD, ∢BAC must be less than ∢ACD (Proposition 3.21(a)).

For remote angle ∢ABC, use the same argument applied to exterior angle ∢BCF, which is congruent to ∢ACD by the vertical angle theorem (Proposition 3.15(a)). ◀

COROLLARY 1. If a triangle has a right or obtuse angle, the other two angles are acute.

The exterior angle theorem will play a very important role in what follows. It was the 16th proposition in Euclid's *Elements*. Euclid's proof had a gap due to reasoning from a diagram. He considered the line $\overleftrightarrow{BM}$ joining B to the midpoint of AC, and he constructed point B′ such that B * M * B′ and BM ≅ MB′ (Axiom C-1). He then assumed from the diagram that B′ lay in the interior of ∢ACD (see Figure 4.4). Since ∢B′CA ≅ ∢A (SAS), Euclid concluded correctly that ∢ACD > ∢A.

The gap in Euclid's argument can easily be filled with the tools we have developed. Since segment BB′ intersects AC at M, B and B′ are on opposite sides of $\overleftrightarrow{AC}$ (by definition). Since BD meets $\overleftrightarrow{AC}$ at C, B and D are also on opposite sides of $\overleftrightarrow{AC}$. Hence B′ and D are on the same side of $\overleftrightarrow{AC}$ (Axiom B-4). Next, B′ and M are on the same side of $\overleftrightarrow{CD}$ since segment MB′ does not contain the point B at which $\overleftrightarrow{MB}′$ meets $\overleftrightarrow{CD}$ (by construction of B′ and Axioms B-1 and B-3). Also, A and M are on the same side of $\overleftrightarrow{CD}$ because segment AM does not contain the

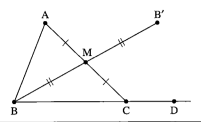

Figure 4.4

point C at which $\overrightarrow{AM}$ meets $\overleftrightarrow{CD}$ (by the definition of midpoint and Axiom B-3). So again, Separation Axiom B-4 ensures that A and B′ are on the same side of $\overleftrightarrow{CD}$. By the definition of "interior" (in Chapter 3, p. 115), we have shown that B′ lies in the interior of ∡ACD. ◄

Many reputable writers mistakenly state that to fill this gap in Euclid one must add an axiom that "lines are infinite in extent"—whatever that may mean. All that is needed are the betweenness axioms and their consequences.

NOTE ON ELLIPTIC GEOMETRY. Figure 3.24, p. 126, shows a triangle on the sphere with both an exterior angle and a remote interior angle that are right angles, so the exterior angle theorem doesn't hold. Our proof of it was based on the alternate interior angle theorem, which can't hold in elliptic geometry because there are no parallels. The proof we gave of Theorem 4.1 breaks down in elliptic geometry because Axiom B-4, which asserts that a line separates the plane into two sides, doesn't hold; we knew points E and D in that proof were distinct because they lay on opposite sides of line *t*. Or, thinking in terms of spherical geometry, where a great circle does separate the sphere into two hemispheres, if points E and D are distinct, there is no contradiction because great circles do meet in two antipodal points.

Euclid's proof of Theorem 4.2 breaks down on the sphere because "lines" are great circles and if segment BM is long enough, the reflected point B′ might lie on it (e.g., if BM is a semicircle, B′ = B).

As a consequence of the exterior angle theorem (and our previous results), you can now prove as exercises the following familiar propositions.

PROPOSITION 4.1 (SAA CONGRUENCE CRITERION). Given AC ≅ DF, ∡A ≅ ∡D, and ∡B ≅ ∡E. Then △ABC ≅ △DEF (Figure 4.5).

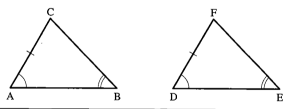

Figure 4.5 SAA.

PROPOSITION 4.2 (HYPOTENUSE-LEG CRITERION). Two right triangles are congruent if the hypotenuse and a leg of one are congruent, respectively, to the hypotenuse and a leg of the other (Figure 4.6).

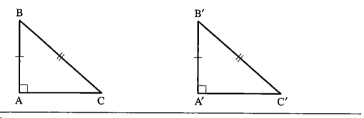

Figure 4.6

PROPOSITION 4.3 (MIDPOINTS). Every segment has a unique midpoint.

Here is a proof that AB has a midpoint, whose steps you are asked to justify in Exercise 5 (see Figure 4.7). You are asked to prove uniqueness of the midpoint in Exercise 6.

PROOF:

(1) Let C be any point not on $\overleftrightarrow{AB}$. (2) There is a unique ray $\overrightarrow{BX}$ on the opposite side of $\overleftrightarrow{AB}$ from C such that ∢CAB ≅ ∢ABX. (3) There is a unique point D on $\overrightarrow{BX}$ such that AC ≅ BD. (4) D is on the opposite side of $\overleftrightarrow{AB}$ from C. (5) Let E be the point at which segment CD intersects $\overleftrightarrow{AB}$. (6) Assume E is not between A and B. (7) Then either E = A, or E = B, or E * A * B, or A * B * E. (8) $\overleftrightarrow{AC}$ is parallel to $\overleftrightarrow{BD}$. (9) Hence, E ≠ A and E ≠ B. (10) Assume E * A * B. (11) Since $\overleftrightarrow{CA}$ intersects side EB of △EBD at a point between E and B, it must also intersect either ED or BD. (12) Yet this is impossible. (13) Hence A is not between E and B. (14) Similarly, B is not between

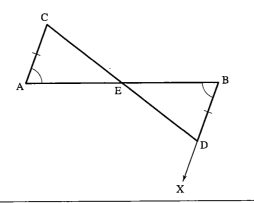

Figure 4.7

A and E. (15) Thus A ∗ E ∗ B. (16) Then ∢AEC ≅ ∢BED. (17) △EAC = △EBD. (18) Therefore, E is a midpoint of AB. ◄

PROPOSITION 4.4 (BISECTORS). (a) Every angle has a unique bisector. (b) Every segment has a unique perpendicular bisector.

Euclid constructed midpoints (I.10) and angle bisectors (I.9) using his previous construction (I.1) of an equilateral triangle on a given segment; we have seen that his proof of I.1 has a gap requiring the circle-circle continuity principle to fill. The construction of the midpoint given above does not depend on equilateral triangles; the construction of the angle bisector follows easily from that. Also, Euclid's proofs of I.9 and I.10 tacitly use betweenness properties—his proof of I.10 requires the crossbar theorem, and his proof of I.9 is based on a diagram where a point he constructs is on the opposite side, from the vertex of the angle, of a certain line he constructs. See the commentary on those proofs in Heath's translation of Euclid.

The next proposition combines Euclid I.18 and I.19.

PROPOSITION 4.5. In a triangle △ABC, the greater angle lies opposite the greater side and the greater side lies opposite the greater angle; i.e., AB > BC if and only if ∢C > ∢A.

The next proposition combines Euclid I.24 and I.25.

PROPOSITION 4.6. Given △ABC and △A'B'C', if we have AB ≅ A'B' and BC ≅ B'C', then ∢B < ∢B' if and only if AC < A'C'.

Measure of Angles and Segments

Thus far in this treatment of geometry, I have refrained from using numbers that measure the sizes of angles and segments—this was in keeping with the spirit of Euclid. After all, for thousands of years, his readers understood what Euclid meant geometrically without using numbers. In accord with modern standards of rigor, Hartshorne in his book has made Euclid's work precise, using congruence classes of segments and angles instead of number measures. That is the correct approach, valid in all Hilbert planes. However, since the treatment of angle measure in Hartshorne's Section 36 requires abstract group theory (his "unwound circle group"), knowledge of which is not presumed for my readers, I must "cop out" and use number measurement as a language for situations where it simplifies the statements.

I also presume that my readers have not necessarily studied the rigorous foundations of real numbers but that they are accustomed to informal talk about them. So although the next theorem refers to real numbers and we only sketch how it is proved, I alert you to the fact that it is not mathematically necessary to bring them in here; mathematically, all that is needed is the ability to do elementary algebra with congruence classes. I only bring in numbers to shorten a long story.[2]

Archimedes' axiom is needed to measure with real numbers—that is why Hilbert called it "the axiom of measurement." Theorem 4.3 below asserts the possibility of measurement and lists its basic properties. In many popular treatments of geometry, a version of this theorem is taken as an axiom (ruler-and-protractor postulates—see, e.g., Moise). The familiar symbol $(\sphericalangle A)°$ denotes the number of degrees in $\sphericalangle A$, and $\overline{AB}$ denotes the length of segment AB with respect to some unit of measurement.

MEASUREMENT THEOREM 4.3. Hypothesis for all but parts (4) and (11): Archimedes' axiom. Hypothesis for parts (4) and (11) as well: Dedekind's axiom.

---

[2] Major Exercise 5, Chapter 5, does use the full power of real numbers for the theory of similar triangles in a real Euclidean plane; again, Hartshorne presents the Hilbert–Enriques approach (using the abstract theory of fields and a crucial proposition about cyclic quadrilaterals), which avoids using real numbers even for that theory. See Hartshorne's Proposition 5.8 and Section 20. The power of Theorem 4.3 is also used for Proposition 9.2, Chapter 9, in Archimedean Hilbert planes. Real numbers are of course used in Chapter 10 on real hyperbolic geometry. For a complete proof of Theorem 4.3, see Borsuk and Szmielew (1960), Chapter 3, Sections 9–10.

A. There is a unique way of assigning a degree measure to each angle such that the following properties hold:

(0) $(\sphericalangle A)°$ is a real number such that $0 < (\sphericalangle A)° < 180°$.

(1) $(\sphericalangle A)° = 90°$ if and only if $\sphericalangle A$ is a right angle.

(2) $(\sphericalangle A)° = (\sphericalangle B)°$ if and only if $\sphericalangle A \cong \sphericalangle B$.

(3) If $\overrightarrow{AC}$ is interior to $\sphericalangle DAB$, then $(\sphericalangle DAB)° = (\sphericalangle DAC)° + (\sphericalangle CAB)°$ (refer to Figure 4.8).

(4) For every real number x between 0 and 180, there exists an angle $\sphericalangle A$ such that $(\sphericalangle A)° = x°$.

(5) If $\sphericalangle B$ is supplementary to $\sphericalangle A$, then $(\sphericalangle A)° + (\sphericalangle B)° = 180°$.

(6) $(\sphericalangle A)° > (\sphericalangle B)°$ if and only if $\sphericalangle A > \sphericalangle B$.

B. Given a segment OI, called a *unit segment*. Then there is a unique way of assigning a length $\overline{AB}$ to each segment AB such that the following properties hold:

(7) $\overline{AB}$ is a positive real number and $\overline{OI} = 1$.

(8) $\overline{AB} = \overline{CD}$ if and only if $AB \cong CD$.

(9) $A * B * C$ if and only if $\overline{AC} = \overline{AB} + \overline{BC}$.

(10) $\overline{AB} < \overline{CD}$ if and only if $AB < CD$.

(11) For every positive real number x, there exists a segment AB such that $\overline{AB} = x$.

NOTE. So as not to mystify you, here is the method for assigning lengths (the method for assigning degrees to angles is similar). We start with a segment OI whose length will be 1. Then any segment obtained by laying off n copies of OI will have length n. By Archimedes' axiom, every other segment AB will have its endpoint B between two points B_{n-1} and B_n such that $\overline{AB_{n-1}} = n - 1$ and $\overline{AB_n} = n$; then $\overline{AB}$ will have to equal $\overline{AB_{n-1}} + \overline{B_{n-1}B}$ by condition (9) of Theorem 4.3, so we may assume $n = 1$ and $B_{n-1} = A$. If B is the midpoint $B_{1/2}$ of AB_1, we set $\overline{AB_{1/2}} = 1/2$; otherwise B lies either in $AB_{1/2}$ or in $B_{1/2}B_1$, say, in $AB_{1/2}$. If then B is the midpoint $B_{1/4}$ of $AB_{1/2}$, we set $\overline{AB_{1/4}} = 1/4$; otherwise B

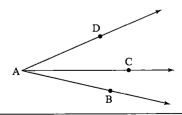

Figure 4.8

lies in $AB_{1/4}$, say, and we continue the process. Eventually B will either be obtained as the midpoint of some segment whose length has been determined, in which case $\overline{AB}$ will be determined to be some dyadic rational number $a/2^n$; or the process will continue indefinitely, in which case $\overline{AB}$ will be the limit of an infinite sequence of dyadic rational numbers; i.e., $\overline{AB}$ will be determined as an infinite decimal with respect to the base 2.

Note conversely that if a Hilbert plane satisfies part B, (7) through (10), then the plane is Archimedean, and if in addition (11) is satisfied, then Dedekind's axiom holds.

DEFINITION. If $(\sphericalangle B)° + (\sphericalangle C)° = 90°$, then $\sphericalangle B$ and $\sphericalangle C$ are called *complements* of each other and are said to be *complementary angles*. It is an easy exercise to show that every acute angle has a complementary angle (Exercise 7).

COROLLARY 2 TO THE EA THEOREM. The sum of the degree measures of any *two* angles of a triangle is less than 180°.

PROOF:

Referring to Figure 4.9, $(\sphericalangle CBD)° > (\sphericalangle A)°$ by the EA theorem. Adding $(\sphericalangle CBA)°$ to both sides of this inequality gives the result. ◄

TRIANGLE INEQUALITY. If $\overline{AB}$, $\overline{BC}$, $\overline{AC}$ are lengths of the sides of a triangle $\triangle ABC$, then $\overline{AC} < \overline{AB} + \overline{BC}$.

PROOF:

(1) There is a unique point D such that $A * B * D$ and $BD \cong BC$ (Axiom C-1 applied to the ray opposite to $\overrightarrow{BA}$). (See Figure 4.9.)

(2) Then $\sphericalangle BCD \cong \sphericalangle BDC$ (Proposition 3.10: base angles of an isosceles triangle).

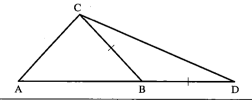

Figure 4.9

(3) $\overline{AD} = \overline{AB} + \overline{BD}$ (Theorem 4.3(9)) and $\overline{BD} = \overline{BC}$ (step 1 and Theorem 4.3(8)); substituting gives $\overline{AD} = \overline{AB} + \overline{BC}$.

(4) $\overrightarrow{CB}$ is between $\overrightarrow{CA}$ and $\overrightarrow{CD}$ (Proposition 3.7); hence $\sphericalangle ACD >$ $\sphericalangle BCD$ (by definition).

(5) $\sphericalangle ACD > \sphericalangle ADC$ (steps 2 and 4; Proposition 3.21(c)).

(6) $AD > AC$ (Proposition 4.5).

(7) Hence $\overline{AB} + \overline{BC} > \overline{AC}$ (Theorem 4.3(10); steps 3 and 6). ◄

Note that in these last two results, the only properties of numbers used were the ability to add and the relationship of addition to order. Numbers provide a more convenient language than the awkward one used by Euclid. Archimedes' axiom and the full power of Theorem 4.3 are certainly not used! For example, Euclid I.20 states the triangle inequality as follows: *In any triangle, two sides taken together in any manner are greater than the remaining one.*

His proof is the same as the one just given, except that he presumes that the reader understands what he means by "two sides taken together." We recognize that as meaning geometric addition of two segment congruence classes. We initially approximate that addition by extending the first segment with a congruent copy of the second one—that is exactly what Euclid's second postulate allows us to do, and you can easily prove Euclid II using Axioms C-1 and B-2. Then Axiom C-3 guarantees that this addition is well-defined for segment congruence classes. It is then routine to verify that this addition has all the familiar algebraic properties and is compatible with the ordering of segment congruence classes (see Major Exercise 9). So numbers are not really needed; they are just convenient and more familiar to beginners than are congruence classes.

The diligent reader is invited to figure out, whenever we use measurement of segments henceforth, how it could be avoided with the algebra of segment congruence classes. In the case of congruence classes of angles, there is a technical difficulty. We could use Proposition 3.19 on "angle addition" to try to define addition of angle congruence classes by juxtaposing the two angles, but that only works in the special case of angles whose degree measures add up to less than 180°. See Hartshorne, Section 36, for the definition and properties of that addition in the general case.

We call the *converse to the triangle inequality* the statement that if a, b, c are lengths such that the sum of any two is greater than the third, then there exists a triangle whose sides have those lengths. This is Euclid I.22, but he of course did not talk about lengths; he talked about segments. His proof has a gap which requires another application of the circle-circle continuity principle. It turns out that the converse to the triangle inequality can be used to prove that principle. The result is the following.

COROLLARY. For any Hilbert plane, the converse to the triangle inequality is equivalent to the circle-circle continuity principle. Hence the converse to the triangle inequality holds in Euclidean planes.

A proof of this equivalence was indicated in Major Exercises 3–5 of Chapter 3, assuming the triangle inequality there, and now we've proved the triangle inequality. The second assertion of this corollary follows from our definition of "Euclidean plane," which includes the circle-circle continuity principle as one of its axioms (see p. 139).

Equivalence of Euclidean Parallel Postulates

We shall now prove the equivalence of Euclid's fifth postulate and Hilbert's Euclidean parallel postulate. Note, however, that we are not proving either or both of the postulates; we are only proving that we *can* prove one *if* we first assume the other. We shall first state Euclid V (all the terms in the statement have now been defined carefully).

EUCLID'S POSTULATE V. If two lines are intersected by a transversal in such a way that the sum of the degree measures of the two interior angles on one side of the transversal is less than 180°, then the two lines meet on that side of the transversal.

THEOREM 4.4. Euclid's fifth postulate ⇔ Hilbert's Euclidean parallel postulate.

PROOF:

First, assume Hilbert's postulate. The situation of Euclid V is shown in Figure 4.10. $(\sphericalangle 1)° + (\sphericalangle 2)° < 180°$ (hypothesis) and $(\sphericalangle 1)° + (\sphericalangle 3)° = 180°$ (supplementary angles, Theorem 4.3(5)). Hence $(\sphericalangle 2)° < 180° - (\sphericalangle 1)° = (\sphericalangle 3)°$. There is a unique ray $\overrightarrow{B'C'}$ such

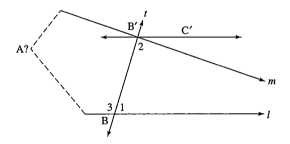

Figure 4.10

that ∢3 and ∢C'B'B are congruent alternate interior angles (Axiom C-4). By Theorem 4.1, $\overleftrightarrow{B'C'}$ is parallel to l. Since $m \neq \overleftrightarrow{B'C'}$, m meets l (Hilbert's postulate). To conclude, we must prove that m meets l on the same side of t as C'. Assume, on the contrary, that they meet at a point A on the opposite side. Then ∢2 is an exterior angle of △ABB'. Yet it is smaller than the remote interior ∢3. This contradiction of Theorem 4.2 proves Euclid V (RAA).

Conversely, assume Euclid V and refer to Figure 4.11, the situation of Hilbert's postulate. Let t be the perpendicular to l through P, and m the perpendicular to t through P. We know that $m \parallel l$ (Corollary 1 to Theorem 4.1). Let n be any other line through P. We must show that n meets l. Let ∢1 be the acute angle n makes with t (which angle exists because $n \neq m$). Then we have (∢1)° + (∢PQR)° < 90° + 90° = 180°. Thus the hypothesis of Euclid V is satisfied. Hence n meets l, proving Hilbert's postulate. ◀

Since Hilbert's Euclidean parallel postulate and Euclid V are logically equivalent in the context of neutral geometry, Theorem 4.4 allows us to use them interchangeably. You will prove as exercises that the

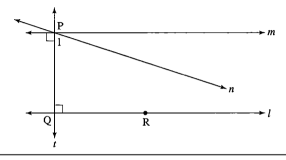

Figure 4.11

following statements are also logically equivalent to the parallel postulate.[3]

PROPOSITION 4.7. Hilbert's Euclidean parallel postulate ⇔ if a line intersects one of two parallel lines, then it also intersects the other.

PROPOSITION 4.8. Hilbert's Euclidean parallel postulate ⇔ converse to the alternate interior angle theorem.

PROPOSITION 4.9. Hilbert's Euclidean parallel postulate ⇔ if t is a transversal to l and m, $l \parallel m$, and $t \perp l$, then $t \perp m$.

PROPOSITION 4.10. Hilbert's Euclidean parallel postulate ⇔ if $k \parallel l$, $m \perp k$, and $n \perp l$, then either $m = n$ or $m \parallel n$.

The next proposition provides a very important consequence of Hilbert's Euclidean parallel postulate. It is *not equivalent to that parallel postulate* without adding further assumptions to our axioms for Hilbert planes, as we shall see later. (Many books state that it is equivalent, but they are assuming other axioms.)

PROPOSITION 4.11. In any Hilbert plane, Hilbert's Euclidean parallel postulate implies that for every triangle $\triangle ABC$,

$$(\sphericalangle A)° + (\sphericalangle B)° + (\sphericalangle C)° = 180°.$$

In words: The angle sum of every triangle is 180° if we assume Hilbert's Euclidean parallel postulate.

PROOF:

Refer to Figure 4.12. By the corollary to the AIA theorem, there is a line through B parallel to line $\overleftrightarrow{AC}$. Since Hilbert's Euclidean parallel postulate is equivalent to the converse to the AIA theorem (Proposition 4.8), the alternate interior angles with respect to the transversals $\overleftrightarrow{BA}$ and $\overleftrightarrow{BC}$ are congruent, as shown. But the three angles at vertex B have degree measures adding to 180°. ◄

We emphasize that this conclusion depends on Hilbert's Euclidean parallel postulate. The simple proof we gave was called by Proclus the *Pythagorean proof* (of the second assertion in Euclid I.32) because it

---

[3] Transitivity of parallelism is also logically equivalent to the parallel postulate (Exercise 10).

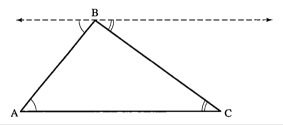

Figure 4.12 Angle sum is 180°.

was known to the Pythagorean school long before Euclid. The next corollary is Euclid's first assertion of I.32.

COROLLARY. Hilbert's Euclidean parallel postulate implies that the degree of an exterior angle to a triangle is equal to the sum of the degrees of its remote interior angles.

PROOF:

Refer again to Figure 4.3 on p. 165. We have

$$(\sphericalangle A)° + (\sphericalangle B)° + (\sphericalangle C)° = 180° = (\sphericalangle ACD)° + (\sphericalangle C)°,$$

so cancel $(\sphericalangle C)°$. ◄

Saccheri and Lambert Quadrilaterals

In this section, we will study certain quadrilaterals that are extremely important in neutral geometry. The results are mainly due to Girolamo Saccheri (1667–1733), who published them in 1733 in a work called *Euclides ab Omni Naevo Vindicatus* (*Euclid Freed of Every Flaw*) or simply *Euclides Vindicatus* (*Euclid Vindicated*). It was so far ahead of its time that it did not receive the appreciation it deserved until 1889, when the geometer Eugenio Beltrami rediscovered it. We will discuss the historical importance of his work in the next chapter. The path Saccheri followed is the correct one. I will often present proofs of his results that are modern simplifications and generalizations, but the ideas are basically his (and his predecessors'—see Rosenfeld (1988), Chapter 2, for the work of his predecessors).

DEFINITION. Quadrilateral ☐ABDC in which the adjacent angles ∢A and ∢B are right angles will be called *bi-right*; we will label such quadrilaterals so that the first two letters denote vertices at which the quadrilateral has right angles. (There may or may not be right angles at one or both of the other vertices as well—we are not assuming anything about them for now.) Side AB joining the right angles will be called the *base* with respect to this labeling; its opposite side CD will be called the *summit*. ∢C and ∢D will be called the *summit angles,* and CA and DB will simply be called the *sides* of the bi-right quadrilateral with respect to this labeling.

An *isosceles bi-right quadrilateral* □ABDC is one whose sides are congruent—i.e., CA ≅ DB—and is called a *Saccheri quadrilateral* (Figure 4.13). Given any segment AB, since perpendiculars can be erected at A and at B (Proposition 3.16) and a segment congruent to a given segment can be laid off on a given ray (Axiom C-1), we see that Saccheri quadrilaterals exist—in fact, they can be constructed on any given base with any given congruence class of the sides.

These quadrilaterals named after Saccheri were studied in the twelfth century by the Iranian poet and mathematician Omar Khayyam, and in the thirteenth century by the Iranian astronomer and mathematician Nasir Eddin (whose work had the similar title *Treatise That Heals the Doubt Raised by Parallel Lines).* These quadrilaterals were also studied later by several Europeans (e.g., Clavius in 1574, Giordano Vitale in 1680). Saccheri developed their significance more deeply. (In what follows, a notation such as Saccheri X stands for Proposition X in Saccheri's treatise.)

PROPOSITION 4.12. (a) (Saccheri I). The summit angles of a Saccheri quadrilateral are congruent to each other. (b) (Saccheri II). The line joining the midpoints of the summit and the base is perpendicular to both the summit and the base.

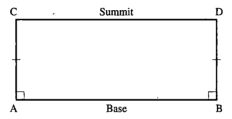

Figure 4.13 Saccheri quadrilateral.

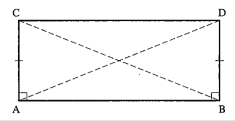

Figure 4.14

PROOF:

(a) By hypothesis and SAS, $\triangle$DBA $\cong$ $\triangle$CAB. Then by SSS, $\triangle$DCB $\cong$ $\triangle$CDA. Hence ⦨C $\cong$ ⦨D by angle addition (Figure 4.14).

(b) (See Figure 4.15.) Let M be the midpoint of the summit and N the midpoint of the base (Proposition 4.3). Then $\triangle$ACM $\cong$ $\triangle$BDM by part (a) and SAS. Hence AM $\cong$ BM (corresponding sides), whence $\triangle$ANM $\cong$ $\triangle$BNM by SSS. By corresponding angles, ⦨ANM $\cong$ ⦨BNM, but since they are supplementary angles, they are by definition right angles. Similarly, we have $\triangle$ACN $\cong$ $\triangle$BDN by SAS and Proposition 3.23, $\triangle$CNM $\cong$ $\triangle$DNM by SSS, so the supplementary angles ⦨CMN and ⦨DMN are congruent. Thus $\overleftrightarrow{MN}$ is perpendicular to both the base and the summit. ◄

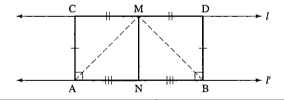

Figure 4.15

PROPOSITION 4.13. In any bi-right quadrilateral □ABDC, ⦨C > ⦨D ⇔ BD > AC. In words: The greater side is opposite the greater summit angle (Figure 4.16).

PROOF:

Assume first BD > AC. Then by definition there is a unique point E between B and D such that AC $\cong$ BE. Then □ABEC is Saccheri,

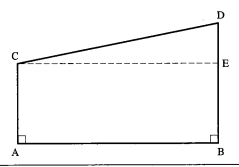

Figure 4.16

so by the previous proposition, ∢ACE ≅ ∢BEC. E is interior to ∢ACD (use Exercise 28). It then follows from the exterior angle theorem and Proposition 3.21 that ∢D < ∢ACE < ∢ACD, as was claimed.

Next, assume that ∢C > ∢D. Suppose that BD is not greater than AC (RAA hypothesis). By Proposition 3.13, either BD < AC or BD ≅ AC. In the former case, reversing the roles of AC and BD, it has been shown that ∢C < ∢D, contradicting our hypothesis. In the latter case, □ABDC is Saccheri, so by the previous proposition, ∢C and ∢D are congruent, contradicting our hypothesis. Hence BD > AC (RAA conclusion). ◄

COROLLARY 1. Given any acute angle with vertex V. Let Y be any point on one side of the angle, let Y' be any point farther out on that side, i.e., V * Y * Y'. Let X, X' be the feet of the perpendiculars from Y, Y', respectively, to the other side of the angle. Then Y' X' > YX. In words: The perpendicular segments from one side of an acute angle to the other increase as you move away from the vertex of the angle (Figure 4.17).

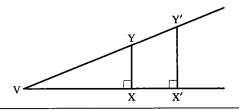

Figure 4.17

PROOF:

By the corollary to the exterior angle theorem, angles ∢VYX and ∢VY'X' are both acute. ∢Y'YX, supplementary to ∢VYX, is therefore obtuse and greater than ∢VY'X'. Now apply Proposition 4.13 to the bi-right quadrilateral □XX'Y'Y. ◄

COROLLARY 2. Euclid V implies Aristotle's axiom.

PROOF:

Refer to Figure 4.18. Let α be the given acute angle and let AB be the test segment for Aristotle's axiom. Let α^* be a complementary angle, so that $\alpha° + \alpha^{*°} = 90°$. On a chosen side of line $\overleftrightarrow{BA}$, lay off angle α^* at A and a 90° angle at B (Axiom C-4). By Euclid V, the rays of those angles not part of $\overleftrightarrow{BA}$ meet at a point C, and by Proposition 4.11 (which assumes Euclid V), ∢C° = $\alpha°$. Let Y be any point such that $C * A * Y$ and let X be the foot of the perpendicular from Y to ray $\overrightarrow{CB}$. By Corollary 1 to Proposition 4.13, YX > AB. ◄

The converse to this corollary does not hold because Aristotle's axiom is also valid in hyperbolic planes (Exercise 13, Chapter 6).

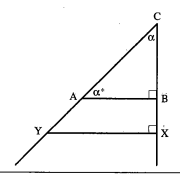

Figure 4.18 YX > AB.

DEFINITION. A quadrilateral with at least three right angles is called a *Lambert quadrilateral*. The remaining angle, about which we are not assuming anything for now, is referred to as the *fourth angle* with respect to the three given right angles. (Now named after J. H. Lambert (1728–1777), these quadrilaterals were studied eight centuries earlier by the Egyptian scientist Ibn al-Haytham and also by Saccheri.)

COROLLARY 3 (SACCHERI III, COROLLARY I). A side adjacent to the fourth angle θ of a Lambert quadrilateral is, respectively, greater than, congruent to, or less than its opposite side if and only if θ is acute, right, or obtuse, respectively. (In Figure 4.19, DB is adjacent to θ and CA is opposite; also, DC is adjacent to θ and BA is opposite.)

PROOF:

This follows from the proposition and trichotomy. ◄

Figure 4.19 Lambert quadrilateral.

OBSERVATION. We can "halve" Saccheri quadrilateral □ABDC in Figure 4.15 to obtain Lambert quadrilateral □NBDM with the fourth angle equal to the summit angle. Conversely, we can "double" Lambert quadrilateral □NBDM by reflecting it across side MN to obtain Saccheri quadrilateral □ABDC with the summit angle equal to the fourth angle of □NBDM. Applying Corollary 3 together with this observation, we obtain the following.

COROLLARY 4 (SACCHERI III). The summit of a Saccheri quadrilateral is, respectively, greater than, congruent to, or less than the base if and only if its summit angle is acute, right, or obtuse, respectively.

NOTE. *Hilbert's Euclidean parallel postulate implies that every Lambert quadrilateral and every Saccheri quadrilateral is a rectangle.* Namely, in Figure 4.19, when a perpendicular is dropped from C to $\overleftrightarrow{BD}$, the foot of that perpendicular must be D; otherwise we would have found a second parallel to $\overleftrightarrow{AB}$ through C (Corollary 1 to the AIA theorem). Thus this Lambert quadrilateral is a rectangle. The assertion about Saccheri quadrilaterals follows by halving.

The next goal is to prove that the behavior of the summit angles and the fourth angles of Saccheri and Lambert quadrilaterals is *uniform* throughout the plane—e.g., if one such quadrilateral has an acute angle, then so do all such quadrilaterals.

UNIFORMITY THEOREM.[4] For any Hilbert plane, if one Saccheri quadrilateral has acute (respectively, right, obtuse) summit angles, then so do all Saccheri quadrilaterals.

The uniformity theorem has a proof which, while elementary, is somewhat lengthy. In order not to exhaust the patience of beginning readers, the proof is indicated in Major Exercises 5–8.

COROLLARY 1. For any Hilbert plane, if one Lambert quadrilateral has an acute (respectively, right, obtuse) fourth angle, then so do all Lambert quadrilaterals. Furthermore, the type of the fourth angle is the same as the type of the summit angles of Saccheri quadrilaterals.

PROOF:

By doubling. ◄

DEFINITION. A Hilbert plane is called *semi-Euclidean*[5] if all Lambert quadrilaterals and all Saccheri quadrilaterals are rectangles. If the fourth angle of every Lambert quadrilateral is acute (respectively, obtuse), we say that *the plane satisfies the acute (respectively, obtuse) angle hypothesis.*

COROLLARY 2. There exists a rectangle in a Hilbert plane iff the plane is semi-Euclidean. Opposite sides of a rectangle are congruent to each other.

COROLLARY 3. In a Hilbert plane satisfying the acute (respectively, obtuse) angle hypothesis, a side of a Lambert quadrilateral adjacent to the acute (respectively, obtuse) angle is greater than (respectively, less than) its opposite side.

COROLLARY 4. In a Hilbert plane satisfying the acute (respectively, obtuse) angle hypothesis, the summit of a Saccheri quadrilateral is

---

[4] Also called the "three musketeers theorem" by historian Jeremy Gray. It shows that the plane is *homogeneous* (geometrically the same everywhere). Saccheri was the first to prove this result in his Propositions V, VI, and VII, but he used an unnecessary continuity argument (see Bonola, 1955, Section 12).

[5] The term "semi-Euclidean" first appeared in the German literature on the foundations of geometry. It is an important name to emphasize that the "right angle hypothesis" does not suffice to prove Euclid V—a further axiom is needed for that, as Hilbert emphasized. Analogous notions of "semihyperbolic" and "semielliptic" planes are discussed and exemplified in Hartshorne's treatise.

greater than (respectively, less than) the base. The midline segment MN is the only common perpendicular segment between the summit line and the base line. If P is any point $\neq$M on the summit line and Q is the foot of the perpendicular from P to the base line, then PQ > MN (respectively, PQ < MN). As P moves away from M along a ray of the summit line emanating from M, PQ increases (respectively, decreases).

These are consequences of Proposition 4.13 and its corollaries.

Angle Sum of a Triangle

The *angle sum* (in degrees) of triangle $\triangle ABC$ is $(\sphericalangle A)° + (\sphericalangle B)° + (\sphericalangle C)°$, by definition. Proposition 4.11 tells us that Hilbert's Euclidean parallel postulate implies that the angle sum of every triangle is 180°, but we are not assuming that postulate here.

SACCHERI'S ANGLE THEOREM (HIS PROPOSITION XV). For any Hilbert plane,
  (a) If there exists a triangle whose angle sum is <180°, then every triangle has an angle sum <180°, and this is equivalent to the fourth angles of Lambert quadrilaterals and the summit angles of Saccheri quadrilaterals being acute.
  (b) If there exists a triangle with angle sum =180°, then every triangle has angle sum =180°, and this is equivalent to the plane being semi-Euclidean.
  (c) If there exists a triangle whose angle sum is >180°, then every triangle has an angle sum >180°, and this is equivalent to the fourth angles of Lambert quadrilaterals and the summit angles of Saccheri quadrilaterals being obtuse.

For the proof of Saccheri's theorem, we need the next lemma (Saccheri VIII).

LEMMA. Let $\square$ABDC be a Saccheri quadrilateral with summit angle class θ. Consider the alternate interior angles $\sphericalangle ACB$ and $\sphericalangle DBC$ with respect to diagonal CB (Figure 4.20).
  (a) $\sphericalangle ACB < \sphericalangle DBC$ iff θ is acute.
  (b) $\sphericalangle ACB \cong \sphericalangle DBC$ iff θ is right.
  (c) $\sphericalangle ACB > \sphericalangle DBC$ iff θ is obtuse.

PROOF:

This is an application of Proposition 4.6 and the work we have just
done. △ACB and △DBC have congruent sides AC and BD (by hypothesis) and have the common side CB congruent to itself. Proposition 4.6 tells us that ⊀ACB is less than, congruent to, or greater
than ⊀DBC according as AB is less than, congruent to, or greater
than CD (those are the sides of the triangles opposite these angles).
But AB is the base and CD is the summit of our Saccheri quadrilateral. The lemma then follows from Corollaries 2 and 4. ◄

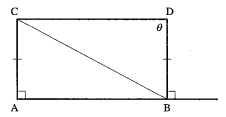

Figure 4.20

PROOF OF SACCHERI'S THEOREM:

Consider first a right triangle △ACB with right angle at A. Erect a
perpendicular to $\overleftrightarrow{AB}$ at B, and on the ray of that perpendicular emanating from B on the same side of $\overleftrightarrow{AB}$ as C, lay off BD ≅ AC so as
to form Saccheri quadrilateral □ABDC (see Figure 4.20). By construction, ⊀DBC is complementary to ⊀CBA. Now apply the lemma.
We conclude that the sum of the degrees of the acute angles in
△ACB is less than, equal to, or greater than 90° iff summit angle θ
is less than, equal to, or greater than 90°. By the uniformity theorem and its corollaries, the conclusion of Saccheri's theorem holds
for right triangles.

   Now let △ACB be arbitrary. By the second corollary to the exterior angle theorem, △ACB has at least two acute angles—say, ⊀A
and ⊀B are acute. Let D be the foot of the perpendicular from C to
$\overleftrightarrow{AB}$. Then A * D * B (by an RAA argument, using the exterior angle
theorem again).

   The angle sum Σ of △ACB is then equal to σ + τ, where σ, τ
is the angle sum of the acute angles in right triangle △ADC, △BDC,
respectively (Figure 4.21). By Saccheri's theorem for right trangles
just proved, σ and τ are either both <90° or both =90° or both
>90°—mutually exclusive cases equivalent to the cases where Σ is
<180°, =180°, or >180°. ◄

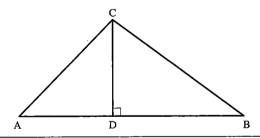

Figure 4.21

We now show that the obtuse angle hypothesis cannot occur if Aristotle's axiom holds. *This is a new result.*

NON-OBTUSE-ANGLE THEOREM. A Hilbert plane satisfying Aristotle's axiom either is semi-Euclidean or satisfies the acute angle hypothesis (so that by Saccheri's angle theorem, the angle sum of every triangle is ≤180°).

PROOF:

Assume on the contrary (using the uniformity theorem) that the fourth angle of every Lambert quadrilateral is obtuse. Since Hilbert's Euclidean parallel postulate implies that Lambert quadrilaterals are rectangles (see note above), that postulate fails in this plane. Hence there is a line l and a point P not on l such that more than one parallel to l passes through P (Figure 4.22). Denote by m the parallel through P obtained by the standard construction of perpendiculars and let n be a second parallel. Let Y be any point on the ray of n from P between m and l and let X be the foot of the perpendicular from Y to m. We claim that Aristotle's assertion fails for acute angle ⋪YPX. Drop a perpendicular from Y to PQ with foot S. We must have P * S * Q because any other position of S on line $\overleftrightarrow{PQ}$ would contradict the parallelism of $\overleftrightarrow{YS}$ with m and l (Corollary 1 to the

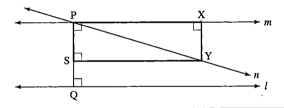

Figure 4.22

AIA theorem). In Lambert quadrilateral □XPSY, ∢Y is obtuse (RAA hypothesis). By Proposition 4.13, YX < SP < PQ. Thus the perpendicular segments YX for ∢YPX, as Y varies on that ray of n, are bounded by fixed segment PQ, contradicting Aristotle's axiom. ◄

COROLLARY. In a Hilbert plane satisfying Aristotle's axiom, an exterior angle of a triangle is greater than or congruent to the sum of the two remote interior angles.

By the EA theorem, that sum is a well-defined angle up to congruence. See Exercise 1(d).

Here is a famous theorem weaker than the non-obtuse-angle theorem because its hypothesis, Archimedes' axiom, is stronger than Aristotle's axiom (Exercise 2, Chapter 5).

SACCHERI–LEGENDRE THEOREM. In an Archimedean Hilbert plane, the angle sum of every triangle is ≤180°.

Direct proofs by Legendre of this theorem that don't invoke the new non-obtuse-angle theorem are indicated in Exercises 15 and 16.

It is natural to generalize the Saccheri–Legendre theorem to polygons other than triangles. For example, let us prove that the angle sum of a quadrilateral ABCD is at most 360°. Break □ABCD into two triangles, △ABC and △ADC, by the diagonal AC (see Figure 4.23). By the Saccheri–Legendre theorem,

$$(∢B)° + (∢BAC)° + (∢ACB)° ≤ 180°$$

and

$$(∢D)° + (∢DAC)° + (∢ACD)° ≤ 180°.$$

Measurement Theorem 4.3(3) gives us the equations

$$(∢BAC)° + (∢DAC)° = (∢BAD)°$$

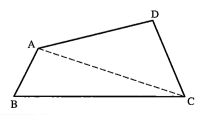

Figure 4.23

and

$$(\measuredangle ACB)° + (\measuredangle ACD)° = (\measuredangle BCD)°.$$

Using these equations, we add the two inequalities to obtain the desired inequality

$$(\measuredangle B)° + (\measuredangle D)° + (\measuredangle BAD)° + (\measuredangle BCD)° \leq 360°.$$

Unfortunately, there is a gap in this simple argument! To get the equations used above, we assumed by looking at the diagram (Figure 4.23) that C was interior to $\measuredangle$BAD and that A was interior to $\measuredangle$BCD. But what if the quadrilateral looked like Figure 4.24? In this case, the equations would not hold. To prevent such a case, we must add a hypothesis; we must assume that the quadrilateral is "convex."

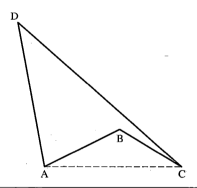

Figure 4.24 Non-convex quadrilateral.

DEFINITION. Quadrilateral □ABCD is called *convex* if it has a pair of opposite sides, e.g., AB and CD, such that CD is contained in one of the half-planes bounded by $\overleftrightarrow{AB}$, and AB is contained in one of the half-planes bounded by $\overleftrightarrow{CD}$.[6]

The assumption made above is now justified by starting with a convex quadrilateral. Thus we have proved the following corollary.

---

[6] It can be proved that this condition also holds for the other pair of opposite sides, AD and BC—see Exercise 28 in this chapter. The use of the word "convex" in this definition does not agree with its use in Exercise 19, Chapter 3; a convex quadrilateral is obviously not a "convex set" as defined in that exercise. However, we can define the *interior* of a convex quadrilateral □ABCD as follows: Each side of □ABCD determines a half-plane containing the opposite side; the interior of □ABCD is then the intersection of the four half-planes so determined. You can then prove that the interior of a convex quadrilateral is a convex set (Exercise 29).

COROLLARY. In an Archimedean Hilbert plane, the angle sum of any *convex* quadrilateral is at most 360°.

NOTE ON ELLIPTIC GEOMETRY. The Saccheri–Legendre theorem is false in elliptic geometry (see Figure 3.24, p. 126). In fact, it can be proved in elliptic geometry that the angle sum of a triangle is always greater than 180° (see Kay, 1969). Since a triangle can have two or three right angles, a *hypotenuse*, defined as a side opposite a right angle, need not be unique, and a *leg*, defined as a side of a right triangle not opposite a right angle, need not exist (and if opposite an obtuse angle, a leg could be longer than a hypotenuse).

NOTE FOR ADVANCED STUDENTS ABOUT NON-ARCHIMEDEAN GEOMETRIES.[7] Non-Archimedean geometries were first considered by Giuseppe Veronese in 1890. Hilbert stated in his *Grundlagen* that they were "of fundamental significance." He provided an algebraic model of a non-Archimedean geometry. Other models were provided by his student Max Dehn and by Friedrich Schur (who also published his own *Grundlagen der Geometrie* in 1909).

There is an algebraic version of Archimedes' property for ordered fields: The ordered field is called *Archimedean* if, given any positive elements t and u, there exists a natural number n such that $nt > u$; consideration of u/t shows that it suffices in this property to take $t = 1$. Hilbert gave an example of a non-Archimedean ordered field F, and other examples have been given since (see Projects 1 and 2). In such a field, there are *infinitesimal* and *infinitely large* elements. A positive element u is called "infinitely large" if it is greater than every natural number; that is the case iff its reciprocal $1/u$ is smaller than the reciprocal $1/n$ of every natural number. An element t is called *infinitesimal* iff its absolute value $|t|$ is smaller than the reciprocal of every natural number.

▓▓ **EXAMPLE 1.** F^2 where F is a non-Archimedean Pythagorean ordered field.

We know that the Euclidean parallel property holds in every model F^2, so by Corollary 2 to Proposition 4.13, Aristotle's axiom holds in this model. But since F is non-Archimedean, so is F^2. Therefore *this model shows that Aristotle's axiom does not imply Archimedes' axiom.*

---

[7] Strange as non-Archimedean geometry may seem, theoretical physicists are applying it to the study of subatomic particles. See Branko Dragovich at http://arxiv.org/PS_cache/math-ph/pdf/0306/0306023.pdf.

▓▓▓ **EXAMPLE 2.** A semi-Euclidean plane in which the Euclidean parallel postulate fails.

Let F be as in the previous example. Let Π be the subplane of F^2 consisting of points (x, y), both of whose coordinates are *infinitesimal*, and lines in F^2 passing through at least two such points. It is straightforward to show that all the 13 axioms for a Hilbert plane still hold when interpreted in Π. Furthermore, the angle sum of every triangle in Π is still 180° because that is the case in the larger plane F^2. However, whenever two lines of Π meet in a point in F^2 whose coordinates are not both infinitesimal, those lines are parallel considered as lines of Π because they do not meet in a point of Π. With the appearance of these new parallels, the Euclidean parallel postulate fails. This example is due to Max Dehn.

Dehn also gave an example, using infinitesimals, of a Hilbert plane in which the fourth angle of every Lambert quadrilateral is *obtuse*.[8] Such examples are important because they contradict the assertion made in some books and articles that "the hypothesis of the obtuse angle" is inconsistent with the first 4 axioms of Euclid. In fact, it is consistent with the 13 axioms for Hilbert planes (which imply those 4 axioms). Many writers claim that to reject the hypothesis of the obtuse angle, one must explicitly assume that, as one popular historian put it, "a line can be extended to any given length" or, as others stated, that "lines are infinite in extent." This claim is erroneous because Euclid's second postulate explicitly assumes the extendability of line segments, which we have proved using Axioms B-2 and C-1. When these writers talk loosely about needing to assume "lines are infinite in extent," they imagine that the only geometries in which "the hypothesis of the obtuse angle" holds are the real spherical and elliptic geometries, where "lines" are topologically circles and have finite length. Some of those writers point to Euclid's proof of the exterior angle theorem, claiming that it tacitly assumes that lines are infinite in extent.

As our discussion on pp. 164–166 showed, what was missing in Euclid's proof was the betweenness axioms, especially the Plane Separation Axiom B-4. Gauss noticed that gap and Pasch filled it. The exterior angle theorem is valid in examples like Dehn's! Saccheri and Legendre both recognized that an additional assumption, acceptable to the ancient

---

[8] "Die Legendre'schen Sätze über die Winkelsumme im Dreieck," *Mathematische Annalen* **53**, 404–439, or see Hartshorne, Exercise 34.14, p. 318 (the infinitesimal neighborhood of a point on a non-Archimedean sphere—the whole sphere is not a Hilbert plane but the infinitesimal neighborhood of a point is).

Greeks, which suffices to reject the hypothesis of the obtuse angle, is Archimedes' axiom (and I showed that the weaker axiom of Aristotle suffices).

Conclusion

In this chapter, we have continued the study of elementary geometry without a parallel postulate (neutral geometry)—specifically, the study of Hilbert planes, which are models of our incidence, betweenness, and congruence axioms. We demonstrated the alternate interior angle (AIA) theorem for arbitrary Hilbert planes, which implies that for every line and every point P not on the line, there exists a parallel line through P, by the standard construction with successive perpendiculars; we do not know in neutral geometry whether that parallel is *unique* or not. We used the AIA theorem to deduce the familiar exterior angle (EA) theorem; from that we deduced further familiar propositions (our Propositions 4.1–4.6) of Euclid, which are valid in arbitrary Hilbert planes.

In the next section, we (unnecessarily!) brought in measurement of segments and angles by real numbers in order to simplify our statements; Archimedes' axiom was used to obtain that. Euclid didn't have any measurement, so many of his statements (such as the triangle inequality) were awkward. We proved his triangle inequality in neutral geometry and showed that its converse is equivalent to the circle-circle continuity principle. We also proved that the angle sum of any *two* angles in a triangle is <180°.

In the next section, we showed that Euclid's fifth postulate is equivalent for Hilbert planes to Hilbert's Euclidean parallel postulate. We also proved it is equivalent to several other familiar statements, such as the converse to the AIA theorem. A subtle point, ignored in most books, was that any one of these equivalent statements implies that the angle sum of every triangle is 180°, but it is not possible to prove the converse for arbitrary Hilbert planes.

The next two sections are rich with less familiar but elementary concepts and results in neutral geometry that appeared in the works of Khayyam, Saccheri, and Lambert (among others). We introduced and studied the important concepts of bi-right, Saccheri, and Lambert quadrilaterals, which will be used extensively in subsequent chapters. The latter two types of quadrilaterals provided our first inkling of non-Euclidean concepts because in Euclidean geometry they are nothing but rectangles.

It is possible, in an arbitrary Hilbert plane, for the angle sum of a triangle to be <180°, =180°, >180°. Our main result was Saccheri's angle theorem that the behavior of that angle sum is *uniform* throughout the plane. Saccheri and Legendre eliminated the case where the summit angles of Saccheri and Lambert quadrilaterals are obtuse, but only by assuming Archimedes' axiom. We proved that the weaker axiom of Aristotle, whose significance was first recognized by Proclus and which is a purely geometric axiom (unlike Archimedes' axiom), suffices to eliminate the case of obtuse angles. In addition, Aristotle's axiom provides the *missing link* between the angle sum of triangles equaling 180° and Euclid's fifth postulate (see Proclus' theorem in Chapter 5).

Review Exercise

Which of the following statements are correct?

(1) If two triangles have the same angle sum, they are congruent.

(2) Euclid's fourth postulate is a theorem in neutral geometry.

(3) Theorem 4.4 shows that Euclid's fifth postulate is a theorem in neutral geometry.

(4) The Saccheri–Legendre theorem tells us that some triangles exist that have angle sums less than 180° and some triangles exist that have angle sums equal to 180°.

(5) The alternate interior angle theorem states that if parallel lines are cut by a transversal, then alternate interior angles are congruent to each other.

(6) It is impossible to prove in neutral geometry that rectangles exist.

(7) The Saccheri–Legendre theorem is false in Euclidean geometry because in Euclidean geometry the angle sum of any triangle is never less than 180°.

(8) According to our definition of "angle," the degree measure of an angle cannot equal 180°.

(9) The notion of one ray being "between" two others is undefined.

(10) It is impossible to prove in neutral geometry that parallel lines exist.

(11) Archimedes' axiom is used to measure segments and angles by real numbers.

(12) An exterior angle of a triangle is any angle that is not in the interior of the triangle.

(13) The SSS criterion for congruence of triangles is a theorem in neutral geometry.

(14) The alternate interior angle theorem implies, as a special case, that if a transversal is perpendicular to one of two parallel lines, then it is also perpendicular to the other.

(15) If a Hilbert plane satisfies Aristotle's axiom, then the fourth angle in a Lambert quadrilateral in that plane cannot be obtuse.

(16) The ASA criterion for congruence of triangles is one of our axioms for neutral geometry.

(17) A Lambert quadrilateral can be "doubled" to form a Saccheri quadrilateral, and a Saccheri quadrilateral can be "halved" to form a Lambert quadrilateral.

(18) If $\triangle ABC$ is any triangle and if a perpendicular is dropped from C to $\overleftrightarrow{AB}$, then that perpendicular will intersect $\overleftrightarrow{AB}$ in a point between A and B.

(19) It is a theorem in neutral geometry that given any point P and any line l, there is at most one line through P perpendicular to l.

(20) It is a theorem in neutral geometry that vertical angles are congruent to each other.

(21) In the sphere interpretation, where "lines" are interpreted to be great circles, Euclid V holds, yet the Euclidean parallel postulate does not.

(22) The gap in Euclid's attempt to prove Theorem 4.2 can be filled using our axioms of betweenness.

Exercises

The exercises that follow are exercises in neutral geometry unless otherwise stated. This means that in your proofs you are allowed to use only those results about Hilbert planes that have been previously demonstrated (including results from previous exercises).

1. (a) State the converse to Euclid V (Euclid's fifth postulate). Prove this converse as a proposition in neutral geometry.

   (b) Prove Corollary 1 to the exterior angle theorem.

   (c) Prove that Hilbert's Euclidean parallel postulate implies that all Saccheri and Lambert quadrilaterals are rectangles and that rectangles exist.

   (d) Prove the corollary to the non-obtuse-angle theorem.

2. The following purports to be a proof in neutral geometry of the SAA congruence criterion. Find the step in the proof that is not valid in neutral geometry and indicate for which special Hilbert planes the proof is valid (see Figure 4.5, p. 167).

Given AC ≅ DF, ∢A ≅ ∢D, ∢B ≅ ∢E. Then ∢C ≅ ∢F since

$$(\sphericalangle C)° = 180° - (\sphericalangle A)° - (\sphericalangle B)°$$
$$= 180° - (\sphericalangle D)° - (\sphericalangle E)° = (\sphericalangle F)°.$$

Hence △ABC ≅ △DEF by ASA (Proposition 3.17).
3. Here is a proof of the SAA criterion (Proposition 4.1) that is valid in neutral geometry. Justify each step (see Figure 4.25).

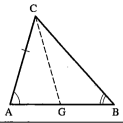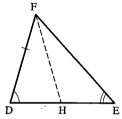

Figure 4.25

(1) Assume side AB is not congruent to side DE. (2) Then AB < DE or DE < AB. (3) If DE < AB, then there is a point G between A and B such that AG ≅ DE. (4) Then △CAG ≅ △FDE. (5) Hence ∢AGC ≅ ∢E. (6) It follows that ∢AGC ≅ ∢B. (7) This contradicts a certain theorem. (8) Therefore, DE is not less than AB. (9) By a similar argument involving a point H between D and E, AB is not less than DE. (10) Hence AB ≅ DE. (11) Therefore, △ABC ≅ △DEF.
4. Prove Proposition 4.2. (Hint: See Figure 4.6. On the ray opposite to $\overrightarrow{AC}$, lay off segment AD congruent to A'C'. First prove △DAB ≅ △C'A'B'; then use isosceles triangles and a congruence criterion to conclude.)
5. Justify each of the 18 steps on p. 167 proving that every segment has a midpoint (Proposition 4.3). Reconstruct Euclid's shorter proof, which uses the existence of an equilateral triangle on any segment (but that existence can't be proved in neutral geometry without a further axiom such as the circle-circle continuity principle).
6. (a) Prove that segment AB has only one midpoint. (Hint: Assume the contrary and use Propositions 3.3 and 3.13 to derive a contradiction, or else derive a contradiction from congruent triangles.)

(b) Prove Proposition 4.4 on the existence of angle bisectors. Prove that the angle bisector is unique.

7. Prove that every acute angle has a complementary angle and that if complements of two acute angles are congruent, then the acute angles are congruent.

8. Prove Proposition 4.5. (Hint: If AB > BC, then let D be the point between A and B such that BD ≅ BC (Figure 4.26). Use isosceles triangle △CBD and exterior angle ∢BDC to show that ∢ACB > ∢A. Use this result and trichotomy of ordering to prove the converse.)

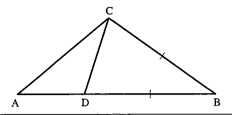

Figure 4.26

9. Here is a sketch of an argument to prove Proposition 4.6. Fill in the details and justify the steps: Given ∢B < ∢B'. Use the hypothesis of Proposition 4.6 to reduce to the case where A = A', B = B', BC ≅ BC', and C is interior to ∢ABC', so that you must show AC < AC' (see Figure 4.27 where D is obtained from the crossbar theorem).

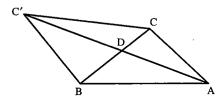

Figure 4.27

The case where C = D being clear, suppose C ≠ D. Proposition 4.5 reduces our task to showing ∢AC'C < ∢ACC' using the hypothesis to show that ∢BC'C ≅ ∢BCC'. In the case where B * D * C (as in Figure 4.27), we have ∢AC'C < ∢BC'C and ∢BCC' < ∢ACC'. In the case where B * C * D, apply the exterior angle theorem twice (see Figure 4.28):

$$\angle ACC' > \angle DCC' > \angle BC'C \cong \angle BCC' > \angle CC'D = \angle AC'C.$$

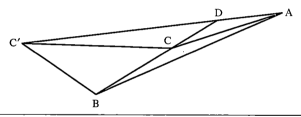

Figure 4.28

The converse implication in Proposition 4.6 follows from the direct implication just shown, using trichotomy.

10. Prove Proposition 4.7. Deduce as a corollary that transitivity of parallelism is equivalent to Hilbert's Euclidean parallel postulate.

11. Prove Proposition 4.8. (Hint: Assume first the converse to the AIA theorem. Let m be the parallel to l through P constructed in the standard way and let n be any parallel to l through P. Use the congruence of alternate interior angles and the uniqueness of perpendiculars to prove $m = n$. Assuming next the parallel postulate, use Axiom C-4 and an RAA argument to establish the converse to the AIA theorem.)

12. Prove Proposition 4.9.

13. Prove Proposition 4.10.

14. The ancient Greek mathematician Heron gave an elegant proof of the triangle inequality different from the one in the text. In order to prove that $\overline{AB} + \overline{AC} > \overline{BC}$, he bisected $\sphericalangle A$. He let the bisector meet BC at point D, which we justify via the crossbar theorem. He then applied the exterior angle theorem and Proposition 4.5 to triangles $\triangle BAD$ and $\triangle CAD$. Fill in the details of this argument.

15. Here is Legendre's lemma—which is needed for his proof found in many texts, based on the Archimedean property of angles—that the angle sum of every triangle is $\leq 180°$. He got the idea for this from Euclid's construction in his (incomplete) proof of the exterior angle theorem (I.16). Given $\triangle ABC$. Let D be the midpoint of BC. Let E be the point on the ray opposite to $\overrightarrow{DA}$ such that $DE \cong DA$. Prove that $\triangle AEC$ has the same angle sum as $\triangle ABC$ and that either $(\sphericalangle AEC)°$ or $(\sphericalangle EAC)°$ is $\leq \frac{1}{2} (\sphericalangle BAC)°$. (Hint: See Figure 4.29. Use congruent triangles to show that $(\sphericalangle EAC)° + (\sphericalangle AEC)° = (\sphericalangle BAC)°$.)

Use Legendre's lemma to give an RAA proof of the Saccheri–Legendre theorem.

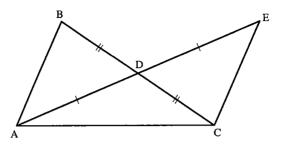

Figure 4.29

16. Here is another proof by Legendre of the Saccheri–Legendre theorem that the angle sum of every triangle is $\leq 180°$ in an Archimedean Hilbert plane (Figure 4.30). Justify the unjustified steps. (1) Let $A_1A_2B_1$ be the given triangle, lay off n copies of segment A_1A_2, and construct a row of triangles $A_jA_{j+1}B_j$, $j = 1, \ldots, n$ congruent to $A_1A_2B_1$, as shown in Figure 4.30. (2) The $B_jA_{j+1}B_{j+1}$, $j = 1, \ldots, n$ are also congruent triangles, the last by construction of B_{n+1}. (3) With angles labeled as in Figure 4.30, $\alpha + \gamma + \delta = 180°$ and we have $\beta + \gamma + \delta$ equal to the angle sum of $A_1A_2B_1$. (4) Assume on the contrary that $\beta > \alpha$. (5) Then $A_1A_2 > B_1B_2$, by Proposition 4.6. (6) Also, $\overline{A_1B_1} + n \cdot \overline{B_1B_2} + \overline{B_{n+1}A_{n+1}} > n \cdot \overline{A_1A_2}$, by repeated application of the triangle inequality. (7) $A_1B_1 \cong B_{n+1}A_{n+1}$. (8) $2\overline{A_1B_1} > n(\overline{A_1A_2} - \overline{B_1B_2})$. (9) Since n was arbitrary, this contradicts Archimedes' axiom. (10) Hence the triangle has angle sum $\leq 180°$.

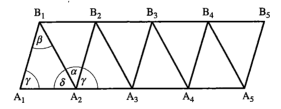

Figure 4.30

17. Prove the following theorems:
    (a) Let γ be a circle with center O and let A and B be two points on γ. The segment AB is called a *chord* of γ; let M be its midpoint. If $O \neq M$, then $\overleftrightarrow{OM}$ is perpendicular to $\overleftrightarrow{AB}$. (Hint: Corresponding angles of congruent triangles are congruent.)
    (b) Let AB be a chord of the circle γ having center O. Prove that the perpendicular bisector of AB passes through the center O of γ.

18. In any Hilbert plane, prove that *every triangle has an inscribed circle*—more specifically, prove that *the three angle bisectors are concurrent* in a point P (called the *incenter*) interior to the triangle which is equidistant from the sides of the triangle—i.e., the perpendiculars dropped from P to the sides are congruent—so that the circle with center P and radius equal to any of those perpendiculars is tangent to the sides of the triangle. (Hint: Show first that two angle bisectors must meet at a point P interior to the triangle; then show by congruent triangles that P is equidistant from the sides and lies on the third angle bisector.)

19. Prove the theorem of Thales that in a semi-Euclidean plane, an angle inscribed in a semicircle is a right angle (see Figure 4.31).

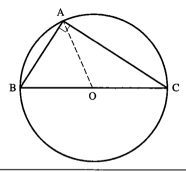

Figure 4.31

20. (a) Find the flaw in the following argument purporting to construct a rectangle. Let A and B be any two points. There is a line l through A perpendicular to $\overleftrightarrow{AB}$ (Proposition 3.16) and, similarly, there is a line m through B perpendicular to $\overleftrightarrow{AB}$. Take any point C on m other than B. There is a line through C perpendicular to l—let it intersect l at D. Then □ABCD is a rectangle.

   (b) In a general Hilbert plane, opposite sides of a parallelogram need not be congruent, as is illustrated by Saccheri and Lambert quadrilaterals in non-semi-Euclidean planes. Prove that in a plane satisfying the Euclidean parallel postulate, opposite sides and opposite angles of a parallelogram are congruent.

21. The sphere, with "lines" interpreted as great circles, is not a model of neutral geometry. Here is a proposed construction of a rectangle on a sphere. Let α, β be two circles of longitude and let them intersect the equator at A and D. Let γ be a circle of latitude in the northern hemisphere intersecting α and β at two other points, B

and C. Since circles of latitude are perpendicular to circles of longitude, the quadrilateral with vertices ABCD and sides the arcs of α, γ, and β and the equator traversed in going from A north to B east to C south to D west to A should be a rectangle. Explain why this construction doesn't work.

22. Given $A * B * C$ and $\overleftrightarrow{DC} \perp \overleftrightarrow{AC}$. Prove that $AD > BD > CD$ (Figure 4.32; use Proposition 4.5).

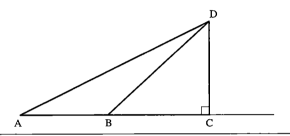

Figure 4.32

23. Given any triangle $\triangle DAC$ and any point B between A and C. Prove that either $DB < DA$ or $DB < DC$. (Hint: Drop a perpendicular from D to $\overleftrightarrow{AC}$ and use the previous exercise.)

24. Recall from Exercise 18, Chapter 3, that a set is called *convex* if whenever points A, B are in the set, the entire segment AB is contained in the set.
    (a) Prove that the interior of a circle is a convex set (the *interior* is the set of all points inside the circle).
    (b) Assume the line-circle continuity principle. Show that if a line passes through a point inside a circle, then it also passes through points outside the circle.

25. Suppose that line l meets circle γ in two points C and D. Prove that:
    (a) Point P on l lies inside γ if and only if $C * P * D$.
    (b) If points A and B are inside γ and on opposite sides of l, then the point E at which AB meets l is between C and D.

26. Look up and state Euclid III.20 and III.32 more precisely. Rewrite his proofs and show that they work in any semi-Euclidean plane.

27. The proof of the uniformity theorem uses the idea of constructing a congruent copy of a Saccheri quadrilateral. Two Saccheri quadrilaterals are defined to be *congruent* if all their corresponding parts are congruent—their bases, their summits, their sides, and their summit angles. For triangles we have an axiom (C-6) and various propositions (ASA, SSS, SAA) which tell us that if three particular corresponding parts are congruent, then the other three correspon-

ding parts are automatically congruent. State and prove one or more analogous propositions for Saccheri quadrilaterals. Explain how to construct a congruent copy.

28. Recall that a quadrilateral □ABCD is formed from four distinct points (called the *vertices*), no three of which are collinear, and from the segments AB, BC, CD, and DA (called the *sides*), which have no intersections except at those endpoints labeled by the same letter. The notation for this quadrilateral is not unique—e.g., □ABCD = □CBAD. Two vertices that are endpoints of a side are called *adjacent*; otherwise the two vertices are called *opposite*. A pair of sides having a vertex in common are called *adjacent*; otherwise the two sides are called *opposite*. The remaining pair of segments AC and BD formed from the four points are called *diagonals* of the quadrilateral; they may or may not intersect at some fifth point. If X, Y, Z are vertices of □ABCD such that Y is adjacent to both X and Z, then ∡XYZ is called an *angle* of the quadrilateral; if W is the fourth vertex, then ∡XWZ and ∡XYZ are called *opposite* angles.

The quadrilaterals of main interest are the *convex* ones. By definition, they are the quadrilaterals such that each pair of opposite sides, e.g., AB and CD, has the property that CD is contained in one of the half-planes bounded by the line through A and B, and AB is contained in one of the half-planes bounded by the line through C and D. Using Pasch's theorem, prove that if one pair of opposite sides has this property, then so does the other pair of opposite sides. Prove, using the crossbar theorem, that the following conditions are equivalent:

(a) The quadrilateral is convex.
(b) Each vertex of the quadrilateral lies in the interior of the opposite angle.
(c) The diagonals of the quadrilateral meet.

Prove that Saccheri and Lambert quadrilaterals are convex.

Draw a diagram of a quadrilateral that is not convex.

29. A convex quadrilateral is not a convex set in the sense of Exercise 18, Chapter 3. However, define the *interior* of a convex quadrilateral to be the intersection of the interiors of its four angles. Prove that the interior of a convex quadrilateral is a convex set and that the point of intersection of the diagonals lies in the interior.

30. State and prove a generalization of Pasch's theorem to Saccheri and Lambert quadrilaterals (or, more generally, to convex quadrilaterals).

31. Prove that there exists a *scalene* triangle—one which is not isosceles.

32. In Figure 4.33, the angle pairs (∡A'B'B″, ∡ABB″) and (∡C'B'B″, ∡CBB″) are called pairs of *corresponding angles* cut off on *l* and *l'* by trans-

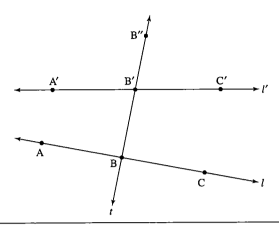

Figure 4.33

versal t. Prove that corresponding angles are congruent if and only if alternate interior angles are congruent.

33. (a) Define a complement of an acute angle without referring to degree measurement.

  (b) Suppose real number measurement of lengths does not exist—i.e., suppose the plane is non-Archimedean. State a version of the triangle inequality for such a plane in terms of addition of segment congruence classes and prove it. Do the same for Corollary 2 to the EA theorem.

  (c) Examples exist of Hilbert planes satisfying the obtuse angle hypothesis (see footnote 8). According to the Saccheri–Legendre theorem, such planes must be non-Archimedean. Now the angle sum of a triangle was defined by adding the real number degree measures of its angles, but in order to obtain such a measurement in Theorem 4.3, Archimedes' axiom was needed. Still, we would like to state that in a Hilbert plane satisfying the obtuse angle hypothesis, the angle sum of every triangle is greater than a "straight" angle, and we would like to prove that statement. Propose a precise definition of that statement and sketch how to prove it.

Major Exercises

1. Fill in the details of the following argument which proves that the *circle-circle continuity principle implies the line-circle continuity principle* (see Figure 4.34; since the circle-circle continuity principle

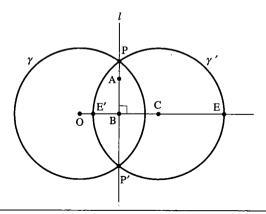

Figure 4.34

holds in a Euclidean plane by definition, this shows that the line-circle continuity principle holds in a Euclidean plane).

Let O be the center of γ. By hypothesis, line l passes through a point A inside γ. The goal is to prove that l intersects γ in two points. The case where l passes through O is easy. Otherwise, let point B be the foot of the perpendicular from O to l. Point C is constructed such that B is the midpoint of OC, and γ' is the circle centered at C having the same radius as γ (γ' is the reflection of γ across l). Prove that the hypothesis of the circle-circle continuity principle is satisfied—specifically that γ' intersects OC in a point E' inside γ and a point E outside γ, so that γ' intersects γ in two points P, P'. Prove that these points lie on the original line l.

2. Prove that *the line-circle continuity principle implies the segment-circle continuity principle and conversely*. (Hint: Use the results in Exercises 22 and 24(b).)

3. (a) Assume the line-circle continuity principle. Prove that there exists a right triangle with a hypotenuse of length c and a leg of length b iff $b < c$. (Hint for the "if" part: Take any point C and any mutually perpendicular lines through C. There exists a point A on one line such that $|AC| = b$. If γ is the circle centered at A of radius c, point C lies inside γ. Show that γ intersects the other line in some point B. Then $\triangle ABC$ is the requisite right triangle.)

   (b) Assume that whenever $b < c$, there exists a right triangle with a hypotenuse of length c and a leg of length b. Prove that this implies the line-circle continuity principle.

4. Let line l intersect circle γ at point A. Let O be the center of γ. If $l \perp \overleftrightarrow{OA}$, we say that l is *tangent* to γ at A; otherwise l is called *secant* to γ.

   (a) Suppose l is secant to γ. Prove that the foot F of the perpendicular t from O to l lies inside γ and that the reflection A′ of A across t is another point at which l meets γ (see Figure 4.35).

   (b) Suppose now that l is tangent to γ at A. Prove that every point B ≠ A lying on l is outside γ, hence A is the unique point at which l meets γ. Prove that all points of γ other than A are on the same side of l. Prove conversely that if a line intersects a circle at only one point, then that line is tangent to the circle.

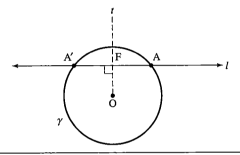

Figure 4.35

The next four major exercises provide a proof of the uniformity theorem. The first two are lemmas needed for the main argument in the third and fourth.

5. Prove Lemma 1. Given a Saccheri quadrilateral □ABDC and a point P between C and D. Let Q be the foot of the perpendicular from P to the base AB (Figure 4.36). Then

   (a) PQ < BD iff the summit angles of □ABDC are acute.

   (b) PQ ≅ BD iff the the summit angles of □ABDC are right angles.

   (c) PQ > BD iff the summit angles of □ABDC are obtuse.

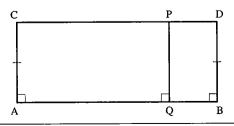

Figure 4.36

(Hint: Apply Proposition 4.13 to the bi-right quadrilaterals □AQPC and □BQPD, using the fact that ⊀QPC and ⊀QPD are supplementary, the definition of a Saccheri quadrilateral, Proposition 4.12(a), and trichotomy.)
6. Prove Lemma 2. Given a Saccheri quadrilateral □ABDC and a point P such that C * D * P. Let Q be the foot of the perpendicular from P to $\overleftrightarrow{AB}$ (Figure 4.37). Then
   (a) PQ > BD iff the summit angles of □ABDC are acute.
   (b) PQ ≅ BD iff the summit angles of □ABDC are right angles.
   (c) PQ < BD iff the summit angles of □ABDC are obtuse.

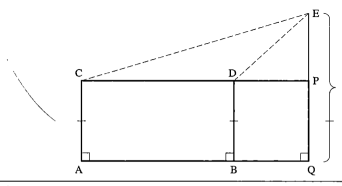

Figure 4.37

(Hints: Suppose PQ ≅ BD. Then □AQPC is Saccheri, so apply part (b) of Lemma 1. Suppose PQ < BD. Then there is a unique point E such that Q * P * E and QE ≅ BD. We then have two more Saccheri quadrilaterals □AQEC and □BQED, each of which has congruent summit angles. To show that ⊀BDC is greater than its supplement ⊀BDP, implying that it is obtuse, use the idea that exterior angle ⊀EDP is greater than remote interior angle ⊀ECD and that C * D * P implies ⊀BDE < ⊀ACE and subtract.

Suppose PQ > BD. Then there is a unique point E such that P * E * Q and QE ≅ BD. We again have two more Saccheri quadrilaterals □AQEC and □BQED, each of which has congruent summit angles. The rest of the argument is similar to the previous case. Finally, show that the other direction of these three cases follows from trichotomy.)
7. Prove the special case of the uniformity theorem where the midline segments of the two given Saccheri quadrilaterals are congruent (Figure 4.38). (Hints: First construct a congruent copy of □A'B'D'C' for which M = M' and N = N', so we can assume these midpoints

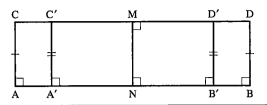

Figure 4.38

coincide, as do the summit and the base lines. Then apply the two lemmas in the preceding exercises.) **Important remark**: In this special case, we have also proved a uniformity result for Lambert quadrilaterals □MNBD and □MNB'D', which have common side MN where there are right angles and common lines containing the sides adjacent to MN.

8. Here is a proof of the general case of the uniformity theorem from the three previous exercises. Your job is to provide justifications for the steps.

PROOF:

The case MN ≅ M'N' having been handled, consider the case M'N' > MN. There is a unique point L such that L ∗ M ∗ N and LN ≅ M'N'. We will construct a Lambert quadrilateral □LNHG with the fourth angle at G, congruent to half of Saccheri quadrilateral □A'B'D'C' (Figure 4.39).

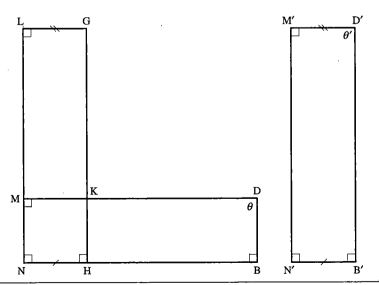

Figure 4.39

On ray $\overrightarrow{NB}$, let H be the point such that NH $\cong$ N'B'. On the same side of $\overleftrightarrow{LN}$ as H and on the perpendicular to $\overleftrightarrow{LN}$ through L, let G be the point such that LG $\cong$ M'D'. Then $\triangle$LNH $\cong$ $\triangle$M'N'B', ⦠GLH $\cong$ ⦠D'M'B', $\triangle$GLH $\cong$ $\triangle$D'M'B', so that ⦠G $\cong$ ⦠D' and, by addition, ⦠NHG $\cong$ ⦠B', which is a right angle.

Since G and L lie on a parallel to $\overleftrightarrow{MD}$, they are on the same side of $\overleftrightarrow{MD}$, and since L is on the opposite side of $\overleftrightarrow{MD}$ from N, G is on the opposite side from H. Let K be the point at which GH meets line $\overleftrightarrow{MD}$, necessarily on ray $\overrightarrow{MD}$ since G and H are on the same side of $\overleftrightarrow{LN}$.

Now apply the important remark from the special case above: ⦠MKH is of the same type as ⦠D. But ⦠MKH is also of the same type as ⦠G. Therefore ⦠G and ⦠D are the same type of angle, and we are done. If M'N' < MN, reverse the roles. ◄

9. Denote by |AB| the congruence class of segment AB (the set of all segments congruent to AB). Then, by definition and Axiom C-2,

$$AB \cong CD \Leftrightarrow |AB| = |CD|.$$

That is the underlying idea of "passing to the quotient"—replacing the equivalence relation $\cong$ with actual *equality* of equivalence classes.

We have already defined an ordering of segments: AB < CD means that there is a point E between C and D such that AB $\cong$ CE. (This seems to depend on the choice of one endpoint C of segment CD; show that it does not.) This ordering induces an ordering of segment congruence classes when we define

$$|AB| < |CD| \Leftrightarrow AB < CD.$$

This definition seems to depend on the choice of representatives of the equivalence classes; using Proposition 3.13, show that it is independent of that choice. Furthermore, show that Proposition 3.13 also yields the following information:

Trichotomy: $a < b$ or $a = b$ or $b < a$, and only one of these possibilities occurs.

Transitivity: $a < b$ and $b < c \Rightarrow a < c$.

Here a, b, c are arbitrary segment congruence classes.

We indicated in the discussion after the triangle inequality how to define addition of congruence classes. Show that Axiom C-3 guarantees that addition is well-defined.

Here are some further properties of addition and order of segment congruence classes that you should verify:

Addition is *commutative*: $a + b = b + a$.
Addition is *associative*: $(a + b) + c = a + (b + c)$.
Subtraction when defined: $a < b$ iff there is a class c such that
$b = a + c$.
Cancellation: $a + c = b + c$ iff $a = b$.
Compatibility of $+$ and $<$: If $a < b$, then for any congruence class
c, $a + c < b + c$.
If A, B, C are collinear, then $A * B * C \Leftrightarrow |AB| + |BC| = |AC|$.

We see that with all these nice properties, the congruence class $|AB|$ of AB behaves just like a measure of length for AB, even though it is not a real number. (The idea for this goes back to Descartes.)

Projects

1. Report on the example of a Pythagorean non-Archimedean ordered field in Hartshorne, Exercise 18.9, p. 163; it is the field $K((t))$ of *formal Laurent series* with coefficients in a Pythagorean field K.
2. Examples of Euclidean non-Archimedean fields: In the previous example, assume now that the coefficient field K is Euclidean. Construct an ascending chain of formal Laurent series fields $K((t_n))$ with $t = t_1$ and $t_{n+1}^2 = t_n$ for any positive integer n. Let F be the union of all those fields, so that an element of F is a formal Laurent series in t_n for some n. Show that F is a Euclidean non-Archimedean field (thus by adjoining iterated square roots of t, one obtains square roots of all positive elements—see Hessenberg and Diller (1967) if you are stymied). For another example, see Hartshorne, Proposition 18.4, p. 161.
3. Report on Euclid's theory of content (area without numbers); use Hartshorne, Chapter 5, as a reference. Indicate which results depend on Archimedes' axiom.
4. Go through all the propositions in Euclid's Books I–IV that we have not discussed and that do not refer to area. With the assistance of Heath's commentaries about them, report on all the flaws found in Euclid's proofs of them and repair those flaws using our axioms for a Euclidean plane and all the results we have proved in the text

and exercises. Be sure to tell which results (if any) are valid in neutral geometry and prove them without using any strictly Euclidean results.

5. Report on interesting theorems about *cyclic quadrilaterals* (quadrilaterals that have a circumscribed circle) in Euclidean planes (use the web or relevant books). Such quadrilaterals are important for developing the theory of similar triangles in planes satisfying Hilbert's Euclidean parallel postulate. Develop a more general theory of cyclic quadrilaterals valid in neutral geometry.

6. Comment on these statements by Edward Nelson, referring to his article *Syntax and Semantics*, accessible online at http://www.math. princeton.edu/%7Enelson/papers/s.pdf:

   (a) In the 1960s infinitesimals rose again, phoenix-like, thanks to the genius of Abraham Robinson, the creator of nonstandard analysis. . . . So do infinitesimals exist or not? This is the wrong question. The question is, as Humpty Dumpty said to Alice, which is to be master—that's all. Mathematics is our invention, and we can have infinitesimals or not, as we choose. The only constraint is consistency.

   (b) But what a constraint that is! Indeed, we have no reason to assume that the axiom systems we use in mathematics are consistent. For all we know, they may lead to a contradiction. Platonists believe otherwise, but to a formalist their arguments carry no conviction.

5

History of the Parallel Postulate

Like the goblin "Puck," [the feat of proving Euclid V] has led me "up and down, up and down," through many a wakeful night: but always, just as I thought I had it, some unforeseen fallacy was sure to trip me up, and the tricksy sprite would "leap out, laughing ho, ho, ho!".

C. L. Dodgson (Lewis Carroll)

Review

Let us summarize what we have done so far. We have discovered certain gaps in Euclid's definitions and postulates for plane geometry. We filled in these gaps and firmed up the foundations for this geometry by presenting (a modified version of) Hilbert's definitions and axioms. We then built a structure of theorems on these foundations. However, the structure thus far erected does not rest on Euclid's parallel postulate, and we called this structure "neutral geometry."

You may feel that to deny the Euclidean parallel postulate would go against common sense. Albert Einstein once said that "common sense is, as a matter of fact, nothing more than layers of preconceived notions stored in our memories and emotions for the most part before age eighteen."

For more than two thousand years, some of the best mathematicians tried to prove Euclid's fifth postulate. What does it mean, according to our terminology, to have a proof? It should not be necessary to assume the parallel postulate as an axiom; one should be able to prove it from

209

the other axioms, so that it would become a theorem in neutral geometry and neutral geometry would encompass all of Euclidean geometry.

In this chapter, we will examine a few illuminating attempts to prove Euclid's parallel postulate (many other attempts are presented in Bonola, 1955; Gray, 1989; and Rosenfeld, 1988). It should be emphasized that most of these attempts were made by outstanding mathematicians of their time, not incompetents. And even though each attempt was flawed, the effort was often not wasted; for, assuming that all but one step can be justified, when we detect the flawed step, we find another statement which to our surprise is equivalent[1] to the parallel postulate. You will have the opportunity to do more of this enjoyable detective work in Exercises 4–8.

Proclus

Proclus (410–485) was the head of the Neoplatonic school in Athens more than seven centuries after Euclid. He was primarily a philosopher, not a mathematician, but his *Commentary on the First Book of Euclid's Elements* is one of the main sources of information on Greek geometry.

Proclus criticized Euclid's fifth postulate as follows: "This ought even to be struck out of the Postulates altogether; for it is a statement involving many difficulties. . . . The statement that since [the two lines] converge more and more as they are produced, they will sometime meet is plausible but not necessary." Proclus offered the example of a hyperbola that approaches its asymptotes as closely as you like without ever meeting them (see Figure 5.1). This example shows that the opposite of Euclid's conclusion can at least be imagined.[2] Proclus adds: "It is then clear from this that we must seek a proof of the present theorem, and that it is alien to the special character of postulates."

Proclus attempted to prove the parallel postulate as follows (see Figure 5.2): Given two parallel lines l and m. Suppose line n cuts m at P. We wish to show n intersects l also (see Proposition 4.7). Let Q be the

---

[1] Actually, the flawed argument only proves that the unjustified statement *implies* the parallel postulate; the converse requires further argument. I do not present any attempts that are uninformative.

[2] Students always object to Figure 5.1 on the grounds that the hyperbola is not "straight." We agreed not to use this word because we don't have a precise definition. A precise definition can be given in differential geometry (see Appendix A).

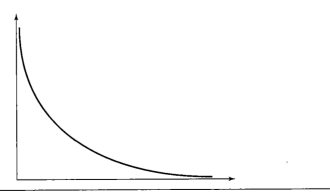

Figure 5.1 Hyperbola with its asymptotes.

foot of the perpendicular from P to *l*. If *n* coincides with $\overleftrightarrow{PQ}$, then it intersects *l* at Q. Otherwise, one ray $\overrightarrow{PY}$ of *n* emanating from P lies between $\overrightarrow{PQ}$ and a ray $\overrightarrow{PX}$ of *m*. Take X to be the foot of the perpendicular from Y to *m*.

Proclus then argued that as the point Y recedes endlessly from P on *n*, segment XY increases without bound, by *Aristotle's axiom*, so eventually XY becomes greater than fixed segment PQ. At that stage, Y must be on the other side of *l*, hence between that position and its starting position, Y must have hit *l*, which means that line *n* intersects *l*.

Proclus' argument is sophisticated, involving motion and betweenness. Moreover, every step in the argument can be shown to be correct (if we assume Aristotle's axiom)—except that the last sentence doesn't follow!

How could one justify the last step? Let us drop a perpendicular YZ from Y to *l*. You might then say that (1) X, Y, and Z are always collinear, and (2) XZ ≅ PQ. Thus, when XY becomes greater than PQ, XY must also be greater than XZ, so that Y must be on the other side of *l*. Here

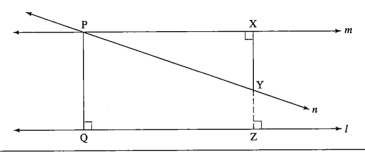

Figure 5.2

the conclusion does indeed follow from statements (1) and (2). The trouble is that there is no justification for these statements!

If this boggles your mind, it may be because Figure 5.2 makes statements (1) and (2) *seem* correct. Recall, however, that we are not allowed to use a diagram to justify a step in a proof. Each step must be proved from stated axioms or previously proved theorems. (We will show later that it is not possible in neutral geometry to prove statements (1) and (2). They can be proved only by using Euclid's parallel postulate or one of its equivalents.)

This analysis of Proclus' faulty argument illustrates how careful you must be in the way you think about parallel lines. You probably visualize parallel lines as railroad tracks, everywhere equidistant from each other, and the ties of the tracks perpendicular to both parallels. This imagery is valid only in Euclidean geometry. Without the parallel postulate, the only thing we can say about two lines that are parallel is that, by the definition of "parallel," they have no point in common. You can't assume they are equidistant; you can't even assume they have a common perpendicular segment. As Humpty Dumpty remarked in *Alice in Wonderland*: "When I use a word it means what I wish it to mean, neither more nor less."

Proclus reported on an earlier attempt to justify Euclid V by the great second-century Greek astronomer Ptolemy. Ptolemy tried to prove the contrapositive (see p. 65) of Euclid V, which is logically equivalent to it. So he started with two parallel lines cut by a transversal. He pointed out correctly that if the interior angles on one side of the transversal add up to <180°, then the interior angles on the other side of the transversal (which are their supplements) add up to >180°. He then argued intuitively that the rays of the parallel lines on one side of the transversal were "no more parallel" than the rays on the opposite side, so this could not happen. If one tries to make that intuitive idea precise, one sees that Ptolemy was tacitly assuming the converse of the AIA theorem, which states that parallel lines are situated symmetrically about any transversal. But we proved in Proposition 4.8 that the converse to the AIA theorem is equivalent to Euclid V in neutral geometry, so Ptolemy was tacitly assuming what he was trying to prove.

According to a medieval Arabic source, Archimedes also wrote a treatise on parallel lines. Unfortunately, it has been lost.[3]

---

[3] In 1906, philologist J. L. Heiberg found an Archimedes manuscript. It was subsequently lost or stolen and then turned up again in 1998. The palimpsest, erased, written over, and even painted over, has been scrutinized using a synchrotron X-ray beam and other technologies to decipher what Archimedes wrote. Do a search on the web for the latest information about this marvelous discovery.

Equidistance

The image of parallel lines as equidistant led to several confused attempts to prove Euclid's parallel postulate. Posidonius (circa 150 B.C.) based his attempt on a different definition of "parallel lines" as two lines for which all the perpendicular segments dropped from either one of them to the other are congruent. Aside from the obvious fallacy of giving the word "parallel" a different meaning, Posidonius could not have proved in neutral geometry that such pairs of lines exist, as we shall later show.

Proceeding more carefully, given a line l and segment PQ of a line perpendicular to l at Q, we can consider the set of all points P' on the same side of l as P such that if Q' is the foot of the perpendicular from P' to l, then PQ $\cong$ P'Q'. Call that set the *equidistant locus (or curve) to l through* P. Christopher Clavius, in 1574, proposed the following axiom as an alternative to Euclid V.

CLAVIUS' AXIOM. For any line l and any point P not on l, the equidistant locus to l through P is the set of all the points on a line through P (which is parallel to l).

The heuristic argument Clavius gave for assuming this axiom was that the equidistant locus has the property that it "lies evenly" with the points on it, hence it must form a line according to Euclid's definition of a line! Centuries earlier, Ibn al-Haytham tried to justify this axiom via a kinematic argument, imagining the rigid segment PQ attached to line l at Q and perpendicular to l; he argued that as Q moved along the (straight) line l, the other end P of the segment had to move along a second (straight) line so long as the segment stayed perpendicular to l. It may be difficult to imagine that the path traced out by P might be curved, but anyhow kinematics is not part of pure geometry.

The following theorem illuminates the status of Clavius' axiom in neutral geometry.

THEOREM. The following three statements are equivalent for a Hilbert plane:
  (a) The plane is semi-Euclidean.
  (b) For any line l and any point P not on l, the equidistant locus to l through P is the set of all the points on the parallel to l through P obtained by the standard construction, i.e., on the

line through P perpendicular to $\overleftrightarrow{PQ}$, where Q is the foot of the perpendicular from P to l.

(c) Clavius' axiom.

PROOF:

(b) $\Rightarrow$ (c) is trivial. Assume (c), let $\square$ABDC be any Saccheri quadrilateral, and let MN be its midline segment. Since M is on the line $\overleftrightarrow{CD}$, Clavius' axiom tells us that M is on the equidistant locus to $\overleftrightarrow{AB}$ through C and D; i.e., $MN \cong CA \cong DB$. So by the corollary to the uniformity theorem of Chapter 4, the plane is semi-Euclidean.

Assume (a). Let m be the parallel to l through P obtained by the standard construction. If P′ is any other point on m and Q′ is the foot of the perpendicular from P′ to l, then $\square$QQ′P′P is a Lambert quadrilateral, hence a rectangle, by (a), so the opposite sides PQ and P′Q′ of this rectangle are congruent (Corollary 3 to Proposition 4.13). Thus P′ lies on the equidistant locus to l through P. Now let P′ $\neq$ P lie on that locus. Then $\square$QQ′P′P is a Saccheri quadrilateral, hence a rectangle, by (a). Thus $\overleftrightarrow{P'P}$ is perpendicular to $\overleftrightarrow{PQ}$ at P. By uniqueness of the perpendicular, P′ lies on m. ◄

As was stated in the note on non-Archimedean geometries at the end of Chapter 4, the Euclidean parallel postulate need not hold in an arbitrary semi-Euclidean non-Archimedean plane, so Clavius' axiom is weaker than the Euclidean parallel postulate and all attempts to prove the parallel postulate using just Clavius' axiom are flawed. Some medieval Arab mathematicians invoked Archimedes' axiom in addition to Clavius' axiom in their flawed attempts (see Chapter 2 of Rosenfeld, 1988), which was the correct idea, as we shall soon show.

Wallis

John Wallis (1616–1703) was the most influential English mathematician before Newton.[4] He made very substantial contributions to the development of calculus, algebra, and analytic geometry.

---

[4] In his 1656 treatise *Arithmetica Infinitorum* (which Newton studied), Wallis introduced the symbol ∞ for "infinity," developed formulas for certain integrals, and presented his famous infinite product formula for π:

$$\frac{\pi}{2} = \frac{2 \cdot 2 \cdot 4 \cdot 4 \cdot 6 \cdot 6 \cdot 8 \cdots}{1 \cdot 3 \cdot 3 \cdot 5 \cdot 5 \cdot 7 \cdot 7 \cdots}$$

Wallis promoted the power of algebra in mathematics, in sharp disagreement with the insistence of Newton's teacher Isaac Barrow on traditional synthetic Euclidean methods.

John Wallis

Wallis was astute enough not to try to prove Euclid's parallel postulate in neutral geometry. Instead, in a treatise on Euclid, which he published in 1693, he proposed a new postulate that he believed to be more plausible. He phrased it as follows:

> Finally (supposing the nature of ratio and of the science of similar figures already known), I take the following as a common notion: to every figure there exists a similar figure of arbitrary magnitude.

In order to make Wallis' postulate precise, we will restrict our attention to triangles instead of arbitrary "figures." We have not developed "the nature of ratio." We can circumvent that difficulty by defining "similar triangles" as follows.

DEFINITION. Two triangles are *similar* if their vertices can be put in one-to-one correspondence in such a way that corresponding angles are congruent (AAA). We use the notation $\triangle ABC \sim \triangle DEF$ to indicate that these triangles are similar when A, B, C correspond, respectively, to D, E, F (see Figure 5.3).

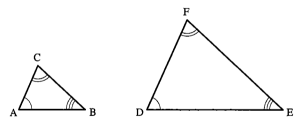

Figure 5.3 Similar triangles.

WALLIS' POSTULATE. Given any triangle △ABC and given any segment DE, there exists a triangle △DEF having DE as one of its sides such that △ABC ~ △DEF.

The intuitive meaning of Wallis' postulate is that you can either magnify or shrink a triangle as much as you like without distortion. Using Wallis' postulate, the Euclidean parallel postulate can be proved as follows.

PROOF:

Given point P not on line *l*, let PQ*lm* be the configuration obtained by the standard parallel construction. Let *n* be any other line through P. We must show that *n* meets *l*. As before, consider a ray of *n* emanating from P that is between a ray of *m* emanating from P and the ray $\overrightarrow{PQ}$. For any point R on this ray, let S be the foot of the perpendicular from R to $\overleftrightarrow{PQ}$ (see Figure 5.4).

Now apply Wallis' postulate to △PSR and segment PQ. It tells us that there is a point T such that △PSR ~ △PQT. We may assume T lies on the same side of $\overleftrightarrow{PQ}$ as R (Figure 5.5)—if not, reflect across $\overleftrightarrow{PQ}$. By the definition of similar triangles, ∢TPQ ≅ ∢RPS. But since

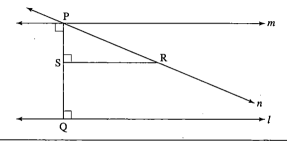

Figure 5.4

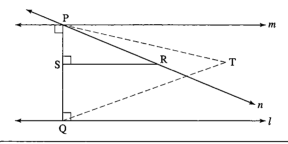

Figure 5.5

these angles have the ray $\overrightarrow{PQ} = \overrightarrow{PS}$ as a common side, and since T lies on the same side of $\overleftrightarrow{PQ}$ as R, the only way they can be congruent is to be equal (by the uniqueness part of Axiom C-4). Thus $\overrightarrow{PR} = \overrightarrow{PT}$, so that T lies on n. Similarly, $\sphericalangle PQT \cong \sphericalangle PSR$, a right angle; hence T lies on l as well. Thus n and l meet at T, and m is the only line through P parallel to l. ◄

There is no reason to believe Wallis' postulate is preferable to Euclid's because you will easily show in Exercise 3(a) that it is equivalent in neutral geometry to Euclid V.

Wallis became publicly engaged in a dispute with the prominent seventeenth-century philosopher John Hobbes after Hobbes published a manuscript in 1655 purporting to square the circle by straightedge and compass and to solve other outstanding geometric problems. Wallis published a reply in the same year in which he pointed out the many errors and rather deplorable state of Hobbes' geometry. Proud Hobbes could not accept Wallis' critique and published an angry attack against him. A bitter, vituperative public verbal battle—encompassing much broader philosophical, political, and religious issues of that time, not just geometry—evolved between them that lasted 20 years. An excellent account of this dispute can be found in the 1999 book by Douglas M. Jesseph, *Squaring the Circle: The War Between Hobbes and Wallis.* There is an informative review of this book at www.maa.org/reviews/squaring.html.

Squaring the circle was a puzzle of widespread popularity among the general population in the late seventeenth century. There were contests open to all, and the March 1686 edition of the *Journal des Savants* even reported that "one young lady positively refused a perfectly eligible suitor simply because he had been unable, within a given time, to produce any new idea about squaring the circle."

You will find a new idea in the last section of Chapter 10.

Saccheri

We next discuss further the remarkable work of the logician and Jesuit priest Girolamo Saccheri (1667–1733), many of whose propositions were proved in Chapter 4.

We saw that the summit angles of his quadrilaterals are congruent to each other, and there are three possible geometries according as those angles are acute, right, or obtuse. Saccheri's idea was to demonstrate that the acute and obtuse angle cases lead to contradictions, leaving the right angle case—the case where the Saccheri quadrilateral is a rectangle—as the only possibility.

By assuming the generally accepted Archimedes' axiom, Saccheri successfully eliminated the case of the obtuse angle (Saccheri–Legendre theorem, Chapter 4). But however hard he tried, Saccheri could not squeeze a contradiction out of "the inimical acute angle hypothesis," as he called it. He was able to deduce many strange results—such as parallel lines having one common perpendicular and then diverging on both sides of the perpendicular, or the possibility of parallel lines diverging in one direction but converging asymptotically in the opposite direction and having a "common perpendicular at infinity" in that direction. He was not able to find a contradiction.

Finally, he exclaimed in frustration: "The hypothesis of the acute angle is absolutely false, because [it is] repugnant to the nature of the straight line!" It is as if a man had discovered a rare diamond but, unable to believe what he saw, announced it was glass. Although he did not recognize it (or was afraid to acknowledge it), Saccheri had discovered the elementary part of non-Euclidean geometry and deserves much acclaim for that discovery.

There is no very serious error in Saccheri's treatise. Moreover, the following remarks by him show that he was aware that his work was not satisfying.

> It is well to consider here a notable difference between the foregoing refutations of the two hypotheses. For in regard to the hypothesis of the obtuse angle the thing is clearer than midday light. . . . But on the contrary, I do not attain to proving the falsity of the other hypothesis, that of the acute angle. . . . I do not appear to demonstrate from the viscera of the very hypothesis, as must be done for a perfect refutation."[5]

---

[5] See the translation of Saccheri's 1733 treatise by G. B. Halsted (Saccheri, 1970, Scholion, p. 233). Saccheri had previously published several versions of his treatise on logic, which Halsted, in his introduction, also lauds as far ahead of his time; for ex-

We will further examine Saccheri's non-Euclidean results in the next chapter. It has been claimed by one anonymous writer that in Saccheri's time, the existence of a valid non-Euclidean geometry was "quite literally, unthinkable—not impossible, not wrong, but *unthinkable*." Well, Saccheri did think about it. Why would a fine logician like Saccheri bother publishing all those correct results in non-Euclidean geometry if he simply believed that such a geometry was "repugnant"? He must have at least sensed that there was something very interesting going on that he couldn't fully understand, and he wanted mathematicians to know about it. By claiming he had vindicated Euclid, his book received the stamp of approval from the Inquisition. Unfortunately, Saccheri died a month after its publication.

Clairaut's Axiom and Proclus' Theorem

Alexis-Claude Clairaut (1713–1765) was a leading French mathematician who made important contributions to differential geometry. Like Wallis, he did not try to prove the parallel postulate in neutral geometry but replaced it in his 1741 text *Éléments de Géométrie* with another axiom.

CLAIRAUT'S AXIOM. Rectangles exist.

In his text, Clairaut tried to make geometry easier for students to understand, presenting it as a practical, common-sense subject. To justify his axiom, Clairaut argued that "we observe rectangles all around us in houses, gardens, rooms, walls."

So why didn't that settle the matter? Perhaps because the game of trying to prove Euclid V had been going on for so many centuries that it became a challenging obsession for mathematicians. Or did mathematicians finally recognize that geometry was not about "physical space"? After all, if you believe that a rectangle can be drawn on the ground, then you cannot also believe that the earth is spherical, because rectangles do not exist on a sphere. If you think you have drawn a "physical rectangle," you could be mistaken because exact measurements are physically impossible. Or did it finally dawn on mathematicians that any postulate proposed to replace Euclid V—no matter how

---

ample, Saccheri was the first to consider the problems of the independence of one postulate from the others and of the consistency of a system of axioms. For an explanation of the "common perpendicular at infinity" to asymptotically parallel lines discovered by Saccheri, see the Conclusion in Chapter 6.

intuitively appealing—was weaker than or logically equivalent to Euclid V and therefore nothing was gained *logically* by the replacement?

Even if one accepts Clairaut's axiom, it does not suffice to demonstrate Euclid's parallel postulate. Our investigations in Chapter 4 show that *Clairaut's axiom holds in a Hilbert plane iff the plane is semi-Euclidean.* As Example 2 in Chapter 4 showed, Euclid V need not hold in a non-Archimedean, semi-Euclidean plane.

Hilbert, in his lectures on geometry after the publication of the first edition of his *Grundlagen,* emphasized that the angle sum of a triangle equaling 180° does not imply Euclid V without a further hypothesis. Dehn provided a non-Archimedean model to show that. Proclus was the first to recognize a correct, purely geometric candidate for that additional hypothesis: *Aristotle's axiom is a missing link.*

Proclus' Theorem. The Euclidean parallel postulate holds in a Hilbert plane if and only if the plane is semi-Euclidean (i.e., the angle sum of a triangle is 180°) and Aristotle's angle unboundedness axiom holds. In particular, the Euclidean parallel postulate holds in an Archimedean semi-Euclidean plane.

Proof:

The last remark follows from the result that Archimedes' axiom implies Aristotle's axiom, which you will prove in Exercise 2.

The "only if" part of the theorem was proved in Chapter 4 (Proposition 4.11 and Corollary 2 to Proposition 4.13). For the "if" part, return to the situation illustrated in Figure 5.6, where m is the parallel to l through P obtained by the standard construction. Let S be the foot of the perpendicular from Y to $\overleftrightarrow{PQ}$. S is on the same side of m as Y and Q because $\overleftrightarrow{SY}$ is parallel to m (Corollary 1 to the AIA theorem). Since the plane is semi-Euclidean, Lambert quadrilateral

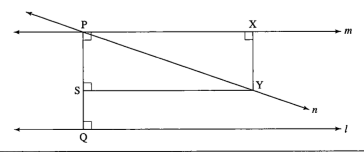

Figure 5.6

□XPSY is a rectangle, hence its opposite sides PS and XY are congruent (Corollary 3 to Proposition 4.13).

We now apply Aristotle's axiom and Proclus' argument: A point Y exists on the given ray of n so that XY > PQ. Then PS, which is congruent to XY, is also > PQ, hence P * Q * S. As before, Y is on the same side of l as S, hence on the opposite side of l from P. By the definition of "opposite side," n meets l at some point between P and Y. ◄

Legendre

Adrien-Marie Legendre (1752–1833), mentioned in Chapter 1, certainly knew of Clairaut's text and rejected Clairaut's axiom because he believed he could prove Euclid V in neutral geometry. He did not know of Saccheri's work and rediscovered (with different proofs) some of Saccheri's main theorems in neutral geometry—the most important one being the Saccheri–Legendre theorem in Chapter 4. Legendre also took Archimedes' axiom for granted. We have already discussed, in Chapter 1, one of Legendre's attempts to prove the parallel postulate, whose flaw we ask you to detect in Exercise 4. Legendre published a collection of his many attempts as late as 1833, the year he died. Here is one of his attempts to prove that the angle sum of any triangle is 180°.

PROOF (SEE FIGURE 5.7):

Suppose, on the contrary, there exists a triangle △ABC having defect $d \neq 0$. By the Saccheri–Legendre theorem, $d > 0$. One of the angles of the triangle—say ∢A—must then be acute (in fact, less than 60°). On the opposite side of $\overleftrightarrow{BC}$ from A, let D be the unique point such that ∢DBC ≅ ∢ACB and BD ≅ AC (Axioms C-1 and C-4). Then △ACB ≅ △DBC (SAS). Also $\overleftrightarrow{BD} \parallel \overleftrightarrow{AC}$ and $\overleftrightarrow{BA} \parallel \overleftrightarrow{DC}$ (by the alternate interior angle theorem, Theorem 4.1), so that D lies in the interior

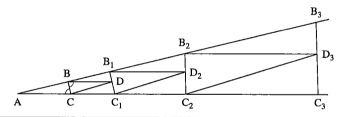

Figure 5.7

of the acute $\sphericalangle A$. Hence there is a line l through D such that l intersects side $\overrightarrow{AB}$ in a point $B_1 \neq A$ and side $\overrightarrow{AC}$ in a point $C_1 \neq A$. Because of the parallel lines, we know that $B_1 \neq B$ and $C_1 \neq C$.

Suppose B_1 is on segment AB. Then A and B_1 would be on the same side of $\overleftrightarrow{BD}$. Since $\overleftrightarrow{BD} \parallel \overleftrightarrow{AC}$, A and C_1 are on the same side of $\overleftrightarrow{BD}$. Thus B_1 and C_1 are on the same side of $\overleftrightarrow{BD}$ (Axiom B-4). But since D lies in the interior of $\sphericalangle A$, $B_1 * D * C_1$ (Proposition 3.7). This contradiction shows that $A * B * B_1$. Similarly, we have $A * C * C_1$. Since $\triangle ACB \cong \triangle DBC$, the defect of $\triangle DBC$ is also d. Therefore, by the additivity of the defect applied to the four triangles into which $\triangle AB_1C_1$ has been decomposed, the defect of $\triangle AB_1C_1$ is greater than or equal to $2d$.

Repeating this construction for $\triangle AB_1C_1$, we obtain $\triangle AB_2C_2$ with defect greater than or equal to $4d$. Iterating the construction n times, we obtain a triangle with defect greater than or equal to $2^n d$, which can be made as large as we like by taking n sufficiently large. But the defect of a triangle cannot be more than $180°$! This contradiction shows that every triangle $\triangle ABC$ has defect 0. ◄

Can you see the flaw? It is easy, because we have justified every step but one, the sentence beginning with "Hence." That is the assumption you were warned on p. 115 not to make. Legendre made the same error as was made many centuries earlier by Simplicius (Byzantine, sixth century), al-Jawhari (Persian, ninth century), Nasir Eddin al-Tusi, and others. He failed to prove in neutral geometry that the defect of every triangle is zero. Nevertheless, Legendre succeeded in proving the following theorem in neutral geometry.

LEGENDRE'S THEOREM (STILL ASSUMING ARCHIMEDES' AXIOM). Hypothesis: For any acute $\sphericalangle A$ and any point D in the interior of $\sphericalangle A$, there exists a line through D and not through A that intersects both sides of $\sphericalangle A$. Conclusion: The angle sum of every triangle is $180°$.

You will easily see from the Klein model in Chapter 7 that the hypothesis of Legendre's Theorem fails in hyperbolic geometry (Figure 7.5). Let us show that *the hypothesis can be proved in Euclidean geometry*. Drop a perpendicular from interior point D to one of the sides of $\sphericalangle A$ and let B be the foot of that perpendicular. Since $\sphericalangle A$ is acute, $(\sphericalangle A)° + (\sphericalangle DBA)° = (\sphericalangle A)° + 90° < 180°$. So $\overrightarrow{BD}$ meets the other side of $\sphericalangle A$, by Euclid V. ◄

For future reference, we name this hypothesis after Legendre.

LEGENDRE'S AXIOM. For any acute angle and any point in the interior of that angle, there exists a line through that point and not through the angle vertex which intersects both sides of the angle.

Just like Saccheri, Legendre wrote that "it is repugnant to the nature of a straight line" for this axiom not to hold.

Lambert and Taurinus

Regarding Euclid V, Johann Heinrich Lambert (1728–1777) wrote:

> Undoubtedly, this basic assertion is far less clear and obvious than the others. Not only does it naturally give the impression that it should be proved, but to some extent it makes the reader feel that he is capable of giving a proof, or that he should give it. However, to the extent to which I understand this matter, that is just a *first* impression. He who reads Euclid further is bound to be amazed not only at the rigor of his proofs but also at the delightful simplicity of his exposition. This being so, he will marvel all the more at the position of the fifth postulate when he finds out that Euclid proved propositions that could far more easily be left unproved.

Lambert studied quadrilaterals having at least three right angles, which are now named after him (though they were studied seven centuries earlier by the Egyptian scientist ibn-al-Haytham). A Lambert quadrilateral can be "doubled" (by reflecting it across an included side of two right angles) to obtain a Saccheri quadrilateral. Lambert was familiar with Saccheri's work. Like Saccheri, Lambert disproved the obtuse angle hypothesis and studied the implications of the "inimical" acute angle hypothesis. He observed that it implied that similar triangles must then be congruent, which in turn implied the existence of an absolute unit of length (see Proposition 6.2, Chapter 6). He called this consequence "exquisite" but did not want it to be true, worrying that the absence of similar, proportional figures "would result in countless inconveniences," especially for astronomers.

He also noticed that the defect of a triangle was proportional to its area (see Chapter 10). He recalled that on a sphere in Euclidean space,

Johann Heinrich Lambert

the angle sum of a triangle formed by arcs of great circles was greater than 180° and that the excess over 180° of the angle sum of the triangle was proportional to the area of the triangle, the constant of proportionality being the square r^2 of the radius of the sphere (see Rosenfeld, 1988, Chapter 1, or Appendix A for the case $r = 1$). If r is replaced by ir $(i = \sqrt{-1})$, squaring introduces a minus sign that converts the excess into the defect in that proportionality. Lambert therefore speculated that the acute angle hypothesis described geometry on a "sphere of imaginary radius."[6]

Fifty years passed before this brilliant idea was further elaborated in a booklet dated 1826 by F. A. Taurinus, who transformed the formulas of spherical trigonometry into formulas for what he called "log-spherical geometry" by substituting ir for r (his formulas are proved by a different method in Theorem 10.4, Chapter 10). When Taurinus first

---

[6] In fact, this idea can be explained in terms of a natural embedding of the non-Euclidean plane in *relativistic* three-space (see Chapter 7). Lambert is known for proving the irrationality of π and of e^x and tan x when x is rational, as well as for important laws he discovered in optics and astronomy. The quote is from B. A. Rosenfeld (1988), p. 100.

notified C. F. Gauss of his work, Gauss replied favorably (see the letter from Gauss on p. 243); but when Taurinus then urged Gauss to publish his own work on this topic, Gauss refused to continue communicating. This rejection threw Taurinus into a state of despair, and he burned the remaining copies of his booklets. Taurinus vacillated over whether such a geometry actually "existed."

Lambert cautiously did not submit his *Theory of Parallels* for publication (it was published posthumously in 1786). It contained an erroneous attempt to disprove the acute angle hypothesis.

Farkas Bolyai

There were so many attempts to prove Euclid V that by 1763 G. S. Klügel was able to submit a doctoral thesis finding the flaws in 28 different supposed proofs of the parallel postulate, expressing doubt that it could be proved. The French encyclopedist and mathematician J. L. R. d'Alembert called this "the scandal of geometry." Mathematicians were becoming discouraged. The Hungarian Farkas Bolyai, who had also tried to prove Euclid V (see Exercise 5), wrote to his son János:

You must not attempt this approach to parallels. I know this way to its very end. I have traversed this bottomless night, which extinguished all light and joy of my life. I entreat you, leave the science of parallels alone. . . . I thought I would sacrifice myself for the sake of the truth. I was ready to become a martyr who would remove the flaw from geometry and return it purified to mankind. I accomplished monstrous, enormous labors; my creations are far better than those of others and yet I have not achieved complete satisfaction. . . . I turned back when I saw that no man can reach the bottom of the night. I turned back unconsoled, pitying myself and all mankind.

I admit that I expect little from the deviation of your lines. It seems to me that I have been in these regions; that I have traveled past all reefs of this infernal Dead Sea and have always come back with broken mast and torn sail. The ruin of my disposition and my fall date back to this time. I thoughtlessly risked my life and happiness—*aut Caesar aut nihil.*[7]

---

[7] The correspondence between Farkas and János Bolyai is from Meschkowski (1964). Farkas Bolyai is credited, along with W. Wallace and P. Gerwien, for having proved the important theorem that polygons of equal area are equidecomposable (see Chapter 10).

Farkas Bolyai

But the young Bolyai was not deterred by his father's warnings, for he had a completely new idea. He assumed that the negation of Euclid's parallel postulate was not absurd, and in 1823 he was able to write to his father:

It is now my definite plan to publish a work on parallels as soon as I can complete and arrange the material and an opportunity presents itself; at the moment I still do not clearly see my way through, but the path which I have followed gives positive evidence that the goal will be reached, if it is at all possible; I have not quite reached it, but I have discovered such wondeful things that I was amazed, and it would be an everlasting piece of bad fortune if they were lost. When you, my dear Father, see them, you will understand; at present I can say nothing except this: that *out of nothing I have created a strange new universe.* All that I have sent you previously is like a house of cards in comparison with a tower. I am no less convinced that these discoveries will bring me honor than I would be if they were completed.

We will explore this "strange new universe" in the following chapters. A century after János Bolyai wrote this letter, the English physicist J. J. Thomson remarked, somewhat facetiously:

> We have Einstein's space, de Sitter's space, expanding universes, contracting universes, vibrating universes, mysterious universes. In fact, the pure mathematician may create universes just by writing down an equation, and indeed if he is an individualist he can have a universe of his own.

In fact, in 1949 the renowned logician Kurt Gödel found a model of the universe that satisfies Einstein's gravitational equations, one in which it is theoretically possible to travel backward in time![8]

Review Exercise

Which of the following statements are correct?

(1) Wallis' postulate implies that there exist two triangles that are similar but not congruent.

(2) A "Saccheri quadrilateral" is a quadrilateral $\square$ABDC such that $\sphericalangle$CAB and $\sphericalangle$DBA are right angles and AC $\cong$ BD.

(3) A "Lambert quadrilateral" is a quadrilateral having at least three right angles.

(4) A quadrilateral that is both a Saccheri and a Lambert quadrilateral must be a rectangle.

(5) A hyperbola comes arbitrarily close to its asymptotes without ever intersecting them.

(6) János Bolyai warned his son Farkas not to work on the parallel problem.

(7) Saccheri succeeded in disproving the "inimical" acute angle hypothesis.

(8) In trying to prove Euclid V, Ptolemy was tacitly assuming the converse to the AIA theorem.

(9) It is a theorem in neutral geometry that if $l \parallel m$ and $m \parallel n$, then $l \parallel n$.

(10) It is a theorem in neutral geometry that every segment has a unique midpoint.

(11) It is a theorem in neutral geometry that if a rectangle exists, then the angle sum of any triangle is 180°.

---

[8] To date, attempts to refute Gödel's model on either mathematical or philosophical grounds have failed. See "On the paradoxical time-structures of Gödel," by Howard Stein, *Journal of the Philosophy of Science*, v. 37, December 1970, p. 589.

(12) It is a theorem in neutral geometry that if l and m are parallel lines, then alternate interior angles cut out by any transversal to l and m are congruent to each other.

(13) Legendre proved in neutral geometry that for any acute $\sphericalangle$A and any point D in the interior of $\sphericalangle$A, there exists a line through D and not through A which intersects both sides of $\sphericalangle$A.

(14) Clairaut showed that Euclid's fifth postulate could be replaced in the logical presentation of Euclidean geometry by the "more obvious" postulate that rectangles exist, yet mathematicians were not appeased by Clairaut's replacement and they continued to try to prove Euclid V.

(15) Lambert guessed that there was such a thing as a "sphere of imaginary radius" on which the acute angle hypothesis was valid.

(16) Gauss responded to Taurinus about his booklet on "log-spherical geometry," telling about his own unpublished work, but when Taurinus urged Gauss to publish it, Gauss did not reply.

(17) Saccheri used the undefined notion of "repugnance" in his attempt to prove Euclid V by an RAA argument.

(18) That Legendre made so many incorrect attempts to prove Euclid V for Archimedean Hilbert planes shows that his work in geometry was worthless.

Exercises

1. Given a right triangle $\triangle$PXY with right angle at X, form a new right triangle $\triangle$PX'Y' that has acute angle $\sphericalangle$P in common with the given triangle, right angle at X', but double the hypotenuse (prove that this can be done); see Figure 5.8. If the plane does not satisfy the obtuse angle hypothesis, prove that the side opposite the acute angle is *at least* doubled, whereas the side adjacent to the acute angle is *at most* doubled. (Hint: Extend side XY far enough to drop a perpendicular Y'Z to $\overleftrightarrow{XY}$. Prove that $\triangle$PXY $\cong$ $\triangle$Y'ZY and apply Corollary 3 to Proposition 4.13, Chapter 4.)

2. Use Exercise 1 and the Saccheri–Legendre theorem to prove that *Archimedes' axiom implies Aristotle's axiom*—i.e., in Figure 5.8, prove that as Y "recedes endlessly" from P, perpendicular segment XY increases without bound. (Hint: Use Archimedes' axiom and the fact that $2^n \to \infty$ as $n \to \infty$.) Does segment PX also increase indefinitely?

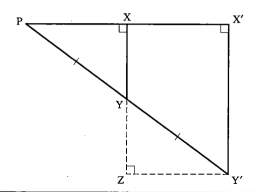

Figure 5.8

3. (a) Prove that Euclid's fifth postulate implies Wallis' postulate (see Figure 5.9). (Hint: Use Axiom C-4 and the fact that in Euclidean geometry the angle sum of a triangle is 180°—Proposition 4.11.)

   (b) Suppose that in the statement of Wallis' postulate we add the assumption AB ≅ DE and replace the word "similar" by "congruent." Prove this new statement in neutral geometry.

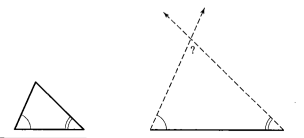

Figure 5.9

4. Reread Legendre's attempted proof of the parallel postulate in Chapter 1. Find the flaw and justify all the steps that are correct. Prove the flawed statement in Euclidean geometry.

5. Find the unjustified assumption in the following "proof" of the parallel postulate by Farkas Bolyai (see Figure 5.10). Given P not on line l, $\overleftrightarrow{PQ}$ perpendicular to l at Q, and line m perpendicular to $\overleftrightarrow{PQ}$ at P. Let n be any line through P distinct from m and $\overleftrightarrow{PQ}$. We must show that n meets l. Let A be any point between P and Q. Let B be the unique point such that A ∗ Q ∗ B and AQ ≅ QB. Let R be the

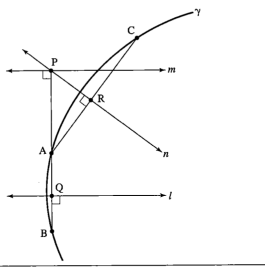

Figure 5.10 Attempted proof by Farkas Bolyai.

foot of the perpendicular from A to n. Let C be the unique point such that $A * R * C$ and $AR \cong RC$. Then A, B, and C are not collinear (else $R = P$); hence there is a unique circle γ passing through them. Since l is the perpendicular bisector of chord AB of γ and n is the perpendicular bisector of chord AC of γ, l and n meet at the center of γ (Exercise 17(b), Chapter 4).

6. The following attempted proof of the parallel postulate is similar to Proclus' but the flaw is different; detect the flaw with the help of Exercise 1 (see Figure 5.11). Given P not on line l, $\overleftrightarrow{PQ}$ perpendicular to l at Q, and line m perpendicular to $\overleftrightarrow{PQ}$ at P. Let n be any line through P distinct from m and $\overleftrightarrow{PQ}$. We must show that n meets l. Let $\overrightarrow{PX}$ be a ray of n between $\overrightarrow{PQ}$ and a ray of m emanating from P

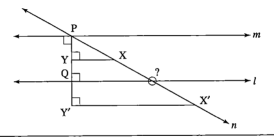

Figure 5.11

and let Y be the foot of the perpendicular from X to $\overleftrightarrow{PQ}$. As X recedes endlessly from P, PY increases indefinitely. Hence, Y even-
tually reaches a position Y′ on $\overrightarrow{PQ}$ such that PY′ > PQ. Let X′ be
the corresponding position reached by X on line n. Now X′ and Y′
are on the same side of l because $\overleftrightarrow{X'Y'}$ is parallel to l. But Y′ and
P are on opposite sides of l. Hence, X′ and P are on opposite sides
of l, so that segment PX′ (which is part of n) meets l.

7. Find the flaw in the following attempted proof of the parallel postulate given by J. D. Gergonne (see Figure 5.12). Given P not on
line l, $\overrightarrow{PQ}$ perpendicular to l at Q, line m perpendicular to $\overleftrightarrow{PQ}$ at P,
and point A ≠ P on m. Let $\overrightarrow{PB}$ be the last ray between $\overrightarrow{PA}$ and $\overrightarrow{PQ}$
that intersects l, B being the point of intersection. There exists a
point C on l such that $Q * B * C$ (Axioms B-1 and B-2). It follows
that $\overrightarrow{PB}$ is not the last ray between $\overrightarrow{PA}$ and $\overrightarrow{PQ}$ that intersects l, and
hence all rays between $\overrightarrow{PA}$ and $\overrightarrow{PQ}$ meet l. Thus m is the only parallel to l through P.

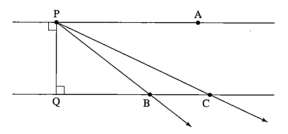

Figure 5.12

8. It was stated at the beginning of this chapter that if all steps but
one of an attempt to prove the parallel postulate are correct, then
the flawed step yields another statement equivalent to Hilbert's parallel postulate. Assuming Aristotle's axiom, show that for Proclus'
attempt, that statement is: Given parallel lines l, m having a common perpendicular and a point Y not lying on l or m, if X (respectively Z) is the foot of the perpendicular from Y to l (respectively
to m), then X, Y, and Z are collinear.

9. You will show in Exercise 16 that the following statement can be
proved in Euclidean geometry: If points P, Q, R lie on a circle with
center O, and if ∢PQR is acute, then $(∢PQR)° = \frac{1}{2}(∢POR)°$. In
neutral geometry, show that this statement implies the existence of
a triangle whose angle sum is 180°.

The remaining exercises in this chapter are exercises in real Euclidean geometry, which means you are allowed to use the parallel postulate and its consequences already established. We will refer to these results in Chapter 7. You are also allowed to use the following result, a proof of which is indicated in the Major Exercises.

PARALLEL PROJECTION THEOREM. Given three parallel lines l, m, and n. Let t and t' be transversals to these parallels, cutting them in points A, B, and C and in points A', B', and C', respectively. Then $\overline{AB}/\overline{BC} = \overline{A'B'}/\overline{B'C'}$ (Figure 5.13).

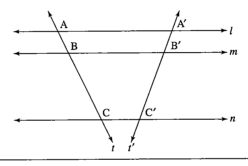

Figure 5.13

10. *Fundamental theorem on similar triangles.* Given $\triangle ABC \sim \triangle A'B'C'$; i.e., given $\angle A \cong \angle A'$, $\angle B \cong \angle B'$, and $\angle C \cong \angle C'$. Then corresponding sides are proportional; i.e., $\overline{AB}/\overline{A'B'} = \overline{AC}/\overline{A'C'} = \overline{BC}/\overline{B'C'}$ (see Figure 5.14). Prove the theorem. (Hint: Let B″ be the point on $\overrightarrow{AB}$ such that $AB'' \cong A'B'$ and let C″ be the point on $\overrightarrow{AC}$ such that $AC'' \cong A'C'$. Use the hypothesis to show that $\triangle AB''C'' \cong \triangle A'B'C'$ and de-

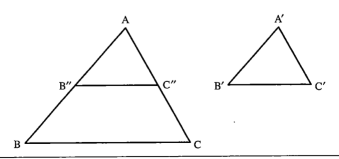

Figure 5.14

duce from corresponding angles that $\overleftrightarrow{B''C''}$ is parallel to $\overleftrightarrow{BC}$. Now apply the parallel projection theorem.)

11. Prove the converse to the fundamental theorem on similar triangles. (Hint: Choose B″ as before. Use Pasch's theorem to show that the parallel to $\overleftrightarrow{BC}$ through B″ cuts AC at a point C″. Then use the hypothesis, Exercise 10, and the SSS criterion to show that we have $\triangle ABC \sim \triangle AB''C'' \cong \triangle A'B'C'$.)

12. *SAS similarity criterion.* If $\angle A \cong \angle A'$ and $\overline{AB}/\overline{A'B'} = \overline{AC}/\overline{A'C'}$, prove that $\triangle ABC \sim \triangle A'B'C'$. (Hint: Same method as in Exercise 11, but using SAS instead of SSS.)

13. Prove the Pythagorean theorem. (Hint: Let CD be the altitude to the hypotenuse; see Figure 5.15. Use the fact that the angle sum of a triangle equals 180° (Proposition 4.11) to show that we have $\triangle ACD \sim \triangle ABC \sim \triangle CBD$. Apply Exercise 10 and a little algebra based on $\overline{AB} = \overline{AD} + \overline{DB}$ to get the result.)

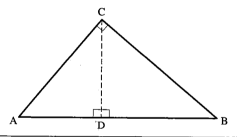

Figure 5.15

14. The fundamental theorem on similar triangles (Exercise 10) allows the trigonometric functions such as sine and cosine to be defined. Namely, given an acute angle $\angle A$, make it part of a right triangle $\triangle BAC$ with right angle at C and set

$$\sin \angle A = (\overline{BC})/(\overline{AB})$$
$$\cos \angle A = (\overline{AC})/(\overline{AB}).$$

These definitions are then independent of the choice of the right triangle used. If $\angle A$ is obtuse and $\angle A'$ is its supplement, set

$$\sin \angle A = +\sin \angle A'$$
$$\cos \angle A = -\cos \angle A'.$$

If $\angle A$ is a right angle, set

$$\sin \angle A = 1$$
$$\cos \angle A = 0.$$

Now, given any triangle $\triangle ABC$, if a and b are the lengths of the sides opposite A and B, respectively, prove the law of sines,

$$\frac{a}{b} = \frac{\sin \sphericalangle A}{\sin \sphericalangle B}.$$

(Hint: Drop altitude CD and use the two right triangles $\triangle ADC$ and $\triangle BDC$ to show that $b \sin \sphericalangle A = \overline{CD} = a \sin \sphericalangle B$; see Figure 5.16.) Similarly, prove the law of cosines,

$$c^2 = a^2 + b^2 - 2ab \cos \sphericalangle C,$$

and deduce the converse to the Pythagorean theorem.

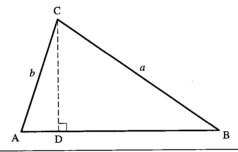

Figure 5.16

15. Given $A * B * C$ and point D not collinear with A, B, and C (Figure 5.17). Prove that

$$\frac{\overline{AB}}{\overline{BC}} = \frac{\overline{AD} \sin \sphericalangle ADB}{\overline{CD} \sin \sphericalangle CDB}$$

$$\frac{\overline{AC}}{\overline{BC}} = \frac{\overline{AD} \sin \sphericalangle ADC}{\overline{BD} \sin \sphericalangle BDC}.$$

(Hint: Use the law of sines to compute $\overline{AB}/\overline{AD}$, $\overline{CD}/\overline{BC}$, and $\overline{BD}/\overline{BC}$ and remember that $\sin \sphericalangle ABD = \sin \sphericalangle CBD$.)

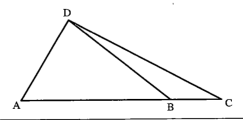

Figure 5.17

16. Let γ be a circle with center O and let P, Q, and R be three points on γ. Prove that if P and R are diametrically opposite, then $\angle$PQR is a right angle, and if O and Q are on the same side of $\overleftrightarrow{PR}$, then $(\angle PQR)° = \frac{1}{2}(\angle POR)°$. (Hint: Again use the fact that the triangular angle sum is 180°. There are four cases to consider, as in Figure 5.18.) State and prove the analogous result when O and Q are on opposite sides of $\overleftrightarrow{PR}$.

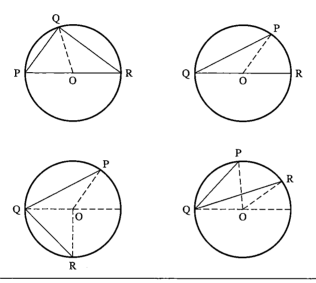

Figure 5.18

17. Prove that if two angles inscribed in a circle subtend the same arc, then they are congruent; see Figure 5.19. (Hint: Apply the previous exercise after carefully defining "subtend the same arc.")

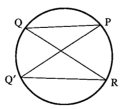

Figure 5.19 $\angle$PQR $\cong$ $\angle$PQ'R.

18. Prove that if $\angle$PQR is a right angle, then Q lies on the circle γ having PR as diameter. (Hint: Use uniqueness of perpendiculars and Exercise 16.)

Major Exercises

These exercises furnish the proof of the parallel projection theorem in Euclidean geometry (p. 232; also see Figure 5.13).

1. Prove the following results about Euclidean parallelograms:
   (a) Opposite sides (and likewise, opposite angles) of a parallelogram are congruent to each other.
   (b) A parallelogram is a rectangle iff its diagonals are congruent, and in that case the diagonals bisect each other.
   (c) A parallelogram has a circumscribed circle iff it is a rectangle. (Hint for the "only if" part: Opposite angles must subtend semicircles.)
   (d) A rectangle is a square iff its diagonals are perpendicular.
2. Let k, l, m, and n be parallel lines, distinct, except that possibly $l = m$. Let transversals t and t' cut these lines in points A, B, C, and D and in A', B', C', and D', respectively (Figure 5.20). If $AB \cong CD$, prove that $A'B' \cong C'D'$. (Hint: Construct parallels to t through A' and C'. Apply Major Exercise 1(a) and the congruence of corresponding angles.)
3. Prove that parallel projection preserves betweenness; i.e., in Figure 5.13, if $A * B * C$, then $A' * B' * C'$. (Hint: Use Axiom B-4).
4. Prove the parallel projection theorem for the special case in which the ratio of lengths $\overline{AB}/\overline{BC}$ is a rational number p/q. (Hint: Divide AB into p congruent segments and BC into q congruent segments so that all $p + q$ segments will be congruent. Use Major Exercise 2, applying it $p + q$ times.)
5. The case where $\overline{AB}/\overline{BC}$ is an irrational number x is the difficult case. Let $\overline{A'B'}/\overline{B'C'} = x'$. The idea is to show that every rational number

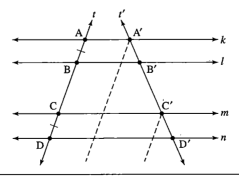

Figure 5.20

p/q less than x is also less than x' (and, by symmetry, vice versa). This will imply $x = x'$ since a real number is the least upper bound of all the rational numbers less than it (see any good text on real analysis). To show this, lay off on $\overrightarrow{BA}$ a segment BD of length $p\overline{CB}/q$ and let D' be the parallel projection of D onto t'. From $p/q < x$, deduce $B * D * A$. Now apply Major Exercises 3 and 4 to show that $p/q < x'$.[9]

6. Given a segment AB of length a with respect to some unit segment OI (see Theorem 4.3). Using straightedge and compass only, show how to construct a segment of length $\sqrt{a}$. (Hint: Extend AB to a segment AC of length $a + 1$; erect a perpendicular through B and let D be one of its intersections with the circle having AC as diameter; apply the theory of similar triangles to show that $\overline{BD} = \sqrt{a}$. Review the construction in Exercise 14, Chapter 1.)

7. Prove that given any line l, two points A and B not on l are on the same side S of l if and only if they lie on a circle contained in S. (Hint: If they are on the same side S, let M be the midpoint and m the perpendicular bisector of AB. Any circle through A and B has its center on m. If $\overleftrightarrow{AB} \parallel l$, take any point P between M and the point where m meets l and use the circle through A, B, and P (see Exercise 10, Chapter 6). Otherwise, if A is closer to l than is B, let the perpendicular from A to l meet m at O. Show that the circle centered at O with radius OA $\cong$ OB lies in S. Be sure to indicate where the hypothesis that *the geometry is Euclidean* is used; see Exercise P-20, Chapter 7.)

8. Let Π be a real Euclidean plane—i.e., a Hilbert plane satisfying Dedekind's axiom and Hilbert's Euclidean parallel postulate. By Theorem 4.3, there exists a real number measure of lengths of segments in Π with respect to some chosen unit segment. Let O be any point in Π, let l, m be two lines through O that are perpendicular, and let r, s be rays of l, m respectively emanating from O. Define a one-to-one mapping φ of Π onto $\mathbb{R}^2$ as follows: $\varphi(O) = (0, 0)$. For any point $P \neq O$, let P', P" be the intersections with l, m of the lines through P that are perpendicular to l, m respectively. Let $x = \overline{P'O}$, $y = \overline{P''O}$, where we define $\overline{OO} = 0$ in the case where P lies on l or m. Then define $\varphi(P) = (\pm x, \pm y)$, where the plus sign is chosen if P', P" lie on rays r, s respectively, and, if not, the appropriate minus sign is chosen for the one or both of them that lie(s) on the

---

[9] This clever method of proof was discovered by the ancient Greek mathematician Eudoxus—see E. C. Zeeman, "Research, Ancient and Modern," *Bulletin of the Institute of Mathematics and Its Applications*, 10 (1974): 272–281, Warwick University, England.

opposite ray. Prove, using the Pythagorean equation, that φ *is an isomorphism* of Π onto $\mathbb{R}^2$ with its structure of Euclidean plane defined in Example 3, p. 139. This result enables us to use coordinates and do analytic geometry in a real Euclidean plane.

Projects

1. Eudoxus was also the founder of theoretical astronomy in antiquity (his work was later refined by Ptolemy). In his model, the universe was bounded by "the celestial sphere," so that the physical interpretation of Euclid's second and third postulates was false! Even Kepler and Galileo believed in an outer limit to the world. It was René Descartes (1596–1650) who promoted the idea that we live in infinite, unbounded Euclidean space. Report on these issues, using Torretti (1978) as one reference.

2. Our treatment of similar triangles in the previous exercises used real numbers. Hilbert, with a later refinement by G. Vaitali, showed that the theory of similar triangles can be fully developed elegantly without real numbers. In that approach, the constants of proportionality come from the intrinsic *field of segment arithmetic.* Report on that development, using Hartshorne, Sections 19, 20, as a reference.

3. Our definition of "Euclidean plane" given in Chapter 3 avoids Dedekind's axiom (which is equivalent to bringing in real numbers), replacing that axiom with the circle-circle continuity principle. What then are the possible Euclidean planes? It turns out they are the models F^2, where F is a Euclidean field. This result is the precise modern formulation of what Descartes, Fermat, Euler *et al.* did when they brought in analytic geometry! If we drop the circle-circle continuity principle from our list of axioms but keep Hilbert's Euclidean parallel postulate, it is still the case that all models have the form F^2, but now we can only assert that F is a Pythagorean field. Curiously, equilateral triangles on arbitrary bases still exist in those models, but Euclid's proof of Euclid I.1 can no longer be used; the result is proved algebraically using the fact that $\sqrt{3}$ is in F. Report on these lovely results, using Hartshorne, Section 21, as a reference.

6

The Discovery of Non-Euclidean Geometry

Out of nothing I have created a strange new universe.

János Bolyai

János Bolyai

It is remarkable that sometimes when the time is right for a new idea to come forth, the idea occurs to several people more or less simultaneously. Thus it was in the eighteenth century with the discovery of the calculus by Newton in England and Leibniz in Germany, and in the nineteenth century with the discovery of non-Euclidean geometry. When János Bolyai (1802–1860) announced privately his discoveries in non-Euclidean geometry, his father Farkas admonished him:

It seems to me advisable, if you have actually succeeded in obtaining a solution of the problem, that, for a two-fold reason, its publication be hastened: first, because ideas easily pass from one man to another who, in that case, can publish them; secondly, because it seems to be true that many things have, as it were, an epoch in which they are discovered in several places simultaneously, just as the violets appear on all sides in springtime.[1]

János Bolyai did publish his discoveries, as a 26-page appendix to a mathematical treatise by his father (the *Tentamen*, 1831). Farkas sent a copy to his friend, the German mathematician Carl Friedrich Gauss (1777–1855), undisputedly the foremost mathematician of his time. Farkas Bolyai had become close friends with Gauss 35 years earlier, when they were both students in Göttingen. After Farkas returned to Hungary, they maintained an intimate correspondence,[2] and when Farkas sent Gauss his own attempt to prove the parallel postulate, Gauss tactfully pointed out the fatal flaw.

Foto: T.D.T.

János Bolyai

---

[1] Quoted in Meschkowski (1964). The title of J. Bolyai's appendix is "The Science of Absolute Space with a Demonstration of the Independence of the Truth or Falsity of Euclid's Parallel Postulate (Which Cannot Be Decided a Priori) and, in Addition, the Quadrature of the Circle in Case of Its Falsity."
[2] For the complete correspondence (in German), see Schmidt and Stäckel (1972).

János was 13 years old when he mastered the differential and integral calculus. His father wrote to Gauss begging him to take the young prodigy into his household as an apprentice mathematician. Gauss never replied to this request (perhaps because he was having enough trouble with his own son Eugene, who had run away from home). Fifteen years later, when Farkas mailed the *Tentamen* to Gauss, he certainly must have felt that his son had vindicated his belief in him, and János must have expected Gauss to publicize his achievement. One can therefore imagine the disappointment János must have felt when he read the following letter to his father from Gauss:

> If I begin with the statement that I dare not praise such a work, you will of course be startled for a moment: but I cannot do otherwise; to praise it would amount to praising myself; for the entire content of the work, the path which your son has taken, the results to which he is led, coincide almost exactly with my own meditations which have occupied my mind for from thirty to thirty-five years. On this account I find myself surprised to the extreme.
>
> My intention was, in regard to my own work, of which very little up to the present has been published, not to allow it to become known during my lifetime. Most people have not the insight to understand our conclusions and I have encountered only a few who received with any particular interest what I communicated to them. In order to understand these things, one must first have a keen perception of what is needed, and upon this point the majority are quite confused. On the other hand, it was my plan to put all down on paper eventually, so that at least it would not finally perish with me.
>
> So I am greatly surprised to be spared this effort, and am overjoyed that it happens to be the son of my old friend who outstrips me in such a remarkable way.[3]

Despite the compliment in Gauss' last sentence, János was bitterly disappointed with the great mathematician's reply; he even imagined that his father had secretly informed Gauss of his results and that Gauss was now trying to appropriate them as his own. A man of fiery temperament, who had fought and won 13 successive duels (unlike Galois, who was killed in a duel at age 20), János never published any of his results in the 14,000 pages of notes he left. A translation of his

---

[3] Wolfe (1945). Gauss did write to Gerling about the appendix a month earlier, saying: "I find all my own ideas and results developed with greater elegance. . . . I regard this young geometer Bolyai as a genius of the first order." That makes it all the more puzzling why Gauss did not help further János' mathematical career.

immortal appendix can be found in J. J. Gray (2004). In 1851, János wrote:

> In my opinion, and as I am persuaded, in the opinion of anyone judging without prejudice, all the reasons brought up by Gauss to explain why he would not publish anything in his life on this subject are powerless and void; for in science, as in common life, it is necessary to clarify things of public interest which are still vague, and to awaken, to strengthen and to promote the lacking or dormant sense for the true and right. Alas, to the great detriment and disadvantage of mankind, only very few people have a sense for mathematics; and for such a reason and pretence Gauss, in order to remain consistent, should have kept a great part of his excellent work to himself. It is a fact that, among mathematicians, and even among celebrated ones, there are, unfortunately, many superficial people, but this should not give a sensible man a reason for writing only superficial and mediocre things and for leaving science lethargically in its inherited state. Such a supposition may be said to be unnatural and sheer folly; therefore I take it rightly amiss that Gauss, instead of acknowledging honestly, definitely and frankly the great worth of the Appendix and the Tentamen, and instead of expressing his great joy and interest and trying to prepare an appropriate reception for the good cause, avoiding all these, he rested content with pious wishes and complaints about the lack of adequate civilization. Verily, it is not this attitude we call life, work and merit.[4]

Gauss

There is evidence that Gauss had anticipated some of J. Bolyai's discoveries—in fact, that Gauss had been working on non-Euclidean geometry since the age of 15, i.e., since 1792 (see Bonola, 1955, Chapter 3). In 1817, Gauss wrote to W. Olbers: "I am becoming more and more convinced that the necessity of our [Euclidean] geometry cannot be proved, at least not by human reason nor for human reason. Perhaps

---

[4] Quoted in L. Fejes Tóth, *Regular Figures* (Macmillan, New York, 1964), pp. 98–99. See the very informative review of Gray's book on Bolyai by Robert Osserman at http://www.ams.org/notices/200509/rev-osserman.pdf. See also an earlier history, *János Bolyai, Appendix*, F. Kártesi, ed., Elsevier, 1987, and the article by E. Kiss in Prékopa and Molnár (2005) that discusses Bolyai's unpublished discoveries in number theory, etc.

in another life we will be able to obtain insight into the nature of space, which is now inattainable." In 1824, Gauss answered F. A. Taurinus, who had attempted to investigate the theory of parallels:

In regard to your attempt, I have nothing (or not much) to say except that it is incomplete. It is true that your demonstration of the proof that the sum of the three angles of a plane triangle cannot be greater than 180° is somewhat lacking in geometrical rigor. But this in itself can easily be remedied, and there is no doubt that the impossibility can be proved most rigorously. But the situation is quite different in the second part, that the sum of the angles cannot be less than 180°; this is the critical point, the reef on which all the wrecks occur. I imagine that this problem has not engaged you very long. I have pondered it for over thirty years, and I do not believe that anyone can have given more thought to this second part than I, though I have never published anything on it.

The assumption that the sum of the three angles is less than 180° leads to a curious geometry, quite different from ours [the Euclidean], but thoroughly consistent, which I have developed to my entire satis-

Carl Friedrich Gauss

faction, so that I can solve every problem in it with the exception of the determination of a constant, which cannot be designated *a priori*. The greater one takes this constant, the nearer one comes to Euclidean geometry, and when it is chosen infinitely large, the two coincide. The theorems of this geometry appear to be paradoxical and, to the uninitiated, absurd; but calm, steady reflection reveals that they contain nothing at all impossible. For example, the three angles of a triangle become as small as one wishes, if only the sides are taken large enough; yet the area of the triangle can never exceed a definite limit, regardless of how great the sides are taken, nor indeed can it never reach it.

All my efforts to discover a contradiction, an inconsistency, in this non-Euclidean geometry have been without success, and the one thing in it which is opposed to our conceptions is that, if it were true, there must exist in space a linear magnitude, determined for itself (but unknown to us). But it seems to me that we know, despite the say-nothing word-wisdom of the metaphysicians, too little, or too nearly nothing at all, about the true nature of space, to consider as *absolutely impossible* that which appears to us unnatural. If this non-Euclidean geometry were true, and it were possible to compare that constant with such magnitudes as we encounter in our measurements on the earth and in the heavens, it could then be determined *a posteriori*. Consequently, in jest I have sometimes expressed the wish that the Euclidean geometry were not true, since then we would have *a priori* an absolute standard of measure.

I do not fear that any man who has shown that he possesses a thoughtful mathematical mind will misunderstand what has been said above, but in any case consider it a private communication of which no public use or use leading in any way to publicity is to be made. Perhaps I shall myself, if I have at some future time more leisure than in my present circumstances, make public my investigations.[5]

It is amazing that, despite his great reputation, Gauss was actually afraid to make public his discoveries in non-Euclidean geometry. He wrote to F. W. Bessel in 1829 that he feared "the howl from the Boeotians" if he were to publish his revolutionary discoveries.[6] He told

[5] Wolfe (1945), pp. 46–47.

[6] An allusion to dull, obtuse individuals. Gauss had more important work to do than to get into a quarrel with them. "Actually, the 'Boeotian' critics of non-Euclidean geometry—conceited people who claimed to have proved that Gauss, Riemann, and Helmholz were blockheads—did not show up before the middle of the 1870s. If you witnessed the struggle against Einstein in the Twenties, you may have some idea of [the] amusing kind of literature [produced by these critics]. . . . Frege, rebuking Hilbert like a schoolboy, also joined the Boeotians. . . . 'Your system of axioms,' he said to Hilbert, 'is like a system of equations you cannot solve'" (Freudenthal, 1962).

H. C. Schumacher that he had "a great antipathy against being drawn into any sort of polemic."

The "metaphysicians" referred to by Gauss in his letter to Taurinus were followers of Immanuel Kant, the supreme European philosopher in the late eighteenth century and much of the nineteenth century. Gauss' discovery of non-Euclidean geometry refuted Kant's position that Euclidean space is *inherent in the structure of our mind*. In his *Critique of Pure Reason* (1781), Kant declared that "the concept of [Euclidean] space is by no means of empirical origin, but is an inevitable necessity of thought." Gauss, in that letter to F. Bolyai, also wrote about " . . . the mistake Kant made in stating that space was merely the *form* of our looking at things."

Another reason that Gauss withheld his discoveries was that he was a perfectionist, one who published only completed works of art. His devotion to perfected work was expressed by the motto on his seal, *pauca sed matura* ("few but ripe"). There is a story that the distinguished mathematician K. G. J. Jacobi often came to Gauss to relate new discoveries, only to have Gauss pull out some papers from his desk drawer that contained the very same discoveries. Perhaps it is because Gauss was so preoccupied with original work in many branches of mathematics, as well as in astronomy, geodesy, and physics (he co-invented an improved telegraph with W. Weber), that he did not have the opportunity to put his results on non-Euclidean geometry into polished form. The few results he wrote down were found among his private papers after his death.

Gauss has been called "the prince of mathematicians" because of the range and depth of his work. (See the biographies by Bell, 1934; Dunnington, 1955; and Hall, 1970.)

Lobachevsky

Another actor in this historical drama came along to steal the limelight from both J. Bolyai and Gauss: the Russian mathematician Nikolai Ivanovich Lobachevsky (1792–1856). He was the first to actually publish an account of non-Euclidean geometry, in 1829. Lobachevsky initially called his geometry "imaginary," then later "pangeometry." His work attracted little attention on the continent when it appeared because it was written in Russian. The reviewer at the St. Petersburg Academy rejected it, and a Russian literary journal attacked Lobachevsky for "the insolence and shamelessness of false new inventions"

Nikolai Ivanovich Lobachevsky

(Boeotians howling, as Gauss predicted). Nevertheless, Lobachevsky courageously continued to publish further articles in Russian and then a treatise in 1840 in German,[7] which he sent to Gauss. In an 1846 letter to Schumacher, Gauss reiterated his own priority in developing non-Euclidean geometry but conceded that "Lobachevsky carried out the task in a masterly fashion and in a truly geometric spirit." At Gauss' secret recommendation, Lobachevsky was elected to the Göttingen Scientific Society. (Why didn't Gauss recommend János Bolyai?)

Lobachevsky openly challenged the Kantian doctrine of space as a subjective intuition. In 1835 he wrote: "The fruitlessness of the attempts made since Euclid's time . . . aroused in me the suspicion that the truth . . . was not contained in the data themselves; that to establish it the aid of experiment would be needed, for example, of astronomical observations, as in the case of other laws of nature." (Gauss privately

[7] For a translation of this paper, see Bonola (1955). For corrections to that translation and an attempt to explain what Lobachevsky and Bolyai did, see Chapter 10 of Jeremy J. Gray's *Ideas of Space: Euclidean, Non-Euclidean and Relativistic*, Oxford University Press, 2nd ed., 1989. Gray has also argued that Gauss' claim to priority in discovering non-Euclidean geometry is unjustified by concrete evidence; see his article "Gauss and Non-Euclidean Geometry" in Prékopa and Molnár (2005).

agreed with this view, having written to Olbers in 1817: "Perhaps we shall come to another insight in another life into the nature of space, which is unattainable for us now. But until then we must not put Geometry on a par with Arithmetic, which exists purely a priori, but rather with Mechanics. . . ." The great French mathematicians J. L. Lagrange (1736–1813) and J. B. Fourier (1768–1830) tried to derive the parallel postulate from the law of the lever in statics.)

Lobachevsky has been called "the great emancipator" by Eric Temple Bell; his name, said Bell, should be as familiar to every schoolboy as that of Michelangelo or Napoleon.[8] Unfortunately, Lobachevsky was not so appreciated in his lifetime; in fact, in 1846 he was fired from the University of Kazan, despite 20 years of outstanding service as a teacher and administrator. He had to dictate his last book in the year before his death, for by then he was blind.

It is amazing how similar are the approaches of J. Bolyai and Lobachevsky and how different they are from earlier work. Both developed the subject much further than Gauss. Both attacked plane geometry via the "horosphere" in hyperbolic three-space (it is the limit of an expanding sphere through a fixed point when its radius tends to infinity). Both showed that geometry on a horosphere, where "lines" are interpreted as "horocycles" (limits of circles), is Euclidean. Both showed that Euclidean spherical trigonometry is valid in hyperbolic geometry, and both constructed a mapping from the sphere to the non-Euclidean plane to derive the formulas of non-Euclidean trigonometry (including the formulas Taurinus discovered—see Chapter 10 for a simpler derivation using a plane model). Both had a constant in their formulas that they could not explain; the later work of Riemann showed it to be the *curvature* of a hyperbolic plane.

It is not entirely accurate to say that J. Bolyai and Lobachevsky "discovered" non-Euclidean geometry. We have seen that Saccheri, Lambert, and Taurinus discovered some basic results in non-Euclidean geometry before them, only these predecessors still doubted that such a geometry was consistent and actually "existed." J. Bolyai and Lobachevsky did believe in its noncontradictory existence, but they did not convincingly establish that. What they did was brilliantly elaborate its properties if it did exist. In an 1865 note on Lobachevsky's work, Arthur Cayley wrote: ". . . it would be very interesting to find a *real* geometric interpretation of Lobachevsky's system of equations." In 1868 Eugenio Beltrami finally found one—see Chapter 7.

[8] Bell (1954, Chapter 14).

Subsequent Developments

It was not until after Gauss' death in 1855, when his correspondence was published, that the mathematical world began to take non-Euclidean ideas seriously. (Yet, as late as 1894, an incorrect attempt to prove Euclid V was published in Arthur Cayley's *Journal of Pure and Applied Mathematics*. Cayley himself never accepted the non-Euclidean geometry of Bolyai–Lobachevsky, though he did work in elliptic geometry.) Some of the best mathematicians (Beltrami, Klein, Poincaré, and Riemann) took up the subject, extending it, clarifying it, and applying it to other branches of mathematics, notably complex function theory. In 1868 Eugenio Beltrami settled once and for all the question of a proof for the parallel postulate. He proved that no proof was possible—by exhibiting a Euclidean model of non-Euclidean geometry. (We will discuss his model in the next chapter.)

Bernhard Riemann, who was a student of Gauss, had the most profound insight into the geometry, not just the logic. In 1854, he built upon Gauss' discovery of the *intrinsic geometry* on a surface in Euclidean three-space. Riemann invented the concept of an abstract geometrical surface that need not be embeddable in Euclidean three-space yet on which the "lines" can be interpreted as geodesics and the intrinsic curvature of the surface can be precisely defined. Elliptic (and, of course, spherical) geometry "exist" on such surfaces that have constant positive curvature, while the hyperbolic geometry of Bolyai and Lobachevsky "exists" on such a *surface of constant negative curvature*. That is the view of geometers today about the "reality" of those non-Euclidean planes. We will describe Gauss and Riemann's idea only in Appendix A, since it is too advanced for the level of this text. Riemann presented the idea of a geometric manifold of arbitrary dimension n, not just $n = 2$ or 3, and defined a notion of curvature for it. He made the revolutionary suggestion that the universe might be finite in extent (as the ancient Greeks believed) but without any boundary if its curvature was slightly positive. A further generalization of that idea provided the geometry for Einstein's general theory of relativity.

Interestingly, a direct relationship between the special theory of relativity and hyperbolic geometry was discovered by the physicist Arnold Sommerfeld in 1909 and elucidated by the geometer Vladimir Varičak in 1912. A model of hyperbolic plane geometry is a sphere of imaginary radius with antipodal points identified in the three-dimensional

Georg Friedrich Bernhard Riemann

space-time of special relativity, vindicating Lambert's idea (see Chapter 7; or Rosenfeld, 1988, pp. 230, 270; or Yaglom, 1979, p. 222 ff.). Moreover, Taurinus' technique of substituting ir for r to go from spherical trigonometry to hyperbolic trigonometry received a structural explanation in 1926–1927 when Élie Cartan developed his theory of Riemannian symmetric spaces: The Euclidean sphere of curvature $1/r^2$ is "dual" to the hyperbolic plane of curvature $-1/r^2$ (see Helgason, 2001).

Non-Euclidean Hilbert Planes

Let us begin our investigation of the particular non-Euclidean plane geometry explored by Saccheri, Lambert, Gauss, J. Bolyai, and Lobachevsky, nowadays called *hyperbolic geometry* (a.k.a. *Lobachevskian* or *Bolyai–Lobachevskian* geometry). To arrive at a correct axiomatization for this geometry, we will proceed along historical lines, not dogmatically. Consider the following.

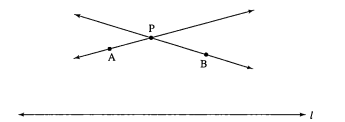

Figure 6.1

NEGATION OF HILBERT'S EUCLIDEAN PARALLEL POSTULATE. There exist a line l and a point P not on l such that at least two distinct lines parallel to l pass through P.

In a plane where such a configuration exists, the entire line l lies in the interior of ⊀APB (Figure 6.1) without meeting either side, which Legendre tacitly assumed to be impossible—that is the flaw in his attempted proof of Euclid V presented in Chapter 1. A Hilbert plane satisfying this negation will be called a *non-Euclidean* Hilbert plane.

To develop an interesting geometry from the consequences of this axiom,[9] we will need to assume more than just the negation of the Euclidean parallel postulate, for there are some non-Euclidean Hilbert planes which are not that important (such as the ones satisfying the obtuse angle hypothesis). One additional assumption is Aristotle's axiom, discussed in Chapters 3–5. Saccheri recognized the importance of that statement for non-Euclidean geometry; it was his Proposition XXI, and he proved it from Archimedes' axiom (Exercise 2, Chapter 5).

BASIC THEOREM 6.1. A non-Euclidean plane satisfying Aristotle's axiom satisfies the acute angle hypothesis. From the acute angle hypothesis alone, the following properties follow: The angle sum of every triangle is <180°, the summit angles of all Saccheri quadrilaterals are acute, the fourth angle of every Lambert quadrilateral is acute, and rectangles do not exist. The summit of a Saccheri quadrilateral is greater than the base. The segment joining the midpoints of the summit and the base is perpendicular to both, is the shortest segment between the base line and the summit line, and is the only common perpendicular segment between those lines. A side adjacent to the acute angle of a Lambert quadrilateral is greater than the opposite side.

---

[9] In previous editions of this book, it was incorrectly called the "hyperbolic axiom."

PROOF:

The non-obtuse-angle theorem in Chapter 4 tells us that in a plane satisfying Aristotle's axiom, the angle sum of every triangle is ≤180°. Proclus' theorem in Chapter 5 tells us that the angle sum cannot equal 180° because we assumed Aristotle's axiom and the plane was non-Euclidean. The only remaining possibility is that it is <180°. In that case, the remaining assertions follow from all our work in Chapter 4. ◄

The negation of Hilbert's Euclidean parallel postulate referred to *some* line l and *some* point P not on l, but we can prove a universal version of that property.[10]

UNIVERSAL NON-EUCLIDEAN THEOREM. In a Hilbert plane in which rectangles do not exist, for every line l and every point P not on l, there are at least two parallels to l through P.

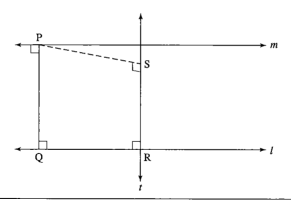

Figure 6.2

PROOF:

Let PQlm be the standard configuration. Let R be another point on l, erect perpendicular t to l through R, and let S be the foot of the perpendicular from P to t (Figure 6.2). Then $\overleftrightarrow{PS}$ is parallel to l since they are both perpendicular to t (Corollary to the AIA theorem). It is a different parallel than m; otherwise S would lie on m and □PQRS would be a rectangle. ◄

---

[10] In previous editions, we called this result the "universal hyperbolic theorem." That name is incorrect because the result is also valid in non-Euclidean planes other than the hyperbolic ones.

COROLLARY. In a Hilbert plane in which rectangles do not exist, for every line l and every point P not on l, there are infinitely many parallels to l through P.

PROOF:

Just vary the point R in the above construction. The nonexistence of rectangles again guarantees that the parallels constructed are distinct. ◄

The Defect

Since the angle sum of every triangle $\triangle ABC$ in a plane as above is $< 180°$, that angle sum is the measure of an angle—namely, an angle constructed by successively juxtaposing the three angles of $\triangle ABC$. The positive measure of the *supplement* of that angle is called the *defect*[11] of the triangle and is denoted $\delta(ABC)$. Thus by definition,

$$(\sphericalangle A)° + (\sphericalangle B)° + (\sphericalangle C)° + \delta(ABC) = 180°.$$

PROPOSITION 6.1 (ADDITIVITY OF THE DEFECT). If D is any point between A and B (Figure 6.3), then

$$\delta(ABC) = \delta(ACD) + \delta(BCD).$$

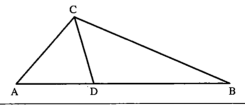

Figure 6.3

PROOF:

This follows immediately from the definition of the defect, from the fact that for the supplementary angles at point D, $(\sphericalangle ADC)° + (\sphericalangle BDC)° = 180°$, and from $(\sphericalangle C)° = (\sphericalangle ACD)° + (\sphericalangle BCD)°$. ◄

---

[11] Hartshorne defines the *defect* as the congruence class of that supplement, not its measure. His definition avoids the use of Archimedes' axiom.

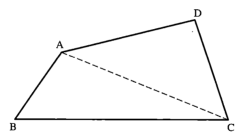

Figure 6.4

REMARK. In Exercise 28 of Chapter 4, we studied the notion of a *convex* quadrilateral. One characterization of convex quadrilaterals is that each vertex lies in the interior of the opposite angle. From that one easily sees (Figure 6.4) that *in a plane satisfying the acute angle hypothesis, the angle sum of every convex quadrilateral is <360°*. The *defect* of a convex quadrilateral is defined to be 360° minus its angle sum.

Similar Triangles

Consider next Wallis' postulate, which cannot hold in a non-Euclidean plane because we saw in Chapter 5 that it implies the Euclidean parallel postulate. The negation of Wallis' postulate asserts that sometimes a triangle similar to a given triangle does not exist. Once again, we can prove a universal version of this statement: Similar noncongruent triangles never exist!

PROPOSITION 6.2 (NO SIMILARITY). In a plane satisfying the acute angle hypothesis, if two triangles are similar, then they are congruent. In other words, AAA is a valid criterion for congruence of triangles.

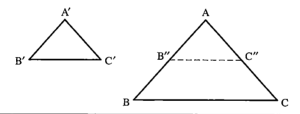

Figure 6.5

PROOF:

Assume on the contrary that there exist triangles $\triangle ABC$ and $\triangle A'B'C'$ which are similar but not congruent. Then no corresponding sides are congruent; otherwise the triangles would be congruent (ASA). Consider the triples (AB, AC, BC) and (A'B', A'C', B'C') of sides of these triangles. One of these triples must contain at least two segments that are larger than the two corresponding segments of the other triple, e.g., AB > A'B' and AC > A'C'. By definition of > there exist points B″ on AB and C″ on AC such that AB″ ≅ A'B' and AC″ ≅ A'C' (see Figure 6.5). By SAS, $\triangle A'B'C' \cong \triangle AB''C''$. Hence, corresponding angles are congruent: ∡AB″C″ ≅ ∡B', ∡AC″B″ ≅ ∡C. By the hypothesis that $\triangle ABC$ and $\triangle A'B'C'$ are similar, we have ∡AB″C″ ≅ ∡B, ∡AC″B″ ≅ ∡C (Axiom C-5). This implies that $\overleftrightarrow{B''C''} \parallel \overleftrightarrow{BC}$ by corresponding angles (AIA theorem and Exercise 32, Chapter 4), so that quadrilateral □BB″C″C is convex (Exercise 28, Chapter 4). By supplementary angles,

$$(∡B)° + (∡BB''C'')° = 180° = (∡C)° + (∡CC''B'')°.$$

It follows that convex quadrilateral □BB″C″C has angle sum 360°. This contradicts the remark after the proof of Proposition 6.1. ◄

A consequence of Proposition 6.2 is that in a plane satisfying the acute angle hypothesis, an angle and a side of an *equilateral* triangle determine one another uniquely. If we assume the circle-circle continuity principle, then we know from Euclid's construction in his first proposition that given any segment, an equilateral triangle exists having that segment as its side. In a hyperbolic plane (studied later in this chapter), for every acute angle $\theta < 60°$, an equilateral triangle exists having θ as its angle; see Chapter 10 for a construction (corollary to the right triangle construction theorem).

Parallels Which Admit a Common Perpendicular

In Chapter 5, in our comment on Proclus' failed attempt to prove Euclid V, the reader was warned not to presume that a pair of parallel lines look like railroad tracks—i.e., not to presume, as Clavius did explicitly, that the set of points on a line through P parallel to a given line l coincides with the equidistant curve to l through P. We saw that Clavius' assumption is equivalent to the plane being semi-Euclidean. Negating that condition, we can prove the following precise result.

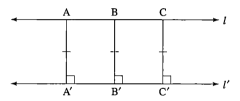

Figure 6.6 AA' ≅ BB' ≅ CC' ≅ . . . .

PROPOSITION 6.3. In a plane in which rectangles do not exist, if *l* ∥ *l'*, then any set of points on *l* equidistant from *l'* has at most two points in it.

PROOF:

Assume, on the contrary, that three points A, B, and C on *l* are equidistant from *l'*. By Axiom B-3, we may assume A ∗ B ∗ C. If A', B', and C' on *l'* are the feet of the perpendiculars from A, B, C, respectively, to *l'*, then AA' ≅ BB' ≅ CC' by the RAA hypothesis. So we obtain three Saccheri quadrilaterals ☐A'B'BA, ☐A'C'CA, and ☐B'C'CB (see Figure 6.6).

We know that the summit angles of any Saccheri quadrilateral are congruent (Proposition 4.12). By transitivity (Axiom C-5), the supplementary angles at B are congruent to each other, hence are right angles. Thus these Saccheri quadrilaterals are rectangles, contradicting our hypothesis that rectangles do not exist. ◄

The proposition states that *at most* two points at a time on *l* can be equidistant from *l'*. It allows the possibility that there are pairs of points (A, B), (C, D), . . . , on *l* such that each pair is equidistant from *l'*— e.g., AA' ≅ BB' and CC' ≅ DD' dropping perpendiculars—but AA' is not congruent to CC'. A diagram for this might be Figure 6.7, which suggests that there is a point of *l* that is closest to *l'*, with *l* diverging

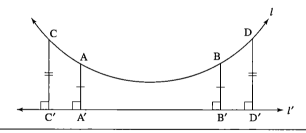

Figure 6.7

from l' symmetrically on either side of this closest point (under the acute angle hypothesis). We will prove that this is indeed the case. (I hope the reader is not too shocked to see line l drawn as being "curved!")

PROPOSITION 6.4. In a Hilbert plane satisfying the acute angle hypothesis, if $l \parallel l'$ and if there exists a pair of points A and B on l equidistant from l', then l and l' have a unique common perpendicular segment MM' dropped from the midpoint M of AB. MM' is the shortest segment joining a point of l to a point of l', and the segments AA' and BB' increase as A, B recede from M.

> *PROOF:*
>
> The common perpendicular segment is obtained by joining the midpoints of the summit and the base of Saccheri quadrilateral □A'B'BA (Proposition 4.12). That it is unique follows from the nonexistence of rectangles. The other assertions follow from the acute angle hypothesis and Propositions 4.5 and 4.13 of Chapter 4. ◄

PROPOSITION 6.5. In a Hilbert plane in which rectangles do not exist, if lines l and l' have a common perpendicular segment MM', then they are parallel and that common perpendicular segment is unique. Moreover, if A and B are any points on l such that M is the midpoint of AB, then A and B are equidistant from l'.

> *PROOF:*
>
> The first statement follows from the corollary to the AIA theorem in Chapter 4 and the nonexistence of rectangles. Suppose now that M is the midpoint of AB, with A and B on l, and let A', B' be the feet of the perpendiculars from A, B to l'. We must prove that AA' ≅ BB' (see Figure 6.8).

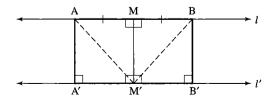

Figure 6.8

Observe that $\triangle\text{AM}'\text{M} \cong \triangle\text{BM}'\text{M}$ (SAS), so that $\text{AM}' \cong \text{BM}'$ and $\sphericalangle\text{AM}'\text{M} \cong \sphericalangle\text{BM}'\text{M}$. Their complementary angles $\sphericalangle\text{A}'\text{M}'\text{A}$ and $\sphericalangle\text{B}'\text{M}'\text{B}$ are then congruent, and we obtain $\triangle\text{AA}'\text{M} \cong \triangle\text{BB}'\text{M}$ (SAA). Hence $\text{AA}' \cong \text{BB}'$. ◀

The preceding propositions give us a good understanding of parallel lines that have a common perpendicular in a plane satisfying the acute angle hypothesis. We know that such parallel lines exist from the standard construction. There remains another possibility for parallel lines in such planes: that there is no pair of points on l equidistant from l' and no common perpendicular between these lines! According to Proposition 4.13 on bi-right quadrilaterals $\square\text{A}'\text{B}'\text{BA}$, l would diverge from l' in one direction and converge toward l' in the opposite direction without meeting it (see Figure 6.9). (Omar Khayyam, trying to prove Euclid V, assumed as a new axiom that this second type of parallel lines could not exist.) As we will discuss in the next section, a further axiom is needed to guarantee that, in certain planes satisfying the acute angle hypothesis, the second type of parallel lines really does exist.

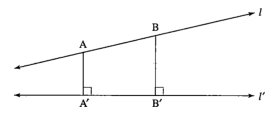

Figure 6.9 BB' > AA'.

Limiting Parallel Rays, Hyperbolic Planes

Saccheri, Gauss, J. Bolyai, and Lobachevsky all took for granted that parallel lines of the second type exist in a very specific manner which we will now describe. Here is the intuitive idea (see Figure 6.10).

Let $PQlm$ be a standard configuration. Consider one ray $\overrightarrow{\text{PS}}$ of m and consider various rays between $\overrightarrow{\text{PS}}$ and $\overrightarrow{\text{PQ}}$. Some of these rays, such as $\overrightarrow{\text{PR}}$, will intersect l, while others, such as $\overrightarrow{\text{PY}}$, will not (universal non-Euclidean theorem). Now imagine R receding endlessly from Q along its ray of l. The master geometers just mentioned all took it for granted that $\overrightarrow{\text{PR}}$ would approach a certain *limiting ray* $\overrightarrow{\text{PX}}$. That ray could not intersect l, for if X were on l, there would exist a point R

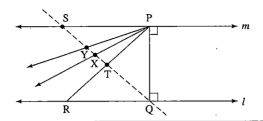

Figure 6.10

further out on l—i.e., $R * X * Q$ (Axiom B-2)—and $\overrightarrow{PX}$ would not be the limit. (Saccheri called $\overrightarrow{PX}$ "the first ray which fails to meet l.") None of the rays $\overrightarrow{PY}$ that lie between $\overrightarrow{PS}$ and $\overrightarrow{PX}$ intersect l, for if one of them did, $\overrightarrow{PX}$ would also have to intersect l by the crossbar theorem. According to Figure 6.10, we could call $\overrightarrow{PX}$ the *left limiting parallel ray* to l emanating from P (in a Euclidean plane, $\overrightarrow{PX}$ would coincide with $\overrightarrow{PS}$). Similarly, there would be a *right limiting parallel ray* to l emanating from P on the opposite side of $\overleftrightarrow{PQ}$.

WARNING It is not possible to prove that limiting parallel rays exist in every plane satisfying the acute angle hypothesis. F. Schur found a non-Archimedean counterexample (the infinitesimal neighborhood of the origin in a non-Archimedean Klein model), and later an Archimedean counterexample was found (the interior of a virtual circle—see Hartshorne, Exercises 39.25–39.31).

ADVANCED THEOREM. *In non-Euclidean planes satisfying Aristotle's axiom and the line-circle continuity principle, limiting parallel rays exist for every line l and point P not on l.*

My proof of this theorem[12] is based on the classification due to W. Pejas of all possible Hilbert planes (see Appendix B). I hope that someday an elementary proof of this theorem will be found that could be presented in a text at this level. János Bolyai foresaw this result when he gave the following *straightedge-and-compass construction of the limiting parallel ray* in such a plane.

J. BOLYAI'S CONSTRUCTION OF THE LIMITING PARALLEL RAY. Let PQlm be a standard configuration. Let R be any point on l different from Q and let S be the foot of the perpendicular from R to m. Then

---

[12] See M. J. Greenberg, "Aristotle's Axiom in the Foundations of Hyperbolic Geometry," *Journal of Geometry,* **33** (1988): 53–57. A proof is sketched at the end of Appendix B.

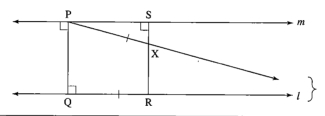

Figure 6.11

□SPQR is a Lambert quadrilateral with acute angle at vertex R (Theorem 6.1). By Corollary 3 to Proposition 4.13, PS < QR. Also, we have PR > QR (hypotenuse greater than leg). By the segment-circle continuity principle, a consequence of the line-circle continuity principle (Major Exercise 2, Chapter 4), the circle with center P and radius congruent to QR will intersect segment SR in a unique point X between S and R. János Bolyai claimed that *ray* $\overrightarrow{PX}$ *is the limiting parallel ray* to *l* emanating from P on the same side of $\overleftrightarrow{PQ}$ as R (see Figure 6.11).

In non-Archimedean examples where the ray constructed by J. Bolyai's method is *not* limiting parallel to *l*, it does have the property that so shocked Saccheri: It has "a common perpendicular with *l* at infinity!"[13]

We will prove below that in a non-Euclidean plane satisfying Dedekind's axiom, limiting parallel rays always exist. However, Bolyai's construction shows that only a very mild quadratic continuity assumption is needed for the existence of limiting parallel rays, not the full power of the real number system! Hilbert's idea was simply to study Hilbert planes in which limiting parallel rays always exist, which finally provides the axiom we need.

DEFINITION. Given a line *l* and a point P not on *l*. Let Q be the foot of the perpendicular from P to *l*. A *limiting parallel ray* to *l* emanating from P is a ray $\overrightarrow{PX}$ that does not intersect *l* and such that for every ray $\overrightarrow{PY}$ which is between $\overrightarrow{PQ}$ and $\overrightarrow{PX}$, $\overrightarrow{PY}$ intersects *l*.

HILBERT'S HYPERBOLIC AXIOM OF PARALLELS. For every line *l* and every point P not on *l*, a limiting parallel ray $\overrightarrow{PX}$ emanating from P exists and it does *not* make a right angle with $\overrightarrow{PQ}$, where Q is the foot of the perpendicular from P to *l*.

[13] See M. J. Greenberg, "On J. Bolyai's Parallel Construction," *Journal of Geometry*, **12**(1) (1979): 45–64.

DEFINITION. A Hilbert plane in which Hilbert's hyperbolic axiom of parallels holds is called a *hyperbolic plane*. Obviously a hyperbolic plane is non-Euclidean.

PROPOSITION 6.6. In a hyperbolic plane, with notation as in the above definition, ⊀XPQ is acute. There is a ray $\overrightarrow{PX'}$ emanating from P, with X' on the opposite side of $\overleftrightarrow{PQ}$ from X, such that $\overrightarrow{PX'}$ is another limiting parallel ray to l and ⊀XPQ ≅ ⊀X'PQ. These two rays, situated symmetrically about $\overrightarrow{PQ}$, are the only limiting parallel rays to l through P.

> *PROOF:*
>
> Let PQ*lm* be the standard configuration and let $\overrightarrow{PS}$ be a ray of *m* with S on the same side of $\overleftrightarrow{PQ}$ as X. If ⊀XPQ were obtuse, $\overrightarrow{PS}$ would lie between $\overrightarrow{PQ}$ and $\overrightarrow{PX}$, hence would intersect l by definition of a limiting parallel ray; but that contradicts *m* being parallel to l. Hence ⊀XPQ is acute. The other limiting parallel ray $\overrightarrow{PX'}$ emanating from P is obtained by reflecting $\overrightarrow{PX}$ across line $\overleftrightarrow{PQ}$. Uniqueness follows from the definition of a limiting parallel ray and the ordering of angles. ◄

DEFINITION. With the above notation, acute angles ⊀XPQ and ⊀X'PQ are called *angles of parallelism* for segment PQ. Lobachevsky denoted their congruence class (or, par abus de langage, any angle congruent to them) by Π(PQ).

Saccheri, in his Proposition XXXII, recognized the existence of this acute angle; Proclus noted that possible existence many centuries earlier.[14] Major Exercise 5 shows that Π(PQ) depends only on the congruence class of PQ.

Hilbert and his followers' development of plane hyperbolic geometry from his hyperbolic axiom is a beautiful tour de force. Although it is all carried out at the same elementary level that we have been working at in this book, the arguments are far too lengthy for our purpose. See Hartshorne, Chapter 7, for all the details.

Instead we will bring in our *deus ex machina*, as classical Greek theatre called it (a god comes down from heaven to save the day): Dedekind's axiom.

THEOREM 6.2. In a non-Euclidean plane satisfying Dedekind's axiom, Hilbert's hyperbolic axiom of parallels holds, as do Aristotle's axiom and the acute angle hypothesis.

[14] See the Morrow edition (1992) of Proclus, p. 290.

PROOF:

For the second part, we know from Chapter 3 that Dedekind's axiom implies Archimedes' axiom, and you showed in Exercise 2, Chapter 5, that Archimedes' axiom implies Aristotle's axiom.[15] The acute angle hypothesis follows from Basic Theorem 6.1.

For the first part, refer again to Figure 6.10 above. To prove rigorously that $\overrightarrow{PX}$ exists, consider the line $\overleftrightarrow{SQ}$ (Figure 6.12). Let Σ_1 be the set of all points T on segment SQ such that $\overrightarrow{PT}$ meets l, together with all points on the ray opposite to $\overrightarrow{QS}$; let Σ_2 be the complement of Σ_1 (so $Q \in \Sigma_1$ and $S \in \Sigma_2$). By the crossbar theorem (Chapter 3), if point T on segment SQ belongs to Σ_1, then the entire segment TQ (in fact, $\overrightarrow{TQ}$) is contained in Σ_1. Hence (Σ_1, Σ_2) is a Dedekind cut. By Dedekind's axiom (Chapter 3), there is a unique point X on $\overleftrightarrow{SQ}$ such that for P_1 and P_2 on $\overleftrightarrow{SQ}$, $P_1 * X * P_2$ if and only if $X \neq P_1$, $X \neq P_2$, $P_1 \in \Sigma_1$, and $P_2 \in \Sigma_2$.

By definition of Σ_1 and Σ_2, rays below $\overrightarrow{PX}$ all meet l and rays above $\overrightarrow{PX}$ do not. We claim that $\overrightarrow{PX}$ does not meet l either. Assume on the contrary that $\overrightarrow{PX}$ meets l in a point U (Figure 6.12). Choose any point V on l to the left of U, i.e., $V * U * Q$ (Axiom B-2). Since V and U are on the same side of $\overleftrightarrow{SQ}$ (Exercise 9, Chapter 3), V and P are on opposite sides, so VP meets SQ in a point Y. We have $Y * X * Q$ (Proposition 3.7), so $Y \in \Sigma_2$, contradicting the fact that $\overrightarrow{PY}$ meets l. It follows that $\overrightarrow{PX}$ is the left limiting parallel ray (we obtain the right limiting parallel ray in a similar manner).

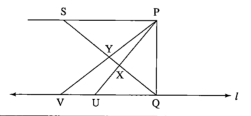

Figure 6.12

To prove symmetry, assume on the contrary that angles $\sphericalangle XPQ$ and $\sphericalangle X'PQ$ are not congruent, e.g., $(\sphericalangle XPQ)° < (\sphericalangle X'PQ)°$. By Axiom C-4, there is a ray between $\overrightarrow{PX'}$ and $\overrightarrow{PQ}$ that intersects l (by

---

[15] In fact, Aristotle's axiom holds in any hyperbolic plane: See Exercise 13. So does the circle-circle continuity principle: See Appendix B or Hartshorne, Corollary 43.4.

the definition of limiting ray) in a point R' such that ∢R'PQ ≅ ∢XPQ. Let R be the point on the opposite side of $\overleftrightarrow{PQ}$ from R' such that R * Q * R' and RQ ≅ R'Q (Axiom C-1). Then △RPQ ≅ △R'PQ (SAS). Hence ∢RPQ ≅ ∢R'PQ, and by transitivity (Axiom C-5), we have ∢RPQ ≅ ∢XPQ. But this is impossible because $\overrightarrow{PR}$ is between $\overrightarrow{PX}$ and $\overrightarrow{PQ}$ (Axiom C-4). ◄

WARNING In the section on incidence geometry, Chapter 2, we called the "hyperbolic parallel property" the property that there is more than one parallel to l through P (the property in the universal non-Euclidean theorem above). Do not confuse that property with the one in Hilbert's hyperbolic axiom of parallels! The latter implies the former, but not conversely, unless additional axioms are assumed (such as Dedekind's or the two axioms in the advanced theorem).

DEFINITION. A non-Euclidean plane satisfying Dedekind's axiom is called a *real hyperbolic plane.*

COROLLARY 1. All the results proved previously in this chapter hold in real hyperbolic planes.

They also hold in general hyperbolic planes—see Hartshorne, Chapter 7.

Engel's theorem in Chapter 10 guarantees that Bolyai's construction gives the limiting parallel ray in a real hyperbolic plane. The construction is also justified for the Klein model at the end of Chapter 7 (pp. 344–345).

COROLLARY 2. A Hilbert plane satisfying Dedekind's axiom is either real Euclidean or real hyperbolic.

More generally, from the advanced theorem, *a Hilbert plane satisfying Aristotle's axiom and the line-circle continuity principle is either Euclidean or hyperbolic.*

Classification of Parallels

We have discussed two types of parallels to a given l. The first type consists of parallels m such that l and m have a common perpendicular; m diverges from l on both sides of the common perpendicular. The second type consists of parallels that approach l asymptotically in one direction (i.e., they contain a limiting parallel ray in that direction) and diverge from l in the other direction. If m is the second type of parallel, Exercises 6 and 7 show that l and m do not have a common perpendicular. We have implied that these two are the only types of parallels, and this is the content of the next theorem.

THEOREM 6.3. In a hyperbolic plane, given m parallel to l such that m does not contain a limiting parallel ray to l in either direction. Then there exists a common perpendicular to m and l (which is unique by Proposition 6.5).

This theorem is proved by Borsuk and Szmielew (1960, p. 291) by a continuity argument, but their proof gives you no idea of how to actually find the common perpendicular. There is an easy way to find it in the Klein and Poincaré models discussed in the next chapter. Hilbert gave a direct construction, which we will sketch. (Project 1 gives another.)

PROOF:

Hilbert's idea is to find two points H and K on l that are equidistant from m, for once these are found, the perpendicular bisector of segment HK is also perpendicular to m (see Basic Theorem 6.1). Choose any two points A and B on l and suppose that the perpendicular segment AA′ from A to m is longer than the perpendicular segment BB′ from B to m (see Figure 6.13). Let E be the point between A′ and A such that A′E $\cong$ B′B. On the same side of $\overleftrightarrow{AA'}$ as B, let $\overrightarrow{EF}$ be the unique ray such that $\sphericalangle$A′EF $\cong$ $\sphericalangle$B′BG, where A * B * G. The key point that will be proved in Major Exercises 2–6 is that $\overrightarrow{EF}$ intersects $\overrightarrow{AG}$ in a point H. Let K be the unique point on $\overrightarrow{BG}$ such that EH $\cong$ BK. Drop perpendiculars $\overleftrightarrow{HH'}$ and $\overleftrightarrow{KK'}$ to m. The upshot of these constructions is that $\square$EHH′A′ is congruent to $\square$BKK′B′ (just divide them into triangles). Hence the corresponding sides HH′ and KK′ are congruent, so that the points H and K on l are equidistant from m, as required. ◄

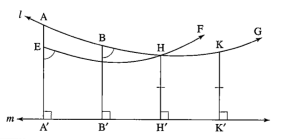

Figure 6.13 Hilbert's construction of the common perpendicular.

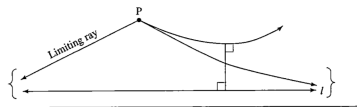

Figure 6.14

SUMMARY. Given a point P not on l, there exist exactly two limiting parallel rays to l through P, one in each direction. There are infinitely many lines through P that do not enter the region between the limiting rays and l. Each such line is divergently parallel to l and admits a unique common perpendicular with l (for one of these lines the common perpendicular will go through P, but for all the rest the common perpendicular will pass through other points).

A NOTE ON TERMINOLOGY. In most books on hyperbolic geometry, the word "parallel" is used only for lines that contain limiting parallel rays. The other lines, which admit a common perpendicular, have various names in the literature: "non-intersecting," "ultraparallel," "hyperparallel," and "superparallel." We will continue to use the word "parallel" to mean "non-intersecting." Following J. Bolyai, a parallel to l that contains a limiting parallel ray to l will be called an *asymptotic parallel*; a parallel to l that admits a common perpendicular to l will be called a *divergently parallel line*. Rays that are limiting parallel will be denoted by a brace in diagrams (see Figure 6.14).

Strange New Universe?

In this chapter, we have only begun to investigate the "strange new universe" of hyperbolic geometry. You can develop much more of this geometry by doing the exercises, major exercises, and projects in this chapter. You will encounter new entities such as asymptotic triangles, lines of enclosure, and ideal and ultra-ideal points at infinity in the projective completion of the hyperbolic plane.

   If you consider this geometry too strange to pursue, you are in for a surprise. You will see in the next chapter that if the undefined terms of hyperbolic geometry are suitably interpreted, hyperbolic geometry can be considered a part of Euclidean geometry!

Meanwhile, notice how we have deepened our understanding of the role of Hilbert's Euclidean parallel postulate P in Euclidean geometry. To simplify, let us work in *real neutral geometry*—the theory of Hilbert planes that satisfy Dedekind's axiom. Any statement S in the language of real neutral geometry that is a theorem in real Euclidean geometry $(P \Rightarrow S)$ and whose negation is a theorem in real hyperbolic geometry $(\sim P \Rightarrow \sim S)$ is equivalent in real neutral geometry to P (by RAA). As an example, take the statement S: "There exist similar non-congruent triangles." Ten other examples are in Exercise 1; you will have the pleasure in that exercise of providing many more.

The angle of parallelism $\Pi(PQ)$ is the key to the deeper results in hyperbolic geometry. Major Exercise 9 shows that it can be any acute angle. It can be used to define segments geometrically, which is impossible in Euclidean geometry (see p. 411, Chapter 9). For example, *Schweikart's segment class* is defined to be the congruence class of a segment whose angle of parallelism is half a right angle (Major Exercise 5 shows that all such segments are congruent). One of the greatest discoveries by J. Bolyai and Lobachevsky is their formula for the measure of $\Pi(PQ)$ (see Theorems 7.2 and 10.2).

An important topic we will sketch in Chapter 10 is the theory of area in hyperbolic planes. It is completely different from the Euclidean theory of area, which is based on squares—there are no squares in hyperbolic planes. The area of a Euclidean triangle can be made as large as you like by taking the base and the height as large as needed. However, *in a hyperbolic plane, the possible areas of triangles are bounded* because it is a fundamental theorem that *the area of a triangle is proportional to its defect* and of course *the defect is bounded* by 180°. But in order to make sense of this strange result, noted by Lambert, one must first clarify what is meant by "area." We defer to other good texts[16] for the details.

The reader's attention is called to Major Exercise 13 of this chapter. That exercise constructs the *projective completion of a hyperbolic plane*, analogous to the construction of the projective completion of an affine plane in Chapter 2, but here we add an entire region at infinity to the hyperbolic plane, not just a line at infinity: The hyperbolic plane

---

[16] See Moise (1990) or Hartshorne. Charles Dodgson (Lewis Carroll) refused to accept such a strange result, not comprehending how the areas of triangles could be bounded when the lengths of their sides are unbounded.

lies inside a *conic at infinity* called the *absolute,* consisting of all the ideal points where asymptotic parallels meet; the *tangent* to the absolute at that meeting point can be considered the "common perpendicular at infinity" whose discovery shocked Saccheri. Outside that conic lie all the ultra-ideal points where divergent parallels meet. This projective completion is the idea behind the Klein model discussed in the next chapter. Another use is the following nice result.

PERPENDICULAR BISECTOR THEOREM. Given any triangle in a hyperbolic plane, the perpendicular bisectors of its sides are concurrent in the projective completion.

Unlike Euclidean planes, those perpendicular bisectors need not be concurrent in an ordinary point; they may meet in the projective completion at an ideal or ultra-ideal point (see Exercises 10 and 11).

NOTE ON OUR AXIOMATIC DEVELOPMENT. By simply negating Hilbert's Euclidean parallel postulate, one allows nonclassical, non-Archimedean Hilbert planes discovered by Dehn, such as semi-Euclidean ones that are not Euclidean and ones satisfying the obtuse angle hypothesis. They can be ruled out by assuming Aristotle's axiom, which reduces us to certain planes satisfying the acute angle hypothesis. Some of those are nonclassical because limiting parallel rays do not exist (my advanced theorem and the second result mentioned in footnote 15 tell us that the line-circle continuity principle is then necessary and sufficient to obtain that existence). Yet all the classical non-Euclidean geometers (Saccheri, Gauss, J. Bolyai, and Lobachevsky) argued intuitively that limiting parallel rays *do* exist. So Hilbert simply took that existence as an axiom and only studied the hyperbolic planes it defines. Hilbert did not wish to bring in the powerful field of real numbers where it was not needed (see the quote by him in Appendix B).

Since Hilbert's development is long and complicated, we invoked Dedekind's axiom to *prove* the existence of limiting parallel rays as well as Aristotle's axiom and the acute angle hypothesis. That is what we called *real plane hyperbolic geometry.* It is less general than Hilbert's theory, which permits coordinatization from arbitrary Euclidean fields (including non-Archimedean ones), not just from the field $\mathbb{R}$ of real numbers. The theory of *real* hyperbolic planes is *categorical*: All its models are isomorphic to the real models in the next chapter (see Hartshorne). But the theory of hyperbolic planes in Hilbert's more general sense is not categorical since not all Euclidean fields are isomorphic (e.g., the constructible field **K** is not isomorphic to ℝ).

Our main tasks in the next chapter will be to prove that axiomatic hyperbolic plane geometry is just as logically secure as plane Euclidean geometry and to reveal how it can be visualized from a Euclidean point of view.

Review Exercise

Which of the following statements are correct?

(1) The negation of Hilbert's Euclidean parallel postulate states that for every line l and every point P not on l there exist at least two lines through P parallel to l.

(2) It is a theorem in neutral geometry that if lines l and m meet on a given side of a transversal t, then the sum of the degrees of the interior angles on that given side of t is less than 180°.

(3) Gauss began working on non-Euclidean geometry when he was 15 years old.

(4) The philosopher Kant taught that our minds could not conceive of any geometry other than Euclidean geometry.

(5) The first mathematician to publish an account of hyperbolic geometry was Lobachevsky.

(6) The crossbar theorem asserts that a ray emanating from a vertex A of △ABC and interior to ∢A must intersect the opposite side BC of the triangle.

(7) It is a theorem in hyperbolic geometry that for any segment AB there exists a square having AB as one of its sides.

(8) In hyperbolic geometry, the summit angles of Saccheri quadrilaterals are always acute.

(9) In hyperbolic geometry, if △ABC and △DEF are equilateral triangles and ∢A ≅ ∢D, then the triangles are congruent.

(10) In hyperbolic geometry, given a line l and a fixed segment AB, the set of all points on a given side of l whose perpendicular segment to l is congruent to AB equals the set of points on a line parallel to l.

(11) In hyperbolic geometry, any two parallel lines have a common perpendicular.

(12) In hyperbolic geometry, the fourth angle of a Lambert quadrilateral is obtuse.

(13) In hyperbolic geometry, some triangles have angle sum less than 180° and some triangles have angle sum equal to 180°.

(14) In hyperbolic geometry, if point P is not on line l and Q is the foot of the perpendicular from P to l, then an angle of parallelism for P with respect to l is the angle that a limiting parallel ray to l emanating from P makes with $\overrightarrow{PQ}$.

(15) J. Bolyai showed how to construct limiting parallel rays using the segment-circle continuity principle.

(16) In hyperbolic geometry, if $l \parallel m$, then there exist three points on m that are equidistant from l.

(17) In hyperbolic geometry, if m is any line parallel to l, then there exist two points on m which are equidistant from l.

(18) In hyperbolic geometry, if P is a point not lying on line l, then there are exactly two lines through P parallel to l.

(19) In hyperbolic geometry, if P is a point not lying on line l, then there are exactly two lines through P perpendicular to l.

(20) In hyperbolic geometry, if $l \parallel m$ and $m \parallel n$, then $l \parallel n$ (transitivity of parallelism).

(21) In hyperbolic geometry, if m contains a limiting parallel ray to l, then l and m have a common perpendicular.

(22) In hyperbolic geometry, if l and m have a common perpendicular, then there is one point on m that is closer to l than any other point on m.

(23) In hyperbolic geometry, if m does not contain a limiting parallel ray to l and if m and l have no common perpendicular, then m intersects l.

(24) In hyperbolic geometry, the summit of any Saccheri quadrilateral is greater than the base.

(25) Every valid theorem of neutral geometry is also valid in hyperbolic geometry.

(26) In hyperbolic geometry, opposite angles of any parallelogram are congruent to each other.

(27) In hyperbolic geometry, opposite sides of any parallelogram are congruent to each other.

(28) In hyperbolic geometry, let $\sphericalangle$P be any acute angle, let X be any point on one side of this angle, and let Y be the foot of the perpendicular from X to the other side. If X recedes without bound from P along its side, then Y will recede without bound from P along its side.

(29) In hyperbolic geometry, if three points are not collinear, there is always a circle that passes through them.

(30) In hyperbolic geometry, there exists an angle and there exists a line that lies entirely within the interior of this angle.

(31) Limiting parallel rays exist in Euclidean planes.

Exercises

1. **This is perhaps the most important exercise in this book.** It is a payoff for all the work you have done. Come back to this exercise as you do subsequent exercises and read further in the book. Your assignment in this exercise is to make a long list of geometric statements that are equivalent to the Euclidean parallel postulate in the sense that they hold in real Euclidean planes and do not hold in real hyperbolic planes. The statements proved in neutral geometry are valid in both Euclidean and hyperbolic planes, so ignore them. To get you started, here are 10 statements that qualify. *They do not say anything about parallel lines*, so you might have been surprised before studying this subject that they are equivalent to the Euclidean parallel postulate.

   Every triangle has a circumscribed circle.
   Wallis' postulate on the existence of similar triangles.
   A rectangle exists.
   Clavius' axiom that the equidistant locus on one side of a line is the set of points on a line.
   Some triangle has an angle sum equal to 180°.
   An angle inscribed in a semicircle is a right angle.
   The Pythagorean equation holds for right triangles.
   A line cannot lie entirely in the interior of an angle.
   Any point in the interior of an angle lies on a segment with endpoints on the sides of the angle.
   Areas of triangles are unbounded.

2. This problem has five parts. In the first part we will construct Saccheri quadrilaterals associated with any triangle △ABC. Then we will apply this construction. Figure 6.15 illustrates the case where the angles of the triangle at A and B are acute; you are invited to draw the figure when one of these angles is obtuse or right.

   (a) Let I, J, K be the midpoints of BC, CA, AB, respectively. Let D, E, F be the feet of the perpendiculars from A, B, C, respectively, to $\overleftrightarrow{IJ}$ (which is called a *medial line*). Prove, in any Hilbert

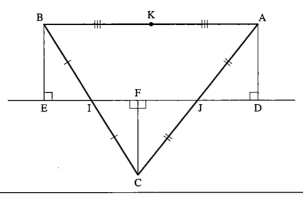

Figure 6.15

plane, that AD ≅ CF ≅ BE, hence that □EDAB is a Saccheri quadrilateral with base ED, summit AB. Show that a triangle and its associated Saccheri quadrilateral have equal content— i.e., that you can dissect the Saccheri quadrilateral region into polygonal pieces and then reassemble these pieces to construct the triangular region.

(b) Prove that the perpendicular bisector of AB is also perpendicular to $\overleftrightarrow{IJ}$. (Hint: Use a result about Saccheri quadrilaterals.) Hence if the plane is hyperbolic, $\overleftrightarrow{IJ}$ is divergently parallel to $\overleftrightarrow{AB}$. Assume now the plane is real, so lengths can be assigned (Theorem 4.3) and the Saccheri–Legendre theorem applies.

(c) Prove that $\overline{ED} = 2\overline{IJ}$. Deduce that $\overline{AB} > 2\overline{IJ}$ (respectively $\overline{AB} = 2\overline{IJ}$) if the plane is hyperbolic (respectively is Euclidean).

(d) Prove that K, F, and C are collinear if and only if AC ≅ BC (isosceles triangle). If that is the case, prove that F is the midpoint of CK iff the plane is Euclidean. If K, F, and C are not collinear and the plane is not Euclidean, prove that $\overleftrightarrow{CF}$ is not perpendicular to $\overleftrightarrow{AB}$ (ray $\overrightarrow{CF}$ does intersect AB at some point G in the case shown, where the angles at A and B are acute, by the crossbar theorem, but CG is not an altitude of the triangle if the plane is not Euclidean).

(e) Show that if the Pythagorean equation holds for all right triangles and if ⟨C is a right angle, then $\overline{AB} = 2\overline{IJ}$ can be proved. Deduce from part (c) that such a plane must be Euclidean. (Use these results to add to your answers to Exercise 1.)

The remaining exercises are in hyperbolic geometry. You can use the results proved in this chapter as well as any results proved in neutral geometry in previous chapters. Do not use any of the

Euclidean results from Exercises 10–18 and the major exercises of Chapter 5.

3. Assume that the parallel lines l and l' have a common perpendicular segment MM'. Prove that MM' is the shortest segment between any point of l and any point of l'. (Hint: In showing MM' < AA', first dispose of the case in which AA' is perpendicular to l' by means of a result about Lambert quadrilaterals and then take care of the other case by Exercise 22, Chapter 4.)

4. Again, assume that MM' is the common perpendicular segment between l and l'. Let A and B be any points of l such that M * A * B and drop perpendiculars AA' and BB' to l'. Prove that AA' < BB'. (Hint: Use Proposition 4.13; see Figure 6.16.)

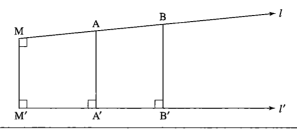

Figure 6.16

5. Given parallel lines l and m. Given points A and B that lie on the opposite side of m from l; i.e., for any point P on l, A and P are on opposite sides of m, and B and P are on opposite sides of m. Prove that A and B lie on the same side of l. (This holds in any Hilbert plane.)

6. Let $\overrightarrow{PY}$ be a limiting parallel ray to l through P and let X be a point on this ray between P and Y (Figure 6.17). It may seem intuitively obvious that $\overrightarrow{XY}$ is a limiting parallel ray to l *through* X, but this requires proof. Justify the steps that have not been justified.

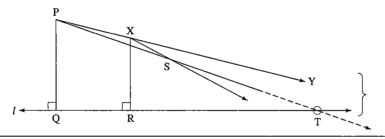

Figure 6.17

PROOF:

(1) We must prove that any ray $\overrightarrow{XS}$ between $\overrightarrow{XY}$ and $\overrightarrow{XR}$ meets l, where R is the foot of the perpendicular from X to l. (2) S and Y are on the same side of $\overleftrightarrow{XR}$. (3) P and Y are on opposite sides of $\overleftrightarrow{XR}$. (4) By Exercise 5, S and Y are on the same side of $\overleftrightarrow{PQ}$. (5) S and R are on the same side of $\overleftrightarrow{XY} = \overleftrightarrow{PY}$. (6) Q and R are on the same side of $\overleftrightarrow{PY}$. (7) Q and S are on the same side of $\overleftrightarrow{PY}$. (8) Thus, $\overrightarrow{PS}$ lies between $\overrightarrow{PY}$ and $\overrightarrow{PQ}$, so it intersects l in a point T. (9) Point X is exterior to $\triangle PQT$. (10) $\overrightarrow{XS}$ does not intersect PQ. (11) Hence $\overrightarrow{XS}$ intersects QT (Proposition 3.9(a)), so $\overrightarrow{XS}$ meets l. ◄

7. Let us assume instead that $\overrightarrow{XY}$ is limiting parallel to l, with P * X * Y. Prove that $\overrightarrow{PY}$ is limiting parallel to l. (Hint: See Figure 6.18. You must show that $\overrightarrow{PZ}$ meets l in a point V. Choose any S such that S * P * Z. Show that SX meets $\overleftrightarrow{PQ}$ in a point U such that U * P * Q. Choose any W such that U * X * W and show that $\overrightarrow{XW}$ is between $\overrightarrow{XY}$ and $\overrightarrow{XR}$ so that $\overrightarrow{XW}$ meets l in a point T. Apply Proposition 3.9(a) to get V.)

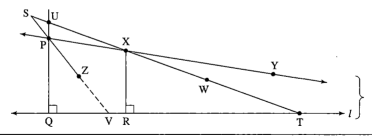

Figure 6.18

8. Let $\overrightarrow{PX}$ be the right limiting parallel ray to l through P and let Q and X' be the feet of the perpendiculars from P and X, respectively, to l (Figure 6.19). Prove that PQ > XX'. (Hint: Use Exercise 6 to

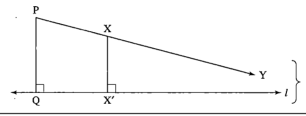

Figure 6.19

show that ∢X'XY is acute and that ∢X'XP is obtuse, so that Proposition 4.13, Chapter 4, can be applied to □PQX'X.) This exercise shows that the distance from X to l decreases as X recedes from P along a limiting parallel ray. In fact, one can prove that the distance from X to l approaches zero (see Major Exercise 11).

9. Assume that the parallel lines l and l' have a common perpendicular $\overleftrightarrow{PQ}$. For any point X on l, let X' be the foot of the perpendicular from X to l'. Prove that as X recedes endlessly from P on l, the segment XX' increases indefinitely; see Figure 6.20. (Hint: We saw that it increases in Exercise 4. Drop a perpendicular XY to the limiting parallel ray between $\overrightarrow{PX}$ and $\overrightarrow{PX'}$. Use the crossbar theorem to show that $\overrightarrow{PY}$ intersects XX' in a point Z. Use Proposition 4.5 to show that XZ ≥ XY. Conclude by applying Aristotle's axiom.)

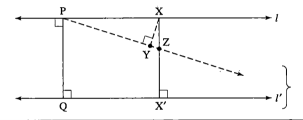

Figure 6.20

10. In Exercise 5, Chapter 5, we saw the elder Bolyai's false proof of the parallel postulate. The flaw in his argument was the assumption that *every* triangle has a circumscribed circle, i.e., that there is a circle passing through the three vertices of the triangle. The idea of the Euclidean proof of this assumption is to show that the perpendicular bisectors of the sides of the triangle meet in a point and that this point is the center of the circumscribed circle. Figure out how Euclid's fifth postulate is used to prove that two of the perpendicular bisectors l and m have a common point (use Proposition 4.10) and then argue by congruent triangles to prove that the third perpendicular bisector passes through that point and that the point is equidistant from the three vertices. (Hint: Join the common point D to the midpoint N of the third side and prove that $\overleftrightarrow{DN}$ is perpendicular to the third side; see Figure 6.21.)

11. Part of the argument in Exercise 10 works for hyperbolic geometry; that is, *if* two of the perpendicular bisectors have a common point, then the third perpendicular bisector also passes through that point.

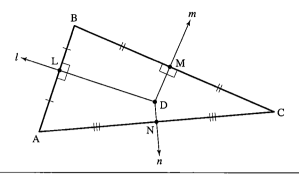

Figure 6.21

In hyperbolic geometry, there will be triangles for which two of the perpendicular bisectors are parallel (otherwise the elder Bolyai's proof would be correct). Moreover, these perpendicular bisectors can be parallel in two different ways. Suppose that they are divergently parallel; that is, suppose that the perpendicular bisectors *l* and *m* have a common perpendicular *t* (see Figure 6.22). Prove that the third perpendicular bisector *n* is also perpendicular to *t*. (Hint: Let A', B', and C' be the feet on *t* of the perpendiculars dropped from A, B, and C, respectively. Let *l* bisect AB at L and be perpendicular to *t* at L' and let *m* bisect BC at M and be perpendicular to *t* at M'. Let N be the midpoint of AC. Show by Proposition 6.5 that

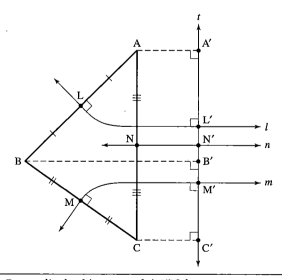

Figure 6.22 Perpendicular bisectors of △ABC have common perpendicular *t*.

AA′ ≅ BB′ and CC′ ≅ BB′. Hence □C′A′AC is a Saccheri quadrilateral with N the midpoint of its summit AC. If N′ is the midpoint of the base A′C′, use Theorem 6.1 to show that $n = \overleftrightarrow{NN'}$ is perpendicular to t and $\overleftrightarrow{AC}$; see Major Exercise 7 for the asymptotically parallel case.)

12. In Theorem 4.1 it was proved in neutral geometry that if alternate interior angles are congruent, then the lines are parallel. Strengthen this result in hyperbolic geometry by proving that the lines are divergently parallel, i.e., that they have a common perpendicular. (Hint: Let M be the midpoint of transversal segment PQ and drop perpendiculars MN and ML to lines m and l; see Figure 6.23. Prove that L, M, and N are collinear by the method of congruent triangles.)

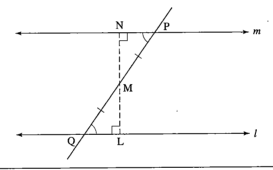

Figure 6.23

13. Prove that Aristotle's axiom holds in a hyperbolic plane. (Hint: For the given acute angle, lay off a segment of parallelism along one side and erect the perpendicular ray at the end of that segment which is limiting parallel to the other side. On that perpendicular ray, lay off the challenge segment AB, at the end of which erect the perpendicular ray that hits the other side of the angle, and from that point of intersection X drop a perpendicular XY to the first side. From the Lambert quadrilateral thus formed, deduce that XY > AB.)

14. Prove that a non-Euclidean Hilbert plane satisfying the important corollary to Aristotle's axiom (stated on p. 133) also satisfies the acute angle hypothesis. (Hint: Find a triangle whose angle sum is <180°.)

15. Comment on the following injunction by Saint Augustine: "The good Christian should beware of mathematicians and all those who make empty prophesies. The danger already exists that the mathematicians have made a covenant with the devil to darken the spirit and to confine man in the bonds of Hell."

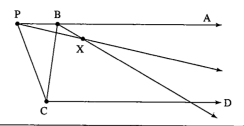

Figure 6.24

Major Exercises

1. Let A, D be the points on the same side of line $\overleftrightarrow{BC}$ such that $\overleftrightarrow{BA} \parallel \overleftrightarrow{CD}$. Then the figure consisting of segment BC (called the *base*) and rays $\overrightarrow{BA}$ and $\overrightarrow{CD}$ (called the *sides*) is called the *biangle* [ABCD with *vertices* B and C (see Figure 6.24). The *interior* of [ABCD is the intersection of the interiors of its *angles* ∢ABC and ∢DCB; if P lies in the interior and X is either vertex, ray $\overrightarrow{XP}$ is called an *interior ray*. We write the relation $\overrightarrow{BA} \mid \overrightarrow{CD}$ when these rays are sides of a biangle and when every interior ray emanating from B intersects $\overrightarrow{CD}$; in that case, we say that $\overrightarrow{BA}$ is *limiting parallel* to $\overrightarrow{CD}$, generalizing the previous definition which required ∢DCB to be a right angle, and we say that the biangle [ABCD is *closed* at B. Given $\overrightarrow{BA} \mid \overrightarrow{CD}$, prove the following generalization of Exercise 6: If P * B * A or if B * P * A, then $\overrightarrow{PA} \mid \overrightarrow{CD}$.

2. *Symmetry of limiting parallelism.* If $\overrightarrow{BA} \mid \overrightarrow{CD}$, then $\overrightarrow{CD} \mid \overrightarrow{BA}$. (In that case, we say simply that biangle [ABCD is *closed*.) Justify the unjustified steps in the proof (see Figure 6.25).

PROOF:

(1) Assume that [ABCD is not closed at C. (2) Then some interior ray $\overrightarrow{CE}$ does not intersect $\overrightarrow{BA}$. (3) Point E, which so far is just a label, can be chosen so that ∢BEC < ∢ECD, by the important corol-

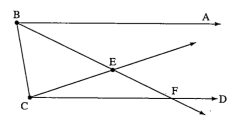

Figure 6.25

lary to Aristotle's axiom, Chapter 3. (4) Segment BE does not intersect $\overrightarrow{CD}$. (5) Interior ray $\overrightarrow{BE}$ intersects $\overrightarrow{CD}$ in a point F, and B * E * F. (6) Since ∢BEC is an exterior angle for △EFC, we have ∢BEC > ∢ECF. (7) Contradiction. (I am indebted to George E. Martin for this simple proof.) ◄

3. *Transitivity of limiting parallelism.* If $\overrightarrow{AB}$ and $\overrightarrow{CD}$ are both limiting parallel to $\overrightarrow{EF}$, then they are limiting parallel to each other. Justify the steps in the proof.

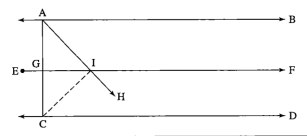

Figure 6.26

PROOF:

(1) $\overleftrightarrow{AB}$ and $\overleftrightarrow{CD}$ have no point in common. (2) Hence there are two cases depending on whether $\overleftrightarrow{EF}$ is between $\overleftrightarrow{AB}$ and $\overleftrightarrow{CD}$ or $\overleftrightarrow{AB}$ and $\overleftrightarrow{CD}$ are both on the same side of $\overleftrightarrow{EF}$. (3) In the case where $\overleftrightarrow{EF}$ is between $\overleftrightarrow{AB}$ and $\overleftrightarrow{CD}$, let G be the intersection of AC with $\overleftrightarrow{EF}$ (see Figure 6.26). We may assume G lies on ray $\overrightarrow{EF}$; otherwise we can consider $\overrightarrow{GF}$. (4) Any ray $\overrightarrow{AH}$ interior to ∢GAB must intersect $\overrightarrow{EF}$ in a point I. (5) $\overrightarrow{IH}$, lying interior to ∢CIF, must intersect $\overrightarrow{CD}$. (6) Hence any ray $\overrightarrow{AH}$ interior to ∢CAB must intersect $\overrightarrow{CD}$, so $\overrightarrow{AB}$ is limiting parallel to $\overrightarrow{CD}$.

    Step (7) is the following sublemma. That this requires such a long proof was overlooked even by Gauss. The proof (for which I am indebted to Edwin E. Moise) uses our hypotheses of limiting parallelism. If we had made the weaker hypothesis of just parallel lines, the sublemma would not follow, as you will show in Exercise K-2(c) of Chapter 7. ◄

SUBLEMMA. If $\overleftrightarrow{AB}$ and $\overleftrightarrow{CD}$ are both on the same side of $\overleftrightarrow{EF}$, then one of them, say $\overleftrightarrow{CD}$, is between $\overleftrightarrow{AB}$ and $\overleftrightarrow{EF}$.

PROOF OF SUBLEMMA:

(1) It suffices to prove there is a line transversal to the three rays $\overrightarrow{AB}$, $\overrightarrow{CD}$, $\overrightarrow{EF}$. (2) In the case where A and F are on the same side of

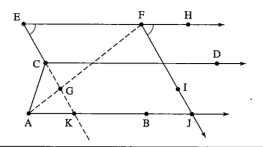

Figure 6.27

$\overleftrightarrow{EC}$, then ray $\overrightarrow{EA}$ is interior to ⦡E. (3) Then $\overrightarrow{EA}$ intersects $\overrightarrow{CD}$, by symmetry. (4) So $\overleftrightarrow{EA}$ is our transversal. (5) In the case where A and F are on opposite sides of $\overleftrightarrow{EC}$, let G be the point at which AF meets $\overleftrightarrow{EC}$ (see Figure 6.27). (6) Choosing H such that E * F * H, we have $\overrightarrow{FH} \mid \overrightarrow{AB}$. (7) ⦡HFG > ⦡E. (8) Therefore there is a ray $\overrightarrow{FI}$ interior to ⦡HFA = ⦡HFG such that ⦡HFI ≅ ⦡E. (9) $\overrightarrow{FI}$ meets $\overrightarrow{AB}$ at a point J. (10) $\overleftrightarrow{FJ} \parallel \overleftrightarrow{EC}$. (11) $\overleftrightarrow{EC}$ intersects side AF and does not intersect side FJ of △AFJ. (12) Hence $\overleftrightarrow{EC}$ intersects AJ and is our transversal. ◄

CONCLUSION OF PROOF OF TRANSITIVITY (SEE FIGURE 6.28):

(8) Then AE intersects $\overleftrightarrow{CD}$ in a point G, which we may assume lies on ray $\overrightarrow{CD}$. (9) Any ray $\overrightarrow{AH}$ interior to ⦡GAB intersects $\overrightarrow{EF}$ in a point I. (10) Since $\overrightarrow{CD}$ enters △AEI at G and does not intersect side EI, it must intersect AI. (11) Therefore, $\overrightarrow{CD}$ is limiting parallel to $\overrightarrow{AB}$. ◄

NOTE 1. The last four steps did not use the hypothesis that $\overrightarrow{CD} \mid \overrightarrow{EF}$; they therefore prove that *any line between two asymptotically parallel lines is asymptotically parallel to both and in the same direction.*

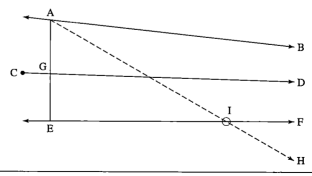

Figure 6.28

NOTE 2. Given rays r and s, define $r \sim s$ to mean that $r \subset s$ or $s \subset r$ or $r \mid s$. Major Exercises 1–3 show that this is an equivalence relation among rays. An equivalence class of rays is called an *ideal point*, or an *end*, and we adopt the convention that it lies on all (and only those) lines containing the rays making up the class. Since a point on a line breaks the line into two opposite rays and opposite rays are not equivalent, we see that *every line has two ends lying on it*. The set of all ideal points was named the *absolute* by Cayley. (This is the beginning of constructing a hyperbolic analogue of the projective completion of an affine plane described in Chapter 2; we continue the construction in Major Exercise 13. The absolute is analogous to the line at infinity of the affine plane, but the absolute could not be a new line because it intersects each old line in two points; it will turn out to be a *conic* in the projective completion.)

If R, S are the vertices of r, s, where $r \mid s$, and Ω is the ideal point determined by these rays, we write $r = P\Omega$ and $s = S\Omega$ and refer to the closed biangle with sides r, s as the *singly asymptotic triangle* $\triangle RS\Omega$. The next two exercises show that these triangles have some properties in common with ordinary triangles. (You can similarly define as an exercise *doubly* (two ideal points) and *triply* (three ideal points) *asymptotic triangles.*)

4. *Exterior angle theorem.* If $\triangle PQ\Omega$ is a singly asymptotic triangle, the exterior angles at P and Q are greater than their respective opposite interior angles. Justify the steps in the proof.

PROOF (SEE FIGURE 6.29):

(1) Given $R * Q * P$. We must show that $\sphericalangle RQ\Omega$ is greater than $\sphericalangle QP\Omega$. (2) Let $\overrightarrow{QD}$ be the unique ray on the same side of $\overleftrightarrow{PQ}$ as ray $Q\Omega$ such that $\sphericalangle RQD \cong \sphericalangle QP\Omega$. (3) If $U * Q * D$, then $\sphericalangle UQP \cong QP\Omega$.

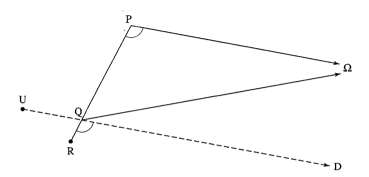

Figure 6.29

(4) By Exercise 12, $\overleftrightarrow{QD}$ is divergently parallel to $\overleftrightarrow{P\Omega}$. (5) Hence $\overrightarrow{QD}$ is between $\overrightarrow{QR}$ and $\overrightarrow{Q\Omega}$. (6) $\sphericalangle RQ\Omega > \sphericalangle QP\Omega$. ◄

5. *Congruence theorem.* If in asymptotic triangles $\triangle AB\Omega$ and $\triangle A'B'\Omega'$ we have $\sphericalangle BA\Omega \cong \sphericalangle B'A'\Omega'$, then $\sphericalangle AB\Omega \cong \sphericalangle A'B'\Omega'$ if and only if $AB \cong A'B'$. Justify the steps in the proof and deduce as a corollary that $PQ \cong P'Q'$ if and only if $\Pi(PQ)° = \Pi(P'Q')°$.

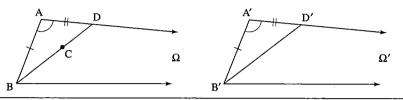

Figure 6.30

PROOF (SEE FIGURE 6.30):

(1) Assume $AB \cong A'B'$ and on the contrary $\sphericalangle AB\Omega > \sphericalangle A'B'\Omega'$. (2) There is a unique ray $\overrightarrow{BC}$ between $B\Omega$ and $\overrightarrow{BA}$ such that $\sphericalangle ABC \cong \sphericalangle A'B'\Omega'$. (3) $\overrightarrow{BC}$ intersects $A\Omega$ in a point D. (4) Let D' be the unique point on $A'\Omega'$ such that $AD \cong A'D'$. (5) Then $\triangle BAD \cong \triangle B'A'D'$. (6) Hence $\sphericalangle A'B'D' \cong \sphericalangle A'B'\Omega'$, which is absurd. (7) Assume conversely that $\sphericalangle AB\Omega \cong \sphericalangle A'B'\Omega'$ and, on the contrary, $A'B' < AB$. (8) Let C be the point on AB such that $BC \cong B'A'$ and let $C\Omega$ be the ray from C limiting parallel to $A\Omega$ (see Figure 6.31). (9) Then $C\Omega$ is also limiting parallel to $B\Omega$. (10) By the first part of the proof, $\sphericalangle BC\Omega \cong \sphericalangle B'A'\Omega'$; hence we have $\sphericalangle BC\Omega \cong \sphericalangle BA\Omega$. (11) But $\sphericalangle BC\Omega > \sphericalangle BA\Omega$, which is a contradiction. ◄

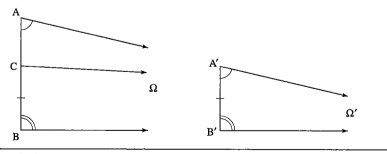

Figure 6.31

6. *Conclusion of the proof of Theorem 6.3.* We wish to show that $\overrightarrow{EF}$ intersects $\overrightarrow{AG}$ (see Figure 6.32). Justify the steps in the proof.

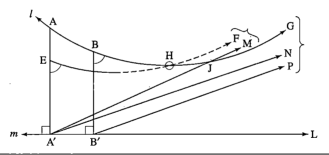

Figure 6.32

PROOF:

(1) Let $\overrightarrow{A'M}$ be limiting parallel to $\overrightarrow{EF}$, $\overrightarrow{A'N}$ limiting parallel to $\overrightarrow{AG}$, and $\overrightarrow{B'P}$ limiting parallel to $\overrightarrow{BG}$. (2) Since $EA' \cong BB'$ and $\sphericalangle A'EF \cong \sphericalangle B'BG$, we have $\sphericalangle EA'M \cong \sphericalangle BB'P$. (3) $\overrightarrow{B'L}$ differs from $\overrightarrow{B'P}$, and $\overrightarrow{A'L}$ differs from $\overrightarrow{A'N}$. (4) $\sphericalangle MA'L \cong \sphericalangle PB'L$. (5) $\overrightarrow{B'P}$ is limiting parallel to $\overrightarrow{A'N}$. (6) Hence $\sphericalangle NA'L$ is smaller than $\sphericalangle PB'L$. (7) It follows that $\overrightarrow{A'M}$ lies between $\overrightarrow{A'N}$ and $\overrightarrow{A'A}$, so it must intersect $\overrightarrow{AG}$ in a point J. (8) J is on the same side of $\overleftrightarrow{EF}$ as A'; hence it is on the side opposite from A. (9) Thus AJ intersects $\overleftrightarrow{EF}$ in a point H, which must be on $\overrightarrow{EF}$ because H is on the same side of $\overleftrightarrow{AA'}$ as J. ◄

Where was the hypothesis of this theorem used?

7. In Exercises 10 and 11 we considered the perpendicular bisectors of the sides of $\triangle ABC$ and showed that (1) if two of them have a common point, the third passes through that point; (2) if two of them have a common perpendicular, the third has that same perpendicular. It follows that if two of them are asymptotically parallel, then any two of them are asymptotically parallel. This result can be strengthened as follows: If perpendicular bisectors l and m are asymptotically parallel in the direction of ideal point Ω, then the third perpendicular bisector n is asymptotically parallel to l and m in the *same direction* Ω. Give the proof and justify each step. The proof is based on the following two lemmas.

LEMMA 6.1. Given $\triangle ABC$. Let l, m, and n be the perpendicular bisectors of sides AB, BC, and AC at their midpoints L, M, and N, respectively. Let $\overline{AC} \geq \overline{AB}$ and $\overline{AC} \geq \overline{BC}$ (AC is the longest side). Then l, m, and n all intersect AC.

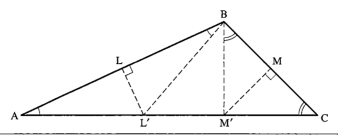

Figure 6.33

PROOF:

(1) $(\angle B)^\circ \geqq (\angle A)^\circ$ and $(\angle B)^\circ \geqq (\angle C)^\circ$. (2) Hence there is a point L′ on AC such that $\angle A \cong \angle L'BA$, and a point M′ on AC such that $\angle C \cong \angle M'BC$ (see Figure 6.33). (3) Then we have $AL' \cong BL'$ and $CM' \cong BM'$. (4) Thus l is the line joining L to L′, and $m = \overleftrightarrow{MM'}$. (5) It follows that all three perpendicular bisectors cut AC. ◄

LEMMA 6.2 No line intersects all three sides of a trebly asymptotic triangle.

PROOF:

(1) Suppose that a line t cuts l at Q and m at P. (2) Then ray $\overrightarrow{PQ}$ of t lies between the rays $P\Omega_2$ and $P\Omega_1$, which are limiting parallel to l (see Figure 6.34). (3) $P\Omega_3$, the other ray through P that is limiting parallel to n, is opposite to $P\Omega_2$. (4) Hence $P\Omega_1$ lies between $\overrightarrow{PQ}$ and $P\Omega_3$. (5) Thus $\overrightarrow{PQ}$ does not intersect n. (6) Similarly, $\overrightarrow{QP}$ does not intersect n. ◄

8. Given any angle $\angle A'OA$. It is a theorem in hyperbolic geometry that there is a unique line l called the *line of enclosure* of this angle such

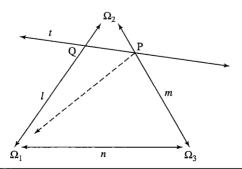

Figure 6.34

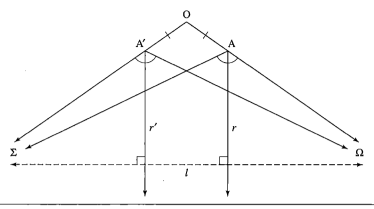

Figure 6.35

that l is limiting parallel to both sides $\overrightarrow{OA'}$ and $\overrightarrow{OA}$. Only the idea of the proof is given here; fill in the details (Hartshorne, Proposition 40.6).

Assume that A and A' are chosen so that $OA \cong OA'$ (see Figure 6.35). Let $A'\Omega$ be the limiting parallel ray to $\overrightarrow{OA}$ through A', and $A\Sigma$ the limiting parallel ray to $\overrightarrow{OA'}$ through A. Let the rays r and r' be the bisectors of $\sphericalangle\Sigma A\Omega$ and $\sphericalangle\Omega A'\Sigma$, respectively. The idea of the proof is to show that the lines m and m' containing these rays are neither intersecting nor asymptotically parallel, so that, by Theorem 6.3, they have a unique common perpendicular l that turns out to be the line of enclosure of $\sphericalangle A'OA$. (See also Exercise K-11, Chapter 7; the advantage of this complicated proof is that it yields a straightedge-and-compass construction.)

9. Use the result of the previous exercise to prove that every acute angle is an angle of parallelism, i.e., given an acute angle $\sphericalangle BOA$, there is a unique line l perpendicular to $\overleftrightarrow{BO}$ and limiting parallel to $\overrightarrow{OA}$. (Hint: Reflect across $\overleftrightarrow{OB}$.)

Alternatively, fill in the details of the following continuity proof of Lobachevsky. First show that there exist perpendiculars to $\overrightarrow{OB}$ that fail to intersect $\overrightarrow{OA}$ by the following argument. In Figure 6.36, B is the foot of the perpendicular from A and $OB \cong BB'$. If the perpendicular at B' intersects $\overrightarrow{OA}$ at A', then

$$\delta OA'B' = \delta OAB + \delta AA'B' = 2\delta OAB + \delta AA'B' > 2\delta OAB.$$

If we iterate this doubling along $\overrightarrow{OB}$ and the perpendicular always hits $\overrightarrow{OA}$, the defects of the resulting triangles will increase indefinitely. So we must eventually arrive at a point where the perpendicular fails to intersect $\overrightarrow{OA}$.

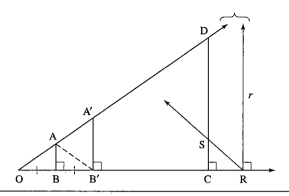

Figure 6.36

Second, apply Dedekind's axiom to obtain "the first" such perpendicular ray r emanating from R.

Finally, show that $r \mid \overrightarrow{OA}$: For any interior ray $\overrightarrow{RS}$, let C be the foot of the perpendicular from S; show that $\overrightarrow{CS}$ hits $\overrightarrow{OA}$ at some point D and apply Pasch's theorem to $\triangle OCD$.

10. Let l and m be divergently parallel lines and let t be their common perpendicular cutting l at Q and m at P (Figure 6.37). Let r be a ray of l emanating from Q and s the ray of m emanating from P on the same side of t as r. Prove that there is a unique point R on r such that the perpendicular to l through R is limiting parallel to s. Prove also that for every point R' on r such that R' * R * Q, the perpendicular to l through R' is divergently parallel to m. (Hint: Use Major Exercises 3 and 9.)

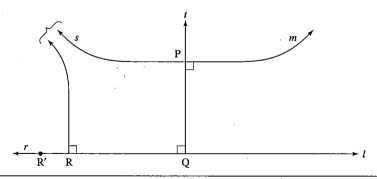

Figure 6.37

11. Let ray r emanating from point P be limiting parallel to line l and let Q be the foot of the perpendicular from P to l (Figure 6.38). Jus-

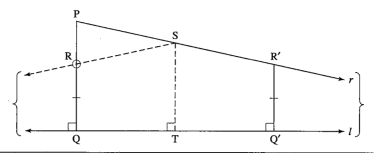

Figure 6.38

tify the terminology "asymptotically parallel" by proving that for any point R between P and Q there exists a point R′ on ray r such that R′Q′ ≅ RQ, where Q′ is the foot of the perpendicular from R′ to l. (Hint: Use Major Exercise 3 and Proposition 6.6 to prove that the line through R that is asymptotically parallel to l in the opposite direction from r intersects r at a point S. Show that if T is the foot of the perpendicular from S to l, the point R′ obtained by reflecting R across line $\overleftrightarrow{ST}$ is the desired point.)

Similarly, show that the lines diverge in the other direction. Use a similar method to prove that the perpendiculars dropped from one line divergently parallel to another are unbounded.

12. Let l and n be divergently parallel lines and PQ their common perpendicular segment. The midpoint S of PQ is called *the symmetry point of l and n*. Let m be the perpendicular to PQ through S. Let Ω and Ω' be the ideal points of l and let Σ and Σ' be the ideal points of n (labeled as in Figure 6.39). By Major Exercise 8, there are unique lines "joining" these ideal points. Prove that (a) $\Omega\Sigma'$ and $\Sigma\Omega'$ meet at S; (b) m is perpendicular to both $\Omega\Sigma$ and $\Omega'\Sigma'$. (Hint: Use Major Exercise 5 and the symmetry part of Proposition 6.6.)

13. *Projective completion of the hyperbolic plane.* The ideal points were defined in Note 2 after Major Exercise 3. By adding them as ends

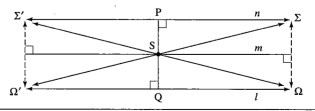

Figure 6.39

to our lines, we ensure that asymptotically parallel lines meet at an ideal point; Major Exercise 11 shows that the lines do converge in the direction of that common end. We need to add more "points at infinity" to ensure that divergently parallel lines will meet. Two divergently parallel lines have a unique common perpendicular t. A third line perpendicular to t can be considered to have "the same direction" as the first two, so all three should meet at the same point, just as in the projective completion of the Euclidean plane. We therefore define the *pole* $P(t)$ to be the set of all lines perpendicular to t and specify that $P(t)$ lies on all those lines and no others; poles of lines are called *ultra-ideal points*. Note here that $t \neq u \Rightarrow P(t) \neq P(u)$ (uniqueness of the common perpendicular, if one exists), unlike the Euclidean case. A "point" of the projective completion $\mathcal{P}$ is defined to be either a point of the hyperbolic plane (called "ordinary") or an ideal point or an ultra-ideal point.

We also add new "lines at infinity" as follows. The *polar* $p(A)$ of an ordinary point A is the set of all poles of lines through A, and the only points incident with $p(A)$ are those poles; polars of ordinary points are called *ultra-ideal lines*. The polar $p(\Omega)$ of an ideal point Ω consists of Ω and all poles of lines having Ω as an end; again, the incidence relation is $\in$, and $p(\Omega)$ is called an *ideal line*. The polar of an ultra-ideal point $P(t)$ is just t. A "line" of $\mathcal{P}$ is defined to be a polar of a point of $\mathcal{P}$. We have defined incidence already. The pole of $p(A)$ is A and of $p(\Omega)$ is Ω.

THEOREM. $\mathcal{P}$ is a projective plane and p is a polarity (an isomorphism of $\mathcal{P}$ onto its dual plane).

Since the ideal points are the only points of $\mathcal{P}$ that lie on their polars, the absolute γ is by definition the conic determined by polarity p, and $p(\Omega)$ is the tangent line to γ at Ω (see Project 2, Chapter 2). If Ω and Σ are the two ends of ordinary line t, then, by definition, the point of intersection of the two tangent lines $p(\Omega)$ and $p(\Sigma)$ is $P(t)$, which gives geometric meaning to the rather abstract $P(t)$. Moreover, the interior of γ is the set of ordinary points since every line through an ordinary point is ordinary and intersects γ twice.

Your exercise is to prove this theorem. To get you started, we show that Axiom I-1 holds for $\mathcal{P}$:

(i)   Two ordinary points A, B lie on ordinary line $\overleftrightarrow{AB}$ and do not lie on any "extraordinary" lines by definition of the latter.

(ii) Given ordinary A and ideal Ω, they are joined by the ordinary line containing ray AΩ.

(iii) Given ideal points Ω and Σ, let A be any ordinary point and consider the rays AΣ and AΩ. If these are opposite, then the line containing them joins Ω and Σ; otherwise, the line of enclosure (Major Exercise 8) of the angle determined by these coterminal rays joins Ω and Σ.

(iv) Given ordinary A and ultra-ideal P(t), the line joining them is the perpendicular to t through A.

(v) Given ideal Ω and ultra-ideal P(t). If Ω lies on t, these points lie on $p(\Omega)$; by the definition of incidence, they do not lie on any other extraordinary line, and they could not lie on an ordinary line u because u would then be both asymptotically parallel to and perpendicular to t. If Ω does not lie on t, let A be a point on t. If ray AΩ is at right angles to t, the line containing AΩ joins Ω to P(t); otherwise, Major Exercise 9 ensures that there is a unique line $u \perp t$ such that AΩ is limiting parallel to u and u joins Ω to P(t).

(vi) Given ultra-ideal points P(t) and P(u), t meets u either at ordinary point A, in which case $p(A)$ is the join, or at ideal point Ω, in which case $p(\Omega)$ is the join, or, by Theorem 6.3, at ultra-ideal point P(m), in which case m (the common perpendicular to t and u) joins P(t) and P(u).

Projects

1. Here is another construction for the common perpendicular between divergently parallel lines l and n. It suffices to locate their symmetry point S, for a perpendicular can then be dropped from S to both lines (Figure 6.39, p. 285). Take any segment AB on l. Construct point C on l such that B is the midpoint of AC and lay off any segment A'B' on n congruent to AB. Let M, M', N, and N' be the midpoints of AA', BB', BA', and CB', respectively. Then the lines $\overleftrightarrow{MM'}$ and $\overleftrightarrow{NN'}$ are distinct and intersect at S. (The proof follows from the theory of *glide reflections*; see Exercises 21 and 22 in Chapter 9; also see Coxeter, 1998, p. 269, where it is deduced from Hjelmslev's midline theorem.) Report on a proof.

2. Report on the development of plane hyperbolic geometry from Hilbert's hyperbolic axiom of parallelism alone, without bringing in Dedekind's axiom, using Hartshorne, Chapter 7, as one reference.

Describe some proofs in your report, particularly a proof of the acute angle hypothesis and the proof of the existence of the *line of enclosure* of an angle.

3. Report on Hilbert's arithmetic of ends, using his *Foundations of Geometry*, Appendix III for reference. Hilbert constructs a field that can be used to coordinatize a general hyperbolic plane and do analytic geometry in it.

4. Report on Joan Richards' study of the resistance to non-Euclidean geometry in late nineteenth-century England; see her *Mathematical Visions: The Pursuit of Geometry in Victorian England*, Chapters 2 and 3, Academic Press, 1988. Your report should discuss the contributions of Hermann von Helmholtz and William Clifford toward enlightening the English on the philosophical implications of Riemann's new ideas about space. Search the library or the web for more information on how Helmholtz and Clifford spread the ideas of non-Euclidean geometry and developed them further, with Clifford's work being a precursor of general relativity 45 years before Einstein—e.g., http://members.aol.com/jebco1st/Paraphysics/twist1.htm

7 Independence of the Parallel Postulate

All my efforts to discover a contradiction, an inconsistency, in this non-Euclidean geometry have been without success. . . .

C. F. Gauss

Consistency of Hyperbolic Geometry

In the previous chapter, you were introduced to hyperbolic geometry and presented with some theorems that must seem very strange to someone accustomed to Euclidean geometry. Even though you may admit that the proofs of these theorems are correct, given our assumptions, you may feel that the basic assumption of hyperbolic geometry— the hyperbolic parallel axiom of Hilbert—is a false assumption. Let's examine what might be meant by saying it's false.

What sort of experiment could I perform to show that the hyperbolic axiom or the negation of Hilbert's Euclidean parallel postulate is false? First of all, I would have to understand what this statement means. What does it mean that l is a "line," that P is a "point" not "on" l, and that there is at most one "parallel" to l through P? I might represent "points" and "lines" with paper, pencil, and straightedge. Suppose I draw the perpendicular from P to l, draw line m through P perpendicular to PQ, and then draw a line n through P, making a very

289

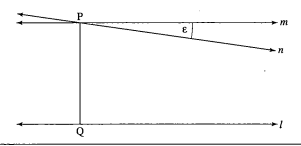

Figure 7.1

small angle ε with m (Figure 7.1). Using Euclidean trigonometry, I can calculate exactly how far out on n I would have to go to get to the point where n is supposed to intersect l, but if ε is small enough, that point might be very far away. Thus I could not physically perform the experiment to prove that the negation of Euclid V is false.

But is geometry about lines that we can draw? Pure geometry is about idealized lines, which are concepts, not objects. The only experiments we can perform on these idealized lines are thought experiments. So the question should be: Can we conceive of a non-Euclidean geometry? Kant said no, that Euclidean geometry is a priori true. At that time, of course, no one had yet conceived of a different geometry. It is in this sense that J. Bolyai and Lobachevsky "created a new universe."

Other questions can be raised. Mathematicians reject many ideas because they either lead to contradictions or do not lead anywhere, i.e., do not prove fruitful, useful, or interesting. Does the hypothesis of the acute angle lead to a contradiction? Saccheri imagined it would and tried to prove Euclid's parallel postulate that way. Is hyperbolic geometry fruitful, useful, or interesting?

Let us postpone the latter question until the end of Chapter 8 (the answer is yes!) and take up the former: Is hyperbolic geometry *consistent?* As was explained in Chapter 2, this is a question in *metamathematics,* i.e., a question outside a mathematical system about the system itself. The question is not about lines or points or other geometric entities; it is a question about the *whole system* of axioms, definitions, and propositions in plane hyperbolic geometry.

If hyperbolic geometry were inconsistent, an ordinary mathematical argument could derive a contradiction. Saccheri tried to do this and failed. Could it be that he wasn't clever enough, that someday some genius will find a contradiction?

On the other hand, can it be proved that hyperbolic geometry is *consistent*—can it be proved that there is no possible way to derive a contradiction?

We might ask the same question about Euclidean geometry: How do we *know* it is consistent? Of course, this was never a burning question before the discovery of non-Euclidean geometry simply because everyone *believed* Euclidean geometry to be consistent since it was supposedly an idealization of physical space. Remarkably enough, if we make this belief an explicit assumption, it is possible to give a proof that hyperbolic geometry is consistent.

METAMATHEMATICAL THEOREM 1. If Euclidean geometry is consistent, then so is hyperbolic geometry.

Granting this result for the moment, we get the following important corollary.

COROLLARY. If Euclidean geometry is consistent, then no proof or disproof of Euclid's parallel postulate from the axioms of neutral geometry will ever be found—Euclid's parallel postulate is independent of the other postulates.

PROOF:

To prove the corollary, assume on the contrary that a proof in neutral geometry of Euclid's parallel postulate exists. Then hyperbolic geometry would be inconsistent since one of its theorems (the negation of Euclid V) contradicts a proved result (recall that neutral geometry is part of hyperbolic geometry). But Metamathematical Theorem 1 asserts that hyperbolic geometry is consistent relative to Euclidean geometry. This contradiction proves that no neutral proof of Euclid's parallel postulate exists (RAA). The hypothesis that Euclidean geometry is consistent ensures that no disproof exists either. ◄

Thus 2000 years of efforts to prove Euclid V were in vain.

Of course, when we say this, we are *assuming* the consistency of the venerable Euclidean geometry. Had Saccheri, Legendre, F. Bolyai, or any of the dozens of other scholars succeeded in proving Euclid V from Euclid's other axioms, with the noble intention of making Euclidean geometry more secure and elegant, they would have instead completely destroyed Euclidean geometry as a consistent body of thought! (I urge you, dear reader, to go over the preceding statements very carefully to make sure you have understood them. If you have not understood, you have missed the main point of this text.)[1] Euclid is "vindicated" by the *failure* of all those dedicated mathematicians who arduously attempted to prove his fifth postulate from his other postulates; the independence of Euclid V (assuming consistency) shows that his insight was profound in assuming a statement that is not "obvious."

In the form given here, Metamathematical Theorem 1 is due to Eugenio Beltrami (1835–1899); a different proof was later given by Felix Klein (1849–1925). Beltrami proved the relative consistency of real hyperbolic geometry in 1868 using differential geometry in a manner influenced by Riemann's new ideas. Klein recognized that projective geometry could be used to give another proof. In 1871 he applied the method to hyperbolic geometry that Arthur Cayley used in 1859 to express distance and angle measure projectively for Euclidean and elliptic geometries. We will discuss their work in the next sections.

To prove Metamathematical Theorem 1, we have to again ask ourselves, what is a "line" in hyperbolic geometry—in fact, what is the hyperbolic plane? The honest answer is that we don't know; it is just an abstraction. A hyperbolic "line" is an undefined term describing an abstract concept that resembles the concept of a Euclidean line except for its parallelism properties. Then how shall we visualize hyperbolic geometry? In mathematics, as in any other field of research, posing the right question is vital.[2]

The question of "visualizing" for us means finding Euclidean objects that represent hyperbolic objects since we are accustomed to seeing diagrams for Euclidean geometry. More precisely, this means finding a Euclidean *model* for hyperbolic geometry. In Chapter 2, we discussed the idea of models for an axiom system; there we showed that the Euclidean parallel postulate is independent of the axioms for incidence geometry by exhibiting three-point and five-point models of incidence geometry that do not satisfy the Euclidean parallel postulate and a four-point model that does. Here we want to know whether the

---

[1] William F. Orr has written the delightful short story "Euclid Alone," about a scientist who believed he had proved Euclidean geometry inconsistent; see http://www.cs. kun.nl/ ~ freek/jordan/euclidalone.html. Some authors state flatly that hyperbolic geometry has been proved consistent because it has Euclidean models; it does not occur to them that the consistency of Euclidean geometry is a hypothesis, not a proven result.

[2] I. I. Rabi, the Nobel Prize–winning physicist, recounted that when he was a boy returning home from school, his mother would usually say, "Did you ask any good questions in school today?" (My thanks to Robert W. Fuller for this anecdote.)

Euclidean parallel postulate is independent of a much larger system of axioms, namely, the axioms for neutral geometry (e.g., the axioms for a Hilbert plane plus Dedekind's axiom). We can show that it is, and by the same method—exhibiting models.

Beltrami's Interpretation

Since Beltrami's work was based on differential geometry, we can only sketch in broad terms what he accomplished. His 1868 paper *Saggio di Interpretazione della Geometria Non-Euclidea* ("Essay on an Interpretation of Non-Euclidean Geometry") has been misrepresented by some popular writers. They claim that he found a model for a region of the real hyperbolic plane only on a certain surface in Euclidean three-space called the *pseudosphere*. He did find that. However, Beltrami also found a model for the entire real hyperbolic plane as a disk in $\mathbb{R}^2$, where hyperbolic lines are represented by Euclidean chords but where the distance function is of course not the usual Euclidean distance.[3] Here is an excerpt from the "sales talk" Beltrami felt he needed to give to begin his groundbreaking study. It shows how controversial these notions still were in 1868.

> In recent times the mathematical public has begun to take an interest in some new concepts which seem destined, if they prevail, to change profoundly the whole complexion of classical geometry.
>
> These concepts are not particularly recent. The master GAUSS grasped them at the beginning of his scientific career, and although his writings do not contain an explicit exposition, his letters confirm that he had always cultivated them and attest his full support for the doctrine of LOBACHEVSKY.
>
> Such attempts at radical innovation in basic principles are encountered not infrequently in the history of ideas. Today they are a natural result of the critical spirit which accompanies all scientific investigation. When these attempts are presented as the fruits of conscientious and sincere investigations, and when they receive the support of a powerful, undisputed authority, it is the duty of men of science to discuss them calmly, avoiding equally both enthusiasm and disapproval. . . .

[3] See p. 11 of the English translation in John Stillwell, *Sources of Hyperbolic Geometry*, American Mathematical Society History of Mathematics Series, vol. 10, 1996. Beltrami's model will be explained in the following sections in a manner that does not require knowledge of differential geometry.

> In this spirit we have sought, to the extent of our ability, to convince ourselves of the results of LOBACHEVSKY's doctrine; then, following the tradition of scientific research, we have tried to find a real substrate for this doctrine, rather than admit the necessity for a new order of entities and concepts.

By "a real substrate" Beltrami meant what we now call a Euclidean model, and having such a model provides a proof of Metamathematical Theorem 1. However, Beltrami did not set out to prove the relative consistency of hyperbolic geometry or the independence of the Euclidean parallel postulate. His purpose was to show that Lobachevsky had not introduced strange new concepts at all, but had merely described the theory of geodesics on surfaces of constant negative curvature, concepts that were familiar to differential geometers.

Eugenio Beltrami

Here is another excerpt, showing that Lambert and Taurinus were on the right track:

> One finds many analogies between the geometries of the sphere and the plane—where the straight lines correspond to geodesics, i.e., great circles—analogies which have been noted in geometry for a long time.

If other analogies, of different type but the same origin, have not been
given equal attention, it is probably because the idea of mapping flex-
ible surfaces onto one another has not become familiar until recently.
. . . We can explain the passage from Euclidean to non-Euclidean
planimetry in terms of the difference between the surfaces of zero curvature and the surfaces of negative curvature.

What Beltrami did was map an abstract complete surface of constant
negative curvature onto a disk in $\mathbb{R}^2$, sending the geodesics of that surface onto chords of the disk, and then observe that the geometry was
hyperbolic.

The curvature of surfaces was first defined and studied in detail by
Gauss. He formulated the concept of a geometry *intrinsic* to a surface,
and his famous *Theorema Egregium* showed that his curvature was intrinsic (see Appendix A). Gauss' student H. F. Minding studied surfaces
of constant negative curvature. He gave the example of the pseudosphere, which is obtained by rotating a curve called a *tractrix* around
its asymptote. It looks like an infinitely long horn. A tractrix is characterized by the property that the tangent line from any point on the
curve to the asymptote has constant length (see Figure 7.2).

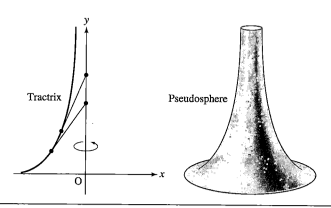

Figure 7.2

As was stated, the pseudosphere is not a model of the entire real
hyperbolic plane but only a model of a horocyclic[4] sector in which
the boundary segments have been identified. Still, the pseudosphere
made a stunning impression in helping people visualize plane hyperbolic geometry at least locally. (Construction of the pseudosphere from

[4] Horocycles will be discussed in Chapters 9 and 10.

a horocyclic sector is analogous to taking a segment in the real Euclidean plane, taking two rays emanating from the endpoints of the segment, perpendicular to and on the same side of it, and then identifying those two rays to form an infinitely long cylinder with a boundary.)

Curiously, in 1839 and 1840 when Minding and in 1857 when Codazzi published their research on surfaces of constant negative curvature, exhibiting the trigonometry on such surfaces, nobody noticed that their formulas were the same as Lobachevsky's until Beltrami made the connection, influenced by Riemann's idea of an abstract geometric surface. In a subsequent 1868 article,[5] Beltrami acknowledged Riemann's ideas and applied them to derive three different models of n-dimensional hyperbolic geometry. In the case where $n = 2$, one of those models is the disk model he exhibited in the previous paper, and the other models (which we will discuss later in this chapter) are now named after Henri Poincaré, who studied them in 1882 and applied them to complex function theory and to quadratic forms. In this second article, Beltrami gives a differential-geometric proof of the result discovered by Wachter in 1816 and shown independently by J. Bolyai and Lobachevsky that a horosphere in hyperbolic three-space has a constant curvature of zero, hence its geometry is Euclidean. Beltrami concludes that the formerly mysterious non-Euclidean geometry of Lobachevsky and J. Bolyai is now transparent from the viewpoint of Riemann. He says:

> Thus all the concepts of non-Euclidean geometry are perfectly matched in the geometry of a space of constant negative curvature. It remains to observe only that whereas the concepts of planimetry receive a true and proper interpretation, because they are *constructible* on a *real* surface, those which embrace three dimensions are susceptible only to an analytic representation. . . . Experience does not seem to accord with the results of this more general geometry. . . . It could be, however, that the triangles we have measured and the portions of space we have observed have been too small, just as measurements on a small portion of the terrestrial surface are insufficiently precise to reveal the sphericity of the globe.

[5] Translated by Stillwell as "Fundamental Theory of Spaces of Constant Curvature," ibid., pp. 35–62. Robert Osserman, in his review of Gray's book on Bolyai, states that overlooking the importance of Beltrami's second article has been "a great historical wrong." See also John Milnor, 1982, "Hyperbolic Geometry: The First 150 Years," *Bulletin of the American Mathematical Society* (N.S.) **6**: 9–24.

Eugenio Beltrami (1835–1899) made other important contributions to differential geometry, analysis, and physics.[6] It was he who, in 1889, resurrected the long-neglected work of Saccheri.

The Beltrami–Klein Model

Felix Klein (1849–1925), in an 1871 article with the translated title "On the So-called Non-Euclidean Geometry,"[7] presented the Beltrami disk model via projective geometry. His formulation is simpler, more general, and more widely known, so the model has come to be named after him. If you read Major Exercise 13 in Chapter 6 on the projective completion of the hyperbolic plane, you will understand the motivation for the Klein model. Instead of constructing the projective plane from the hyperbolic plane, Klein does the reverse. We will not follow Klein exactly because our purpose is to construct a model within a Euclidean plane, which itself has a completion to a projective plane. But the basic idea is Klein's.

Felix Klein was a master of many branches of mathematics and a very influential teacher. His history of nineteenth-century mathematics shows how familiar he was with all aspects of the subject. Klein's famous inaugural address in 1872, his *Erlanger Programme*, made the study of groups of transformations and their invariants the key to geometry (see Chapter 9); this work emerged out of his collaboration with Sophus Lie. Klein's lectures on non-Euclidean geometry, published in 1928 after his death, are masterpieces of exposition. His work on complex function theory was a major mathematical contribution, summarized in the four volumes he wrote jointly with Robert Fricke. Poincaré, who competed with Klein in the study of automorphic functions, named the groups which occur in that theory after Klein; those groups are still an active area of research today.

While Abel had shown that the general polynomial equation of degree 5 could not be solved by radicals, Klein used the symmetry group of the icosahedron and elliptic modular functions to solve it. In topology, there is a compact nonorientable surface called the "Klein bottle"; it cannot be embedded in Euclidean three-space without crossing itself. Another surface for the study of which he is famous is the "Klein

---

[6] See, for example, http://www-groups.dcs.st-and.ac.uk/~history/Mathematicians/Beltrami.html and the references therein.

[7] Stillwell, ibid., pp. 63–111.

Felix Klein

quartic." Klein also very actively worked to improve mathematical teaching.[8]

For the Klein model, we fix once and for all a circle γ in a Euclidean plane (Cayley called γ the "absolute"). If O is the center of γ and OR is a radius, the *interior* of γ by definition consists of all points X such that OX < OR. In Klein's model, the points in the interior of γ represent the points of the hyperbolic plane.

Recall that a chord of γ is a segment AB joining two points A and B on γ. We wish to consider the segment without its endpoints, which we will call an *open chord* and denote by A)(B. In Klein's model the open chords of γ represent the lines of the hyperbolic plane. The relation "lies on" is represented in the usual sense: P lies on A)(B means that P lies on the Euclidean line $\overleftrightarrow{AB}$ and P is between A and B. The hyperbolic relation "between" is represented by the usual Euclidean relation "between." This much is easy. The representation of "congruence" is much more complicated, and we will

---

[8] See http://wwwgroups.dcs.stand.ac.uk/~history/Mathematicians/Klein.html and the references therein for more detail about the life and work of Klein. See the last section of Chapter 8 for more on the Klein quartic.

discuss it later on in this chapter (The Projective Nature of the Beltrami–Klein Model).

It is immediately clear from Figure 7.3 that the negation of Hilbert's Euclidean parallel postulate holds in this representation.

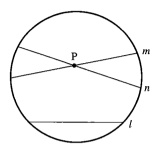

Figure 7.3

Here the two open chords m and n through P are both parallel to the open chord l—for what does "parallel" mean in this representation? The definition of "parallel" states that two lines are parallel if they have no point in common. In Klein's representation, this becomes: Two open chords are parallel if they have no point in common (in the definition of "parallel," replace the word "line" by "open chord"). The fact that the three chords, when extended, may meet outside the circle γ is irrelevant—points outside of γ do not represent points of the hyperbolic plane. So let us summarize the Beltrami–Klein proof of the relative consistency of hyperbolic geometry as follows.

First, a glossary is set up to "translate" the five undefined terms ("point," "line," "lies on," "between," and "congruent") into their interpretations in the Euclidean model (we have done this for the first four terms). All the defined terms are then interpreted by "translating" all occurrences of undefined terms. For instance, the defined term "parallel" was interpreted by replacing every occurrence of the word "line" in the definition by "open chord." Once all the defined terms have been interpreted, we have to interpret the axioms of the system. Incidence Axiom 1, for example, has the following interpretation in the Klein model.

INCIDENCE AXIOM 1 (KLEIN). Given any two distinct points A and B in the interior of circle γ. There exists a unique open chord l of γ such that A and B both lie on l.

We must prove that this is a theorem in Euclidean geometry (and similarly, prove the interpretations of all the other axioms). Once all the interpreted axioms have been proved to be theorems in Euclidean geometry, any proof of a contradiction within hyperbolic geometry could be translated by our glossary into a proof of a contradiction in Euclidean geometry. From our assumption that Euclidean geometry is consistent, it follows that no such proof exists. Thus if Euclidean geometry is consistent, so is hyperbolic geometry.

We must now backtrack and prove that the interpretations of the axioms of hyperbolic geometry in the Klein model are theorems in Euclidean geometry. Let us prove Axiom I-1 (Klein) stated above.

PROOF:

Given A and B interior to γ. Let $\overleftrightarrow{AB}$ be the Euclidean line through them (see Figure 7.4). This line intersects γ in two distinct points C and D. Then A and B lie on the open chord C)(D, and, by Axiom I-1 for Euclidean geometry, this is the only open chord on which they both lie. ◄

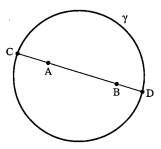

Figure 7.4

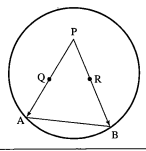

Figure 7.5 Limiting parallel rays.

In the second step of the proof, we used the line-circle continuity principle of Euclidean geometry, which states that a line passing through the interior of a circle intersects the circle in two distinct points. This can be proved from the circle-circle continuity principle (see Major Exercise 1, Chapter 4). Verifications of the interpretations of the other incidence axioms, the betweenness axioms, and Dedekind's axiom (if the Euclidean plane is real) are left as exercises; the congruence axioms are verified later in this chapter.

. One nice aspect of the Klein model is that it is easy to visualize the limiting parallel rays (see Figure 7.5). Let P be a point interior to γ and not on the open chord A)(B. A and B are points on the circle and therefore do not represent points in the hyperbolic plane; they represent *ideal points* and are called the *ends* of the hyperbolic line represented by A)(B. Then the limiting parallel rays to A)(B from P are represented by the segments PA and PB with the endpoints A and B omitted. It is clear that any ray between these limiting parallel rays intersects the open chord A)(B, whereas all other rays emanating from P do not. The symmetry and transitivity of limiting parallelism are utterly obvious in the Klein model, as is the fact that every angle has a *line of enclosure* (given ∢QPR, if A is the end of $\overrightarrow{PQ}$ and B is the end of $\overrightarrow{PR}$, then A)(B is the line of enclosure of ∢QPR). Thus four fundamental theorems of axiomatic real hyperbolic geometry, whose proofs were fairly difficult, are perfectly clear in the Klein model. Compare Theorem 6.2 and Major Exercises 2, 3, and 8, Chapter 6.

Let us conclude this section by considering the interpretation in the Klein model of "congruence," the subtlest part of the model. One method of interpretation is to use a system of numerical measurement of angle degrees and segment lengths. Two angles would then be interpreted as congruent if they had the same number of degrees, and two segments would be interpreted as congruent if they had the same length. The catch is that Euclidean methods of measuring degrees and lengths cannot be used. If we use Euclidean length, for example, then every line (i.e., open chord) would have a finite length less than or equal to the length of a diameter of γ. This would invalidate the interpretations of Axioms B-2 and C-1, which ensure that lines are infinitely long.

We will further discuss the matter in this chapter (in the sections Perpendicularity in the Beltrami–Klein Model and The Projective Nature of the Beltrami-Klein Model), but first let's consider the Poincaré models, in which congruence of angles is easier to describe.

The Poincaré Models

A disk model due to Henri Poincaré (1854–1912) also represents points of the hyperbolic plane by the points *interior* to a Euclidean circle γ, but lines are represented differently. First, all open chords that pass through the center O of γ (i.e., all *open diameters l* of γ) represent lines. The other lines are represented by *open arcs of circles orthogonal to γ*. More precisely, let δ be a circle orthogonal to γ (at each point of intersection of γ and δ the radii of γ and δ through that point are perpendicular). Then intersecting δ with the interior of γ gives an open arc m, which by definition represents a hyperbolic line in the Poincaré model. So we will call *Poincaré line*, or "P-line," either an open diameter l of γ or an open circular arc m orthogonal to γ (see Figure 7.6).

A point interior to γ "lies on" a Poincaré line if it lies on it in the Euclidean sense. Similarly, "between" has its usual Euclidean interpretation (for A, B, and C on an open arc coming from an orthogonal circle δ with center P, B is *between* A and C if $\overrightarrow{PB}$ is between $\overrightarrow{PA}$ and $\overrightarrow{PC}$).

The interpretation of *congruence for segments* in the Poincaré model is complicated, being based on a way of measuring length that is dif-

Henri Poincaré

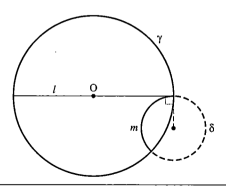

Figure 7.6

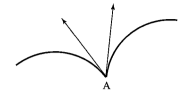

Figure 7.7

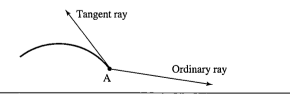

Figure 7.8

ferent from the usual Euclidean way, just as in the Klein model (see p. 320). *Congruence for angles* has the usual Euclidean meaning, however, and this is the main advantage of the Poincaré model over the Klein model.[9] Specifically, if two directed circular arcs intersect at a point A, the number of degrees in the *angle* they make is by definition the number of degrees in the angle between their tangent rays at A (see Figure 7.7). Or, if one directed circular arc intersects an ordinary ray at A, the number of degrees in the *angle* they make is by definition the number

[9] Technically, we say that the Poincaré model is *conformal*—meaning it represents angles accurately—while the Klein model is not. Another example of a conformal model is Mercator's map of the surface of the earth.

of degrees in the angle between the tangent ray and the ordinary ray at A (see Figure 7.8).

Having interpreted all the undefined terms of hyperbolic geometry in the Poincaré model, we get (by substitution) interpretations of all the defined terms. For example, two Poincaré lines are *parallel* if and only if they have no point in common. Then all the axioms of hyperbolic geometry get translated into statements in Euclidean geometry, and it will be shown in the section Inversion in Circles, Poincaré Congruence later in this chapter that these interpretations are theorems in Euclidean geometry. Hence the Poincaré model furnishes another proof that if Euclidean geometry is consistent, so is hyperbolic geometry.

The limiting parallel rays in the Poincaré model are illustrated in Figure 7.9. Here we have chosen l to be an open diameter A)(B; the rays are circular arcs that meet $\overleftrightarrow{AB}$ at A and B and are tangent to this line at those points. You can see how these rays approach l asymptotically as you move out toward the ideal points represented by A and B.

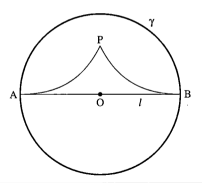

Figure 7.9 Limiting rays.

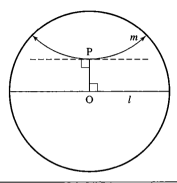

Figure 7.10 Divergent parallels.

Figure 7.10 illustrates two parallel Poincaré lines with a common perpendicular. The diagram shows how m diverges from l on either side of the common perpendicular PO.

Figure 7.11 illustrates a Lambert quadrilateral. You can see that the fourth angle is acute. By adding the mirror image of this Lambert quadrilateral, we get a diagram illustrating a Saccheri quadrilateral (Figure 7.12).

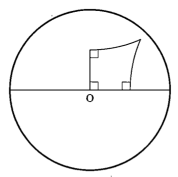

Figure 7.11 Lambert quadrilateral.

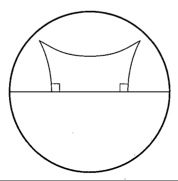

Figure 7.12 Saccheri quadrilateral.

You may be surprised that we have two different models of hyperbolic geometry, one due to Klein and the other to Poincaré. (There is a third model, also due to Poincaré, and a fourth model on one sheet of a hyperboloid in three-space will be described later in this chapter.) Yet you may have the feeling that these models are not "essentially different." In fact, these models are *isomorphic* in the technical sense that one-to-one correspondences can be set up between the "points" and "lines" in one model and the "points" and "lines" in the other so as to preserve the relations of incidence, betweenness, and congruence. Such isomorphism is illustrated in Figure 7.13. We start with the Klein model and consider, in Euclidean three-space, a sphere of the same radius sitting on the plane of the Klein model and tangent to it at the origin. We project upward orthogonally the entire Klein model onto the lower hemisphere of this sphere; by this projection, the chords in the Klein model become arcs of circles orthogonal to the equator. We then project stereographically from the north pole of the sphere onto the original plane. The equator of the sphere will project onto a circle larger than the one used in the Klein model, and the lower hemisphere will project stereographically onto the inside of this circle. Under these successive transformations, the chords of the Klein model will be mapped one-to-one onto the diameters and orthogonal arcs of the Poincaré model. In this way the isomorphism of the models may be established.

One can actually prove that *all possible models of real hyperbolic geometry are isomorphic to one another,* i.e., that the axioms for real hyperbolic geometry are *categorical.* The same is true for real Euclidean geometry. The categorical nature of real Euclidean geometry is established by introducing Cartesian coordinates into the real Euclidean plane. Analogously, the categorical nature of real hyperbolic geometry

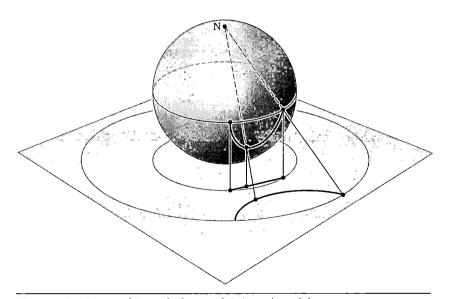

Figure 7.13 Isomorphism of Klein and Poincaré models.

is established by introducing Beltrami coordinates into the real hyperbolic plane (for which real hyperbolic trigonometry must first be developed).[10]

In the other Poincaré model mentioned here, the points of the hyperbolic plane are represented by the points of one of the Euclidean half-planes determined by a fixed Euclidean line. If we use the Cartesian model for the Euclidean plane, it is customary to make the x-axis the fixed line and then to use for our model the upper half-plane consisting of all points (x, y) with $y > 0$. Hyperbolic lines are represented in two ways:

1. As rays emanating from points on the x-axis and perpendicular to the x-axis;
2. As semicircles in the upper half-plane whose center lies on the x-axis (see Figure 7.14).

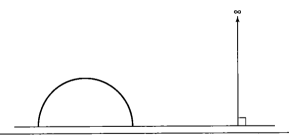

Figure 7.14 P-lines in upper half-plane model.

Incidence and betweenness have the usual Euclidean interpretations. This model is conformal also (degrees of angles are measured in the Euclidean way). Measurement of lengths will be discussed later.

To establish isomorphism with the previous models, choose a point E on the equator of the sphere in Figure 7.13 and let Π be the plane tangent to the sphere at the point diametrically opposite to E. Stereographic projection from E to Π maps the equator onto a line in Π and the lower hemisphere onto the lower half-plane determined by this line. Notice that the points on this line represent ideal points. However, one ideal point is missing: The point E got lost in the stereographic projection. It is customary to imagine an ideal "point at infinity" ∞ that corresponds to E; it is the common end of all the vertical rays.

Like Gauss, Henri Poincaré made profound discoveries in many branches of mathematics and physics. He even started a new branch

---

[10] See Chapter 10 as well as Borsuk and Szmielew (1960, Chapter 6).

of mathematics, algebraic topology, inventing the *fundamental group* and other concepts. Settling his famous conjecture about the three-sphere is one of the seven millennium problems for the solution of which the Clay Institute is offering a million-dollar prize. He was a pioneer in the currently very active field of dynamics and chaos theory, as well as in function theory in several complex variables. He used his models of hyperbolic geometry to discover new theorems about automorphic functions of a complex variable. There is a widely published account of his two experiences, while on vacation, of suddenly realizing that the transformations he had used to define Fuchsian functions and the arithmetic transformations of ternary quadratic forms are identical to those of hyperbolic geometry.[11] This epiphany solidified the acceptance of hyperbolic geometry by the mathematics community and led to very important further research still ongoing today (see the last section of Chapter 8). Poincaré made major contributions to several branches of mathematical physics, particularly celestial mechanics. He was nearly a co-discoverer with Einstein of the theory of relativity in physics. Poincaré is also important as a philosopher of science (Chapter 8 has a discussion of his conventionalist philosophy of mathematics).[12]

Perpendicularity in the Beltrami–Klein Model

The Klein model is not conformal. Congruence of angles is interpreted differently from the usual Euclidean way and will be explained later in this chapter. Here we will describe only those angles that are congruent to their supplements, namely, right angles.

Let l and m be open chords of y. To describe when $l \perp m$ in the Klein model, there are two cases to consider:

CASE 1. One of l and m is a diameter. Then $l \perp m$ in the Klein sense if and only if $l \perp m$ in the Euclidean sense (see Figure 7.15).

CASE 2. Neither l nor m is a diameter. In this case, we associate with l a certain point P(l) outside of γ called the *pole* of l and defined

---

[11] See, for example, "Mathematical Creation," in vol. 4 of *The World of Mathematics*, J. R. Newman, ed., Allen & Unwin Ltd., London, 1960, pp. 2041–2050.

[12] See   http://www-groups.dcs.st-and.ac.uk/~history/Mathematicians/Poincare.html, http://www.utm.edu/research/iep/p/poincare.htm and the references therein for more detail about the life and work of Henri Poincaré, who was the cousin of the president of France, Raymond Poincaré. Also see *The Poincaré Conjecture* by D. O'Shea (Walker, 2007).

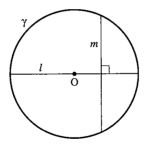

Figure 7.15

as follows. Let t_1 and t_2 be the tangents to γ at the endpoints of l. Then by definition $P(l)$ is the unique point common to t_1 and t_2 (t_1 and t_2 are not parallel because l is not a diameter); see Figure 7.16.

It turns out that *l is perpendicular to m in the sense of the Klein model if and only if the Euclidean line extending m passes through the pole of l.*

This description of perpendicularity will be justified later. We can use it to see more easily why divergently parallel lines have a common perpendicular. We are given two parallel lines that are not asymptotically parallel. In the Klein model, this means that we are given open chords l and m that do not intersect and do not have a common end. How do we find their common perpendicular k? Let's discuss case 2, leaving case 1 as an exercise. By the above description of perpendicularity, if k were perpendicular to both l and m, the extension of k would have to pass through the pole of l and the pole of m. Hence to construct k, we need only join these poles by a

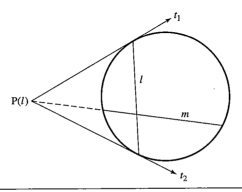

Figure 7.16 m is Klein perpendicular to l.

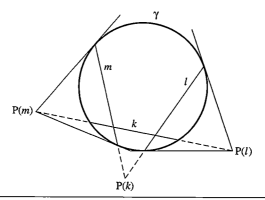

Figure 7.17 k is Klein perpendicular to l and m.

Euclidean line and take k to be the open chord of γ cut out by this line (Figure 7.17).[13]

We will use the language of the projective completion, Major Exercise 13, Chapter 6, to describe the behavior of pairs of lines in the Klein model. Let us call the points inside circle γ (which represent all the points in the hyperbolic plane) *ordinary points*. We already called the points on the circle γ *ideal points*. Let us call the points outside γ *ultra-ideal points*. Finally, for every diameter of γ, let us add the point "at infinity" such that all the Euclidean lines parallel in the Euclidean sense to this diameter meet in this point at infinity in the projective completion of the Euclidean plane (see Chapter 2). These points at infinity will also be called *ultra-ideal*. We can then say that two Klein lines "meet" at an ordinary point, an ideal point, or an ultra-ideal point, depending on whether they are intersecting, asymptotically parallel, or divergently parallel, respectively. The ultra-ideal point at which divergently parallel Klein lines l and m "meet" is the pole $P(k)$ of their common perpendicular k (see Figure 7.17).

This language is suggestive of further theorems in hyperbolic geometry. For example, we know that two ordinary points determine a unique line, and we have seen that two ideal points also determine a unique line, the line of enclosure of Major Exercise 8, Chapter 6. We can ask the same question about two points that are ultra-ideal or about two points of different species. For example, an ordinary point and an ideal

---

[13] If l and m did have a common end Ω, the Euclidean line joining $P(l)$ and $P(m)$ would be tangent to γ at Ω. That is why Saccheri claimed that asymptotically parallel lines have "a common perpendicular at infinity," and this he found repugnant.

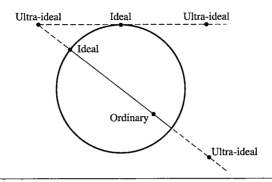

Figure 7.18

or ultra-ideal point always determine a unique ordinary line, but two ultra-ideal points may or may not (see Figure 7.18). Let us translate back from this language, say, in the case of an ordinary point O and an ultra-ideal point P(l) that is the pole of a Klein line l. What is the Klein line "joining" O to P(l)? It is the unique Klein line m through O that is perpendicular in the sense of the Klein model to the line l (see Figure 7.16). We leave the other cases for exercises.

  If you did most of the exercises in hyperbolic geometry in Chapter 6, deriving results without having reliable diagrams to guide you, the Klein and Poincaré models must come as a great relief. It is a useful exercise to take an absurd diagram like Figure 6.22 and draw those divergently parallel perpendicular bisectors of the triangle more accurately in one of the models. It is amazing that J. Bolyai and Lobachevsky were able to visualize hyperbolic geometry without such models, especially since they worked in three dimensions. They must have had non-Euclidean eyesight.

A Model of the Hyperbolic Plane from Physics

This model comes from the theory of special relativity. In Cartesian three-space $\mathbb{R}^3$, with coordinates denoted x, y, and t (for *time*), distance will be measured by the *Minkowski metric*

$$ds^2 = dx^2 + dy^2 - dt^2.$$

Then with respect to the Minkowski metric, the surface of equation

$$x^2 + y^2 - t^2 = -1$$

is a "sphere" centered at the origin O = (0, 0, 0) of imaginary radius
$i = \sqrt{-1}$. (As was mentioned in Chapter 5, Lambert was the first to
wonder whether such a model existed.) In Euclidean terms, it is a two-
sheeted *hyperboloid* (surface of revolution obtained by rotating the hyperbola $t^2 - x^2 = 1$ around the x-axis). We choose the sheet Σ: $t \geq 1$
as our model. It looks like an infinite bowl (see Figure 7.19). Analogously with our interpretation of "lines" on a sphere in Chapter 2, Exercise 10(c), "lines" are interpreted to be the sections of Σ cut out by
planes through O; thus a "line" is one branch of a *hyperbola* on Σ.

Here is an isomorphism of Σ with the Beltrami–Klein model Δ. The
plane $t = 1$ is tangent to Σ at the point C = (0, 0, 1). Let Δ be the unit
disk centered at C in this plane. Projection from O gives a one-to-one
correspondence between the points of Δ and the points of Σ (i.e., point
P of Δ corresponds to the point P′ at which ray $\overrightarrow{OP}$ pierces Σ). Similarly, each chord m of Δ lies on a unique plane Π through O, and m

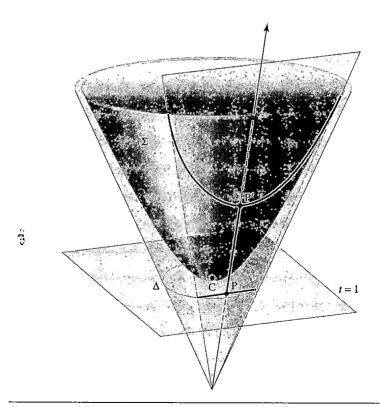

Figure 7.19 Isomorphism of Klein and hyperboloid models.

corresponds to the section m' of Σ cut out by Π. This isomorphism of incidence models can be used to interpret betweenness and congruence on Σ. Alternatively, they can be defined in terms of the measurement of arc length induced on Σ by the Minkowski metric; then further argument is needed to verify that our correspondence is indeed an isomorphism of models of hyperbolic geometry. Another justification of Σ as a model of the hyperbolic plane will be given analytically in Chapter 10 (see the discussion of Weierstrass coordinates in the section Coordinates in the Hyperbolic Plane).

NOTE. From the point of view of Einstein's special relativity theory, Σ can be identified with the set of plane uniform motions, and the hyperbolic distance can be identified with the relative velocity of one motion with respect to the other. A glossary can be set up to translate every theorem of hyperbolic geometry into a theorem of relativistic kinematics, and conversely. See Yaglom (1979, p. 225 ff.). See also Chapter 10 of Ramsay and Richtmyer (1995) for a more detailed discussion of this model and its relation to special relativity.

Inversion in Circles, Poincaré Congruence

In order to define congruence in the Poincaré models and verify the axioms of congruence, we must study inversion in a Euclidean circle; when interpreted in the model, this transformation turns out to be reflection across a line in the hyperbolic plane. *This theory is part of Euclidean geometry* and is called *inversive geometry*. It originated with Apollonius in ancient Greece and was developed much further by Jakob Steiner in the 1820s and by August Möbius in the 1850s, among others.

Steiner was a purist about using only synthetic methods in geometry, considering that calculation replaces thinking whereas geometry stimulates thinking. While our development will be primarily synthetic, we will not be so austere as Steiner and will occasionally use coordinate methods available to us in $\mathbb{R}^2$. This is justified by Major Exercise 8, Chapter 5, which showed, using the Pythagorean equation, that a real Euclidean plane must be isomorphic to $\mathbb{R}^2$. (Almost everything we do works just as well in F^2, where F is any Euclidean field.) We will use the results on similarity and circles proved for a real Euclidean plane in the exercises toward the end of Chapter 5.

DEFINITION. Let γ be a circle of radius r, center O. For any point $P \neq O$ the *inverse* P' of P with respect to γ is the unique point P' on ray $\overrightarrow{OP}$ such that $(\overline{OP})\,(\overline{OP'}) = r^2$ (see Figure 7.20).

The following properties of inversion are immediate from the definition.

PROPOSITION 7.1. (a) P = P′ if and only if P lies on the circle of inversion γ. (b) If P is inside γ, then P′ is outside γ; if P is outside γ, then P′ is inside γ. (c) (P′)′ = P.

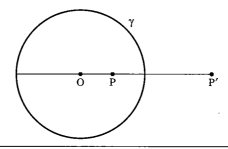

Figure 7.20

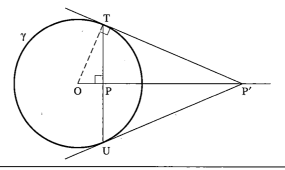

Figure 7.21

DEFINITION. If TU is a chord of circle γ which is not a diameter, then the *pole* of TU is the point of intersection of the tangents to γ at T and U (see Figure 7.21). That point exists because TU is a transversal to those tangents satisfying the hypothesis of Euclid V.

The next two propositions tell how to construct the inverse point with a straightedge and compass.

PROPOSITION 7.2. Suppose P ≠ O is inside γ. Let TU be the chord of γ through P, which is perpendicular to $\overleftrightarrow{OP}$. Then the inverse P' of P is the pole of chord TU (see Figure 7.21).

> *PROOF:*
>
> Suppose the tangent to γ at T cuts $\overrightarrow{OP}$ at point P'. Right triangle △OPT is similar to right triangle △OTP' (since they have ∢TOP in common and the angle sum is 180°). Hence corresponding sides are proportional (Exercise 10, Chapter 5). Note that $\overline{OT} = r$, so we get $(\overline{OP})/r = r/(\overline{OP'})$, which shows that P' is inverse to P. Reflecting across $\overleftrightarrow{OP}$ (Major Exercise 2, Chapter 3), we see that the tangent to γ at U also passes through P', so P' is indeed the pole of TU. ◄

PROPOSITION 7.3. If P is outside γ, let Q be the midpoint of segment OP. Let σ be the circle with center Q and radius $\overline{OQ} = \overline{QP}$. Then σ cuts γ in two points T and U, $\overleftrightarrow{PT}$ and $\overleftrightarrow{PU}$ are tangent to γ, and the inverse P' of P is the intersection of TU and OP (see Figure 7.22).

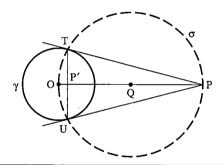

Figure 7.22

> *PROOF:*
>
> By the circle-circle continuity principle (Chapter 3), σ and γ do meet in two points T and U. Since ∢OTP and ∢OUP are inscribed in semicircles of σ, they are right angles (Exercise 16, Chapter 5); hence $\overleftrightarrow{PT}$ and $\overleftrightarrow{PU}$ are tangent to γ. If TU meets OP in a point P', then P is the inverse of P' (Proposition 7.2); hence P' is the inverse of P in γ. ◄

The next proposition shows how to construct the Poincaré line joining two ideal points—the line of enclosure of ∢TOU in Figure 7.23.

PROPOSITION 7.4. Let T and U be points on γ that are not diametrically opposite and let P be the pole of TU. Then we have PT ≅ PU,

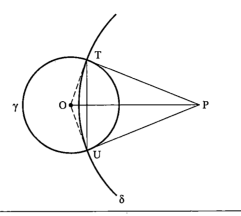

Figure 7.23

$\angle PTU \cong \angle PUT$, $\overleftrightarrow{OP} \perp \overleftrightarrow{TU}$, and the circle δ with center P and radius $\overline{PT} = \overline{PU}$ cuts γ orthogonally at T and U (see Figure 7.23).

PROOF:

By the definition of pole, $\angle OTP$ and $\angle OUP$ are right angles; so by the hypotenuse-leg criterion, $\triangle OTP \cong \triangle OUP$. Thus $PT \cong PU$ and $\angle OPT \cong \angle OPU$. The base angles $\angle PTU$ and $\angle PUT$ of the isosceles triangle $\triangle TPU$ are then congruent, and the angle bisector $\overrightarrow{PO}$ is perpendicular to the base TU. The circle δ is then well defined because $\overline{PT} = \overline{PU}$ and δ cuts γ orthogonally by our hypothesis that $\overleftrightarrow{PT}$ and $\overleftrightarrow{PU}$ are tangent to γ. ◄

LEMMA 7.1. Given that point O does not lie on circle δ. (a) If two lines through O intersect δ in pairs of points (P_1, P_2) and (Q_1, Q_2), respectively, then $(\overline{OP_1})(\overline{OP_2}) = (\overline{OQ_1})(\overline{OQ_2})$. This common product is called the *power* of O with respect to δ when O is outside δ, and minus this product is called the power of O when O is inside δ. (b) If O is outside δ and a tangent to δ from O touches δ at point T, then $(\overline{OT})^2$ equals the power of O with respect to δ.

PROOF:

(a) Since angles that are inscribed in a circle and subtend the same arc are congruent (Exercise 17, Chapter 5), we have

$$\angle P_2P_1Q_2 \cong \angle P_2Q_1Q_2$$

$$\angle P_1Q_2Q_1 \cong \angle P_1P_2Q_1$$

(see Figure 7.24). It follows that $\triangle OP_1Q_2$ and $\triangle OQ_1P_2$ are similar, so that $(\overline{OP_1})/(\overline{OQ_1}) = (\overline{OQ_2})/(\overline{OP_2})$, as asserted.

(b) Let C be the center of δ and let line OC cut δ at P_1 and P_2, with $O * P_1 * C * P_2$. By the Pythagorean theorem,

$$(\overline{OT})^2 = (\overline{OC})^2 - (\overline{CT})^2$$
$$= (\overline{OC} - \overline{CT})(\overline{OC} + \overline{CT})$$
$$= (\overline{OC} - \overline{CP_1})(\overline{OC} + \overline{CP_2})$$
$$= (\overline{OP_1})(\overline{OP_2})$$

(see Figure 7.25.). ◄

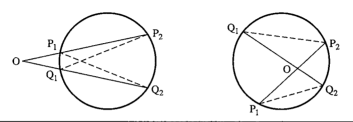

Figure 7.24

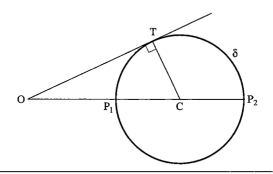

Figure 7.25

PROPOSITION 7.5. Let P be any point that does not lie on circle γ and that does not coincide with the center O of γ, and let δ be a circle through P. Then δ cuts γ orthogonally if and only if δ passes through the inverse point P' of P with respect to γ.

PROOF:

Suppose first that δ passes through P'. Then the center C of δ lies on the perpendicular bisector of PP' (Exercise 17, Chapter 4); hence $\overline{CO} > \overline{CP}$ (Exercise 27, Chapter 4) and O lies outside δ. Therefore, there is a point T on δ such that the tangent to δ at T passes through O (Proposition 7.3). Lemma 7.1(b) then gives $(\overline{OT})^2 = (\overline{OP})(\overline{OP'}) = r^2$, so that T also lies on γ and δ cuts γ orthogonally.

Conversely, let δ cut γ orthogonally at points T and U. Then the tangents to δ at T and U meet at O, so that O lies outside δ. It follows that $\overrightarrow{OP}$ cuts δ again at a point Q. By Lemma 7.1(b), we have $r^2 = (\overline{OT})^2 = (\overline{OP})(\overline{OQ})$, so that Q = P', the inverse of P in γ. ◀

COROLLARY. Let P be as in Proposition 7.5. Then the locus of the centers of all circles δ through P orthogonal to γ is the line l, which is the perpendicular bisector of PP'. If P is inside γ, then l is a line in the exterior of γ. Conversely, let l be any line in the exterior of γ, let C be the foot of the perpendicular from O to l, let δ be the circle centered at C which is orthogonal to γ (constructed as in Proposition 7.3), and let P be the intersection of δ with OC; then l is the locus of the centers of all circles orthogonal to γ that pass through P.

PROOF:

Any δ orthogonal to γ must pass through P and P', so its center C must be equidistant from P and P'. The locus of all such C is the perpendicular bisector of PP'. As we have seen, the center of any circle δ orthogonal to γ lies outside γ. We leave the converse as an easy exercise. ◀

Proposition 7.5 can be used to construct the P-line joining two points P and Q inside γ that do not lie on a diameter of γ. First, construct the inverse point P', using Proposition 7.2. Then construct the circle δ determined by the three noncollinear points P, Q, and P' (use Exercise 10, Chapter 6). By Proposition 7.5, δ will be orthogonal to γ; intersecting δ with the interior of γ gives the desired P-line. This verifies the interpretation of **Axiom I-1** for the Poincaré disk model. The verification is even simpler for the Poincaré upper half-plane model: Given two points P and Q that do not lie on a vertical ray, let the perpendicular bisector of Euclidean segment PQ cut the x-axis at C. Then the semicircle centered at C and passing through P and Q is the desired P-line.

We could also have verified the interpretations of the incidence axioms, the betweenness axioms, and Dedekind's axiom by using isomorphism with the Klein model (where the verifications are trivial). The advantage of the argument we gave is that it provides an explicit *construction*, and constructions are a main theme of this section.

We turn now to the congruence axioms. Since angles are measured in the Euclidean sense in the Poincaré models, the interpretation of **Axiom C-5** is trivially verified. Consider **Axiom C-4**, the laying off of a congruent copy of a given angle at some vertex A (for the disk model). If A is the center of γ, the angle is formed by diameters and the laying off is accomplished in the Euclidean way. If A is not the center O of γ, then the verification is a matter of finding a unique circle δ through A that is orthogonal to γ and tangent to a given Euclidean line l that passes through A and not through O (since the tangents determine the angle measure). By Proposition 7.5, δ must pass through the inverse A' of A with respect to γ. The center C of δ must lie on the perpendicular bisector of chord AA' (Exercise 17, Chapter 4); call this bisector m. If δ is to be tangent to l at A, then C must also lie on the perpendicular n to l at A. So δ must be the circle whose center is the intersection C of m and n and whose radius is CA (see Figure 7.26).

To define congruence of segments in the disk model, we introduce the following definition of length.

DEFINITION. Let A and B be points inside γ and let P and Q be the ends of the P-line through A and B. We define the *cross-ratio* (AB, PQ) by

$$(AB, PQ) = \frac{\overline{(AP)}\,\overline{(BQ)}}{\overline{(BP)}\,\overline{(AQ)}}$$

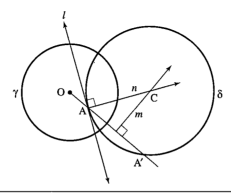

Figure 7.26

where, for example, $\overline{AP}$ is the Euclidean length of the Euclidean segment AP). We then define the *Poincaré length* $d(AB)$ by

$$d(AB) = |\log(AB, PQ)|.$$

IMPORTANT REMARK. This definition makes no sense for Euclidean planes coordinatized by an arbitrary Euclidean field F because there is no log function defined for such fields as there is for any Euclidean subfield of $\mathbb{R}$. The only reason for introducing the log here is for the length to be additive, as is customary and as will soon be proved. The logarithm function converts multiplication into addition, but why is it necessary to do that? We could just as well have length be multiplicative, as it would be if we simply used the cross-ratio and dispensed with the logarithm, with one proviso: The order in which we write the letters (AB, PQ) matters. It doesn't matter when we bring in the absolute value of the log, as we will soon show. So to use (AB, PQ) as our multiplicative P-length for P-segment AB, we must specify that on the circular arc which is the P-line joining A to B, A lies between P and B. Then B lies between A and Q, and by this convention the multiplicative P-length is also equal to (BA, QP), as a little algebra shows. The multiplicative length is denoted $\mu(AB)$.[14] *We leave it to the reader to verify that everything we do with the additive version of P-length works equally well with the multiplicative definition just given when the formulas are adjusted appropriately; hence our results are also valid for arbitrary Euclidean fields.*

P-length $d(AB)$ does not depend on the order in which we write P and Q: If (AB, PQ) $= x$, then we have (AB, QP) $= 1/x$, and therefore $|\log(1/x)| = |-\log x| = |\log x|$. Furthermore, since (AB, PQ) = (BA, QP), we see that $d(AB)$ also does not depend on the order in which we write A and B.

We may therefore interpret the Poincaré segments AB and CD to be *Poincaré-congruent* if $d(AB) = d(CD)$. With this interpretation, **Axiom C-2** is immediately verified.

Suppose we fix the point A on the P-line from P to Q and let point B move continuously from A to P, where $Q * A * B * P$, as in Figure 7.27. The cross-ratio (AB, PQ) will increase continuously from 1 to ∞ since $(\overline{AP})/(\overline{AQ})$ is constant, $\overline{BP}$ approaches zero, and $\overline{BQ}$ approaches $\overline{PQ}$. If we fix B and let A move continuously from B to Q, we get the

---

[14] Robin Hartshorne introduced this very valuable notion, which will be exploited in Appendix B—see his Section 39. He described the cross-ratio as "magic" because one cannot visualize it geometrically. It is the fundamental invariant for coordinate projective geometry (Exercise 68, Chapter 9).

same result. It follows immediately that for any Poincaré ray $\overrightarrow{CD}$, there is a unique point E on $\overrightarrow{CD}$ such that $d(CE) = d(AB)$, where A and B are given in advance. This verifies **Axiom C-1**.

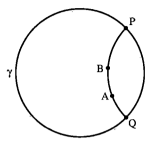

Figure 7.27

REMARK. The argument just given is valid only for the real Euclidean plane because it appeals to the intermediate value theorem for continuous functions of a real variable. An argument that works for arbitrary Euclidean planes will be given after the sublemma following Theorem 7.1.

We next verify **Axiom C-3**. This will follow immediately from the additivity of the Poincaré length, which asserts that if A ∗ C ∗ B in the sense of the disk model, then $d(AC) + d(CB) = d(AB)$. To prove this additivity, label the ends so that Q ∗ A ∗ B ∗ P. Then the cross-ratios (AB, PQ), (AC, PQ), and (CB, PQ) are all greater than 1 (because $\overline{AP} >$ $\overline{BP}$, $\overline{BQ} > \overline{AQ}$, etc.); their logs are thus positive, and we can drop the absolute value signs. We have

$$d(AC) + d(CB) = \log(AC, PQ) + \log(CB, PQ)$$
$$= \log[(AC, PQ)(CB, PQ)],$$

but (AC, PQ)(CB, PQ) = (AB, PQ), as can be seen by canceling terms.

Finally, to verify **Axiom C-6 (SAS)**, we must study the effect of inversions on the objects and relations in the disk model.

DEFINITION. Let O be a point and k a positive number. The *dilation* with *center* O and *ratio* k is the transformation of the Euclidean plane that fixes O and maps a point P ≠ O onto the unique point P* on $\overrightarrow{OP}$ such that $\overline{OP^*} = k(\overline{OP})$ (so that points are moved radially from O a distance k times their original distance).

LEMMA 7.2. Let δ be a circle with center C ≠ O and radius s. Under the dilation with center O and ratio k, δ is mapped onto the circle δ^*

with center C^* and radius ks. If Q is a point on δ, the tangent to δ^* at Q^* is parallel to the tangent to δ at Q.

PROOF:

Choose rectangular coordinates so that O is the origin. Then the dilation is given by $(x, y) \rightarrow (kx, ky)$. The image of the line having equation $ax + by = c$ is the line having equation $ax + by = kc$; hence the image is parallel to the original line. In particular, $\overleftrightarrow{CQ}$ is parallel to $\overleftrightarrow{C^*Q^*}$, and their perpendiculars at Q and Q^*, respectively, are also parallel. If δ has equation $(x - c_1)^2 + (y - c_2)^2 = s^2$, then δ^* has equation $(x - kc_1)^2 + (y - kc_2)^2 = (ks)^2$, from which the lemma follows. ◄

REMARK. The argument just given uses analytic geometry for the first time. It is quicker than a synthetic argument, which can also be given.

PROPOSITION 7.6. Let γ be a circle of radius r and center O, δ a circle of radius s and center C. Assume that O lies outside δ; let p be the power of O with respect to δ (see Lemma 7.1). Let $k = r^2/p$. Then the image δ' of δ under inversion in γ is the circle of radius ks whose center is the image C^* of C under the dilation from O of ratio k. If P is any point on δ and P' is its inverse in γ, then the tangent t' to δ' at P' is the reflection across the perpendicular bisector of PP' of the tangent to δ at P (see Figure 7.28).

PROOF:

Since O is outside δ, $\overrightarrow{OP}$ either cuts δ in another point Q or is tangent to δ at P (in which case let Q = P). Then

$$\frac{\overline{OP'}}{\overline{OQ}} = \frac{\overline{OP'}}{\overline{OQ}} \cdot \frac{\overline{OP}}{\overline{OP}} = \frac{r^2}{p},$$

which shows that P' is the image of Q under the dilation from O of ratio $k = r^2/p$. Hence $\delta^* = \delta'$. By Lemma 7.2, the tangent t' to δ' at P' is parallel to the tangent u to δ at Q. Let t be tangent to δ at P. By Proposition 7.4, t and u meet at a point R such that $\angle RQP \cong \angle RPQ$. Then t and t' meet at a point S such that $\angle SP'P \cong \angle SPP'$ by transitivity of congruence and corresponding angles of parallel lines in a Euclidean plane. Since $\triangle PSP'$ is an isosceles triangle (base angles are congruent), S lies on the perpendicular bisector of PP'. Hence t' is the reflection of t across this perpendicular bisector. ◄

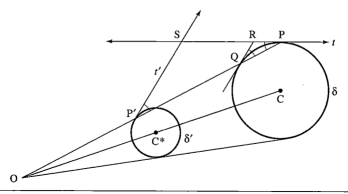

Figure 7.28

COROLLARY. Circle δ is orthogonal to circle γ if and only if δ is mapped onto itself by inversion in γ.

PROOF:

If δ is orthogonal to γ and P lies on δ, then $p = (\overline{OP})(\overline{OP'}) = r^2$ (Proposition 7.5 and Lemma 7.1), so $k = 1$ and $\delta = \delta'$. Conversely, if $\delta = \delta'$, then $p = r^2$ and δ passes through the inverse P' of P in γ, so that by Proposition 7.5, δ is orthogonal to γ. ◄

LEMMA 7.3. Let O be the center of circle γ, let P and Q be two points that are not collinear with O, and let P' and Q' be their inverses in γ. Then $\triangle POQ$ is similar to $\triangle Q'OP'$ (Figure 7.29).

PROOF:

The triangles have $\sphericalangle POQ$ in common and we have $(\overline{OP})(\overline{OP'}) = r^2 = (\overline{OQ})(\overline{OQ'})$. Thus the SAS similarity criterion is satisfied (Exercise 12, Chapter 5). ◄

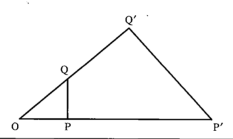

Figure 7.29

PROPOSITION 7.7. Let l be a line not passing through the center O of circle γ. The image of l under inversion in γ is a punctured circle with missing point O. The diameter through O of the completed circle δ is (when extended) perpendicular to l (see Figure 7.30).

> *PROOF:*
>
> Let A be the foot of the perpendicular from O to l, P be any other point on l, and A' and P' their inverses in γ. By Lemma 7.3, $\triangle$OP'A' is similar to $\triangle$OAP. Hence $\sphericalangle$OP'A' is a right angle, so that P' must lie on the circle δ having OA' as a diameter (Exercise 18, Chapter 5). Conversely, if we start with any point P' on δ other than O and let $\overrightarrow{OP'}$ cut l in P (using Euclid V), then reversing the above argument shows that P' is the inverse of P in γ. ◄

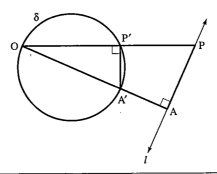

Figure 7.30

NOTE. A line through O is transformed into itself by inversion in γ, by the definition of inversion.

PROPOSITION 7.8. Let δ be a circle passing through the center O of γ. The image of δ minus O under inversion in γ is a line l not through O; l is parallel to the tangent to δ at O.

> *PROOF:*
>
> Let A' be the point on δ diametrically opposite to O, let A be its inverse in γ, and let l be the line perpendicular to $\overrightarrow{OA}$ at A (see Figure 7.30). By the proof of Proposition 7.7, inversion in γ maps l onto δ minus O; hence, it must map δ minus O onto l (Proposition 7.1(c)). ◄

Reflection in a Euclidean line preserves the magnitude but reverses the sense of directed angles (angles whose rays have a specified order). The next proposition generalizes this to inversions.

PROPOSITION 7.9. A directed angle of intersection of two circles is preserved in magnitude but reversed in sense by an inversion. The same applies to the angle of intersection of a circle and a line or of two lines.

PROOF:

Suppose that circles δ and σ intersect at point P with tangents l and m there. Let P' be the inverse of P in γ, let δ' and σ' be the images of δ and σ under inversion in γ, and let l' and m' be their respective tangents at P'. The first assertion then follows from the fact that l' and m' are the reflections of l and m across the perpendicular bisector of PP' (Proposition 7.6). The other cases follow from Propositions 7.7 and 7.8. ◄

The next proposition shows that inversion preserves the cross-ratio used to define Poincaré length.

PROPOSITION 7.10. If A, B, P, Q are four points distinct from the center O of γ and A', B', P', Q' are their inverses in γ, then we have $(AB, PQ) = (A'B', P'Q')$.

PROOF:

By Lemma 7.3, we see that $(\overline{AP})/(\overline{OA}) = (\overline{A'P'})/(\overline{OP'})$ and that $(\overline{AQ})/(\overline{OA}) = (\overline{A'Q'})/(\overline{OQ'})$, whence:

$$(1) \qquad \frac{\overline{AP}}{\overline{AQ}} = \frac{\overline{AP}}{\overline{OA}} \cdot \frac{\overline{OA}}{\overline{AQ}} = \frac{\overline{OQ'}}{\overline{OP'}} \cdot \frac{\overline{A'P'}}{\overline{A'Q'}}.$$

Similarly,

$$(2) \qquad \frac{\overline{BQ}}{\overline{BP}} = \frac{\overline{OP'}}{\overline{OQ'}} \frac{\overline{B'Q'}}{\overline{B'P'}}.$$

Multiplying equations (1) and (2) gives the result. ◄

PROPOSITION 7.11. Let circle δ be orthogonal to circle γ. Then inversion in δ maps γ onto γ and maps the interior of γ onto itself. Inversion in δ preserves incidence, betweenness, and congruence in the sense of the Poincaré disk model inside γ.

PROOF:

The corollary to Proposition 7.6 tells us that γ is mapped onto it-self. Suppose that P is inside γ and P' is its inverse in δ. Let C be the center and s the radius of δ. Let $\overrightarrow{CP}$ cut γ at Q and Q', so that by Proposition 7.5 $\overline{(CQ)}\,\overline{(CQ')} = s^2 = \overline{(CP)}\,\overline{(CP')}$. Since P lies between Q and Q', we have the inequalities CQ < CP < CQ'. Taking the reciprocal reverses inequalities, and we get $s^2/\overline{CQ} > s^2/\overline{CP} > s^2/\overline{CQ'}$, which is the same as CQ' > CP' > CQ. Thus P' lies between Q and Q' and therefore is inside γ.

By Propositions 7.6, 7.8, and 7.9, inversion in δ maps any circle σ orthogonal to γ either onto another circle σ' orthogonal to γ or onto a line σ' orthogonal to γ, i.e., a line through the center O of γ. The line σ joining O to C is mapped onto itself, and any other line σ through O is mapped onto a circle σ' punctured at C, which is orthogonal to γ (by Propositions 7.7 and 7.9). In all these cases, the above argument shows that the part of σ inside γ maps onto the part of σ' inside γ. Hence P-lines are mapped onto P-lines.

If A and B are inside γ and P and Q are the ends of the P-line through A and B, then inversion in δ maps P and Q onto the ends of the P-line through A' and B'. By Proposition 7.10, $d(AB) = d(A'B')$, so congruence of segments is preserved. Proposition 7.9 shows that congruence of angles is also preserved. Furthermore, Poincaré betweenness is also preserved because B is between A and D if and only if A, B, and D are Poincaré-collinear and $d(AD) = d(AB) + d(BD)$. ◄

NOTE. If, in the statement of Proposition 7.11, δ is taken to be a line through O and "inversion in δ" is replaced by "reflection across δ," then the conclusion of Proposition 7.11 still holds—see Major Exercise 2, Chapter 3. Proposition 7.11 shows that in the P-model, inversion is the interpretation of hyperbolic reflection for P-lines that are not diameters of γ. Combining these two cases of P-lines, we call either of these two transformations *P-reflections*.

THEOREM ON THE CONSTRUCTION OF P-REFLECTIONS. For any two points A, B in the disk, a unique P-line δ can be constructed such that the P-reflection in δ interchanges A and B. The intersection of δ with the P-line joining A and B is their P-midpoint.

PROOF:

We start with the case where neither A nor B is O. Let A', B' be the inverses of A, B in γ. Let us work backward. Suppose that δ is the desired circle orthogonal to γ. Then inversion in δ must interchange A′ and B′ as well as interchanging A and B. Hence the center C of δ must be the intersection of the E-lines $\overleftrightarrow{AB}$ and $\overleftrightarrow{A'B'}$ if they are distinct and meet; then Proposition 7.3 tells how to construct the unique circle δ orthogonal to γ with center C. What if those E-lines are equal? They will be equal only when A and B lie on a diameter d, in which case the center C of δ must lie on the extension $\overline{d}$ of that diameter. Construct the E-lines α, β perpendicular to d at A, B, respectively, and let them intersect γ on a chosen side of d at A″, B″, respectively. Then C will be the intersection of $\overline{d}$ with $\overleftrightarrow{A''B''}$ since inversion in δ must interchange A″ and B″. What if those E-lines are parallel? That will happen only when A and B are equidistant from O; then reflection in the diameter perpendicular to d will interchange A and B. Similarly, in the case where $\overleftrightarrow{AB}$ and $\overleftrightarrow{A'B'}$ are distinct but parallel, we must have AO $\cong$ BO, so A and B will be interchanged by reflection in the diameter perpendicular to $\overleftrightarrow{AB}$. Finally, if, say, A = O, then C will again be the intersection of $\overline{d}$ with $\overleftrightarrow{A''B''}$. ◄

We come finally to the verification of the **SAS axiom**. We are given two Poincaré triangles △ABC and △XYZ inside γ such that $\sphericalangle$A $\cong$ $\sphericalangle$X, $d(AC) = d(XZ)$, and $d(AB) = d(XY)$ (Figure 7.31). We must prove that the triangles are Poincaré-congruent. We first reduce to the case where A = X = O (the *center* of γ): By the theorem just proved, if A $\neq$ O, there is a unique circle ε orthogonal to γ such that inversion in ε maps A to O; by Proposition 7.11, inversion in ε maps the Poincaré triangle △ABC onto a Poincaré-congruent Poincaré triangle △OB′C′. In the same way,

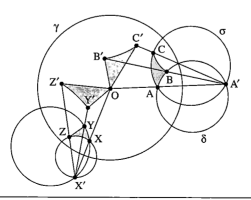

Figure 7.31 Proof of SAS for the Poincaré model.

Poincaré triangle $\triangle$XYZ can be mapped by inversion onto a Poincaré-congruent Poincaré triangle $\triangle$OY′Z′ (see Figure 7.31).

LEMMA 7.4. If $d(OB) = d$, then $\overline{OB} = r(e^d - 1)/(e^d + 1)$, where e is the base of the natural logarithm and r is the radius of γ.

PROOF:

If P and Q are the ends of the diameter of γ through B, labeled so that Q ∗ O ∗ B ∗ P, then $d = \log(OB, PQ)$. Exponentiating both sides of this equation gives

$$e^d = (OB, PQ) = \frac{\overline{OP}}{\overline{OQ}} \cdot \frac{\overline{BQ}}{\overline{BP}} = \frac{\overline{BQ}}{\overline{BP}} = \frac{r + \overline{OB}}{r - \overline{OB}},$$

and solving this equation for $\overline{OB}$ gives the result. ◄

COROLLARY. OB is P-congruent to OC iff they are Euclidean-congruent.

Returning to the proof of SAS, we have shown that we may assume that A = X = O. By Lemma 7.4 and the SAS hypothesis, we see that $\overline{OB} = \overline{OY}$, $\overline{OC} = \overline{OZ}$, and $\sphericalangle$BOC $\cong$ $\sphericalangle$YOZ. Hence a suitable Euclidean rotation about O—combined, if necessary, with reflection in a diameter—will map Euclidean triangle $\triangle$OBC onto Euclidean triangle $\triangle$OYZ. This transformation maps γ onto itself, and the orthogonal circle through B and C onto the orthogonal circle through Y and Z, preserving Poincaré length and angle measure. Hence the Poincaré triangles $\triangle$OBC and $\triangle$OYZ are Poincaré-congruent. ◄

NOTE. We have verified SAS in the Poincaré disk model by *superposition*, which was Euclid's idea! More precisely, we verified SAS by "rigidly moving" one triangle onto the other via a sequence of P-reflections. In fact, we have proved the following strong result (using Proposition 7.11).

THEOREM 7.1. Two triangles in the Poincaré disk model are P-congruent if and only if one can be mapped onto the other by a succession of P-reflections.

Let us call a transformation of the Poincaré disk model which is a composition of P-reflections a *P-rigid-motion*. Such a transformation preserves incidence, betweenness, and P-congruence in the model. These motions will be studied in greater detail in Chapter 9. For now, we need the following result.

SUBLEMMA. (a) For any two points A, B in the Poincaré disk model, there is a P-rigid-motion sending A to B. (b) For any three noncollinear points A, B, B′, there is a P-rigid-motion fixing A and sending P-ray $\overrightarrow{AB}$ to P-ray $\overrightarrow{AB'}$.

PROOF:

(a) In fact, we previously proved that there is a P-reflection that interchanges A and B. (b) If A ≠ O, let R be the P-reflection sending A to O and let R(B) = C, R(B′) = C′. The P-rays $\overrightarrow{AB}$ and $\overrightarrow{AB'}$ are mapped by R to P-rays emanating from O, which are just part of Euclidean rays emanating from O that form a Euclidean angle with vertex O. If S is the Euclidean reflection across the Euclidean bisector of ⦩COC′, then S interchanges $\overrightarrow{OC}$ and $\overrightarrow{OC'}$. Then the P-rigid-motion RSR fixes A and sends P-ray $\overrightarrow{AB}$ to P-ray $\overrightarrow{AB'}$. ◄

Let us use this sublemma to give a **verification of Axiom C-1** for the P-model, which does not appeal to continuity (hence is valid in arbitrary Euclidean planes, not just the real Euclidean plane). We are given a P-segment AB, a point A′, and a P-ray r emanating from A′. C-1 requires us to find a point B′ on r such that we have AB ≅ A′B′ (P-congruence). By (a), there is a P-rigid-motion T sending A to A′. By (b), there is a P-rigid-motion S fixing A′ and sending P-ray $T(\overrightarrow{AB})$ to r. Let B′ = ST(B). Then AB ≅ A′B′ because P-rigid-motions map any P-segment onto a P-congruent P-segment (a consequence of Proposition 7.10 and the corollary to Lemma 7.4). ◄

We have verified the axioms for a Hilbert plane in the Poincaré disk model. It follows that all propositions and theorems valid in Hilbert planes are valid in this model. It is, however, an interesting exercise to verify some of those propositions in the model by direct Euclidean constructions. For example, in the sublemma above, the P-rigid-motions mentioned can actually be taken to be P-reflections. We've shown that for part (a). For part (b), we can use the P-reflection across the P-angle-bisector t of P-angle ⦩BAB′. In Exercise P-4 you are asked to construct t (we did that in the special case where A = O).

It remains to verify Hilbert's hyperbolic axiom of parallels for the Poincaré disk model. Just use the isomorphism with the Klein model, where the verification is trivial. For a direct construction of the P-limiting parallel rays, see Exercise P-10.

Having verified that the Poincaré disk model is indeed a model of plane hyperbolic geometry within a Euclidean plane, *we have proved*

Metamathematical Theorem 1: If plane Euclidean geometry is consistent, then so is plane hyperbolic geometry.[15]

Let us next study what P-circles look like in the Poincaré disk model.

PROPOSITION 7.12. A P-circle is a Euclidean circle in the disk, and conversely, but the P-center differs from the Euclidean center except when the center is O.

> **PROOF:**
>
> Consider first the case where the P- or E-center of the circle is O. The result follows from the corollary to Lemma 7.4. Next, suppose the P-center of the P-circle δ is $A \neq O$. Let R be the P-reflection interchanging A and O. Then $R(\delta)$ is a P-circle with P-center O, hence a Euclidean circle with E-center O by the first case. By Proposition 7.6, $\delta = R(R(\delta))$ is a Euclidean circle with Euclidean center C not equal to A (Proposition 7.6 tells us that C is the image of O under a certain dilation from the center A' of ε, not the image A of O under R).
>
> Conversely, let δ be a Euclidean circle inside the disk, having Euclidean center $O' \neq O$. Let the line m joining O to O' intersect δ at points A, B (m is both a P-line and an E-line). Let M be the P-midpoint of AB and let R be the inversion (P-reflection) interchanging M and O. Then $R(\delta)$ is a Euclidean circle with diameter $R(A)R(B)$. Since O is the P-midpoint of this diameter, it is also the Euclidean midpoint (corollary to Lemma 7.4), so O is the Euclidean center of $R(\delta)$, and $R(\delta)$ is a P-circle with P-center O. Hence we see that $\delta = R(R(\delta))$ is a P-circle with P-center M. ◀

COROLLARY. The circle-circle continuity principle holds in the Poincaré disk model within a Euclidean plane.

> **PROOF:**
>
> P-circles are Euclidean circles inside the disk, and the inside of a circle is the same. So the result follows from that principle in the Euclidean plane. ◀

NOTE. This corollary furnishes a proof that the circle-circle (hence line-circle) continuity principle holds in all hyperbolic planes once it is

---

[15] The converse to this metatheorem has also been proved recently—see Project 1 and the application of polar coordinates in Chapter 10.

proved that every hyperbolic plane is isomorphic to a Poincaré disk model within a Euclidean plane. Hartshorne, Section 43, proves the latter using Hilbert's Euclidean field of ends. It would be interesting to have a direct, synthetic proof of those principles.

CONSTRUCTION OF THE P-CIRCLE WITH GIVEN P-CENTER AND P-RADIUS, AS WELL AS ITS E-CENTER.

Given distinct points A, B in the disk. We wish to construct the P-circle ε with P-center A passing through B. We have shown that it is a Euclidean circle, so we need only find its E-center C and construct the E-circle ε with E-center C and E-radius CB. If A = O, then C = O, so suppose A $\neq$ O. Then the diameter d of γ through A is a P-line through the P-center of ε and so must be orthogonal to ε; that means it passes through the E-center C of ε. If B does not lie on d, construct the E-circle δ orthogonal to γ through A and B (the E-circle through A, B, and A'); it cuts out the P-line joining A to B, which cuts out a P-diameter of ε, so ε is orthogonal to δ at B. That means the tangent t to δ at B passes through C. Hence C is the point where d meets t (see Figure 7.32).

Suppose B lies on d. Construct the P-perpendicular δ to d at A: It is cut out of the disk by the E-circle through A and A' whose center is the midpoint of AA'. Construct the inverse B' of B in δ (as in Proposition 7.3). Then B' has the same P-distance from A as B, and so lies on ε. Hence the E-center C of ε is the E-midpoint of BB'. ◄

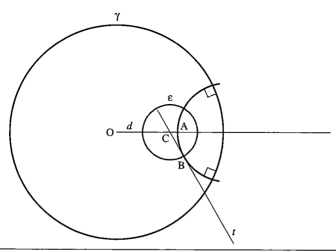

Figure 7.32 Construction of E-center C of P-circle ε.

EUCLIDEAN CHARACTERIZATION OF THE P-CENTER. Since P-circles are just E-circles ε inside γ, we can ask what the P-center of ε is from the Euclidean point of view. In any Hilbert plane, the center of a circle ε is characterized as the point of concurrence of all the lines that intersect ε orthogonally. Knowing the interpretation of "line" in the Poincaré model, we see that the P-center A of ε is the point of concurrence inside γ of all E-circles that intersect *both* ε and γ orthogonally and of the unique diameter of γ, which intersects ε orthogonally. (Those E-circles and that diameter extended have another point of concurrence outside γ—namely, the inverse A′ of A in γ.)

We will now apply the Poincaré model to determine the formula of J. Bolyai and Lobachevsky for the angle of parallelism. Let $\Pi(d)$ denote the number of *radians* in the angle of parallelism corresponding to the Poincaré distance d (the number of radians is $\pi/180$ times the number of degrees).

THEOREM 7.2. In the Poincaré disk model, the formula for the angle of parallelism is $e^{-d} = \tan[\Pi(d)/2]$.[16]

In this formula, e is the base for the natural logarithm. The trigonometric tangent function is defined analytically as sin/cos, where the sine and cosine functions are defined by their Taylor series expansions (given in Chapter 10). The tangent is *not* to be interpreted as the ratio of opposite to adjacent for a right triangle in the hyperbolic plane!

PROOF:

By definition of the angle of parallelism, d is the Poincaré distance $d(PQ)$ from some point P to some Poincaré line l, and $\Pi(d)$ is the number of radians in the angle that a limiting parallel ray to l through P makes with $\overrightarrow{PQ}$. We may choose l to be a diameter of γ and Q to be the center of γ, so that P lies on the perpendicular diameter. A limiting parallel ray through P is then an arc of a circle δ orthogonal to γ such that δ is tangent to l at one end Σ. The tangent line to δ at P therefore meets l at some interior point R that is the pole of chord $P\Sigma$ of δ, and, by Proposition 7.4, $\sphericalangle RP\Sigma$ and $\sphericalangle R\Sigma P$ both have the same number of radians β (see Figure 7.33). Let $\alpha = \Pi(d)$, which is the number of radians in $\sphericalangle RPQ$. Since 2β is the number of radians in $\sphericalangle PRQ$ (exterior to $\Delta PR\Sigma$), we get $\alpha + 2\beta = \pi/2$, or

---

[16] Theorem 7.2 uses real numbers, of course. Hartshorne has a version of it, valid in arbitrary hyperbolic planes, which uses his multiplicative length and his algebraic version of the tangent function: See his Proposition 41.9.

$\beta = \pi/4 - \alpha/2$. The Euclidean distance $\overline{PQ}$ is $r \tan \beta$, so that, by the proof of Lemma 7.4,

$$e^d = \frac{1 + \tan \beta}{1 - \tan \beta}.$$

Using the formula for β and the trigonometric identity

$$\tan \left(\frac{\pi}{4} - \frac{\alpha}{2} \right) = \frac{1 - \tan(\alpha/2)}{1 + \tan(\alpha/2)},$$

we get the desired formula after some algebra. ◄

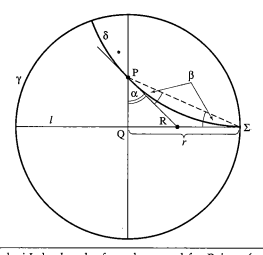

Figure 7.33 Bolyai-Lobachevsky formula proved for Poincaré model.

We have developed only enough of the geometry of inversion in circles to verify the axioms in the Poincaré disk model. You will find further developments in the exercises and in Chapters 9 and 10. Inversion has many other applications in geometry, notably in Feuerbach's famous theorem on the nine-point circle of a triangle, the problem of Apollonius (Hartshorne, Section 38), and the construction of linkages that change linear motion into curvilinear motion (see Kay, 1969, and Pedoe, 1970).

The Projective Nature of the Beltrami–Klein Model

Although the Klein model is also located on an open disk in a Euclidean plane, it can best be understood via the projective completion of that Euclidean plane, which is also the model's projective completion as a hyperbolic plane (see Major Exercise 13, Chapter 6). We know now that the Klein interpretation is a model of hyperbolic plane geometry because it is isomorphic to the Poincaré disk model, as we showed.

To be more explicit, consider the unit sphere Σ in Cartesian three-dimensional space given by the equation $x_1^2 + x_2^2 + x_3^2 = 1$. Let γ be the unit circle in the equatorial plane of Σ, determined by the equation $x_3 = 0$ and the equation for Σ. We will represent both the Poincaré disk and the Klein disk by the set Δ of points inside γ, and we will take as our isomorphism F the composite of two mappings: If N is the north pole $(0, 0, 1)$ of Σ, first project Δ onto the southern hemisphere of Σ stereographically from N. Then project orthogonally back upward to the disk Δ (see Figure 7.34).

The isomorphism F will be considered to go from the Poincaré model to the Klein model. By an easy exercise in similar triangles, you can show that F is given in coordinates by

$$F(x_1, x_2, 0) = \left(\frac{2x_1}{1 + x_1^2 + x_2^2}, \frac{2x_2}{1 + x_1^2 + x_2^2}, 0 \right).$$

Or, if we ignore the third (zero) coordinate and use the single complex coordinate $z = x_1 + ix_2$, then F is given by

$$F(z) = \frac{2z}{1 + |z|^2}.$$

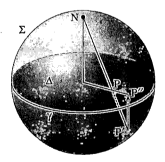

Figure 7.34 Isomorphism F via the sphere.

It is clear that F maps a diameter of γ onto the same diameter (but moving the points on the diameter out toward the circle). Let δ be a circle orthogonal to γ and cutting γ at points P and Q. We claim that F maps the Poincaré line with ends P and Q onto the open chord P)(Q. In fact, *if A is on the arc of δ from P to Q inside γ, then $F(A)$ is the point at which $\overrightarrow{OA}$ hits chord* PQ (see Figure 7.35).

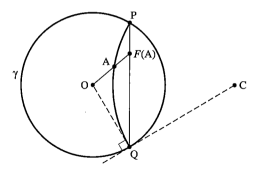

Figure 7.35 Isomorphism F within the disk.

PROOF:

We can prove this as follows. Suppose the center C of δ has coordinates (c_1, c_2). By Proposition 7.3, the points P and Q are the intersections with γ of the circle having CO as diameter. After simplifying, the equation of this circle turns out to be

(1) $$x_1^2 - c_1 x_1 + x_2^2 - c_2 x_2 = 0.$$

Combining this equation with the equation $x_1^2 + x_2^2 = 1$ for γ gives the equation

(2) $$c_1 x_1 + c_2 x_2 = 1$$

for the line joining P to Q (called the *polar* of C with respect to γ). Since δ is orthogonal to γ, $\sphericalangle OQC$ is a right angle, and the Pythagorean theorem gives

(3) $$\overline{CQ}^2 = \overline{CO}^2 - \overline{OQ}^2 = c_1^2 + c_2^2 - 1$$

for the square of the radius of δ. Hence δ is the circle

$$(x_1 - c_1)^2 + (x_2 - c_2)^2 = c_1^2 + c_2^2 - 1,$$

which simplifies to

(4) $$x_1^2 + x_2^2 = 2c_1 x_1 + 2c_2 x_2 - 1.$$

If now $A = (a_1, a_2)$ lies on δ and $F(A) = (b_1, b_2)$ is its image under F, we have for $j = 1, 2$,

(5) $$b_j = 2a_j/(1 + a_1^2 + a_2^2),$$

(6) $$b_j = a_j/(c_1 a_1 + c_2 a_2).$$

It follows that

(7) $c_1 b_1 + c_2 b_2 = 1,$

and hence $F(A)$ lies on the polar of C, as asserted. ◄

We now use the isomorphism F to define congruence in the Klein model. Two segments (respectively, two angles) are interpreted to be *Klein-congruent* if their inverse images under F in the Poincaré model are Poincaré-congruent (as was defined before). With this interpretation, the verification of the congruence axioms is immediate. (It follows from this interpretation that the Klein model is conformal only at O.)

Next, let us *justify the previous description of perpendicularity in the Klein model.* According to the above definition, two Klein lines l and m are Klein-perpendicular if and only if their inverse images $F^{-1}(l)$ and $F^{-1}(m)$ are perpendicular Poincaré lines. There are three cases to consider.

▨ **CASE 1.** Both l and m are diameters. In this case, it is clear that perpendicularity has its usual Euclidean meaning.

▨ **CASE 2.** Only l is a diameter. Then $F^{-1}(l) = l$. The only way $F^{-1}(m)$, an arc of an orthogonal circle δ, can be perpendicular to l is if the Euclidean line extending l passes through the center C of δ (see Figure 7.36). In that case, the extension of l is the perpendicular bisector of chord m (Exercise 17, Chapter 4). Conversely, if l is perpendicular to m in the Euclidean sense, l bisects m, and hence the extension of l goes through C and l is then perpendicular to arc $F^{-1}(m)$.

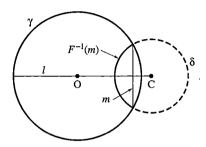

Figure 7.36

CASE 3. Neither l nor m is a diameter. Then $F^{-1}(l)$ and $F^{-1}(m)$ are arcs of circles δ and σ orthogonal to γ. Suppose δ is orthogonal to σ. By Proposition 7.4, the centers of these circles are the poles $P(l)$ and $P(m)$ of l and m since these circles meet γ at the ends of l and m. Let P and Q be the ends of m. Inversion in δ interchanges P and Q since this inversion maps both γ and σ onto themselves (corollary to Proposition 7.6). But if P and Q are inverse in δ, the Euclidean line joining them has to pass through the center $P(l)$ of δ (see Figure 7.37).

Conversely, if the extension of m passes through $P(l)$, then P and Q are inverse to each other in δ (since points on γ are mapped onto γ by inversion in δ). By Proposition 7.5, σ is orthogonal to δ. ◄

Next, let us describe the interpretation of reflections in the Klein model. In both Euclidean and hyperbolic geometries, the *reflection* in a line m is the transformation R_m of the plane, which leaves each point of m fixed and transforms a point A not on m as follows. Let M be the foot of the perpendicular from A to m. Then, by definition, $R_m(A)$ is

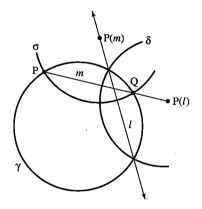

Figure 7.37

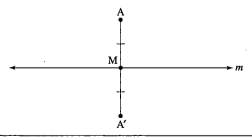

Figure 7.38 A′ is the reflection of A across m.

the unique point A' such that A' $*$ M $*$ A and A'M $\cong$ MA (Figure 7.38). In Major Exercise 2, Chapter 3, you showed that reflection preserves incidence, betweenness, and congruence in any Hilbert plane.

Returning to the Klein model, assume first that m is not a diameter of γ and let P be its pole. To drop a Klein perpendicular from A to m, we draw the line joining A and P. Let it cut m at M and let t be the chord of γ cut out by this Euclidean line. Let Q be the pole of t and draw the line joining Q and A. Let this line cut γ at Σ and Σ' and let n be the open chord $\Sigma)(\Sigma'$. Draw the line joining Σ' and M and let it cut γ again at point Ω. If we now join Ω and Q, we obtain a line that cuts t at A' and γ again at Ω' (see Figure 7.39).

CONTENTION. The point A' just constructed is the reflection in the Klein model of A across m. The Euclidean lines extending $\Omega\Sigma$ and $\Omega'\Sigma'$ meet at P, and $\Omega\Sigma'$ meets $\Omega'\Sigma$ at point M.

One justification for this construction is given in Major Exercise 12, Chapter 6. Here is another. Start with divergently parallel Klein lines $l = \Omega\Omega'$ and $n = \Sigma\Sigma'$ and their common perpendicular t. Let l meet t in A' and n meet t in A and let M be the midpoint of AA' in the sense of the model. Let m be the Klein line through M Klein-perpendicular to t; m is obtained by joining M to the pole Q of t. Ray MΣ' is limiting parallel to n. If we reflect across m, then n is mapped onto the line

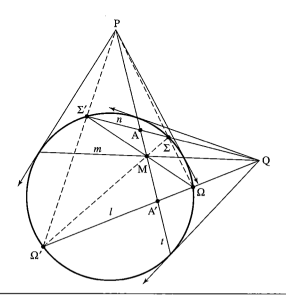

Figure 7.39 A' is the Klein reflection of A across m.

through A' Klein-perpendicular to t, namely, the line l. The end Σ' is mapped onto the end of l on the same side of t as Σ', namely, the point Ω'. Hence ray $M\Sigma'$ is mapped onto ray $M\Omega'$. Now reflect across t; Ω' is sent to Ω, so $M\Omega'$ is mapped to $M\Omega$. But successive reflections in the Klein-perpendicular lines m and t combine to give the 180° rotation about M. Hence $M\Omega$ is the ray opposite to $M\Sigma'$. Similarly, $M\Sigma$ is the ray opposite to $M\Omega'$. Since reflection in m sent Σ' to Ω' and Σ to Ω, $\Sigma'\Omega'$ and $\Sigma\Omega$ must both be Klein-perpendicular to m and their Euclidean extensions meet at the pole P of m.

Second, let us describe the Klein reflection for the case in which m is a diameter of γ. In this case, P is a point at infinity, t is perpendicular to m in the Euclidean sense, and M is the Euclidean midpoint of chord t (since a diameter perpendicular to a chord bisects it). Chord $\Omega\Sigma$ was shown to be perpendicular to diameter m in the argument above, so Ω is the Euclidean reflection of Σ across m. Hence $\overleftrightarrow{Q\Omega}$ is the Euclidean reflection of $\overleftrightarrow{Q\Sigma}$, and we deduce that A' *is the ordinary Euclidean reflection of* A *across diameter* m (see Figure 7.40).

In order to describe the Klein reflection more succinctly, let us return to the notion of cross-ratio (AB, CD) defined by the formula

$$(AB,\ CD) = \frac{\overline{AC}}{\overline{AD}} \cdot \frac{\overline{BD}}{\overline{BC}}.$$

DEFINITION. If A, B, C, and D are four distinct *collinear* points in the Euclidean plane such that (AB, CD) = 1, we say that C and D are *harmonic conjugates* with respect to AB and that ABCD is a *harmonic tetrad*. By symmetry of the cross-ratio, A and B are then also harmonic conjugates with respect to CD.

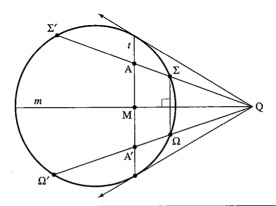

Figure 7.40 A' is the Euclidean reflection of A across diameter m.

Another way to write the condition for a harmonic tetrad is $\overline{AC}/\overline{AD} = \overline{BC}/\overline{BD}$. Since C and D are distinct, one must be inside segment AB and the other outside (so that "C and D divide AB internally and externally in the same ratio"). Moreover, given AB, then C and D determine each other uniquely. For example, suppose A * C * B and let $k = \overline{AC}/\overline{CB}$. If $k < 1$, then D is the unique point such that D * A * B and $\overline{DB} = \overline{AB}/(1 - k)$, whereas if $k > 1$, then D is the unique point such that A * B * D and $\overline{DB} = \overline{AB}/(k - 1)$; see Figure 7.41. The case where $k = 1$ is indeterminate, for there is no point D outside AB such that $\overline{AD} = \overline{BD}$. Thus the midpoint M of AB has no harmonic conjugate. This exception can be removed by completing the Euclidean plane to the real projective plane by adding a "line at infinity" (see Chapter 2). Then the harmonic conjugate of M is defined to be the "point at infinity" on $\overleftrightarrow{AB}$.

There is a nice way of constructing the harmonic conjugate of C with respect to AB with a straightedge alone: Take any two points I and J collinear with C but not lying on $\overleftrightarrow{AB}$. Let $\overleftrightarrow{AJ}$ meet $\overleftrightarrow{BI}$ at point K and let $\overleftrightarrow{AI}$ meet $\overleftrightarrow{BJ}$ at point L. *Then* $\overleftrightarrow{AB}$ *meets* $\overleftrightarrow{KL}$ *at the harmonic conjugate* D *of* C (Figure 7.42).

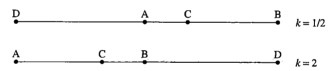

Figure 7.41

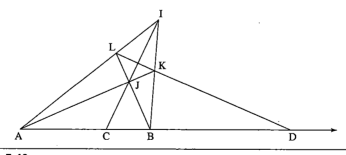

Figure 7.42

We will justify this *harmonic construction* momentarily. Meanwhile, as a device to help remember the construction, "project" line $\overleftrightarrow{ID}$ to infinity. Then our figure becomes Figure 7.43. Since $\Box A'B'K'L'$ is now a parallelogram, we see that C' is the midpoint of A'B' and its harmonic conjugate is the "point at infinity" D' on $\overleftrightarrow{A'B'}$. (This mnemonic device can be turned into a proof based on projective geometry; see Eves, 1972, Chapter 6.)

If you will now refer back to Figure 7.39, where the Klein reflection A' of A was constructed, you will see that A' *is the harmonic conjugate of A with respect to* MP. Just relabel the points in Figure 7.39 by the correspondences I-Σ', J-Σ, K-Ω, L-Ω', A-P, B-M, C-A, and D-A' to obtain a figure for constructing the harmonic conjugate.

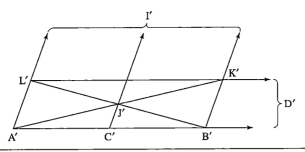

Figure 7.43

DEFINITION. Let m be a line and P a point not on m. A transformation of the Euclidean plane called the *harmonic homology with center P and axis m* is defined as follows. Leave P and every point on m fixed. For any other point A, let the line t joining P to A meet m at M. Assign to A the unique point A' on t, which is the harmonic conjugate of A with respect to MP.

With this definition we can restate our result.

THEOREM 7.3. Let m be a Klein line that is not a diameter of γ and let P be its pole. Then reflection across m is interpreted in the Klein model as restriction to the interior of γ of the harmonic homology with center P and with axis the Euclidean line extending m. If m is a diameter of γ, then reflection across m has its usual Euclidean meaning.

To justify the harmonic construction, we need the notion of a *perspectivity*. This is the mapping of a line l onto a line n obtained by projecting from a point P not on either line (Figure 7.44). It assigns to point A on l the point A' of intersection of $\overleftrightarrow{PA}$ with n. (Should $\overleftrightarrow{PA}$ be parallel to n, the image of A is the point at infinity on n.) P is called the *center* of this perspectivity.

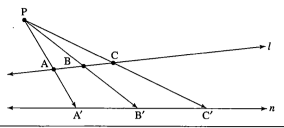

Figure 7.44 Perspectivity with center P.

LEMMA 7.5. A pespectivity preserves the cross-ratio of four collinear points; i.e., if A, B, C, and D are four points on line l and A', B', C', and D' are their images on line n under the perspectivity with center P, then (AB, CD) = (A'B', C'D').

PROOF:

By Exercise 15, Chapter 5, we have

$$\frac{\overline{AC}}{\overline{BC}} = \frac{\overline{AP} \sin \sphericalangle APC}{\overline{BP} \sin \sphericalangle BPC}$$

and

$$\frac{\overline{BD}}{\overline{AD}} = \frac{\overline{BP} \sin \sphericalangle BPD}{\overline{AP} \sin \sphericalangle APD},$$

which give

$$(AB,\ CD) = \frac{(\sin \sphericalangle APC)(\sin \sphericalangle BPD)}{(\sin \sphericalangle BPC)(\sin \sphericalangle APD)}.$$

But sin $\sphericalangle APC$ = sin $\sphericalangle A'PC'$, sin $\sphericalangle BPD$ = sin $\sphericalangle B'PD'$, and so on, so we obtain the same formula for (A'B', C'D').

Now refer back to Figure 7.42. Let $\overleftrightarrow{IJ}$ meet $\overleftrightarrow{KL}$ at point M. Using the perspectivity with center I, Lemma 7.5 gives us the relation (AB, CD) = (LK, MD), whereas using the perspectivity with center J, we get (AB, CD) = (KL, MD). But (KL, MD) = 1/(LK, MD), by the definition of cross-ratio. Hence (AB, CD) is its own reciprocal, which means that (AB, CD) = 1; i.e., ABCD is a harmonic tetrad, as asserted. This justifies the harmonic construction previously given. ◀

Next, we will apply Theorem 7.3 to calculate the length of a segment in the Klein model. According to our general procedure, length in the Klein model is defined by pulling back to the Poincaré model via the inverse of the isomorphism F and using the definition of length already given there. Thus the length $d'(AB)$ of a segment in the Klein model is given by $d'(AB) = d(ZW) = |\log(ZW, PQ)|$, where $A = F(Z)$, $B = F(W)$, and P and Q are the ends of the Poincaré line through Z and W. By our earlier result illustrated in Figure 7.35, P and Q are also the ends of the Klein line through A and B.

The next theorem shows how to calculate $d'(AB)$ directly in terms of A, B, P, and Q. In its proof we will need the remark "the cross-ratio (AB, PQ) is preserved by any Klein reflection." This is clear if we are reflecting in a diameter of γ. Otherwise, by Theorem 7.3, the Klein reflection is a harmonic homology whose center R lies outside γ. A reflection in the hyperbolic plane preserves collinearity, so for any Klein line l, the mapping of l onto its Klein reflection n is just the perspectivity with center R. Therefore, Lemma 7.5 ensures that the cross-ratio is preserved.

THEOREM 7.4. If A and B are two points inside γ and P and Q are the ends of the chord of γ through A and B, then the *Klein length* of segment AB is given by the formula

$$d'(AB) = \frac{1}{2} \, |\log(AB, PQ)|.$$

PROOF:

We saw in the verification of the SAS axiom for the Poincaré disk model that any Poincaré line can be mapped onto a diameter by an inversion in a suitable orthogonal circle. Proposition 7.10 guarantees that cross-ratios are preserved by inversions. The transformation of the Klein model that corresponds to this inversion under our isomorphism F is a harmonic homology (Theorem 7.3), and this preserves cross-ratios of collinear points by the above remark. Hence we may assume that A and B lie on a diameter.

Let $A = F(Z)$ and $B = F(W)$, so that, by definition, we have $d'(AB) = d(ZW)$. After a suitable rotation (which preserves cross-ratios), we may assume that the given diameter is the real axis. Its ends P and Q then have complex coordinates -1, $+1$. If Z and W have real coordinates z and w, then

$$(ZW, PQ) = \frac{1 + z}{1 - z} \cdot \frac{1 - w}{1 + w},$$

$$(AB, PQ) = \frac{1 + F(z)}{1 - F(z)} \cdot \frac{1 - F(w)}{1 + F(w)}.$$

But

$$1 - F(z) = 1 - \frac{2z}{1 + |z|^2} = \frac{1 - 2z + |z|^2}{1 + |z|^2}$$

$$1 + F(z) = \frac{1 + 2z + |z|^2}{1 + |z|^2}$$

$$\frac{1 + F(z)}{1 - F(z)} = \frac{1 + 2z + |z|^2}{1 - 2z + |z|^2}.$$

Since z is real, $z = \pm|z|$ and we get

$$\frac{1 + F(z)}{1 - F(z)} = \left(\frac{1 + z}{1 - z}\right)^2.$$

From this and the formula obtained from it by substituting w for z, it follows that $(AB, PQ) = (ZW, PQ)^2$, and taking logarithms of both sides proves the theorem. ◀

NOTE. In the proof just given, and earlier, we have used real and complex numbers. However, everything we have done so far in this section is valid over an arbitrary Euclidean field F, not just over $\mathbb{R}$, once one makes the following observation: Complex numbers were only used as a shorthand to abbreviate more complicated formulas involving two real variables (the real and imaginary parts of these complex numbers). That shorthand can also be used over F by formally adjoining a symbol i whose square is -1 and manipulating the elements $a + bi$ with $a, b \in F$ in the same way complex numbers are manipulated—i.e., for those who know abstract algebra, by working in the field $F(i)$.

Finally, let us apply our results to **justify J. Bolyai's construction of the limiting parallel ray** (Chapter 6). We are given a Klein line l and a point P not on it. Point Q on l is the foot of the Klein perpendicular t from P to l, and m is the Klein perpendicular to t through P. Let R be any other point on l, and S, the foot on m of the Klein perpendicular from R. Bolyai's construction is based on the contention that if the limiting parallel ray to l from P in the direction $\overrightarrow{QR}$ meets RS at X, then PX is Klein-congruent to QR.

Let T and M be the poles of t and m. Let Ω and Ω' be the ends of l. If we join these ends to M, the intersections Σ and Σ' with γ will be the ends of the Klein reflection n of l across m.

As Figure 7.45 shows, the collinear points Ω, X, P, and Σ' are in perspective with the collinear points Ω, R, Q, and Ω' (in that order), the center of the perspectivity being M. By Lemma 7.5, such a perspectivity preserves cross-ratios, so that we have (XP, $\Omega\Sigma'$) = (RQ, $\Omega\Omega'$). Theorem 7.4 tells us that d'(XP) = d'(RQ), justifying Bolyai's contention. (In the case where m is a diameter of γ, M is a point at infinity; then instead of Lemma 7.5 we use the parallel projection theorem (preceding Exercises 10–18, Chapter 5) to deduce the above equality of cross-ratios, or move the figure by a harmonic homology so that m is not a diameter.) ◄

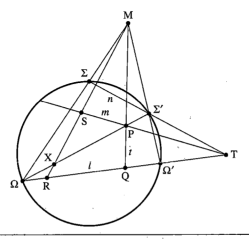

Figure 7.45 Bolyai's construction in the Klein model.

NOTE: The method used to prove Theorems 7.2 and 7.4 is very useful for solving other problems in the Klein and Poincaré models. The idea is that the figure being studied can be moved, by a succession of hyperbolic reflections, to a special position where one or more of the hyperbolic lines is represented by a diameter of the absolute circle γ and one point is the center O of γ. The movement to this special position does not alter the geometric properties of the figure, and in that special position, elementary arguments and calculations based on Euclidean geometry can be used to solve the problem.

For example, if P, P' ≠ O, then the statement OP ≅ OP' has the same truth value whether interpreted in the Euclidean, Poincaré, or Klein sense (according to Lemma 7.4 and Theorem 7.4), and ∢POP' has the same measure in all three senses. In particular, a hyperbolic circle with hyperbolic center O is represented in both models by a Euclidean circle with Euclidean center O.

You will see some nice applications of this method in Exercises K-15, K-17 through K-20, and P-5, and in Chapters 9 and 10. The general study of geometric motions is in Chapter 9.

Let us apply this method to **verify the Bolyai–Lobachevsky formula in the Klein model**: Let $\alpha = \Pi(AB)$. Take A to be the center of the unit Klein disk and also the vertex of angle α. The end of the side of α not containing B is a point Ω on the unit circle. ΩBA is a Euclidean right triangle and also a singly asymptotic Klein right triangle with right angle at B since AB is a segment of a diameter PQ. The Euclidean length $\overline{AB}$ is equal to $\cos \alpha$ since $\overline{A\Omega} = 1$. If $P * A * B * Q$, we calculate the cross-ratio

$$(AB, PQ) = \frac{\overline{BQ}}{\overline{BP}} = \frac{1 - \cos \alpha}{1 + \cos \alpha}.$$

But by a trigonometric formula, this is equal to $\tan(\alpha/2)$. ◄

Conclusion

We have proved that Euclid V cannot be proved from the axioms of neutral geometry by studying three isomorphic models of plane hyperbolic geometry within a Euclidean plane—one named after Klein and the other two after Poincaré (although Beltrami found all three of them first). Thus the attempts over 2000 years to prove Euclid V from his other axioms had to fail (if Euclidean geometry is consistent). These models make the "strange new universe" of hyperbolic geometry much less strange and help us visualize it. In the process of verifying the hyperbolic axioms in the Poincaré models, we have learned a good deal of *inversive geometry* (much more will be found in the P-exercises) and some more *projective geometry* for the Klein model (see the H-exercises and the H-exercises).

Review Exercise

Which of the following statements are correct?

(1) Although 2000 years of efforts to prove the parallel postulate as a theorem in neutral geometry have been unsuccessful, it is still possible that someday some genius will succeed in proving it.

(2) If we add to the axioms of neutral geometry the elliptic parallel postulate (that no parallel lines exist), we get another consistent geometry called elliptic geometry.

(3) All the ultra-ideal points in the Klein model are points in the Euclidean plane outside γ.

(4) Both the Klein and Poincaré models are "conformal" in the sense that congruence of angles has the usual Euclidean meaning.

(5) In the Poincaré model, "lines" are represented by all open diameters of a fixed circle γ and by all open arcs inside γ of circles intersecting γ.

(6) For any chord A)(B whatever of circle γ, the tangents to γ at the endpoints A and B of the chord meet in a unique point called the *pole* of that chord.

(7) In the Poincaré model, two Poincaré lines are interpreted as "perpendicular" if and only if they are perpendicular in the usual Euclidean sense.

(8) In the Klein model, two open chords are interpreted to be "perpendicular" if and only if they are perpendicular in the usual Euclidean sense.

(9) Inversion in a given circle maps all circles onto circles.

(10) Ultra-ideal points have no representation in the Poincaré models.

(11) Four points in the Euclidean plane form a harmonic tetrad if they are collinear and their cross-ratio equals 1.

(12) If point O is outside circle δ and a tangent from O to δ touches δ at point T, then the power of O with respect to δ is equal to the square of the distance from O to T.

(13) Let point P lie on circle δ and let P' and δ' be their inverses in another circle γ such that γ does not pass through P or the center of δ. Then the tangent to δ' at P' is parallel to the tangent to δ at P.

(14) The inverse of the center of a circle δ is the center of the inverted circle δ'.

(15) In order for the midpoint M of segment AB to have a harmonic conjugate with respect to AB, for all A and B, a Euclidean plane must be extended to a projective plane by adding a line of points at infinity.

(16) If a geometric statement in real hyperbolic geometry holds when interpreted in the Klein or Poincaré models, then that statement is a theorem in hyperbolic geometry.

(17) If O is the center of the Poincaré disk and ε is a P-circle whose P-center A is not O, then the E-center C of ε is E-between O and A.

The following exercises will be divided into four categories: (1) K-exercises, on the Klein model; (2) P-exercises, on the Poincaré models and on circles; (3) H-exercises, on harmonic tetrads and theorems of Menelaus, Ceva, Gergonne, and Desargues; (4) projects. The K-exercises and P-exercises are extremely important for a visual understanding of plane hyperbolic geometry.

K-Exercises

K-1. Verify the interpretations of the incidence axioms, the betweenness axioms, and Dedekind's axiom (if the Euclidean plane is real) for the Klein model.

K-2. (a) Let l be a diameter of γ and let m be an open chord of γ that does not meet l and whose endpoints differ from the endpoints of l. Draw a diagram showing the common perpendicular k to l and m in the Klein model. (Hint: Use the pole of m and the case 1 description of perpendicularity.)

(b) Let l and m be intersecting open chords of γ. It is a valid theorem in hyperbolic geometry that for any two intersecting nonperpendicular lines there exists a third line perpendicular to one of them and asymptotically parallel to the other (see Major Exercise 9, Chapter 6). Draw the two lines in the Klein model that are perpendicular to l and asymptotically parallel to m (on the left and right, respectively). This shows that the angle of parallelism can be any acute angle whatever. Explain.

(c) In the Euclidean plane, any three parallel lines have a common transversal. Draw three parallel lines in the Klein model that do *not* have a common transversal.

K-3. (a) In the Klein model, an ideal point and an ordinary point always determine a unique Klein line. Translate this back into a theorem in hyperbolic geometry about limiting parallel rays.

(b) Suppose the ultra-ideal points P(l) and P(m) are poles of Klein lines l and m, respectively. You saw in Figure 7.18 that the Euclidean line joining P(l) and P(m) need not cut through the circle γ and hence need not determine a Klein line. Show that the only case in which there is a Klein line joining P(l) and P(m) is when lines l and m are divergently parallel.

(c) Suppose the ultra-ideal point P(l) is the pole of a Klein line l and Ω is an ideal point; Ω is uniquely determined by a ray r in the direction of Ω. State the necessary and sufficient conditions on r and l in order that P(l) and Ω determine a Klein line. Translate this into a theorem in hyperbolic geometry.

K-4. Given chords l and m of γ that are not diameters. Suppose the line extending m passes through the pole of l. Prove that the line extending l passes through the pole of m. (Hint: Use either equation (2), in the last section of this chapter, or the theory of orthogonal circles.)

K-5. Use the Klein model to show that in the hyperbolic plane there exists a pentagon with five right angles and there exists a hexagon with six right angles. (Hint: Begin with two lines having a common perpendicular. Locate the poles of these two lines, then draw an appropriate line through each of the poles, etc.) Does there exist, for all $n \geq 5$, an n-sided polygon with n right angles?

K-6. Justify the following construction of the Klein reflection A′ of A across m, which is simpler than the one in Figure 7.39. Let Λ be an end of m and let P be the pole of m. Join Λ to A and let this line cut γ again at Φ. Join Φ to P and let this line cut γ at Φ'. Then A′ is the intersection of $\overleftrightarrow{AP}$ with $\Lambda\Phi'$ (see Figure 7.46).

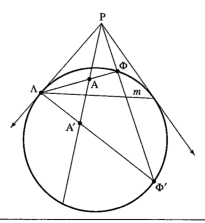

Figure 7.46 Simpler construction of the Klein reflection.

K-7. Given a segment AA′ in the Klein model. Show how to construct its hyperbolic midpoint with straightedge and compass (see Figures 7.39 and 7.40).

K-8. Construct triangles in the Klein model such that the perpendicular bisectors of the sides are (a) divergently parallel and (b) asymptotically parallel. (See Exercise 11 and Major Exercise 7, Chapter 6.)

K-9. Prove the formula

$$F(z) = \frac{2z}{1 + |z|^2}$$

for the isomorphism F of the Poincaré model onto the Klein model (see Figure 7.34). What is the formula for the inverse isomorphism? Angle measure in the Klein model is defined so that F preserves angle measure; draw the diagram which illustrates this.

K-10. Let A = (0, 0), let B = (0, $\frac{1}{2}$), and let l be the diameter of γ cut out by the x-axis.

  (a) Find the Klein length $d'(AB)$.

  (b) Find the coordinates of the point M on segment AB that represents its midpoint in the Klein model.

  (c) Find the equation of the equidistant curve to l through B. Show that it is an arc of an ellipse.

K-11. Let Ω and Ω' be distinct ideal points and A an ordinary point. Let P be the pole of chord $\Omega\Omega'$ and let Euclidean ray $\overrightarrow{AP}$ cut γ at Σ. Prove that AΣ represents the bisector of $\sphericalangle\Omega A\Omega'$ in the Klein model (see Figure 7.47). Apply this result to justify the construction of the line of enclosure given in Major Exercise 8, Chapter 6. (Hint: Use Proposition 6.6.)

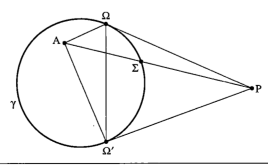

Figure 7.47 Angle bisector in Klein model.

K-12. In Exercise 18, Chapter 4, you proved the theorem that the angle bisectors of a triangle in hyperbolic geometry (in fact, in neutral geometry) are concurrent. Using the construction of angle
bisectors given in the previous exercise and the glossary of the
Klein model, translate this theorem into a famous theorem in
Euclidean geometry due to Brianchon (see Figure 7.48). This
gives a hyperbolic proof of a Euclidean theorem (for a Euclidean proof, see Coxeter and Greitzer, 1967, p. 77).

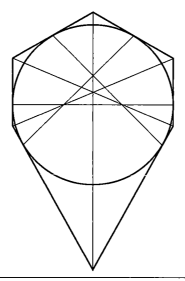

Figure 7.48 Brianchon's theorem.

K-13. It is a theorem in hyperbolic geometry that inside every trebly
asymptotic triangle $\Delta\Sigma\Omega\Lambda$ there is a unique point G equidistant
from all sides, which is the point of concurrences of the alti-
tudes. Show that in the Klein model this theorem is a consequence of Gergonne's theorem in Euclidean geometry, which asserts that if the inscribed circle of $\triangle$PQR touches the sides at
points Λ, Σ, and Ω, then segments PΣ, QΩ, and RΛ are concurrent (see Figure 7.49 and Exercise H-9). Show that
$(\measuredangle\Lambda G\Sigma)° = 120°$ in the sense of degree measure for the Klein
model. (Hint: To take care of the special case where one side of
$\Delta\Sigma\Omega\Lambda$ is a diameter, apply a harmonic homology to transform
to the case where Gergonne's theorem applies.)

Note that G is the hyperbolic center of the inscribed hyperbolic circle of $\Delta\Sigma\Omega\Lambda$.

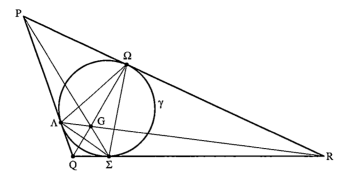

Figure 7.49 Gergonne point G is Klein incenter of trebly asymptotic triangle.

K-14. In order to express the Klein length $d'(AB) = \frac{1}{2}\,|\log(AB, PQ)|$ in
terms of the coordinates (a_1, a_2) of A and (b_1, b_2) of B, prove
that with a suitable ordering of the ends P and Q of the Klein
line through A and B, you have the formula

$$(AB, PQ)$$

$$= \frac{a_1b_1 + a_2b_2 - 1 - \sqrt{(a_1 - b_1)^2 + (a_2 - b_2)^2 - (a_1b_2 - a_2b_1)^2}}{a_1b_1 + a_2b_2 - 1 + \sqrt{(a_1 - b_1)^2 + (a_2 - b_2)^2 - (a_1b_2 - a_2b_1)^2}}\,.$$

(Hint: If A and B have complex coordinates z and w, then P and
Q have complex coordinates $tz + (1 - t)w$ and $uz + (1 - u)w$,
where t and u are roots of a quadratic equation $Dx^2 + 2Ex + F = 0$ expressing the fact that P and Q lie on the unit circle. Find the
coefficients D, E, and F and show that

$$(AB, PQ) = \frac{t(1 - u)}{u(1 - t)} = \frac{E + F - \sqrt{E^2 - DF}}{E + F + \sqrt{E^2 - DF}}\,.$$

K-15. Use the formula for Klein length given in Theorem 7.4 to derive
a proof of the Bolyai–Lobachevsky formula in Theorem 7.2 for
the Klein model. (Hint: Take the vertex of the angle of parallelism α to be the center O of the absolute and show that the
Klein distance d' corresponding to α is given by

$$d' = \frac{1}{2}\log\frac{1 + \cos\alpha}{1 - \cos\alpha}.$$

Then use a half-angle formula from trigonometry.)

K-16. (a) Show that a Cartesian line l of equation $Ax + By + C = 0$ is a secant of the unit circle if and only if

$$A^2 + B^2 - C^2 > 0.$$

We will denote the expression on the left of this inequality by $|l|^2$.

(b) Prove that if $P' = (x', y')$ is the Klein reflection of point $P = (x, y)$ across l, then

$$x' = \frac{|l|^2 x - 2A(Ax + By + C)}{|l|^2 + 2C(Ax + By + C)},$$

$$y' = \frac{|l|^2 y - 2B(Ax + By + C)}{|l|^2 + 2C(Ax + By + C)}.$$

(Hint: Use Theorem 7.3. In the case where $C = 0$, the Euclidean reflection is easy to calculate. If $C \neq 0$, the pole L of l has coordinates $(-A/C, -B/C)$, according to equation (2) in the last section of this chapter; you must calculate the coordinates of the point M where line $\overleftrightarrow{LP}$ meets l and then calculate the coordinates of the harmonic conjugate P' of P with respect to L and M.)

(c) Suppose l is a secant of the unit circle and let line $l' \neq l$ be another secant, having equation $A'x + B'y + C' = 0$. Show that the algebraic criterion for the Klein lines cut out by l and l' to be Klein-perpendicular is $AA' + BB' - CC' = 0$. (Hint: If $C \neq 0$, use the coordinates of the pole of l.)

(d) Let l and l' be secants as above. The determinant criterion for them to be parallel is

$$D = \det \begin{pmatrix} A & B \\ A' & B' \end{pmatrix} = AB' - BA' = 0.$$

The Klein lines they cut out will a fortiori then be Klein-parallel. Suppose now $D \neq 0$. Then the Klein lines cut out by these secants are Klein-parallel iff the point at which the secants meet is not inside the unit circle. Show that an algebraic equation on those coefficients which is necessary and sufficient for them to intersect on the unit circle (so that the Klein lines they cut out are asymptotically parallel) is

$$(BC' - B'C)^2 + (AC' - A'C)^2 = D^2.$$

Find an algebraic inequality on the coefficients of l and l' which is necessary and sufficient for them to intersect inside the unit circle. (Hint: Knowledge of Cramer's rule in two-dimensional linear algebra is helpful here to solve for the co-ordinates of the point of intersection. The quantities being squared on the left side of this equation are also subdeterminants of the 2×3 matrix of coefficients.)

K-17. The line perpendicular to the bisector of $\sphericalangle A$ at A is called the *external bisector* of $\sphericalangle A$ (because its rays emanating from A bisect the two supplementary angles to $\sphericalangle A$). You proved (in Exercise 18, Chapter 4) that the (internal) bisectors of the angles of $\triangle ABC$ concur in the center I of the inscribed circle—this is a theorem in neutral geometry.

(a) Prove that in Euclidean geometry the internal bisector of $\sphericalangle A$ is concurrent with the external bisectors of $\sphericalangle B$ and $\sphericalangle C$.

(b) Deduce from the Klein model that in hyperbolic geometry, the internal bisector of $\sphericalangle A$ is "concurrent" with the external bisectors of $\sphericalangle B$ and $\sphericalangle C$ in a point which may be ordinary, ideal, or ultra-ideal (see Figure 7.50). (Hint for part (a): Use the facts that the bisector of an angle is the locus of interior points equidistant from the sides and that external bisectors are not parallel. Hint for part (b): Take I to be the center O of the absolute γ and notice, using K-11, that the hyperbolic internal bisectors, being diameters of γ, coincide with the Euclidean internal bisectors. Hence the hyperbolic external bisectors, being perpendicular to the diameters of γ, coincide with the Euclidean external bisectors.)

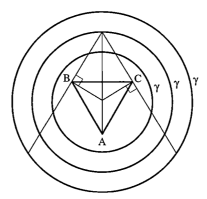

Figure 7.50 Three possible positions of the absolute.

K-18. It is a theorem in Euclidean geometry that the altitudes of an acute triangle are concurrent and the lines containing the altitudes of an obtuse triangle are concurrent (see Problem 8, Chapter 9). Applying this theorem to the Klein model, deduce that in hyperbolic geometry the altitudes of an acute triangle are concurrent and that the lines containing the altitudes of an obtuse triangle are "concurrent" in a point that may be ordinary, ideal, or ultra-ideal. (Hint: Place the triangle so that one vertex is O; show that the Klein lines containing the altitudes then coincide with the Euclidean perpendiculars from the vertices to the opposite sides. Use the crossbar and exterior angle theorems to verify that for acute triangles the point of concurrence is ordinary.)

K-19. (a) Prove, using analytic geometry, that the medians (segments from a vertex to the midpoint of the opposite side) of a triangle in a Euclidean plane are concurrent (simplifying the algebra by placing one vertex at the origin). Show that this result can be proved synthetically from the converse to Desargues' theorem (see Exercise H-10 and Project 1, Chapter 2) in the projective completion of the Euclidean plane. (Hint: Each medial line joining midpoints is parallel to the line containing the third side, so the point of intersection of these lines in the projective completion lies on the line at infinity.)

(b) Show that this theorem also holds in hyperbolic geometry by a special position argument in the Klein model.[17] (Hint: If O is the hyperbolic midpoint of AB, it is also the Euclidean midpoint; if J, I are the hyperbolic midpoints of AC, BC, use Exercise 2(b), Chapter 6, to show that $\overleftrightarrow{JI}$ is Euclidean-parallel to $\overleftrightarrow{AB}$—that is, $\overleftrightarrow{JI}$ "meets" $\overleftrightarrow{AB}$ in the harmonic conjugate at infinity of O with respect to A and B. The result then follows from the converse to harmonic construction in Figure 7.51.) Alternatively, as in part (a), apply the converse to Desargues' theorem to △ABC and △IJK formed by the midpoints of the sides of △ABC. Use perpendicular bisector concurrence in the projective completion and Exercise 2(b), Chapter 6, to show that the points

[17] Bachmann (1959, p. 74), has proved the concurrence of the medians for arbitrary Hilbert planes and, more generally, for his "metric" planes, using his calculus of reflections—based on Hjelmslev's methods. Hartshorne used those methods to prove, for arbitrary Hilbert planes, that if two altitudes of a triangle meet, then the third altitude is concurrent with them. See his Theorem 43.15, p. 430.

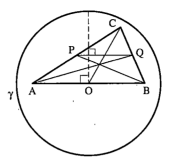

Figure 7.51 Concurrence of medians in Klein model.

of intersection in the projective completion of corresponding sides of these two triangles lie on the polar of the point
of concurrence of the perpendicular bisectors.

(c) Refer to Figure 6.15 and Exercise 2, Chapter 6. If △ABC is
not isosceles, then according to part (d) of that exercise, the
perpendiculars dropped from vertices of the triangle to medial lines are not altitudes of △ABC if the plane is hyperbolic. Show, using the Klein model, that those perpendiculars are also concurrent in the projective completion. (Hint:
Apply the converse to Desargues' theorem to △ABC and to
the triangle in the projective completion formed by the poles
of the medial lines of △ABC. To show that the hypothesis
of the converse to Desargues' theorem is satisfied by these
two triangles, again apply perpendicular bisector concurrence in the projective completion to deduce that the poles
of the perpendicular bisectors of △ABC are collinear.) This
gives us a new *special point* of a triangle in the hyperbolic
plane!

K-20. (a) In any Hilbert plane, let □ABCD have both pairs of opposite sides congruent. Prove that both pairs of opposite angles are congruent and that the lines containing opposite
sides have a common perpendicular (use Exercise 12, Chapter 6), in particular are parallel. Such a quadrilateral will be
called a *symmetric parallelogram*. Prove that □ABCD is a
symmetric parallelogram iff the diagonals bisect each other.
Let S be their common midpoint. Prove that the diagonals
of a symmetric parallelogram are perpendicular iff all four
sides are congruent, and in that case □ABCD has an inscribed circle with center S. Prove that the diagonals of a

symmetric parallelogram are congruent iff all four angles are congruent (in a semi-Euclidean plane, that happens iff □ABCD is a rectangle), and in that case □ABCD has a circumscribed circle with center S. □ABCD is called a *regular 4-gon* if all four sides and all four angles are congruent.

(b) Prove that in a Euclidean plane, every parallelogram is symmetric, whereas in a hyperbolic plane, there exist parallelograms that are not symmetric.

(c) Suppose that □ABCD is a symmetric parallelogram in a hyperbolic plane, with S the midpoint of its diagonals. Show that for each pair of opposite sides, S is the symmetry point for the lines containing those sides, in the sense of Major Exercise 12, Chapter 6. In the Klein model, suppose also that S = O. Show that □ABCD is a Euclidean parallelogram, that it is a Euclidean rectangle iff all four angles are Klein-congruent, and that it is a Euclidean square iff it is a hyperbolic regular 4-gon.

(d) In a Euclidean plane, use the results above about parallelograms to give a synthetic proof of the following theorem: In △ABC, let B′, C′ be the midpoints of AC, AB, respectively. BB′ and CC′ meet at a point G, by the crossbar theorem. Let L, M be the midpoints of BG, CG, respectively. Then BL ≅ LG ≅ GB′ and CM ≅ MG ≅ GC′. In words: G is two-thirds of the distance from each vertex to the opposite midpoint. Deduce from this that the three medians of △ABC are concurrent. (Hint: Show that □LMB′C′ is a parallelogram.)

K-21. It has been shown by Jenks that in hyperbolic geometry, "betweenness," "congruence," and "asymptotic parallelism" can all be defined in terms of incidence alone. (An important consequence of this observation is that every collineation of the hyperbolic plane is a motion; see Chapter 9). Here are his observations (draw diagrams in the Klein model to see what is going on). First, three distinct lines *a, b, c* form an asymptotic triangle *abc* if and only if for any point P on any one of them—say, on *a*—there exists a unique line *p* ≠ *a* through P which is parallel to both *b* and *c* (*p* is called an *asymptotic transversal* through P). Second, *a | b* (*a* asymptotically parallel to *b*) if and only if there exists a line *c* such that *a, b, c* form an asymptotic triangle. Third, given three points P, Q, R on a line *m*, P * Q * R if and only if given any *a* ≠ *m* through P, *b* ≠ *m* through R, and c such that a, b, c form an asymptotic triangle, every line through Q meets at least one of the sides of abc. Fourth, segment PQ on a is congruent to segment P'Q' on a' if and only if either (1) a | a' and both are asymptotically parallel to the join of the meets of the asymptotic transversals through P and P' and through Q and Q', or (2) both are asymptotically parallel to some line a" on which lies a segment P"Q" congruent with both PQ and P'Q' in the sense of (1). Justify (1) by drawing the diagram in the Klein model and then applying Lemma 7.5 and Theorem 7.4 (see Blumenthal and Menger, 1970, p. 220).

P-Exercises

P-1. Using the glossary for the Poincaré disk model, translate the following theorems in hyperbolic geometry into theorems in Euclidean geometry:

    (a) If two triangles are similar, then they are congruent.

    (b) If two lines are divergently parallel, then they have a common perpendicular and the latter is unique.

    (c) The fourth angle of a Lambert quadrilateral is acute.

P-2. State and prove the analogue of Proposition 7.6 when O lies inside δ and the power p of O with respect to δ is negative.

P-3. Let δ be a circle with center C and α a circle not through C having center A. Let A' be the inverse of A in δ and let circle α' be the image of α under inversion in δ. Prove that A' is the inverse of C in α' and hence that A' is not the center of α'. (Hint: Show that any circle β through A' and C is orthogonal to α' by observing that the image β' of β under inversion in δ is a line orthogonal to α.)

P-4. Construct the P-angle-bisector t of a P-angle ∢BAB' in the case where A ≠ O. (Hint: Choose B and B' so that P-segments AB and AB' are P-congruent. Then find the construction of the P-perpendicular-bisector of P-segment BB' previously worked out in the text.)

P-5. We have proved that P-circles in the disk are E-circles, and conversely. We also showed how to construct the E-center of a P-circle given its P-center. Show conversely, given an E-circle in the disk and its E-center, how to construct its P-center. (Hint: It comes down to constructing the P-midpoint of a certain E-segment that is contained in a diameter of the disk.)

P-6. In the hyperbolic plane with some given unit of length, the distance d for which the angle of parallelism $\Pi(d)° = 45°$ is called *Schweikart's constant*. Schweikart was the first to notice that if $\triangle ABC$ is an isosceles right triangle with base BC, then the length of the altitude from A to BC is bounded by this constant, which is the least upper bound of the lengths of all such altitudes. Prove that for the length function we have defined for the Poincaré disk model, Schweikart's constant equals $\log(1 + \sqrt{2})$ (see Figure 7.52). (Hint: Schweikart's constant is the Poincaré length d of segment OP in Figure 7.52. Show that the Euclidean length of OP is $\sqrt{2} - 1$ and apply Lemma 7.4 to solve for d.)

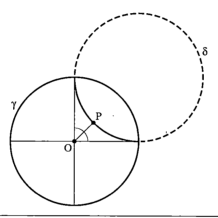

Figure 7.52 Schweikart's segment OP.

P-7. Let α be a circle with center A and radius of length r, and β a circle with center B and radius of length s. Assume $A \neq B$ and let C be the unique point on $\overleftrightarrow{AB}$ such that $\overline{AC^2} - \overline{BC^2} = r^2 - s^2$. The line through C perpendicular to $\overleftrightarrow{AB}$ is called the *radical axis* of the two circles.

(a) Prove (e.g., by introducing coordinates) that C exists and is unique and that for any point P different from A and B, P lies on the radical axis if and only if $\overline{PA^2} - \overline{PB^2} = r^2 - s^2$.

(b) For any point X outside both α and β, let T be a point of α such that $\overleftrightarrow{XT}$ is tangent to α at T; similarly let U on β be a point of tangency for $\overleftrightarrow{XU}$. Prove that $\overline{XT} = \overline{XU}$ if and only if X lies on the radical axis of α and β.

(c) Prove that if α and β intersect in two points P and Q, $\overleftrightarrow{PQ}$ is their radical axis.

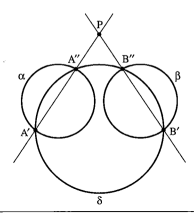

Figure 7.53

   (d) Prove that if α and β are tangent at point C, the radical axis is the common tangent line through C.

   (e) Let X be a point outside both α and β. Prove that X lies on the radical axis of α and β if and only if X has the same power with respect to α and β (see Lemma 7.1).

P-8. Given two nonintersecting, nonconcentric circles α and β with centers A and B, respectively. Justify the following straightedge-and-compass construction of the radical axis of α and β. Draw any circle δ that cuts α in two points A' and A" and cuts β in two points B' and B". If $\overleftrightarrow{A'A''}$ and $\overleftrightarrow{B'B''}$ intersect in a point P, then P lies on the radical axis; the latter is therefore the perpendicular to $\overleftrightarrow{AB}$ through P. (Hint: Draw tangents PS, PT, and PU from P to δ, α, and β and apply Exercises P-7(b) and P-7(c) to show that $\overline{PT} = \overline{PS} = \overline{PU}$. See Figure 7.53.)

P-9. Use Exercise P-7 to verify by a straightedge-and-compass construction that in the Poincaré model two divergently parallel Poincaré lines have a common perpendicular. (Hint: There are four cases to consider, depending on whether the Poincaré line is a diameter of γ or an arc of a circle α orthogonal to γ and depending on whether radical axes intersect or not. One case is illustrated in Figure 7.54. In the case where the radical axes are parallel, use the fact that the perpendicular bisector of a chord of a circle passes through the center of the circle (Exercise 17, Chapter 4).)

P-10. Given any Poincaré line l and any Poincaré point P not on l. Construct the two rays from P in the Poincaré model that are limiting parallel to l. (If l is an arc of a circle α orthogonal to γ

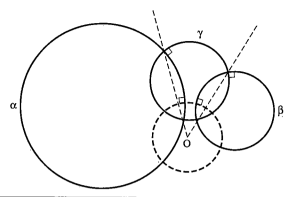

Figure 7.54 Common Poincaré perpendicular of divergent parallels.

and intersecting γ at A_1 and A_2, then the problem amounts to constructing a circle β_i through P that is orthogonal to γ and tangent to α at A_i for each of $i = 1, 2$. See Figure 7.55 and use Proposition 7.5.)

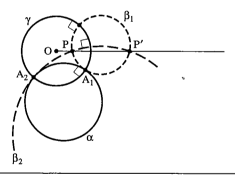

Figure 7.55 Poincaré limiting parallel rays.

P-11. We define three types of *coaxal pencils of circles* as follows:

(1) Given a line t and a point C on t. The corresponding *tangent coaxal pencil* consists of all circles tangent to t at C.

(2) Given two points A and B. The corresponding *intersecting coaxal pencil* consists of all the circles that pass through both A and B, and A and B are the *limiting points* of this pencil.

(3) Given a circle γ and a line t not meeting γ. The corresponding *non-intersecting coaxal pencil* consists of γ and all other circles δ such that t is the radical axis of γ and δ.

Prove the following:

(a) Any two nonconcentric circles belong to a unique coaxal pencil.

(b) Given a coaxal pencil C. All pairs of circles belonging to C have the same radical axis, and the centers of all circles in C lie on a line perpendicular to this radical axis called the *line of centers* of C. (Hint: See Exercise P-7.)

P-12. Given circle γ with center O. For any point $P \neq O$, if P' is the inverse of P in γ, then the line through P' that is perpendicular to $\overleftrightarrow{OP}$ is called the *polar* of P with respect to γ and will be denoted $p(P)$. When P lies outside γ, its polar joins the points of contact of the two tangents to γ from P (see Figure 7.22). When P lies on γ, its polar is the tangent to γ at P, and this is the only case in which P lies on $p(P)$. Prove the following duality property. B lies on $p(A)$ if and only if A lies on $p(B)$. (Hint: If B lies on $p(A)$, let B' be the foot of the perpendicular from A to $\overleftrightarrow{OB}$. See Figure 7.56. Show that $\triangle OAB'$ is similar to $\triangle OBA'$ and deduce that B' is the inverse of B in γ. For the significance of this operation of polar reciprocation for the theory of conics, see Coxeter and Greitzer, 1967, Chapter 6.)

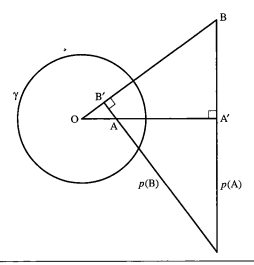

Figure 7.56 Polar reciprocation for a circle.

P-13. Given an acute angle in the Poincaré model. Construct the unique Poincaré line that is perpendicular to a given side of this angle and limiting parallel to the other. This shows that the angle of parallelism can be any acute angle whatever. (Hint: If both Poincaré lines are arcs of orthogonal circles α and β, let P' be the intersection with γ of the part of α containing the given ray and let P be the other intersection with γ of $\overleftrightarrow{P'B}$, B being the center of β; see Figure 7.57. Show that P and P' are the inverse points in circle β, then find the point of intersection of the tangents to γ at P and P'. Compare with Major Exercise 9, Chapter 6.)

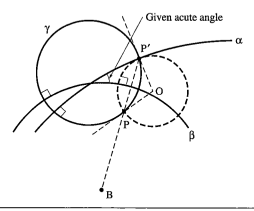

Figure 7.57 Construction of Poincaré segment of parallelism.

P-14. Prove the following:
   (a) The set of all circles orthogonal to two given circles γ and δ tangent at C is the tangent coaxal pencil through C whose line of centers is the common tangent t to γ and δ.
   (b) The set of all circles orthogonal to two given non-intersecting non-concentric circles γ and δ is the intersecting coaxal pencil whose line of centers is the radical axis t of γ and δ and whose limiting points are the two points at which every member of this pencil cuts the line joining the centers of γ and δ.
   (c) The set of all circles orthogonal to two given circles γ and δ intersecting at A and B is the non-intersecting non-concentric coaxal pencil whose line of centers is $\overleftrightarrow{AB}$ and whose radical axis is the perpendicular bisector of AB (see Figure 7.58).
P-15. Given three circles α, β, and γ. Is there always a fourth circle δ orthogonal to all three of them? If so, is δ unique? (Hint: Consider the radical axes of the three pairs of circles obtained from the three given circles; the center of δ must lie on all three radical axes and must lie outside the three circles.)
P-16. Given a circle γ with center O.
   (a) Given $P \neq O$ and P' its inverse in γ. Prove that inversion in γ maps the pencil of lines through P' onto the intersecting

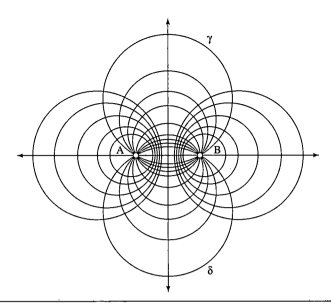

Figure 7.58 Orthogonal coaxal pencils.

coaxal pencil of circles through O and P and maps the orthogonal pencil of concentric circles centered at P' onto the non-intersecting coaxal pencil of circles whose radical axis is the perpendicular bisector of OP.

(b) Given a line l through O. Prove that inversion in γ maps the pencil of lines parallel to l onto the pencil of circles tangent to l at O.

P-17. The *inversive plane* is obtained from the Euclidean plane by adjoining a single point at infinity ∞, which by convention lies on every Euclidean line but does not lie on any Euclidean circle. By a "circle" we mean either an ordinary Euclidean circle or a line in the inversive plane. Two parallel Euclidean lines meet at ∞ when extended to inversive lines; as "circles" they will be considered to be tangent at ∞. Given an ordinary circle γ with center O, define the inverse of O in γ to be ∞. By inversion in a "circle" we mean either inversion in an ordinary circle or reflection across a line. Prove the following:

(a) Inversion in a given "circle" maps "circles" onto "circles."

(b) If A and B are inverse to each other in a "circle" α, and if under inversion in another "circle" β they map to A', B', α', then A' and B' are inverse to each other in α'. (Hint for part (b): Show that any "circle" γ' through A' and B' is orthogonal to α' by observing that inversion preserves orthogonality—use Propositions 7.5 and 7.9.)

P-18. In addition to the tangent, intersecting, and non-intersecting coaxal pencils of circles defined in Exercise P-11, define three further pencils of "circles" in the inversive plane as follows:

(4) All the circles having a given point as center

(5) All the lines passing through a given ordinary point

(6) A given line and all lines parallel to it

Furthermore, given a coaxal pencil of circles, we will consider its radical axis as one more "circle" belonging to the pencil. Prove the following:

(a) Two distinct "circles" belong to a unique pencil of "circles."

(b) A pencil of "circles" is invariant as a set under inversion in any "circle" in the pencil. (Hint for part (b): The statement is obvious for the three new types of pencils just introduced. For the three coaxal types, use the two preceding exercises.)

P-19. Construct a regular 4-gon in the Poincaré disk model. (Hint: Choose a point A ≠ O on the line $y = x$; let B (respectively, D) be its reflection across the x-axis (respectively, the y-axis) and let C be obtained from A by 180° rotation about O. Show that □ABCD is a regular 4-gon. Note that as A approaches O, ⊀A approaches a right angle, while as A moves away toward the ideal end of ray $\overrightarrow{OA}$, ⊀A approaches the zero angle.)

P-20. Use the Poincaré model to show that in the hyperbolic plane, there exist two points A, B lying on the same side S of a line l such that no circle through A and B lies entirely within S. This shows that the result in Major Exercise 7, Chapter 5, is another statement equivalent to Euclid's parallel postulate. (Hint: Proposition 7.12.)

H-Exercises

Once again, these are exercises for a Euclidean plane.

H-1. Let M be the midpoint of AB, let $r = \overline{MA}$, and let C, D on $\overleftrightarrow{AB}$ lie on the same side of M, with A, B, C, D distinct. Then C and D are harmonic conjugates with respect to AB if and only if we have $r^2 = (\overline{MD})(\overline{MC})$.

H-2. If γ and δ are orthogonal circles, if AB is a diameter of γ, and if δ cuts $\overleftrightarrow{AB}$ in points C and D, then C and D are harmonic

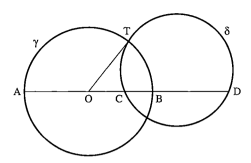

Figure 7.59

conjugates with respect to AB; conversely, if a diameter of one circle is cut harmonically by a second circle when the diameter is extended, then the two circles are orthogonal (see Figure 7.59). (Hint: If T is a point of intersection of γ and δ, use Lemma 7.1 to show that the circles are orthogonal if and only if $(\overline{OT})^2 = (\overline{OC})(\overline{OD})$. Now apply Exercise H-1.)

H-3. Given three collinear points A, B, and C. Prove that the fourth harmonic point D is the inverse of C in the circle having AB as diameter. (Hint: Use Exercise H-2 and Proposition 7.5.)

H-4. *Sensed magnitudes.* Given two points A, B. Assign arbitrarily an order (i.e., a direction) to $\overleftrightarrow{AB}$. Then the length of AB will be considered positive or negative according to whether the direction from A to B is the positive or negative direction on the line. We will denote this signed length by $\underline{AB}$, so that we have $\underline{AB} = -\underline{BA}$. If C is a third point on the directed line $\overleftrightarrow{AB}$, we define the signed ratio in which C divides AB to be $\underline{AC}/\underline{CB}$.

(a) Prove that this signed ratio is independent of the direction assigned to the line and that point C is uniquely determined by this ratio. (Note that C would not be uniquely determined by the unsigned ratio.)

(b) Given parallel lines l and m. Let transversals t and t' cut l and m in B, C and B', C', respectively and let t meet t' at point A. Prove that $\underline{AB}/\underline{BC} = \underline{AB'}/\underline{B'C'}$ (use the fundamental theorem on similar triangles, Chapter 5).

H-5. *Theorem of Menelaus.* Given $\triangle ABC$ and points D on $\overleftrightarrow{BC}$, E on $\overleftrightarrow{CA}$, and F on $\overleftrightarrow{AB}$ that do not coincide with any of the vertices of the triangle. Define the *linearity number* by the relation $[ABC/DEF] = (\underline{AF}/\underline{FB})\,(\underline{BD}/\underline{DC})\,(\underline{CE}/\underline{EA})$. Then a necessary and sufficient condition for D, E, and F to be collinear (Figure 7.60) is that $[ABC/DEF] = -1$. (Hint: If D, E, and F lie on a line l, let the parallel m to l through A cut $\overleftrightarrow{BC}$ at G. Use Exercise H-4 to get $\underline{CE}/\underline{EA} = \underline{CD}/\underline{DG}$ and $\underline{AF}/\underline{FB} = \underline{GD}/\underline{DB}$ and deduce that the linearity number is -1. Conversely, use Exercise H-4 to show that $\overleftrightarrow{EF}$ cannot be parallel to $\overleftrightarrow{BC}$. If these lines meet at D', use the first part of the proof and the hypothesis to show that $\underline{BD}/\underline{DC} = \underline{BD'}/\underline{D'C}$ and apply Exercise H-4(a).)

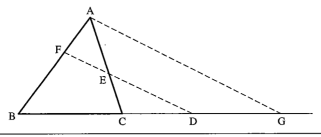

Figure 7.60

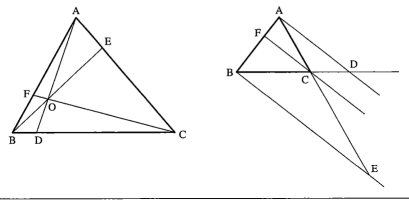

Figure 7.61

H-6. *Theorem of Ceva.* Given $\triangle ABC$ and a third point D (respectively, E, F) on $\overleftrightarrow{BC}$ (respectively, on $\overleftrightarrow{AC}$, $\overleftrightarrow{AB}$). Then the three lines $\overleftrightarrow{AD}$, $\overleftrightarrow{BE}$, and $\overleftrightarrow{CF}$ are either concurrent or parallel if and only if [ABC/DEF] = +1 (see Figure 7.61). (Hint: Suppose that the three lines meet at O; apply Menelaus' theorem to $\triangle ADB$ and $\triangle ADC$ to obtain two different expressions for $\underline{OD}/\underline{AO}$, then divide one expression by the other to see that the linearity number is +1. If the three lines are parallel, apply Exercise H-4(b). Conversely, if the linearity number is +1 and the three lines are not parallel, let $\overleftrightarrow{BE}$ and $\overleftrightarrow{CF}$, for example, meet at O and let $\overleftrightarrow{AO}$ meet $\overleftrightarrow{BC}$ at D'. Use the first part of the proof and the hypothesis to show that $\underline{BD}/\underline{DC} = \underline{BD'}/\underline{D'C}$ and apply Exercise H-4(a).)

H-7. Given four collinear points A, B, C, and D. Define their *signed cross-ratio* $(\underline{AB}, \underline{CD})$ by $(\underline{AB}, \underline{CD}) = (\underline{AC}/\underline{CB})/(\underline{AD}/\underline{DB})$.

    (a) Prove that ABCD is a harmonic tetrad if and only if we have $(\underline{AB}, \underline{CD}) = -1$.

    (b) Prove that signed cross-ratios are preserved by perspectivities and parallel projections (see Lemma 7.5 and the parallel projection theorem preceding Exercise 10, Chapter 5).

H-8. Prove that ABCD is a harmonic tetrad if and only if we have that $1/\underline{AB} = \frac{1}{2}(1/\underline{AC} + 1/\underline{AD})$.

H-9. Suppose the inscribed circle of $\triangle ABC$ touches sides BC, CA, and AB at D, E, and F, respectively. Prove that AD, BE, and CF are concurrent in a point G called the *Gergonne point* of $\triangle ABC$; see Figure 7.62. (Hint: By Execise 18, Chapter 4, the center I of the inscribed circle lies on all three angle bisectors; this gives three pairs of congruent right triangles that can be used to verify the criterion of Ceva's theorem.)

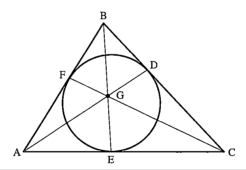

Figure 7.62 Gergonne point.

H-10. Use the theorem of Menelaus to prove Desargues' theorem as stated in Project 1, Chapter 2. (Hint: Referring to Figure 2.10, apply Menelaus' theorem to $\triangle BCP$, $\triangle CAP$, and $\triangle ABP$ and then multiply the three equations to get [ABC/RST] $= -1$. Now apply Menelaus' theorem once more.) By the principle of duality, this also proves the converse to Desargues' theorem, which is its dual. Combining these two results yields the following theorem, which actually holds in any projective plane coordinatized by a division ring: *Two triangles are in perspective from a point if and only if they are in perspective from a line.*

H-11. Use the theorem of Menelaus to prove Pappus' theorem as stated in Project 3, Chapter 2. (Hint: Referring to Figure 2.13 and using · to denote the intersection of lines, let $N = BA' \cdot B'A$, let $L = CA' \cdot C'A$, and let $M = BC' \cdot B'C$. Pappus asserts that L, M, N are collinear. Consider the generic case where the points $U = BC' \cdot B'A$, $V = CA' \cdot B'A$, and $W = CA' \cdot C'B$ form a triangle. Apply Menelaus' theorem to this triangle and the five triples of collinear points LC'A, CMB', A'BN, CBA, and A'C'B' to obtain five linearity numbers that are equal to -1. Do the algebra required to show the sixth linearity number needed to prove Pappus' assertion is also equal to -1. The nongeneric case requires a separate argument.)

H-12. Apply Ceva's theorem to prove that (a) the medians of a triangle are concurrent, (b) the altitudes of a triangle are concurrent. Apply Menelaus' theorem to prove Pascal's theorem for a circle: *If a hexagon is circumscribed by a circle and the three pairs of opposite sides intersect (if necessary, when extended), then those three points of intersection are collinear.* (Refer to Coxeter and Greitzer, 1967, Section 3.8.)

Projects

1. A model of a Euclidean plane can be constructed within a hyperbolic plane Π, not just within hyperbolic space on the horosphere. The idea is to fix a point O and take as "points" all the points of Π, but interpret "lines" as the hyperbolic lines through O plus all the (singly) equidistant curves for those lines, with "incidence" interpreted as a point lying on that "line" in the hyperbolic plane. Betweenness is induced by the betweenness relation in Π. It is straightforward to verify the incidence and betweenness axioms and the Euclidean parallel property for this interpretation. The more difficult part is to define a suitable congruence relation and to verify the congruence axioms. (This model shows that Clavius and the Greek and Islamic geometers before him, who proposed that equidistant curves were straight lines, may have unconsciously been working within a hyperbolic plane!) See if you can work that out. One method is given in Chapter 10.

To be more concrete, take Π to be the Poincaré disk model within a Euclidean plane and O to be the center of the disk. Then "lines" are diameters of the circle γ, which is the rim of the disk, plus arcs · in the disk of Euclidean circles intersecting γ at the ends of a diameter. Figure out appropriate measures of angles and segment lengths so as to satisfy the congruence axioms. One would then have a Euclidean model within a hyperbolic model within a Euclidean plane—not philosophically significant, but a possible guide to understanding the abstract problem.

2. Report on the determination of angle and segment measure in a real projective plane relative to a distinguished conic called the absolute, which may be real, imaginary, or degenerate. The resulting geometries, which include the elliptic, hyperbolic, and parabolic among several others (such as the Galilean and Minkowskian), are called *Cayley–Klein geometries* since Arthur Cayley had the original idea and Felix Klein brought it to fruition. When the absolute is real, the metrical formulas are exactly those obtained for hyperbolic geometry, and when the absolute is imaginary, the formulas are the same as those of elliptic geometry. One reference, if you read German, is Klein (1968). A brief reference in English is Chapter 12, Section 6, in vol. II of *Fundamentals of Mathematics: Geometry*, H. Behnke, F. Bachmann, H. Kunle, and K. Fladt, eds., MIT Press, Cambridge, MA, 1974. A more detailed account of all nine plane Cayley–Klein geometries, with applications to physics, can be found in I. M. Yaglom, *A Simple Non-Euclidean Geometry and Its Physical Basis*, Springer, New York, 1979. If you do a search on the web, you will find many articles applying Cayley–Klein geometries to quantum physics and to engineering, e.g., http://www.parcellular.fsnet.co.uk/Spin5%20master.htm.

3. If you have a facility with computer drawing of diagrams in Poincaré models of the real hyperbolic plane, draw an accurate version in one of those models for the atrocious Figure 6.22 illustrating perpendicular bisectors of a triangle that do not meet in an ordinary point but that themselves have a common perpendicular. Also, draw the case where the perpendicular bisectors meet at an ideal point. I would love to see those drawings if you can do them (please send them to me c/o W. H. Freeman & Co., 41 Madison Avenue, New York, NY 10010).

Here are some online references for drawing in Poincaré models:

http://cgm.cs.mcgill.ca/labdocs/CinderellaManual/Texts/
    Introduction.html
http://cs.unm.edu/~joel/NonEuclid/

8

Philosophical Implications, Fruitful Applications

The value of non-Euclidean geometry lies in its ability to liberate us from preconceived ideas in preparation for the time when exploration of physical laws might demand some geometry other than the Euclidean.

G. F. B. Riemann

What Is the Geometry of Physical Space?

We have shown that if Euclidean geometry is consistent, so is hyperbolic geometry, since we can construct models for it within Euclidean geometry. Conversely, it can be proved that if hyperbolic geometry is consistent, so is Euclidean geometry, via the model described at the end of the Coordinates in the Real Hyperbolic Plane section in Chapter 10. Traditionally, the Euclidean plane has been modeled by the

371

points on the horosphere in hyperbolic three-space, with the horocycles acting as "lines" and with the metric induced from the hyperbolic metric, but now we have a model of the Euclidean plane within the hyperbolic plane. *Thus, the two geometries are equally consistent.*

You may grant now that, logically speaking, hyperbolic geometry deserves to be put on an equal footing with Euclidean geometry. But you may also feel that hyperbolic geometry is just an amusing intellectual pastime, whereas Euclidean geometry accurately represents the physical world we live in and is therefore far more important. Let's examine this idea a little more closely.

Certainly, engineering and architecture are evidence that Euclidean geometry is extremely useful for ordinary measurement of distances that are not too large. However, the representational accuracy of Euclidean geometry is less certain when we deal with larger distances. For example, let us interpret a "line" physically as the path traveled by a light ray. We could then consider three widely separated light sources forming a physical triangle. We would want to measure the angles of this physical triangle in order to verify whether the sum is 180° or not (such an experiment would presumably settle the question of whether space is Euclidean or hyperbolic).

F. W. Bessel, a friend of Gauss, performed such a measurement, using the angle of parallax of a distant star. The results were inconclusive. Why? Because any physical experiment involves experimental error. Our instruments are never completely accurate. Suppose the sum did turn out to be 180°. If the error in our measurement were at most 1/100 of a degree, we could conclude only that the sum was between 179.99° and 180.01°. We could never be sure that it actually was 180°.

Suppose, on the other hand, that measurement gave us a sum of 179°. Although we could conclude only that the sum was between 178.99° and 179.01°, we would be certain that the sum was less than 180°. In other words, the only conclusive result of such an experiment would be that space is hyperbolic![1] The inconclusiveness of Bessel's experiment shows only that if space is hyperbolic, the defects of terrestrial triangles are extremely small.

To repeat the point: Because of experimental error, a physical experiment can never prove conclusively that space is Euclidean—it can prove only that space is non-Euclidean.

The discussion can be made more subtle. We must question the nature of our instruments—aren't they designed on the basis of Euclid-

---

[1] If the measurement gave us a sum of 181° with error at most .01°, we would conclude that space is elliptic.

Albert Einstein

ean assumptions? We must question our interpretation of "lines"—couldn't light rays travel on curved paths? We must question whether space, especially space of cosmic dimensions, cannot be described by geometries other than these two.

The latter is in fact our present scientific attitude. According to Einstein, space and time are inseparable and the geometry of space-time is affected by matter, so that light rays are indeed curved by the gravitational attraction of masses. Space is no longer conceived of as an empty Newtonian box whose contours are unaffected by the rocks put into it. The problem is much more complicated than Euclid or Lobachevsky ever imagined—neither of their geometries is adequate for our present conception of space. This does not diminish the historical importance of non-Euclidean geometry. Einstein said, "To this interpretation of geometry I attach great importance, for should I not have been acquainted with it, I never would have been able to develop the theory of relativity."[2]

[2] See George Gamow (1956), which tells how Einstein developed a geometry appropriate to general relativity from the ideas of Georg Friedrich Bernhard Riemann (1826–1866).

Here is the famous response of Poincaré to the question of which geometry is true:

If geometry were an experimental science, it would not be an exact science. It would be subjected to continual revision. . . . *The geometrical axioms are therefore neither synthetic a priori intuitions nor experimental facts. They are conventions.* Our choice among all possible conventions is guided by experimental facts; but it remains free, and is only limited by the necessity of avoiding every contradiction, and thus it is that postulates may remain rigorously true even when the experimental laws which have determined their adoption are only approximate. In other words, *the axioms of Geometry* (I do not speak of those of arithmetic) *are only definitions in disguise.* What then are we to think of the question: Is Euclidean Geometry true? It has no meaning. We might as well ask if the metric system is true and if the old weights and measures are false; if Cartesian coordinates are true and polar coordinates false. *One geometry cannot be more true than another: it can only be more convenient.* [italics added][3]

Essay Topic 18 at the end of this chapter provides a vivid example due to Poincaré illustrating this *conventionalist* philosophy with regard to physics and the hyperbolic plane.

You may think that Euclidean geometry is the most convenient—it is for ordinary engineering, but not for the theory of relativity. Moreover, R. K. Luneburg contends that visual space, the space mapped on our brains through our eyes, is most conveniently described by hyperbolic geometry.[4]

Philosophers are still arguing about Poincaré's philosophy of conventionalism. One school, which includes Newton, Helmholtz, Russell, and Whitehead, contends that space has an intrinsic metric or standard of measurement. The other school, which includes Riemann, Poincaré, Clifford, and Einstein, contends that a metric is stipulated by convention. The discussion can become very subtle (see Torretti, 1978, Chapter 4).

What Is Mathematics About?

The preceding discussion sheds new light on what geometry, and in general, mathematics, is about. Geometry is not about light rays, but

---

[3] H. Poincaré (1952), p. 50.
[4] R. K. Luneburg (1947), and his article in the *Optical Society of America Journal*, October 1950, p. 629. See also the articles by O. Blank in that same journal, December 1958, p. 911, and March 1961, p. 335, and the explanation in Trudeau (1987), pp. 251–254.

the path of a light ray is one possible *physical interpretation* of the un-defined geometric term "line." Bertrand Russell once said that "math-ematics is the subject in which we do not know what we are talking about nor whether what we say is true." This is because certain prim-itive terms, such as "point," "line," and "plane," are undefined and could just as well be replaced with other terms without affecting the validity of results. Instead of saying "two points determine a unique line," we could just as well write "two alphas determine a unique beta." Despite this change in terms, the proofs of all our theorems would still be valid because correct proofs do not depend on diagrams; they de-pend only on stated axioms and the rules of logic. Thus, geometry is a purely *formal* exercise in deducing certain conclusions from certain formal premises. Mathematics makes statements of the form "if . . . then"; it does not say anything about the meaning or truthfulness of the hypotheses. The primitive notions (such as "point" and "line") ap-pearing in the hypotheses are implicitly defined by these axioms, by the rules, as it were, that tell us how to play the game.[5]

To illustrate how radically different this view of mathematics is, ob-serve the following interaction (Torretti, 1987, p. 235). Gottlob Frege (1848–1925), who is considered the founder of modern mathematical logic, wrote to Hilbert:

> I give the name of axioms to propositions which are true, but which are not demonstrated because their knowledge proceeds from a source which is not logical, which we may call space intuition. The truth of the axioms implies of course that they do not contradict each other. That needs no further proof.

Frege has stated the traditional view. Hilbert replied:

> Since I began to think, to write and to lecture about these matters, I have always said exactly the contrary. If the arbitrarily posited axioms do not contradict one another or any of their consequences, they are true and the things defined by them exist. That is for me the criterion of truth and existence.

Hilbert knew that Euclidean and hyperbolic geometries are equally con-sistent, so it follows that for him they "exist" and are both "true." The discovery that Euclidean geometry was not "absolute truth" had a lib-erating effect on mathematicians, who now feel free to invent any set

---

[5] For a clear exposition of this viewpoint, which is due to Hilbert, see Hempel (1945).

of axioms they wish and deduce conclusions from them. In fact, this freedom may account for the great increase in the scope and generality of modern mathematics. Georg Cantor, the daring founder of infinite set theory, with its cardinal and ordinal numbers, said: "The essence of mathematics is its freedom." In a 1961 address, Jean Dieudonné remarked on Gauss' discovery of non-Euclidean geometry:

> [It] was a turning point of capital significance in the history of mathematics, marking the first step in a new conception of the relation between the real world and the mathematical notions supposed to account for it; with Gauss' discovery, the rather naive point of view that mathematical objects were only "ideas" (in the Platonic sense) of sensory objects became untenable, and gradually gave way to a clearer comprehension of the much greater complexity of the question, wherein it seems to us today that mathematics and reality are almost completely independent, and their contacts more mysterious than ever.[6]

The Controversy about the Foundations of Mathematics

It would be misleading to say that mathematics is just a formal game played with symbols and having no broader significance. Mathematicians do not arbitrarily make up axioms—it is unlikely that anyone would ever develop a geometry in which it is assumed that nonsupplementary right angles are never congruent to each other. Axioms must lead to interesting and fruitful results. Of course, some axioms that appear uninteresting may turn out to have surprising consequences—this was the case with the hyperbolic axiom, which was virtually ignored during the lifetimes of Gauss, Bolyai, and Lobachevsky. If, however, axiom systems do not bear interesting results, they become neglected and eventually forgotten.

Arguing against the description of mathematics as a "formal game," R. Courant and H. Robbins (1941) insist that "a serious threat to the very life of science is implied in the assertion that mathematics is nothing but a system of conclusions drawn from definitions and postulates that must be consistent but otherwise may be created by the free will of the mathematician. If this description were accurate, mathematics

---

[6] J. Dieudonné, *L'Oeuvre Mathématique de C. F. Gauss,* Poulet-Malassis Alençon: L'Imprimerie Alençonnaise, 1961.

could not attract any intelligent person. It would be a game with definitions, rules and syllogisms, without motivation or goal."

And Hermann Weyl has remarked:

> The constructions of the mathematical mind are at the same time free and necessary. The individual mathematician feels free to define his notions and to set up his axioms as he pleases. But the question is, will he get his fellow mathematicians interested in the constructs of his imagination? We can not help feeling that certain mathematical structures which have evolved through the combined efforts of the mathematical community bear the stamp of a necessity not affected by the accidents of their historical birth.[7]

Axiom systems that are fruitful can also be controversial in the mathematical world, as are the axioms for infinite sets developed by Georg Cantor, E. Zermelo, and others. A controversy occurs because some outstanding mathematicians (such as Weyl, L. E. J. Brouwer, and Errett Bishop in the case of infinite sets) simply do not *believe* all these axioms. If axioms were truly meaningless formal statements, how could there be any controversy about them? Is there any controversy about the rules of chess? It would seem that the formalist viewpoint—the view that mathematics is just a formal game—is a dodge to avoid having to face the difficult philosophical and psychological problem of the nature of mathematical creations or discoveries. Just what is asserted when a mathematician claims that something exists? When the Pythagoreans discovered that the hypotenuse of an isosceles right triangle was not commensurable with the leg, they tried to keep this discovery secret, calling such lengths "irrational." Nowadays we aren't upset over numbers like $\sqrt{2}$. Similarly, mathematicians have accommodated themselves to "imaginary" numbers such as $i = \sqrt{-1}$, exploited by J. Cardan.[8]

The most "fundamentalist" position on the philosophy of mathematics is that of Leopold Kronecker, a leading German mathematician in the late nineteenth century. According to Kronecker, "God created

---

[7] From H. Weyl, "A Half-Century of Mathematics," *American Mathematical Monthly,* **58** (1951): 523–553.

[8] Jacques Hadamard has said about Cardan: "It would naturally be expected that the discovery of imaginaries, which seems nearer to madness than to logic and which, in fact, has illuminated the whole mathematical science, would come from such a man whose adventurous life was not always commendable from the moral point of view, and who from childhood suffered from fantastic hallucinations. . . ." (Hadamard, 1945). For the history and importance of $\sqrt{-1}$, see Nahin (1998).

the whole numbers—all else is man-made." In particular, Kronecker re-pudiated Georg Cantor's theory of transfinite cardinal and ordinal num-bers. Hilbert later defended Cantor, proclaiming that "no one shall ex-pel us from the paradise which Cantor has created for us." Subsequently Kronecker was portrayed as the nasty reactionary whose rejection of Cantor's revolutionary new ideas drove Cantor to the insane asylum (see Bell, 1961); this may be a myth, and the philosophical issues un-derlying the Kronecker–Cantor controversy are far from settled.

In the twentieth century, Cantor's set theory, made precise by the Zermelo–Fraenkel (ZF) axioms, became the new "absolute truth" that was the foundation for all of mathematics. However, there was some controversy about one axiom, the axiom of choice (AC), and there was so much uncertainty about another idea of Cantor's that it was called a "hypothesis"—the continuum hypothesis (CH). The first on Hilbert's famous 1900 list of 23 problems was to prove or disprove CH. Forty years later, Kurt Gödel created a model of the other ZF axioms in which both AC and CH were true; that demonstrated the impossibility of dis-proving them. History repeated itself when, in 1963, models were cre-ated[9] in which either AC or CH or both were false. Thus AC and CH are independent of the other ZF axioms and of each other. There ex-ists an equally valid non-Cantorian set theory, just as there is an equally valid non-Euclidean geometry.

One mystery about mathematics is perhaps the most compelling of all. If mathematical creations are merely arbitrary fancies, how is it that some turn out to have physical applications—for example, applications that enable us to calculate orbits well enough to put men on the moon? When the Greeks developed the theory of ellipses, they had no inkling that it would have applications to a "space race."[10]

These questions and viewpoints are not intended to confuse you, but to point up the fact that mathematics is alive, ever changing, and incomplete. Moreover, according to a metamathematical theorem of Kurt Gödel, mathematics is forever destined to remain incomplete. He proved that there will always be valid mathematical statements that cannot be demonstrated from systems of axioms that are broad enough to include arithmetic (see DeLong, 2004). In other words, Gödel provided a formal demonstration of the inadequacy of formal demonstrations!

[9] By Paul J. Cohen; see P. J. Cohen and R. Hersh, "Non-Cantorian Set Theory," *Scientific American*, **217** (December 1967).

[10] See E. Wigner, "The Unreasonable Effectiveness of Mathematics," *Communications in Pure and Applied Mathematics*, **13** (1960): 1 ff.

Kurt Gödel

Perhaps the following remarks by René Thom are an appropriate reaction to Gödel's incompleteness theorem:

The mathematician should have the courage of his private convictions; he would then affirm that mathematical structures have an existence independent of the human mind that thinks about them. The form of this existence is undoubtedly different from the concrete and material existence of the external world, but it is nevertheless subtly and profoundly linked to objective existence. For how else explain—if mathematics is merely a gratuitous game, the random product of our cerebral activities—its indisputable success in describing the universe? Mathematics is encountered—not only in the rigid and mysterious laws of physics—but also, in a more hidden but still indubitable manner, in the infinitely playful succession of forms of the animate and inanimate world, in the appearance and destruction of their symmetries. That's why the Platonic hypothesis of Ideas informing the universe is—despite appearances—the most natural and philosophically the most economical. But, at any instant, mathematicians have only an incomplete and

> fragmentary vision of this world of Ideas . . . , we have to recreate it in our consciousness by a ceaseless and permanent reconstruction. . . . With this confidence in the existence of an ideal universe, the mathematician will not overly worry about the limits of formal procedures, he will be able to forget the problem of consistency. For the world of Ideas infinitely exceeds our operational possibilities, and the ultima ratio of our faith in the truth of a theorem resides in our intuition—a theorem being above all, according to a long-forgotten etymology, the object of a vision.[11]

The Meaning

In the first edition of this book, I ended this chapter with that inspiring quote from Thom (the founder of "catastrophe theory"). Further inquiry into these questions prompts me to a more somber conclusion. Namely, there is at present no intelligible account of what the statements of pure mathematics are about.

My claim that the formalist viewpoint is a dodge is substantiated by the following revealing admission by Jean Dieudonné:[12]

> On foundations we believe in the reality of mathematics, but of course when philosophers attack us with their paradoxes we rush to hide behind formalism and say "Mathematics is just a combination of meaningless symbols," and then we bring out Chapters 1 and 2 on set theory. Finally we are left in peace to go back to our mathematics and do it as we have always done, with the feeling each mathematician has that he is working with something real. This sensation is probably an illusion, but is very convenient. That is Bourbaki's attitude toward foundations.

An article by Reuben Hersh[13] forcefully demonstrates the philosophical plight of the working mathematician, who is "a Platonist on weekdays and a formalist on Sundays." Hersh contends that the tension caused by holding contradictory views on the nature of his work must

---

[11] R. Thom, "'Modern' Mathematics: An Educational and Philosophic Error?" *American Scientist*, November 1971, p. 695 ff. The translation here is my own from the original (in *L'Age de Science*, **III** (3): 225).

[12] "The work of Nicholas Bourbaki," *American Mathematical Monthly*, **77** (1970): 134–145.

[13] "Some Proposals for Reviving the Philosophy of Mathematics," *Advances in Mathematics* **31** (1979): 31–50.

affect the self-confidence of a person who is supposed, above all things, to hate contradiction.

Dieudonné admits that the Platonic view is probably an illusion. In a very interesting essay, Gabriel Stolzenberg[14] argues that the illusion consists in being taken in by a present-tense *language of objects and their properties*, a language that has the appearance—but only that— of being meaningful. The psychological act of accepting this appearance produces a notion of "reality" so strong that it becomes very difficult to step aside and question it.

We have already seen examples of such illusion. If one believes that points and lines in the plane are "real objects," then they either satisfy Euclid's fifth postulate or they don't (with the corollary belief that Euclidean geometry is either "true" or "false"). Similarly, if sets are "real objects," then they either satisfy Cantor's continuum hypothesis or they don't (Gödel believed that they don't).

The fundamental illusion, according to Stolzenberg (and Brouwer before him), is the belief that a mathematical statement can be "true" without anyone being able to know it. This belief is so strong that only the few constructivist mathematicians have been willing to give it up. They contend that "$\mathscr{S}$ is true" is a signal to announce a state of knowing, which one has attained by means of an act of proof. Stolzenberg (1978) claims (p. 265):

> What one "sees" or "discovers" at the conclusion of an act of proof is that a certain structure (which is constructed in the course of the proof) displays a certain form: a form of the type that, according to the conventions of mathematical language use that have been established, entitles anyone who observes it to say "$\mathscr{S}$ is true." But "$\mathscr{S}$ is true" is merely what one *says*, not what one *sees*; the expression itself is merely the "brand name" for the type of thing that one sees at the conclusion of the proof. And it is a type of thing that may be seen only by constructing a proof—not because we need to use the proof as "a ladder" to get ourselves into a position to see it but rather because what one sees is "in" the structure that is created by the act "of making the proof."

An interesting consequence of this position is that "the knower" is brought into the philosophy of mathematics (just as "the observer" has

---

[14] "Can an Inquiry into the Foundations of Mathematics Tell Us Anything Interesting about Mind?" in *Psychology and Biology of Language and Thought*, Essays in Honor of Eric Lenneberg, G. Miller and Elizabeth Lenneberg, eds. (New York: Academic Press, 1978, 221–269).

been brought into the philosophy of physics by Heisenberg's uncertainty principle).

If indeed the philosophical impasse is the result of a linguistic illusion, then deep insights are needed to develop a new language system.

On the other hand, the Platonic "illusion" has shown itself to be very valuable heuristically (e.g., Gödel credited the Platonic viewpoint for his insights). An intelligible justification for Platonic heuristics may someday be found (just as one was found in the twentieth century by the logician Abraham Robinson for the "illusory" infinitesimals used in the seventeenth century by the founders of the calculus). Physics has continued to advance despite the even worse mess in its philosophical foundations, so the proverbial "working mathematicians" will have no trouble continuing to ignore the irritating question of the *meaning* of their theorems.

Paul J. Cohen recently challenged Hilbert's optimism that mathematical reasoning from axioms could resolve all mathematical questions. Cohen asserted: "The vast majority of questions even in elementary number theory, of reasonable complexity, are beyond the reach of any such reasoning."[15]

The Fruitfulness of Hyperbolic Geometry for Other Branches of Mathematics, Cosmology, and Art

While the philosophy of mathematics may be at an impasse, mathematics itself certainly is not. In particular, hyperbolic geometry, which was considered a dormant subject when I first started teaching it, is very much alive and has turned out to have extraordinary applications to other branches of mathematics and to models of the shape of the universe. All the applications are too extensive to explain in detail here, so we briefly introduce six of them to show you how important it is. This is just intended to whet your interest for further study, so don't be disturbed if you only get a rough idea of these topics.

---

[15] See his article "Skolem and pessimism about proof in mathematics," *Philosophical Transactions of the Royal Society of London, Series A,* **363** (2005): 2407–2418. Cohen also stated: "Even if the formalist position is adopted, in actual thinking about mathematics one can have no intuition unless one assumes that models exist and that the structures are real . . . . there is a reality to mathematics, but axioms cannot describe it."

▨ 1. UNIFORMIZATION OF COMPACT CONNECTED ORIENTABLE SURFACES

These surfaces (meaning two-dimensional topological manifolds) were long known to be just spheres with handles, the number of handles being the *genus g* of the surface *S*. The surface can be "unrolled" into its *universal covering space U*, which is *simply connected* (any loop can be shrunk to a point). Topologically, *U* is either the sphere (if $g = 0$) or the plane $\mathbb{R}^2$, and in the latter case, *S* is a quotient space of $\mathbb{R}^2$ under the action of a discontinuous fixed-point-free group G.

Uniformization has to do with a deeper structure, a canonical *geometry of constant curvature* on *S*. For $g = 0$ it is obviously *spherical geometry* (if we considered nonorientable surfaces as well, *elliptic geometry* would occur). For $g = 1$, the surface *S* is topologically a *torus*. The covering map $U \rightarrow S$ transports the Euclidean geometry on $\mathbb{R}^2$ to a Riemannian geometry on *S* of constant curvature 0, endowing *S* with the structure of a *flat torus*; the group G is generated by two independent translations, which are isometries of the Euclidean plane. (Of course, this geometry on the torus is not the same as its geometry as a doughnut sitting in $\mathbb{R}^3$, whose curvature is not constant.)

For all $g > 1$, $\mathbb{R}^2$ must be replaced with the Poincaré disk **D**, to which it is homeomorphic, but which, as we know, models hyperbolic geometry. Group G is a discrete group of hyperbolic isometries of **D**, and the covering map induces on *S* a Riemannian geometry of *constant negative curvature*, creating a *hyperbolic surface*.

This great result is just the beginning of a long study. Some references are Stillwell's relatively informal *Geometry of Surfaces* (1992),[16] Beardon (1983), and Hubbard (2006). For the case $g = 2$, where *S* is obtained by identifying sides of a regular octagon in **D**, see http://mathworld.wolfram.com/UniversalCover.html, which includes a pretty illustration.

Uniformization was originally studied as part of complex function theory on Riemann surfaces, but that approach does not generalize to three dimensions.

▨ 2. THURSTON'S GEOMETRIZATION CONJECTURE FOR THREE-DIMENSIONAL MANIFOLDS

William Thurston conjectured (and proved in special cases) that three-manifolds also have geometric structures which enable us to classify them. In two dimensions we saw that there are three possible geometries; in three dimensions there are

---

[16] See my customer review on amazon.com. Isometries will be studied in the next chapter.

eight, and two of them involve hyperbolic geometry. Thurston emphasized the importance of *hyperbolic manifolds*. A corollary of his conjecture is the Poincaré conjecture about the three-sphere, for a proof of which the Clay Institute is offering a million dollar prize, and for his proof of it that is still being checked as of this writing, Grigory Perelman was awarded a Fields Medal in 2006. See O'Shea (2007).

Here are some online links for information about these topics:

http://en.wikipedia.org/wiki/Geometrization_conjecture
http://www.msri.org/publications/books/Book31/files/cannon.pdf#se
    arch=%22applications%20of%20hyperbolic%20geometry%22
http://www.ams.org/notices/200402/fea-anderson.pdf
http://www.math.sunysb.edu/~jack/PREPRINTS/tpc.pdf#search=
    %22John%20Milnor%20Poincare%20Conjecture%22

See also Thurston (1997), Casson and Bleiler (1988), Ratcliffe (2nd ed., 2006), W. Thurston, "Three dimensional manifolds, Kleinian groups and hyperbolic geometry," *Bulletin of the American Mathematical Society* (N.S.) **6** (1982): 357–381, and http://en.wikipedia.org/wiki/William_Thurston.

Jeffrey Weeks' program Snap Pea, used for studying hyperbolic three-manifolds, is at http://geometrygames.org/SnapPea/index.html.

Not Knot is a guided tour of computer-animated hyperbolic space. It proceeds from the world of knots to their complementary spaces—what's not a knot. Profound theorems of recent mathematics show that most knot complements carry the structure of hyperbolic geometry: http://www.geom.uiuc.edu/graphics/pix/Video_Productions/Not_Knot/.

3. MODULAR FORMS, SHIMURA–TANIYAMA THEOREM, AND FERMAT'S LAST THEOREM   This topic is a confluence of ideas from analysis, number theory, algebraic geometry, and hyperbolic geometry. The proof of the above theorems (by Andrew Wiles and Richard Taylor in the semistable case, sufficient to deduce Fermat's last theorem (FLT) as a corollary because of earlier work by Gerhard Frey, Jean-Pierre Serre, and Ken Ribet) was one of the greatest accomplishments of twentieth-century mathematics. Much very important research on the subject of modular forms is ongoing.

In Chapter 9, we will show that in the upper half-plane model **H** of the hyperbolic plane, the group $PSL(2, \mathbb{R})$ of 2×2 real matrices with determinant 1 modulo sign acts on **H** as the group of orientation-preserving isometries. The *modular group* Γ is the subgroup $PSL(2, \mathbb{Z})$ of those matrices with integer entries. In analytic number theory, the action of this group and its *congruence subgroups* is studied.

Of fundamental importance in algebraic number theory and algebraic geometry are *elliptic curves*. Barry Mazur's version of the Shimura–Taniyama theorem states that every elliptic curve defined over $\mathbb{Q}$ admits a *hyperbolic uniformization* that is periodic with respect to a congruence subgroup of Γ. Mazur wrote: "It is the confluence of two uniformizations, the Euclidean one, and the hyperbolic one of arithmetic type, that puts an exceedingly rich geometric structure on an arithmetic elliptic curve, and that carries deep implications for arithmetic questions."

Two sources of information about these topics are *A First Course in Modular Forms*, by F. Diamond and J. Shurman (2006), and *Modular Forms and Fermat's Last Theorem*, edited by G. Cornell, J. Silverman, and G. Stevens. See also:

> http://en.wikipedia.org/wiki/Taniyama%E2%80%93Shimura_
> theorem
> http://www.pbs.org/wgbh/nova/transcripts/2414proof.html
> http://www.ams.org/notices/199911/comm-darmon.pdf#search=
> %22Shimura-Taniyama%20Theorem%22

Arakelov geometry is a new type of geometry that arose out of the study of Diophantine algebraic geometry. A 2002 article by Yuri Manin and Matilde Marcolli at http://arxiv.org/PS_cache/hep-th/pdf/0201/ 0201036.pdf states: "We show that the relation between hyperbolic geometry and Arakelov geometry at arithmetic infinity involves exactly the same geometric data as the Euclidean AdS3 holography of *black holes*."

4. FUCHSIAN AND KLEINIAN GROUPS This is an older subject that goes back to the competition between Klein and Poincaré but which is still very active, related to *fractal geometry*. The book *Indra's Pearls* by David Mumford, Caroline Series, and David Wright (Cambridge University Press, 2002) is a beautifully illustrated introductory treatment of the subject. See also:

> http://en.wikipedia.org/wiki/Fuchsian_group
> http://en.wikipedia.org/wiki/Kleinian_group

▓▓▓▓ **5. THE GEOMETRY OF THE UNIVERSE** Einstein's description of the universe is via a four-dimensional space-time geometry. When we speak here of "the geometry of the universe," we are considering a three-dimensional slice of space-time at a particular instant. Possible models to describe this geometry are developed from three-dimensional spherical, Euclidean, or hyperbolic geometries. If you search the web for this topic, you will find an enormous number of web pages about it. Here are a few recommendations.

The latest edition of Jeffrey Weeks' *The Shape of Space* is ostensibly written at a high school level but covers very sophisticated topics such as the eight geometries relevant to Thurston's geometrization conjecture and programs of astronomical observation that might tell us the precise topology and geometry of the universe. It is reviewed at http://www.maa.org/reviews/shapeofspace.html.

Some introductory web references:

> The beautifully illustrated 1999 *Scientific American* article "Is Space Finite?" at http://cosmos.phy.tufts.edu/~zirbel/ast21/sciam/IsSpaceFinite.pdf#search=%22Is%20Space%20Finite%3F%22
>
> The also well-illustrated 2005 article about the *Poincaré dodecahedral universe* at http://luth2.obspm.fr/~luminet/physworld.pdf#search=%22Dodecahedral%20universe%22
>
> Weeks' "flight simulator" for tilings of three-dimensional spherical, flat, and hyperbolic spaces at http://geometrygames.org/CurvedSpaces/index.html
>
> The 1998 Cornish and Weeks article "Measuring the Shape of the Universe" at http://www.ams.org/notices/199811/cornish.pdf#search=%22shape%20of%20the%20universe%22
>
> See also http://en.wikipedia.org/wiki/Shape_of_the_universe

▓▓▓▓ **6. TESSELLATIONS OF THE HYPERBOLIC PLANE, THE ART OF M. C. ESCHER** A *tessellation* is a covering of the plane by a figure repeated over and over again, with overlaps only along the boundary of the figure. Another word for a tessellation is a *tiling*, and the generating figure is a *tile*.

The simplest figure is a regular polygon (all sides and all angles congruent), possibly giving rise to a *regular tessellation*, defined as an edge-to-edge covering of the plane by copies of a regular polygon in which the same number of polygons meet at each vertex. In the Euclidean plane, the only possible such tessellations are by equilateral triangles, squares, or hexagons.

But in the hyperbolic plane there are infinitely many different regular tessellations. Each regular tessellation has its Schläfli symbol $\{p, q\}$, which indicates that q regular p-gons surround each vertex. A necessary and sufficient condition that a hyperbolic tessellation $\{p, q\}$ exists is that $(p - 2)(q - 2) > 4$; thus each such $\{p, q\}$ has a *dual* tessellation $\{q, p\}$. The proof of this criterion is based on the fact that a regular p-gon with angle θ has angle sum $p\theta < 180(p - 2)$. For every such θ, a regular p-gon exists with angle θ. To fit exactly q of these about a point, we must have $q\theta = 360$. Substituting this in the previous inequality leads to the necessity. For sufficiency, when θ satisfies that equation, one must show that repeatedly reflecting the regular p-gon across its edges generates the tessellation. For example, using equilateral triangles $p = 3$, q must be at least 7, unlike the Euclidean tessellation $\{3, 6\}$.

Triangle tessellations of the hyperbolic plane are very important. Start with any triangle and try to generate a tessellation by repeatedly reflecting across the edges. This will actually generate a tessellation if and only if the angles of the triangle, in radian measure, have the form π/p, π/q, π/r, where p, q, r are integers satisfying the inequality

$$\frac{1}{p} + \frac{1}{q} + \frac{1}{r} < 1,$$

asserting that the angle sum is $< \pi$. Notation for such a triangle is (p, q, r).

A famous example is the $\{7, 3\}$ regular tessellation generated by $(2, 3, 7)$. Klein discovered that 336 triangles in this tessellation aggregate to form a regular 14-gon (see Figure 8.1), and the surface obtained from suitably identifying sides of this 14-gon has genus 3 and represents a quartic curve in $P^2(\mathbb{C})$. A beautifully illustrated discussion of the *Klein quartic* is at http://math.ucr.edu/home/baez/klein.html; the mathematics and history are explained by Jeremy Gray at http://www.msri.org/publications/books/Book35/files/gray.pdf, which also shows how hyperbolic tessellations are related to hyperbolic surfaces (topic 1). The key theorem about this relationship is due to Poincaré in 1882 (see Stillwell, ibid., p. 180).

In the hyperbolic plane, one can also consider tessellations in which some of the vertices are *ideal points*. For example, under the action of the *modular group PSL*(2, $\mathbb{Z}$), the Poincaré upper half-plane is tessellated by singly asymptotic triangles—(3, 3, ∞) triangles in an extension of the above notation (Figure 8.2). See http://en.wikipedia.org/wiki/Modular_group, which leads us back to topic 4.

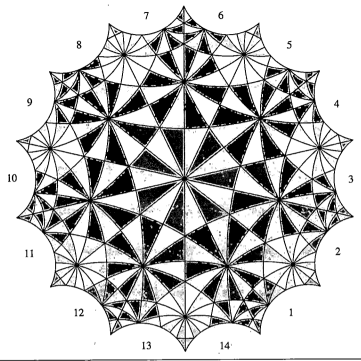

Figure 8.1 Regular 14-gon generated by (2, 3, 7) triangles.

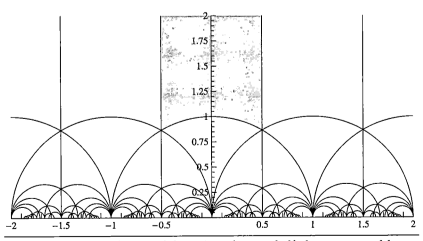

Figure 8.2 The tessellation of the Poincaré upper half-plane generated by the fundamental domain of the modular group.

M. C. Escher's *Circle Limit* drawings are based on hyperbolic tessellations. The great geometer H. S. M. Coxeter corresponded with Escher and wrote about his art—see, e.g., http://www.mathaware.org/mam/03/essay1.html and the references therein.

If you search the web, you will find a great many sites about tessellations. You will find many lovely illustrations, such as at

http://www.hadron.org/~hatch/HyperbolicTesselations/
http://www2u.biglobe.ne.jp/~hsaka/mandara/index.html
http://aleph0.clarku.edu/~djoyce/poincare/poincare.html
http://mcs.open.ac.uk/tcl2/nonE/nonE.html
http://comp.uark.edu/~strauss/papers/hypcomp.pdf

Review Exercise

Which of the following statements are correct?

(1) It is impossible to verify by physical experiments whether hyperbolic geometry is true because hyperbolic geometry is not about physical entities.

(2) If we interpret the undefined terms of geometry physically, e.g., by interpreting "line" as "path of a light ray in empty space," then it makes sense to ask whether this interpretation is a model of Euclidean geometry; however, due to experimental error, physical experiments could never prove conclusively that it is a model.

(3) Hyperbolic geometry is consistent if and only if Euclidean geometry is consistent.

(4) Poincaré maintained that it was meaningless to ask which geometry is "true," and that it only makes sense to ask which geometry is more "convenient" for physics.

(5) The most convenient geometry for astrophysics is neither Euclidean nor hyperbolic geometry but a more complicated geometry of space-time developed by Einstein out of ideas from Riemann.

(6) The Klein and Poincaré models, although they appear to be different, are actually isomorphic to each other.

(7) Hyperbolic geometry, although equally as consistent as Euclidean geometry, has no application to other branches of mathematics or to other sciences.

Some Topics for Essays

1. Comment on this quotation from Albert Einstein: "As far as the mathematical theorems refer to reality, they are not sure, and as far as they are sure, they do not refer to reality." (See Hempel, 1945, for a development of this theme.)

2. Report on the debate about the philosophy of *conventionalism*, using Grünbaum (1968), Poincaré (1952), and Nagel (1939) as sources.

3. Report on the use of hyperbolic geometry to describe binocular vision, referring to Luneburg and Blank (see note 4 in this chapter).

4. It can be said that the discovery of non-Euclidean geometry led to the extensive modern development of mathematical logic. Elaborate on this statement using DeLong (2004), Chapters 1 and 2, as a source.

5. Jacques Hadamard said: "Practical application is found by not looking for it, and one can say that the whole progress of civilization rests on that principle. . . . It seldom happens that important mathematical researches are *directly* undertaken in view of a given practical use: they are inspired by the desire which is the common motive of every scientific work, the desire to know and understand."[17]

   Along the same lines, David Hilbert maintained that in spite of the importance of the applications of mathematics, these must never be made the measure of its value. And the mathematician Jacobi said that "the glory of the human spirit is the sole aim of all science."

   Nevertheless, Lobachevsky believed that "there is no branch of mathematics, however abstract, that may not someday be applied to phenomena of the real world."

   Comment on these viewpoints.

6. Read the "Socratic Dialogue on Mathematics" in Renyi (1967) and discuss the following questions therein:

   (a) "Is it not mysterious that one can know more about things which do not exist than about things which do exist?"

   (b) "How do you explain that, as often happens, mathematicians living far from each other and having no contact independently discover the same truths?"

7. Comment on the following statement by Michael Polanyi (1964; see especially Chapter 6, Sections 9–11):

---

[17] Hadamard (1945); see especially Chapter 9.

We can now turn to the paradox of a mathematics based on a system of axioms which are not regarded as self-evident and indeed cannot be known to be mutually consistent. To apply the utmost ingenuity and the most rigorous care to prove the theorems of logic or mathematics while the premises of these inferences are cheerfully accepted, without any grounds being given for doing so . . . might seem altogether absurd. It reminds one of the clown who solemnly sets up in the middle of the arena two gateposts with a securely locked gate between them, pulls out a large bunch of keys, and laboriously selects one which opens the lock, then passes through the gate and carefully locks it after himself—while all the while the whole arena lies open on either side of the gateposts where he could go round them unhindered.

8. Comment on the following statements:

There is a scientific taste just as there is a literary or artistic one. . . . Concerning the fruitfulness of the future result—about which, strictly speaking, we most often do not know anything in advance—[the] sense of beauty can inform us and I cannot see anything else allowing us to foresee. . . . Without knowing anything further we *feel* that such a direction of investigation is worth following. . . . Everybody is free to call or not to call that a feeling of beauty. This is undoubtedly the way the Greek geometers thought when they investigated the ellipse, because there is no other conceivable way. (Hadamard, 1945.)

We dwell on mathematics and affirm its statements for the sake of its intellectual beauty. . . . For if this passion were extinct, we would cease to understand mathematics; its conceptions would dissolve and its proofs carry no conviction. Mathematics would become pointless and lose itself in a welter of insignificant tautologies. . . . (Polanyi, 1964.)

We all believe that mathematics is an art. The author of a book or the lecturer in a classroom tries to convey the structural beauty of mathematics to his readers, to his listeners. In this attempt he must always fail. Mathematics is logical, to be sure; each conclusion is drawn from previously derived statements. Yet the whole of it, the real piece of art, is not linear; worse than that, its perception should be instantaneous.[18]

9. Comment on the following statements. G. H. Hardy (1940) said:

For me, and I suppose for most mathematicians, there is another reality, which I will call "mathematical reality"; and there is no sort of

[18] Emil Artin, "Review of *Algèbre* by N. Bourbaki," *Bulletin of the American Mathematical Society*, **59** (1953): 474.

argument about the nature of mathematical reality among either mathematicians or philosophers. . . . A man who could give a convincing account of mathematical reality would have solved very many of the most difficult problems of metaphysics. . . . I believe that mathematical reality lies outside us, that our function is to discover or *observe* it, and that the theorems which we prove, and which we describe grandiloquently as our "creations," are simply the notes of our observations. This view has been held, in one form or another, by many philosophers of high reputation from Plato onwards. . . .

Heinrich Hertz, the discoverer of radio waves, said:

One cannot escape the feeling that these mathematical formulas have an independent existence and an intelligence of their own, that they are wiser than we are, wiser even than their discoverers, that we get more out of them than was originally put into them.

10. Comment on the following remarks by Kurt Gödel:

I don't see any reason why we should have less confidence in this kind of perception, i.e., in mathematical intuition, than in sense perception, which induces us to build up physical theories and to expect that future sense perceptions will agree with them and, moreover, to believe that a question not decidable now has meaning and may be decided in the future. The set theoretical paradoxes are hardly any more troublesome for mathematics than deceptions of the senses are for physics. . . . Evidently the "given" underlying mathematics is closely related to the abstract elements contained in our empirical ideas. It by no means follows, however, that the data of this second kind [mathematical intuitions], because they cannot be associated with actions of certain things upon our sense organs, are something purely subjective, as Kant asserted. Rather, they, too, may represent an aspect of objective reality. But as opposed to the sensations, their presence in us may be due to another kind of relationship between ourselves and reality.[19]

Gödel in this passage speaks primarily of *set theoretical intuition.* As far as geometric intuition is concerned, the following, according to Gödel, would have to be added:

Geometrical intuition, strictly speaking, is not mathematical, but rather a priori physical intuition. In its purely mathematical aspect our Eu-

---

[19] K. Gödel, "What Is Cantor's Continuum Problem?" in Benacerraf and Putnam's *Philosophy of Mathematics,* 2nd ed. (Englewood Cliffs, N.J.: Prentice-Hall, 1964), p. 271.

clidean space intuition is perfectly correct, namely, it represents correctly a certain structure existing in the realm of mathematical objects. Even physically it is correct "in the small."[20]

11. Comment on the following quotation from Rolf R. Loehrich:

The communication of a new mathematical system or game meets with peculiar obstacles. Each mathematician has a preferred game. A new game may not capture his interest if it is significantly different from those he has been accustomed to play. . . .

A mathematical system is hardly ever presented axiomatized at its inception. Successful axiomatization is a fruition of an *exercitium cogitandi*. Once a system is axiomatized, mathematical activity can be played as a game, as a manipulation of symbols by virtue of rule-systems thought of as invented, but this does not assert that the mathematician who invented or presumably discovered the system meant to play a game. . . . Roberts and I are convinced that there is what might be adequately referred to as a mathematical universe. We believe that, with the complex instrumentations and empirical data set forth in *Exercitium Cogitandi*, the ontological value of confrontations belonging to this universe can be determined with a high degree of accuracy. . . . If this is true, then indeed a mathematician may think of himself as an explorer of the mathematical universe, and any new mathematical system functions as the inception of a possible creation of a universe which comprehends any of the other universes.[21]

12. Write an essay on the development of geometry in ancient Greece, using the resources of your school library. You may be particularly interested in the female mathematician Hypatia.

13. Comment on the following remarks about the true role of logic in mathematics:

If logic is the hygiene of the mathematician, it is not his source of food; the great problems furnish the daily bread on which he thrives. We have learned to trace our entire science back to a single source, constituted by a few signs and by a few rules for their use; this is an unquestionable stronghold, inside which we could hardly confine ourselves without risk of famine, but to which we are always free to retire in case of uncertainty or external danger. (A. Weil, "The Future of Mathematics," *American Mathematical Monthly*, **57** (1950): 295–306.)

---

[20] Private communication to the author, October 1973.

[21] R. R. Loehrich (with L. G. Roberts), *Exercitium Cogitandi*, vol. II (Oxford: Center for Medieval and Renaissance Studies, 1978).

Thus, in a sense, mathematics has been most advanced by those who distinguished themselves by intuition rather than by rigorous proofs. (Felix Klein.)

Discovery, after all, is more important in science than strict deductive proof. Without discovery there is nothing for deduction to attack and reduce to order. (E. T. Bell, *Development of Mathematics*, 2nd ed. McGraw-Hill, New York, 1945, p. 83.)

14. Report on Imre Lakatos' critique of the formalist philosophy of mathematics and his ideas on how mathematics is discovered, as presented in his book *Proofs and Refutations: The Logic of Mathematical Discovery* (New York: Cambridge University Press, 1976). Here are some pertinent Lakatos quotes:

Euclid has been the evil genius particularly for the history of mathematics and for the teaching of mathematics, both on the introductory and the creative levels. . . . The two activities of guessing and proving are rigidly separated in the Euclidean tradition. . . . It was the infallibilist philosophical background of Euclidean method that bred the authoritarian traditional patterns in mathematics, that prevented publication and discussion of conjectures, that made impossible the rise of mathematical criticism. . . . The discovery of non-Euclidean geometries (by Lobatschewsky in 1829 and Bolyai in 1831) shattered infallibilist conceit. . . . There is no infallibilist logic of scientific discovery, one which would infallibly lead to results; there is a fallibilist logic of discovery which is the logic of scientific process.

15. Write a detailed report on the theory of area in hyperbolic geometry using Moise (1990), Chapter 24, as a reference.

16. Report on Bertrand Russell's doctoral dissertation *An Essay on the Foundations of Geometry* (Dover reprint, 1956). Show how Russell very capably refutes theories of geometry due to Kant and other philosophers, but then proclaims his own incorrect notion of space (that was later refuted by Einstein). See also the critique in Torretti (1978), Chapter 4.

17. Report on Chapter 3 of Roberto Torretti's sublime treatise *Philosophy of Geometry from Riemann to Poincaré* (1978). This chapter is on the foundations of geometry. Here is one important quote:

The fact that these semi-circles [in the Poincaré upper half-plane model] behave exactly like Euclidean lines with regard to every logical consequence of Hilbert's axioms [for neutral geometry] bespeaks a deep analogy between them, which can come as a shock only to the mathematically uneducated. To maintain that *line* means something entirely different in Bolyai-Lobachevsky geometry and in Euclidean geometry is not more reasonable than to say that *heart* has a completely different meaning in the anatomy and physiology of elephants and in that of frogs.

---

18. To further illustrate his contention that it is meaningless to ask which geometry is "true," Poincaré invented a "universe" U occupying the interior of a sphere S of radius R in Euclidean space, in which the following physical laws hold:
    (a) At any point P inside S, the absolute temperature T is directly proportional to $R^2 - r^2$, where r is the distance from P to the center of S.
    (b) The length, width, and height of an object vary directly with the absolute temperature of the object.
    (c) All objects in U instantaneously take on the temperatures of their locations.
    (d) Light travels along the shortest path from one point to another.
    Show that an inhabitant of U could not detect his change in temperature and size as he moves about with a thermometer or a tape measure, and that he could never reach the boundary S of his universe, so would consider it infinitely far away. Poincaré showed that the shortest path in U joining point A to point B is the smaller arc of the circle through A and B that cuts S orthogonally. Hence, if an inhabitant interprets "straight line segment" in his universe to be the path of a light ray, he would conclude that the "true" geometry of his world was hyperbolic. In other words, this is a region of Euclidean space that, because of different and undetectable physical laws, appears to its inhabitants to be non-Euclidean. Comment, using Poincaré (1952) as a reference, as well as Torretti (1978) and Grünbaum (1968).
19. Albert Einstein stated: "The concept 'true' does not tally with the assertions of pure geometry, because by the word 'true' we are eventually in the habit of designating always the correspondence with a 'real' object; geometry, however, is not concerned with the relation of the ideas involved in it to objects of experience, but only with the logical connection of these ideas among themselves." Read as much of his *Relativity* (2005 edition, for example) as you can and comment on this quote by him as well as anything else he says that interests you.

9 Geometric Transformations

> *I have spent a lifetime applying Klein's program to differential geometry.*
>
> W. Blaschke

Klein's Erlanger Programme

In 1872, a year after his decisive publication of the projective models for non-Euclidean geometries, Felix Klein was appointed (at age 23) to a chair at the University of Erlangen. He submitted an inaugural research proposal describing a new unifying principle for classifying the various geometries that were rapidly being developed and for discovering relationships between them. This *Erlanger Programme* has had an enormous impact on all of mathematics to the present day.[1]

The key notion, according to Klein, involves the group of all automorphisms of a mathematical structure. In Chapter 2, we defined the concept of an isomorphism of one model onto another, and in Chapter 7 we used a specific isomorphism to relate the Klein and Poincaré

---

[1] For an English translation of Klein's lecture, see the *Bulletin of the New York Mathematical Society*, **2** (1893): 215–249. And see the project at the end of this chapter.

models of the hyperbolic plane. An isomorphism mapping a given model onto itself is called an *automorphism* of that model; thus an automorphism is a one-to-one mapping (or transformation) of each basic set of objects in the model onto itself that preserves the basic relations among the objects. In geometry, an automorphism is often called a *symmetry*.

The importance of the group of automorphisms was first recognized in connection with the problem of solving an algebraic equation by radicals. Évariste Galois (1811–1832) showed that a solution by radicals was possible if and only if the group of automorphisms of the field extension generated by the roots of the equation is a solvable group. This implies Abel's particular discovery that the general equation of degree 5 cannot be solved by radicals. Klein later discovered a relation between the group of rotations of an icosahedron and the roots of the quintic equation that explained why the latter can be solved by elliptic functions.[2]

Here is an example of the simplest type of geometric automorphism.

EXAMPLE 1. Consider models of incidence geometry (Chapter 2). The basic sets of objects are the sets of points and lines, and the only basic relation is incidence of a point and a line. An automorphism T will therefore map each point P and each line l onto a point P′ and a line $l′$ such that P lies on l if and only if P′ lies on $l′$. By Axiom I-1, a line is determined by any two points lying on it, so T is determined as a mapping of lines once its effect on the points is known, namely,

$$T(\overleftrightarrow{PQ}) = \overleftrightarrow{P'Q'}.$$

Since T preserves incidence and is one-to-one on the set of lines, it has the property that three points O, P, Q are collinear if and only if their images O′, P′, Q′ are collinear. Hence an automorphism of a model of incidence geometry is called a *collineation*.

For example, in the three-point model, every permutation of the three noncollinear points is a collineation. However, for the seven-point projective plane (Figure 9.1), you can show that, of the 7! = 5040 permutations of the points, only 168 are collineations (Exercise 1).

It is important to note that an automorphism not only preserves the basic relations but also *all* the relations that can be defined from them. For example, a collineation of an incidence plane preserves parallelism $(l \parallel m \Rightarrow l' \parallel m')$.

---

[2] See Klein's *Lectures on the Icosahedron*, 2nd English edition (New York: Dover Books, 2003).

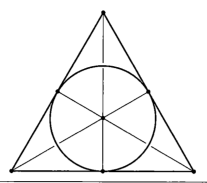

Figure 9.1 Seven-point projective plane.

Groups

Transformations of a set onto itself can be multiplied by first applying one transformation T and then another transformation S; thus the composite transformation ST is defined by the equation

(0) $$ST(x) = S(T(x))$$

for all x in the set.

With this multiplication, the set $\mathcal{G}$ of all automorphisms of a structure has itself the stucture of a *group*, which means that the following properties hold:

1. $S, T \in \mathcal{G} \Rightarrow ST \in \mathcal{G}$.
2. $I \in \mathcal{G}$ (where I is the *identity* transformation that leaves all the objects fixed; the identity transformation satisfies $IT = T = TI$ for all $T \in \mathcal{G}$).
3. $T \in \mathcal{G} \Rightarrow T^{-1} \in \mathcal{G}$ (where the *inverse* T^{-1} of T is characterized by the equations $TT^{-1} = I = T^{-1}T$).
4. $S(TU) = (ST)U$ for all $S, T, U \in \mathcal{G}$ (this *associative law* is an immediate consequence of the definition (0) of multiplication).

To illustrate these properties, let us consider rotations about a point O, which will be rigorously defined later but can now be thought of as transformations that turn the entire plane through a certain angle about O. If T is the rotation through $t°$ clockwise and S the rotation through $s°$ clockwise, then ST is the rotation through $(s + t)°$ clockwise. T^{-1} is the rotation through $t°$ counterclockwise. I can be thought of as the rotation through $0°$.

WARNING The product ST is not, in general, equal to the product TS in the opposite order, as the next example shows.

EXAMPLE 2. Consider the equilateral triangle $\triangle ABC$ situated symmetrically about its centroid O in Figure 9.2. If we let T be the rotation through 120° counterclockwise about O and let S be the reflection across the altitude $\overleftrightarrow{AO}$, then TS leaves C fixed and interchanges A and B (in fact, TS is the reflection across $\overleftrightarrow{CO}$); whereas ST leaves B fixed and interchanges A and C (ST is the reflection across $\overleftrightarrow{BO}$).

If two transformations, S, T happen to have the property $ST = TS$, we say that they *commute*, and a collection of transformations in which every pair commutes is called *commutative* (or Abelian, after the great Norwegian mathematician N. H. Abel). For instance, any two rotations about the same point O commute.

The more structure a geometry has, the smaller its group of automorphisms. Neutral geometry is incidence geometry with the additional relations of betweenness and congruence; hence the group of automorphisms of a neutral geometry is the *subgroup* of those collineations T for which betweenness and congruence are *invariant*, i.e., for which

$$A * B * C \Rightarrow A' * B' * C'$$

$$AB \cong CD \Rightarrow A'B' \cong C'D'$$

(we will systematically use X' *to denote the image of any object* X— *point, line, circle, etc.—under a transformation denoted* T*)*. We have not assumed T preserves congruence of angles because this can be proved: If $\sphericalangle ABC \cong \sphericalangle DEF$, we can assume by Axiom C-1 that $AB \cong DE$ and $BC \cong EF$, so that $AC \cong DF$ (SAS); since T preserves congruence of segments, $\triangle A'B'C' \cong \triangle D'E'F'$ (SSS), hence $\sphericalangle A'B'C' \cong \sphericalangle D'E'F'$. Notice also that if a transformation preserves betweenness, it must be a collineation (by Axioms B-1 and B-3).

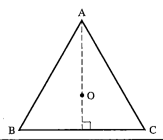

Figure 9.2

The principal objective of this chapter will be to explicitly determine all the automorphisms of Euclidean and hyperbolic planes and to classify them according to their geometric properties, particularly their invariants.

We say that a property or relation is "invariant" under a transformation or group of transformations if the property or relation still holds after the transformations are applied; a geometric figure is "invariant" if it is mapped onto itself by the transformations.

"Invariance" and "group" are the unifying concepts in Klein's *Erlanger Programme*. Groups of transformations had been used in geometry for many years, but *Klein's originality consisted in reversing the roles, in making the group the primary object of interest and letting it operate on various geometries, looking for invariants.* For example, the group $PSL(2, \mathbb{R})$ of 2-by-2 projective transformations with real coefficients (see Proposition 9.26) operates on both the hyperbolic plane and the real projective line; for the latter operation, the cross-ratio of four points is the fundamental invariant, whereas for the former operation, the length of a segment (which is calculated by means of cross-ratios in the Klein and Poincaré models) is the fundamental invariant.

Klein classified the following geometries as subgeometries of real plane projective geometry:

1. Affine geometry is the study of invariants of the subgroup of those projective transformations (called *affine transformations*) that leave the line at infinity invariant (points at infinity may be moved by affine transformations, but they stay on the line at infinity).
2. Hyperbolic geometry is the study of invariants of the subgroup of those projective transformations that leave a given real conic ("the absolute") invariant.
3. Elliptic geometry is the study of invariants of the subgroup of those projective transformations that leave a given imaginary conic invariant.
4. Parabolic geometry is the study of invariants of the subgroup of those affine transformations (called *similarities*) that leave invariant the two imaginary circular points at infinity (see Exercise 72).
5. Euclidean geometry is the study of invariants of the subgroup of those similarities (called *motions*) that preserve length (which is defined in terms of an arbitrarily chosen unit segment).

During the two decades preceding Klein's address, Cayley and Sylvester had developed a general theory of algebraic invariants together with a systematic procedure for determining generators and relations for them (see J. Dieudonné and J. Carrell, *Invariant Theory, Old and New*, New York: Academic Press, 1971). Klein proposed to translate geometric problems in projective geometry into algebraic problems in invariant theory, where such problems could be solved by the known algebraic methods (for a readable explanation of this program, see Part 3 of Klein's *Geometry*, which is Part 2 of his *Elementary Mathematics from an Advanced Standpoint*, New York: Dover, 2004).

Klein's idea of looking for various actions or representations of a group and their invariants has proved to be fruitful in many branches of mathematics and physics, not just in geometry.

In physics, for example, the invariance of Maxwell's equations for electromagnetism under Lorentz transformations suggested to Minkowski a new geometry of space-time whose group of automorphisms is the Lorentz group; this was the beginning of relativity theory, for which Einstein at one point considered the name *Invariantentheorie*. In atomic physics, the regularities revealed in the periodic table are a direct consequence of invariance under rotations. In elementary particle physics, considerations of invariance and symmetry have led to several nontrivial predictions. Physical entities such as energy, momentum, electrical charge, and spin have all been defined in terms of invariants, each one for a specific group of transformations. Each law of conservation in physics corresponds to a certain group of transformations. E. Wigner has said that in the future we may well "derive the laws of nature and try to test their validity by means of the laws of invariance rather than to try to derive the laws of invariance from what we believe to be the laws of nature."[3]

In this chapter, we will explore the insights Klein's point of view gives to plane Euclidean and hyperbolic geometries. From our axioms we will deduce a description of all possible motions, showing how they are built up from reflections (see Table 9.1, p. 430). Then we will show how to calculate using these transformations in terms of the coordinates in our models. We will implement Klein's program by replacing congruence axioms with group axioms. Finally, we will apply group-theoretic methods to questions of symmetry.

[3] E. Wigner, "Invariance in Physical Theory," *Proceedings of the American Philosophical Society*, **93** (1949): 521–526.

Applications to Geometric Problems

Here are some examples[4] of geometric problems that can easily be solved using transformations; the solutions will use certain properties of reflections, rotations, translations, and dilations, which will be demonstrated in the following sections. The purpose in discussing these problems at this time is to illustrate concretely the power of transformation techniques. *You will better comprehend the solutions after you study the theory that follows*, and I suggest that you then reread these solutions and test your understanding with Exercises 73–78.

PROBLEM 1. Given two points A, B on the same side of line l. Find the point C on l such that $\overrightarrow{CA}$ and $\overrightarrow{CB}$ make congruent angles with l (if l were a mirror, ACB would be the path of a ray of light traveling from A to B by reflecting in l).

SOLUTION. (See Figure 9.3.) Let B′ be the reflection of B across l. Then C is the intersection of AB′ with l.

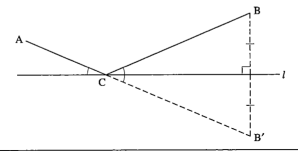

Figure 9.3

PROBLEM 2. Point Q is called a *center of symmetry* for figure F if whenever AA′ is a segment having Q as midpoint and A is in F, then A′ also belongs to F. Show that a figure can only have zero, one, or infinitely many centers of symmetry.

SOLUTION. Q is a center of symmetry if and only if the figure is invariant under the half-turn (180° rotation) H_Q about Q. A triangle

---

[4] Several hundred more examples will be found in the monumental three-volume treatise by I. M. Yaglom, *Geometric Transformations*, Washington, DC: Mathematical Association of America, 1962.

has zero, a circle has one, and a line has infinitely many centers of symmetry. Suppose figure F has at least two centers Q and Q'. Then $H_Q(Q') = Q''$ is a third center, $H_{Q'}(Q'')$ is a fourth center, etc.

NOTE. The preceding problems were stated and solved in neutral geometry. For the remaining problems, we will assume the geometry to be Euclidean.

PROBLEM 3. Let L, M, N be the respective midpoints of sides AB, BC, CA of $\triangle$ABC. Let O_1, O_2, O_3 be the *circumcenters* (i.e., the centers of the circumscribed circles) of triangles $\triangle$ALN, $\triangle$BLM, $\triangle$CMN, respectively, and let P_1, P_2, P_3 be the *incenters* (i.e., the centers of the inscribed circles) of these same triangles. Show that we have the relation $\triangle O_1O_2O_3 \cong \triangle P_1P_2P_3$.

SOLUTION. (See Figure 9.4.) Observe that each of the three triangles is obtained from each of the others by a translation—e.g., translating $\triangle$ALN in direction $\overrightarrow{AB}$ through distance $\overline{AL} = \overline{LB}$ gives $\triangle$LBM. This translation carries the circumscribed circle (and its center) of one triangle onto the circumscribed circle (and its center) of the other; similarly for the inscribed circles. Hence we not only have $O_1O_2 \cong AL \cong P_1P_2$, etc., giving $\triangle O_1O_2O_3 \cong \triangle P_1P_2P_3$, but we also see that corresponding sides of these two triangles are parallel.

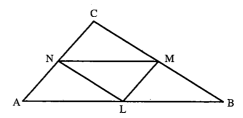

Figure 9.4 L, M, N are midpoints of the sides.

PROBLEM 4. Given an acute-angled triangle, find the inscribed triangle of minimum perimeter (Fagnano's problem).

SOLUTION. Consider $\triangle$XYZ inscribed as in Figure 9.5(a). Reflect X across $\overleftrightarrow{AB}$ to point X_1 and across $\overleftrightarrow{AC}$ to point X_2. Then the perimeter of $\triangle$XYZ is equal to the length of the polygonal path X_1ZYX_2. If we fix X, this length will be minimized when Z and Y are chosen to lie on

X_1X_2, and then $\overline{X_1X_2}$ will equal the perimeter of $\triangle XYZ$. We have that $AX_1 \cong AX \cong AX_2$ and $\sphericalangle X_1AX_2 \cong 2\sphericalangle A$. If we now vary X, the summit angle of isosceles triangle $\triangle X_1AX_2$ remains constant in measure and the base $\overline{X_1X_2}$ varies in direct proportion to $\overline{AX}$ (in fact, trigonometry gives us $\overline{X_1X_2} = 2\overline{AX}$ sin $\sphericalangle A$). Hence the minimum perimeter is achieved when $\overline{AX}$ is a minimum, and that occurs when X is the foot of the altitude from A (Figure 9.5(b)). By the same argument, Y and Z must then also be the feet of the altitudes from B and C. Hence the unique inscribed triangle of minimum perimeter is the *orthic* or *pedal triangle* formed by the feet of the altitudes of $\triangle ABC$.

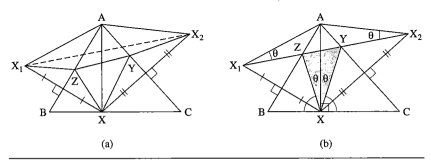

(a)   (b)

Figure 9.5

▨▨▨ **PROBLEM 5.** Given three parallel lines, find an equilateral triangle whose vertices lie on them.

▨▨▨ **SOLUTION.** Choose any point A on the first line l. Rotate the second line m about A through 60° to a new line m'. Let C be the intersection of m' with the third line n and let B be the point on m obtained by rotating C about A through 60° in the opposite direction. Then $\triangle ABC$ is a solution.

▨▨▨ **PROBLEM 6.** For any triangle $\triangle ABC$, construct equilateral triangles on the sides of $\triangle ABC$, exterior to it. Show that the centers of these triangles also form the vertices of an equilateral triangle (Napoleon's theorem).

▨▨▨ **SOLUTION.** Call the centers O_1, O_2, and O_3 and consider the rotations R_1, R_2, and R_3 through 120° counterclockwise about O_1, O_2, and O_3, respectively; then $R_1(A) = B$, $R_2(B) = C$, and $R_3(C) = A$. Now R_2R_1 is the clockwise rotation through 120° about the point O'_3 of the intersection of two lines, one through O_1 and the other through O_2, each making an angle of $60°$ with $\overleftrightarrow{O_1O_2}$, so that $\triangle O_1O_2O_3'$ is equilateral. Since R_3^{-1} is also a clockwise rotation through $120°$ taking A into C, we must have $R_3^{-1} = R_2R_1$ and $O_3' = O_3$.

PROBLEM 7. Given a circle κ and a point P on κ. Find the locus κ' of midpoints M of all chords PA of κ through P.

SOLUTION. (See Figure 9.6.) Since κ' is obtained from κ by dilation of center P and ratio $\frac{1}{2}$, κ' is the circle with diameter OP, O being the center of κ.

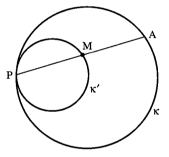

Figure 9.6

PROBLEM 8. Given any triangle $\triangle ABC$, consider its *circumcenter* O (point of concurrence of the perpendicular bisectors of the sides), its *centroid* G (point of concurrence of the medians), and its *orthocenter* H (point of concurrence of the altitudes). You showed O exists in Exercise 10, Chapter 6. You showed in Exercise K-20(d), Chapter 7, that G exists and lies two-thirds of the distance from each vertex to the midpoint of the opposite side; thus the dilation T of center G and ratio $-\frac{1}{2}$ maps $\triangle ABC$ onto the *medial triangle* $\triangle A'B'C'$. The problem we pose now is to show that H exists, that O, G, and H lie on a line (called the *Euler line* of $\triangle ABC$), and that G lies two-thirds of the distance from H to O.

SOLUTION. Dilation T^{-1} maps $\triangle ABC$ onto $\triangle A_1B_1C_1$ having sides parallel to the respective sides of $\triangle ABC$ and twice as long (Figure 9.7). $\triangle ABC$ is then the medial triangle of $\triangle A_1B_1C_1$, and the altitudes of $\triangle ABC$ are the perpendicular bisectors of $\triangle A_1B_1C_1$, hence are concurrent in a point H.

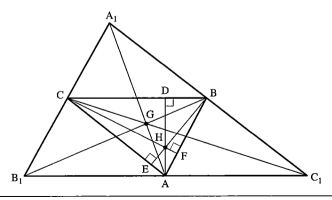

Figure 9.7 The orthocenter H of $\triangle ABC$ is the circumcenter of $\triangle A_1B_1C_1$.

The original dilation T, being a similarity, preserves perpendicularity, hence maps the orthocenter H of $\triangle ABC$ onto the orthocenter of the medial triangle $\triangle A'B'C'$, which is O; since G is the center of T and $-\frac{1}{2}$ the ratio, the conclusion follows from the definition of dilation.

▓▓ PROBLEM 9. Let H be the orthocenter, O the circumcenter, L, M, N the midpoints of the sides, D, E, F the feet of the altitudes of $\triangle ABC$. Show that L, M, N, D, E, F and the midpoints of segments HA, HB, HC all lie on a circle whose center U lies on the Euler line and is the midpoint of HO (*the nine-point circle* of $\triangle ABC$).

▓▓ SOLUTION. Consider the dilation T of center H and ratio 2. If we show that T maps all nine points onto the circumscribed circle κ of $\triangle ABC$, the conclusion will follow from Lemma 7.2, Chapter 7, applied to dilation T^{-1} of ratio $\frac{1}{2}$ (T^{-1} maps κ onto a circle of half the radius and whose center is the midpoint of OH). Clearly T maps the midpoints of HA, HB, HC onto A, B, C on κ.

Let P be the point on κ diametrically opposite to A (see Figure 9.8). Since $\sphericalangle ACP$ is inscribed in a semicircle, $\overleftrightarrow{PC} \perp \overleftrightarrow{AC}$, hence $\overleftrightarrow{PC}$ is parallel to altitude $\overleftrightarrow{BH}$. Similarly $\overleftrightarrow{PB} \parallel \overleftrightarrow{CH}$. Thus $\square PCHB$ is a parallelogram, hence the midpoint of diagonal HP coincides with the midpoint L of BC. This shows $T(L) = P$ on κ (and similarly for $T(M)$ and $T(N)$).

Let ray $\overrightarrow{HD}$ meet κ at D'. Since $\sphericalangle AD'P$ is inscribed in a semicircle of κ, $\overleftrightarrow{D'P} \perp \overleftrightarrow{AD'} = \overleftrightarrow{AD} \perp \overleftrightarrow{DL}$; i.e., $\overleftrightarrow{D'P} \parallel \overleftrightarrow{DL}$, which implies that D is the midpoint of HD' (since L is the midpoint of HP). Thus $T(D) = D'$ on κ (and similarly for $T(E)$ and $T(F)$).

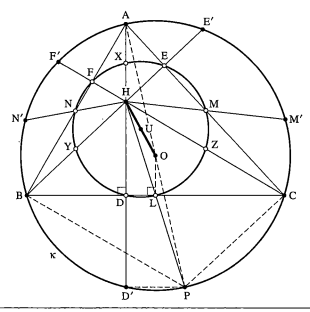

Figure 9.8 Nine-point circle.

Motions and Similarities

Henceforth the word "automorphism" will be used only for an auto-morphism of a neutral geometry, i.e., for a transformation that pre-serves incidence, betweenness, and congruence.

DEFINITION. A transformation T of the entire plane onto itself is called a *motion*[5] or an *isometry* if length is invariant under T, i.e., if for every segment AB, $\overline{AB} = \overline{A'B'}$, or equivalently, AB $\cong$ A'B'.

PROPOSITION 9.1. (a) Every motion is an automorphism. (b) The mo-tions form a subgroup of the group of automorphisms.

PROOF:
 (a) Let T be a motion. If AB $\cong$ CD, then

$$\overline{A'B'} = \overline{AB} = \overline{CD} = \overline{C'D'},$$

---

[5] Some authors call these transformations *rigid motions*. The term "motion" as we use it here does not mean continuous movement of a physical body as in common usage, although it is suggested by the latter.

so that $A'B' \cong C'D'$. That T also preserves betweenness follows from Theorem 4.3(9), which says that $A * B * C$ if and only if we have $\overline{AC} = \overline{AB} + \overline{BC}$.

(b) You must verify properties 1 through 3 in the definition of a group, which is an easy exercise. ◄

PROPOSITION 9.2. If T is an automorphism of an Archimedean Hilbert plane, then $\sphericalangle A \cong \sphericalangle A'$ for every angle. Thus T preserves angle measure.

PROOF:

Given an automorphism T, define a possibly new measure of $\sphericalangle A$ to be the measure $(\sphericalangle A')°$ of its image under T. You will show in Exercise 3 that this new measure satisfies all the basic properties 1 through 6 in Theorem 4.3 (the plane must be Archimedean for measure by real numbers to exist). But that theorem says there is a *unique* degree measure with these properties. Hence $(\sphericalangle A)° = (\sphericalangle A')°$. ◄

WARNING If the Hilbert plane is non-Archimedean, we have not constructed any numerical measure of angles and there are examples of automorphisms T for which $\sphericalangle A$ is not congruent to $\sphericalangle A'$ (see Exercise 69). Such automorphisms are not of interest geometrically (e.g., Lemma 9.2 below and its consequences are false for them). So *in what follows, whenever we speak of automorphisms of a Hilbert plane, we will assume that we are working in an Archimedean plane.*

COROLLARY 1. If $\triangle A'B'C'$ is the image of $\triangle ABC$ under an automorphism, then $\triangle A'B'C'$ is similar to $\triangle ABC$.

PROOF:

Corresponding angles are congruent. ◄

COROLLARY 2. Every automorphism of an Archimedean plane satisfying the acute angle hypothesis is a motion.

PROOF:

Proposition 6.2 says $\triangle ABC \cong \triangle A'B'C'$, hence $\overline{AB} = \overline{A'B'}$. ◄

Also, as we have already observed using the Klein model (Exercise K-21, Chapter 7), every collineation of a hyperbolic plane onto itself is a motion.

Because of Corollary 1, an automorphism of a Euclidean plane is called a *similarity;* by definition, it is a collineation that preserves angle measure. An example of a similarity that is not a motion is the *dilation*[6] with center O and ratio $k \neq 0$: If $k > 0$ (respectively, $k < 0$), this transformation T fixes O and maps any other point P onto the unique point P' on ray $\overrightarrow{OP}$ (respectively, on the ray opposite to $\overrightarrow{OP}$) such that

$$\overline{OP'} = |k|\overline{OP}.$$

If we introduce Cartesian coordinates with origin O, this transformation is represented by

$$(x, y) \rightarrow (kx, ky).$$

Hence, if A, B have coordinates (a_1, a_2), (b_1, b_2), we have

$$(\overline{A'B'})^2 = (ka_1 - kb_1)^2 + (ka_2 - kb_2)^2 = k^2(\overline{AB})^2;$$

i.e., $\overline{A'B'} = |k|\overline{AB}$. From this you can show that T preserves betweenness and congruence (Exercise 4), which is the "if" part of the next proposition.

PROPOSITION 9.3. A transformation T of a Euclidean plane is a similarity if and only if there is a positive constant k such that

$$\overline{A'B'} = k\overline{AB}$$

for all segments AB.

PROOF:

Given similarity T and segment AB, choose any point C not collinear with A, B and consider $\triangle A'B'C'$ similar to $\triangle ABC$. By the fundamental theorem on similar triangles (Exercise 10, Chapter 5), there is a positive constant k such that the ratio of corresponding sides of these triangles is equal to k. If D is any other point, the same argument applied to $\triangle ACD$ (or $\triangle BCD$ if A, C, D are collinear) gives $\overline{C'D'} = k\overline{CD}$. And if D, E lie on $\overleftrightarrow{AB}$, the same argument applied to $\triangle CDE$ gives $\overline{D'E'} = k\overline{DE}$. Thus k is the proportionality constant for all segments. ◄

The proof just given, together with Exercise 4, shows the following.

---

[6] This notion of dilation is more general than the one on p. 321, where only the case $k > 0$ was considered. A dilation is also called a *homothety* or *central similarity*. Note that similarities are characterized as collineations that preserve circles (see Theorem 9.3, p. 447).

COROLLARY. A one-to-one transformation T of a Euclidean plane onto itself is an automorphism if and only if for every triangle $\triangle ABC$, we have $\triangle ABC \sim \triangle A'B'C'$.

We can conclude from these results that hyperbolic planes have *invariant* distance functions $AB \to \overline{AB}$, whereas Euclidean planes do not. According to Klein's viewpoint, any function or relation that is not invariant under the group of automorphisms of a structure is not an intrinsic part of the theory of that structure; it is only part of the theory of the new structure described by those transformations that do leave it invariant. So if we want distance to be a part of Euclidean geometry, we would have to redefine Euclidean geometry as the study of invariants of the group of Euclidean *motions* only. Klein suggested the name *parabolic geometry* for the study of invariants of the full group of Euclidean similarities. We will not adopt this terminology.

NOTE. The existence of similarities which are not motions proves that *there cannot be any "absolute" (geometrically defined) unit of length in Euclidean geometry* because any possible such length would have to be invariant under all automorphisms. This is one difference with hyperbolic geometry that intrigued Lambert and Gauss.

Reflections

The most fundamental type of motion from which we will generate all others is the *reflection* R_m across line m, its *axis* (see Major Exercise 2, Chapter 3). We will denote the image of a point A under R_m by A^m. Reflecting across m twice sends every point back where it came from, so $R_m R_m = I$ or $R_m = (R_m)^{-1}$. A transformation that is equal to its own inverse and that is not the identity is called an *involution*. The 180° rotation about a point is another example of an involution. (You will show in Exercise 9 that there are no other involutions.)

A *fixed point* of a transformation T is a point A such that $A' = A$. The fixed points of a reflection R_m are the points lying on m. We will use fixed points to classify motions.

LEMMA 9.1. If an automorphism T fixes two points A, B, then it is a motion and it fixes every point on line $\overleftrightarrow{AB}$.

PROOF:

Since $\overline{AB} = \overline{A'B'}$, the constant k in Proposition 9.3 in the Euclidean case is equal to 1. Let C be a third point on $\overleftrightarrow{AB}$. Consider the case A * B * C (the other two cases are treated similarly). Then A * B * C' and $\overline{AC} = \overline{AC'}$. By Axiom C-1, C = C'. ◄

LEMMA 9.2. If an automorphism fixes three noncollinear points, then it is the identity.

PROOF:

If A, B, C are fixed, then by Lemma 9.1 so is every point on the lines joining these three points. If D is not on those three lines, choose any E between A and B. By Pasch's theorem, line $\overleftrightarrow{DE}$ meets another side of △ABC in a point F. Since E and F are fixed, Lemma 9.1 tells us D is fixed. ◄

PROPOSITION 9.4. If an automorphism fixes two points A, B and is not the identity, then it is the reflection across line $\overleftrightarrow{AB}$.

PROOF:

Lemma 9.1 ensures that every point of $\overleftrightarrow{AB}$ is fixed. Let C be any point off AB and let F be the foot of the perpendicular from C to $\overleftrightarrow{AB}$. Since automorphisms preserve angle measure, they preserve perpendicularity, so C' must lie on $\overleftrightarrow{CF}$. Lemma 9.2 ensures that C' ≠ C, and since $\overline{CF} = \overline{C'F}$, C' is the reflection of C across $\overleftrightarrow{AB}$. ◄

The next result shows that "motion" is the precise concept that justifies Euclid's idea of superimposing one triangle on another.

PROPOSITION 9.5. △ABC ≅ △A'B'C' if and only if there is a motion sending A, B, C, respectively, onto A', B', C' and that motion is unique.

PROOF:

Uniqueness follows from Lemma 9.2, for if T and T' had the same effect on A, B, C, then $T^{-1}T'$ would fix these points, hence $T^{-1}T' = I$ and $T = T'$. It's clear that a motion maps △ABC onto a congruent triangle (SSS). So we will assume conversely that △ABC ≅ △A'B'C' and construct the motion. We may assume, say, A ≠ A' and let t be the perpendicular bisector of AA'. Then reflection across t sends A to A' and B, C to points B^t, C^t. If the latter are B', C', we're done, so assume that B' ≠ B^t. We have

$$A'B' \cong AB \cong A'B^t.$$

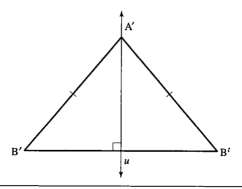

Figure 9.9

Let u be the perpendicular bisector of $B'B^t$, so that R_u sends B^t to B′ (Figure 9.9). This reflection fixes A′ because if A′, B^t, B′ are collinear, A′ is the midpoint of $B'B^t$ and lies on u, whereas if they are not collinear, u is the perpendicular bisector of the base of isosceles triangle $\triangle B'A'B^t$ and u passes through the vertex A′.

Thus, the composite R_uR_t sends the pair (A, B) to the pair (A′, B′). If it also sends C to C′, we're done; otherwise let C″ be its effect on C. Then

$$A'C' \cong AC \cong A'C''$$

$$B'C' \cong BC \cong B'C'',$$

so that $\triangle A'B'C' \cong \triangle A'B'C''$. An easy argument with congruent triangles (see Figure 9.10) shows that C′ is the reflection of C″ across $v = \overleftrightarrow{A'B'}$. Thus $R_vR_uR_t$ is the motion we seek. ◄

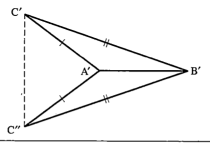

Figure 9.10

COROLLARY. Every motion is a product of at most three reflections.

This was shown in the course of the proof (where we consider the identity a "product of zero reflections" and a reflection a "product of one reflection").

We are next going to examine products of two reflections $T = R_l R_m$. If l meets m at a point A, T is called a *rotation* about A. If l and m have a common perpendicular t, T is called a *translation along t*. Finally, in the hyperbolic plane only, if l and m are asymptotically parallel in the direction of an ideal point Ω, T is called a *parallel displacement* or *limit rotation* about Ω. These cases are mutually exclusive, but by convention the identity motion will be considered to be a rotation, translation, and parallel displacement (this is the case $l = m$).

Proposition 9.4 showed the importance of fixed points in describing motions. Another important tool is invariant lines: We say that line l is *invariant* under T if $l' = l$. This does not imply that all the points on l are fixed; it only implies that if a point on l is moved by T, it is moved to another point on l. For example, the only lines besides m that are invariant under a reflection R_m are the lines perpendicular to m (Exercise 7).

Rotations

PROPOSITION 9.6. Let $l \perp m$, let A be the point of intersection of l and m, and let $T = R_l R_m$. Then for any point B $\neq$ A, A is the midpoint of BB'.

PROOF:

The assertion is clear if B lies on either l or m, so assume it does not (Figure 9.11).

Let C be the foot of the perpendicular from B to m. Then B' is on the opposite side of both l and m from B, and C' is on the opposite side of l from C. From the congruence $\angle BAC \cong \angle B'AC'$ we deduce that these must be vertical angles, hence A, B, B' are collinear. Since AB $\cong$ AB', A is the midpoint. ◄

The motion T in Proposition 9.6 can be described as the 180° rotation about A; we will call it the *half-turn* about A and denote it H_A. The image of a point P under H_A will be denoted P^A.

COROLLARY. H_A is an involution, and its invariant lines are the lines through A.

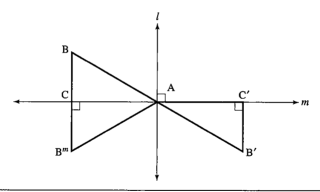

Figure 9.11

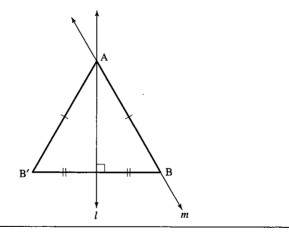

Figure 9.12

PROPOSITION 9.7. A motion $T \neq I$ is a rotation if and only if T has exactly one fixed point.

PROOF:

Suppose T has only one fixed point, A, and choose $B \neq A$. Let l be the perpendicular bisector of BB'. Since $AB \cong AB'$, A lies on l, and the motion $R_l T$ fixes both A and B. If $R_l T = I$, then $T = R_l$, which contradicts the hypothesis that T has only one fixed point. Hence if $m = \overleftrightarrow{AB}$, Proposition 9.4 implies $R_l T = R_m$, so that $T = R_l R_m$ and T is a rotation about A (see Figure 9.12).

Conversely, given rotation $T = R_l R_m$ about A, assume on the contrary that point $B \neq A$ is fixed. Then $B^l = B^m$, so that joining this point to B gives a line perpendicular to both l and m, which is impossible. ◄

NOTE ON ELLIPTIC GEOMETRY. This last argument breaks down in an elliptic plane because there it is possible for intersecting lines to have a common perpendicular. In fact, each point P has a line l called its *polar* such that l is perpendicular to every line through P (see Figure 9.13).

In the elliptic plane, the half-turn H_p about P is the same as the reflection R_l across l. (Lemmas 9.1 and 9.2 are also false in the elliptic plane.) It can be shown that rotations are the only motions of an elliptic plane (see Ewald, 1971, p. 50). In the sphere model, with antipodal points identified, the motions are represented by Euclidean rotations about lines through the center of the sphere (Artzy, 1965, p. 181).

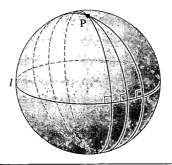

Figure 9.13

If you reread the first part of the proof of Proposition 9.7 and refer to Figure 9.12, you will see that we have also proved the following proposition, which is the first case of the fundamental theorem on three reflections (p. 428).

PROPOSITION 9.8. If T is a rotation about A and m is any line through A, then there is a unique line l through A such that $T = R_l R_m$. If l is not perpendicular to m, then for any point B ≠ A,

$$(\sphericalangle BAB')° = 2d°,$$

where d is the number of degrees in the acute angle made by intersecting lines l and m. (So, for example, to express a 90° rotation about A in the form $R_l R_m$, you must choose lines l and m through A that make a 45° angle.)

WARNING The rotation $R_l R_m$ is not the same as the rotation $R_m R_l$ unless $l \perp m$. Intuitively, one of these rotations is the "clockwise rotation" through $2d°$ about A, while the other is the "counterclockwise rotation" through $2d°$. For a more rigorous argument, note that

$$(R_m R_l)(R_l R_m) = R_m (R_l^2) R_m = R_m I R_m = R_m^2 = I,$$

so that $R_m R_l$ is the inverse of $R_l R_m$. In Exercise 9, you will show that the only rotation equal to its inverse is a half-turn.

PROPOSITION 9.9. Given a point A, the set of rotations about A is a commutative group.

> *PROOF:*
>
> The identity is a rotation about A by definition, and we have just shown that the inverse of a rotation about A is a rotation about A. We must show that the product TT' of rotations about A is a rotation about A. Let $T' = R_l R_m$. By Proposition 9.8, there is a unique line k through A such that $T = R_k R_l$. Then
>
> $$TT' = (R_k R_l)(R_l R_m) = R_k (R_l^2) R_m = R_k I R_m = R_k R_m,$$
>
> which is a rotation about A. To prove commutativity, apply Proposition 9.8 again to get a unique line n such that $T^{-1} = R_n R_m$. Then $T = R_m R_n$ and $T'T = (R_l R_m)(R_m R_n) = R_l (R_m^2) R_n = R_l R_n$. Since $TT' = R_k R_m$, $R_k(TT')R_m = R_k^2 R_m^2 = I$. But we also have $R_k(T'T)R_m = R_k(R_l R_n)R_m = (R_k R_l)(R_n R_m) = TT^{-1} = I$. Hence $TT' = T'T$ by canceling on the right and the left. ◄

WARNING Rotations about different points never commute (unless at least one rotation is the identity). For if T is a rotation about A and T' is a rotation about B, $T'T$ sends A to A'', whereas TT' sends A to $(A'')'$. Furthermore, the product of such rotations may or may not be a rotation (Exercise 10).

Translations

We turn next to translations $T = R_l R_m$, where l and m have a common perpendicular t. The geometric properties of translations in hyperbolic planes are different from those in Euclidean planes (unlike rotations, which behave the same in both geometries).

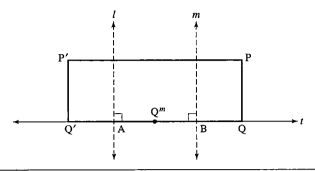

Figure 9.14

PROPOSITION 9.10. Let $l \perp t$ at A, $m \perp t$ at B, $T = R_l R_m$. If Q lies on t, then $\overline{QQ'} = 2(\overline{AB})$. If P does not lie on t, then P' lies on the same side of t as P, and $\overline{PP'} = 2(\overline{AB})$ if the plane is Euclidean, $\overline{PP'} > 2(\overline{AB})$ if the plane is hyperbolic.

PROOF:

(See Figure 9.14.) We will prove the assertion about $\overline{QQ'}$ when $A * B * Q$, leaving the other cases as an exercise. If $\overline{BQ} < \overline{AB}$, then $A * Q^m * B$ and

$$\overline{QQ'} = \overline{Q'A} + \overline{AB} + \overline{BQ}$$
$$= \overline{Q^mA} + \overline{AB} + \overline{BQ^m}$$
$$= 2\overline{AB}.$$

If $\overline{BQ} = \overline{AB}$, then $Q' = A$ and $\overline{QQ'} = 2(\overline{AB})$. If $\overline{BQ} > \overline{AB}$, then we have $Q^m * A * B$ and $\overline{Q^mA} = \overline{Q^mB} - \overline{AB} = \overline{BQ} - \overline{AB}$. Hence we have $Q^m * A * Q' * Q$, which results in

$$2\overline{BQ} = \overline{QQ^m} = \overline{QQ'} + 2\overline{Q'A}$$
$$= \overline{QQ'} + 2(\overline{BQ} - \overline{AB}),$$

which gives $\overline{QQ'} = 2\overline{AB}$.

If P does not lie on t, then P and P^m lie on a line perpendicular to m, hence parallel to t, and thus are on the same side of t; similarly, P^m and P' are on the same side of t; so, by Axiom B-4, P and P' are on the same side of t. Let Q be the foot of the perpendicular from P to t. Since T preserves perpendicularity, Q' is the foot of the perpendicular from P' to t, and since T is a motion, $P'Q' \cong PQ$. Thus, $\square PQQ'P'$ is a Saccheri quadrilateral. In Euclidean geometry, it's a rectangle and its opposite sides are congruent, so $\overline{PP'} = \overline{QQ'} = 2(\overline{AB})$; in hyperbolic geometry, the summit is larger than the base (Theorem 6.1, Chapter 6), so we have $\overline{PP'} > \overline{QQ'} = 2(\overline{AB})$. ◄

COROLLARY. If a translation has a fixed point, then it is the identity motion.

PROPOSITION 9.11. If T is a translation along t and m is any line perpendicular to t, then there is a unique line $l \perp t$ such that $T = R_l R_m$.

PROOF:

Let m cut t at Q and let l be the perpendicular bisector of QQ'. Then $R_l T$ fixes Q. Let P be any other point on m, so that as before, $\square PQQ'P'$ is a Saccheri quadrilateral. Since l is perpendicular to the base QQ' at its midpoint, l is also perpendicular to the summit PP' at its midpoint (Proposition 4.12(b)), so that P is the reflection of P' across l. Thus $R_l T$ fixes every point on m, whence $R_l T = R_m$, and

$$T = (R_l)^2 T = R_l R_m.$$

As for uniqueness, if $T = R_k R_m$, then

$$R_l = T R_m = R_k$$

and $l = k$. ◄

PROPOSITION 9.12. Given a line t, the set of translations along t is a commutative group.

The proof is the same as the proof of Proposition 9.9, using Proposition 9.11 in place of Proposition 9.8. ◄

PROPOSITION 9.13. Let $T \neq I$ be a translation along t. If the plane is Euclidean, the invariant lines of T are t and all lines parallel to t. If the plane is hyperbolic, t is the only invariant line.

PROOF:

It's clear that t is invariant. In the Euclidean case, if $u \parallel t$, then T is also a translation along u (Proposition 4.9), so u is invariant. In both cases, if u meets t at A, then A' lies on t and A' $\neq$ A, so A' does not lie on u and u is not invariant. Suppose that in the hyperbolic case u is invariant and parallel to t. Choose any P on u; then $u = \overleftrightarrow{PP'}$. But we have already seen that P and P' are equidistant from t, whence u and t have a common perpendicular m (Proposition 6.4). We've shown that m is not invariant, and since T preserves perpendicularity, m' is also perpendicular to $t = t'$ and $u = u'$. This contradicts the uniqueness of the common perpendicular in hyperbolic geometry (Proposition 6.5). ◄

PROPOSITION 9.14. Given a motion T, a line t, and a point B on t. Then T is a translation along t if and only if there is a unique point A on t such that T is the product of half-turns $H_A H_B$.

PROOF:

Let m be the perpendicular to t through B. If T is a translation along t, then by Proposition 9.11 there is a unique line $l \perp t$ such that $T = R_l R_m$. If l meets t at A, then $H_A H_B = (R_l R_t)(R_t R_m) = R_l(R_t^2)R_m = R_l R_m = T$. Reverse the argument to obtain the converse. ◄

Half-Turns

Having shown that the product of two half-turns is a translation, we now naturally ask: What is the product of three half-turns? Once again, the answer depends on whether the geometry is Euclidean or hyperbolic.

PROPOSITION 9.15. In a Euclidean plane, the product $H_A H_B H_C$ of three half-turns is a half-turn. In a hyperbolic plane, the product is a half-turn only when A, B, C are collinear; if they are not, the product can be a rotation, a translation, or a parallel displacement.

PROOF:

Suppose that A, B, C are collinear, lying on t, and that l, m, n are the respective perpendiculars to t through these points. Then

$$H_A H_B H_C = (R_l R_t)(R_t R_m)(R_n R_t)$$

$$= R_l(R_t^2)R_m R_n R_t$$

$$= (R_l R_m R_n)R_t$$

$$= R_k R_t,$$

where the line $k \perp t$ such that $R_k R_n = R_l R_m$ is furnished by Proposition 9.11. If k meets t at D, we have shown $H_A H_B H_C = H_D$.

Suppose that A, B, C are not collinear, that $t = \overleftrightarrow{AB}$, that $l \perp t$ at A, and that $m \perp t$ at B. We may assume C lies on m (otherwise replace B by the foot of the perpendicular from C to t and replace A by the point furnished by Proposition 9.14). Let u be the perpendicular to m through C (Figure 9.15). Then

$$H_A H_B H_C = (R_l R_t)(R_t R_m)(R_m R_u) = R_l(R_t^2)(R_m^2)R_u = R_l R_u.$$

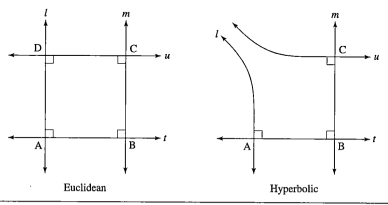

Euclidean   Hyperbolic

Figure 9.15

In the Euclidean case, l meets u at a point D and $l \perp u$ (Propositions 4.7 and 4.9), so $H_A H_B H_C = H_D$. In the hyperbolic case, l and u may meet, be divergently parallel, or be asymptotically parallel; if they do meet at point D, then $H_A H_B H_C$ is a rotation about D, but it is not a half-turn because $\sphericalangle$D is the fourth angle of a Lambert quadrilateral. ◄

COROLLARY. In a Euclidean plane, the product of two translations along different lines is again a translation, and the set of all translations along all lines is a commutative group (proof left for Exercise 13).

NOTE. In the hyperbolic case, Proposition 9.15 can be strengthened. Suppose A, B, C are not collinear. Then $H_A H_B H_C$ is (1) a rotation that is not a half-turn iff A, B, C lie on a circle, (2) a translation iff A, B, C lie on an equidistant curve, or (3) a parallel displacement iff A, B, C lie on a horocycle. Another way of expressing these three possibilities is that the perpendicular bisectors of $\triangle$ABC (1) are concurrent, (2) have a common perpendicular, or (3) are asymptotically parallel in the same direction (see Exercise 70).

Ideal Points in the Hyperbolic Plane

We next study the effect of motions in the hyperbolic plane on ideal points. An *ideal point* Ω is by definition an equivalence class of rays, where rays are in the same class if one is contained in the other or if they are limiting parallel to each other (Major Exercises 2 and 3 of Chapter 6 ensure that this situation does define an equivalence relation).

Now limiting parallelism is defined in terms of incidence and betweenness, hence motions preserve the relation of limiting parallelism. Thus it makes sense to choose any ray r from the class Ω, consider its image r' under T, and define the image Ω' to be the class of r'. If we have $\Omega = \Omega'$ (which means that either r' is limiting parallel to r or one ray contains the other), we say Ω is an *ideal fixed point* of T.

Given a line t containing a ray r, the class of r and the class of the opposite ray are called the two *ends* of t and are said to *lie on* t. Two ideal points Ω, Σ lie on a unique line $\Omega\Sigma$ (namely, if rays $r \in \Omega$, $s \in \Sigma$ emanate from the same point and are not opposite, then $\Omega\Sigma$ is the line of enclosure of the angle formed by r and s—see Major Exercise 8, Chapter 6). We say that Ω and Σ are *on the same side* of line t if neither of them is an end of t and if line $\Omega\Sigma$ is parallel to t. This defines a transitive relation on the set of ideal points off t.

PROPOSITION 9.16

    (a) The ends of m are the only ideal fixed points of the reflection R_m and any translation $(\neq I)$ along m.

    (b) If a rotation has an ideal fixed point, then it is the identity.

    (c) If $(\Omega, \Sigma, \Lambda)$ and $(\Omega', \Sigma', \Lambda')$ are any triples of ideal points, then there is a unique motion sending one triple onto the other.

PROOF:

    (a) It is clear that R_m and any translation $T \neq I$ along m fix the ends Σ, Ω of m. If any other ideal point Λ were fixed, then the line $\Sigma\Lambda$ would be invariant; but T has no invariant lines other than m (Proposition 9.13), and the only other invariant lines of R_m are the perpendiculars to m, whose ends are interchanged by R_m.

    (b) If a rotation about A fixes Ω, then ray $A\Omega$ will be invariant; but Propositions 9.6 and 9.8 imply that only the identity rotation has an invariant ray emanating from A.

    (c) There is a unique point B on $\Sigma\Omega$ such that $\sphericalangle\Lambda B\Omega$ is a right angle (Major Exercise 10, Chapter 6). Let B' be the point on $\Sigma'\Omega'$

such that $\sphericalangle\Lambda'B'\Omega'$ is a right angle. Let A be any point $\neq B$ on $B\Lambda$ and let C be any point $\neq B$ on $B\Omega$. By Axiom C-1, there are unique points A' on $B'\Lambda'$, and C' on $B'\Omega'$, such that

$$AB \cong A'B' \quad \text{and} \quad CB \cong C'B'.$$

Then $\triangle ABC \cong \triangle A'B'C'$ (SAS), so by Proposition 9.5, a unique motion T effects this congruence. Clearly T sends $(\Omega, \Sigma, \Lambda)$ to $(\Omega', \Sigma', \Lambda')$. Conversely, any such motion must send (A, B, C) onto (A', B', C'), so by Proposition 9.5, T is unique. ◄

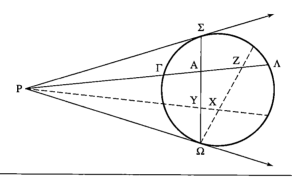

Figure 9.16

NOTE. Part (c) of this proposition can be visualized nicely in terms of the Klein model. Ideal points are represented by points on the absolute (the unit circle). There is a unique motion T mapping any triple of points $(\Sigma, \Omega, \Lambda)$ on the absolute onto any other. The effect of T on the other points can be described as follows (see Figure 9.16).

If P is the pole of chord $\Omega\Sigma$, then line $P\Lambda$ is Klein-perpendicular to $\Omega\Sigma$ at some point A. Then the image A' of A must be the intersection of $\Omega'\Sigma'$ with $\Lambda'P'$, where P' is the pole of $\Omega'\Sigma'$. Take any other point B on $\Omega\Sigma$, say, $\Omega * B * A$. By Theorem 7.4, the image B' of B is the unique point between Ω' and A' such that cross-ratios are preserved:[7]

$$(AB, \Omega\Sigma) = (A'B', \Omega'\Sigma').$$

---

[7] The mapping of $\Omega\Sigma$ onto $\Omega'\Sigma'$ given by this equality of cross-ratios is called a *projectivity*. It can be described more geometrically by a sequence of at most three perspectivities (see Figure 7.44); this is essentially the "fundamental theorem of projective geometry" (Ewald, 1971, Theorem 5.9.5, p. 226).

Let Γ be the other intersection of $P\Lambda$ with the absolute—its image Γ' is the other intersection of $P'\Lambda'$ with the absolute. As previously, we can use cross-ratios to determine the image of any point on $\Gamma\Lambda$. Finally, given any other point X (ideal or ordinary), represent it as the intersection of two lines, one being XP, which cuts $\Omega\Sigma$ at some point Y, and the other being $X\Omega$ (or $X\Sigma$), which cuts $\Gamma\Lambda$ at some point Z. Then X' is the intersection of P'Y' and $Z'\Omega'$(or $Z'\Sigma'$).

This construction describes the motion T in terms of incidence alone. It suggests the conjecture that *every collineation of the hyperbolic plane is a motion;* this conjecture was demonstrated by Karl Menger and his students.[8] In the Euclidean plane, there are lots of collineations that are not motions or similarities (see Exercise 34 on affine transformations).

Parallel Displacements

We next study parallel displacements about an ideal point Σ.

PROPOSITION 9.17. Given a parallel displacement $T = R_l R_m$, where l and m are asymptotically parallel in the direction of ideal point Σ. Then

(a) T has no ordinary fixed points.
(b) Let k be any line through Σ and A any point on k; then Σ lies on the perpendicular bisector h of AA' and $T = R_h R_k$.
(c) T has no invariant lines.
(d) The only ideal fixed point of T is Σ.
(e) The set of parallel displacements about Σ is a commutative group.
(f) A motion with exactly one ideal fixed point is a parallel displacement.

PROOF:

(a) Assume A is fixed. Then the line joining A to $A^m = A^l$ is perpendicular to both l and m, contradicting the hypothesis.

(b) Σ lies on two perpendicular bisectors l and m of $\triangle AA^m A'$, so by Major Exercise 7, Chapter 6, Σ also lies on the third perpendicular bisector h. Then $R_h T$ fixes A and Σ. By Proposition 9.16(b),

---

[8] See L. Blumenthal and K. Menger (1970, p. 220). See also K. Menger, "The New Foundation of Hyperbolic Geometry," in J. C. Butcher (ed.), *A Spectrum of Mathematics* (Oxford: Auckland and Oxford University Presses, 1971), p. 86. The idea of the proof is given in Exercise K-21, Chapter 7. You can add this result to your list for Exercise 1, Chapter 6.

R_hT cannot be a rotation about A. By Proposition 9.4, it must be a reflection, and by Proposition 9.16(a), it has to be the reflection across the line k joining A to Σ (see Figure 9.17).

(c) Suppose line t were invariant under T. Choose any point A lying on t and let h, k be as in part (b). Then $h \perp T = \overleftrightarrow{AA'}$, so t is invariant under R_h too. Hence t is invariant under $R_k = R_hT$, which means either $t \perp k$ or $t = k$. But the asymptotically parallel lines h and k cannot have a common perpendicular or be perpendicular to each other.

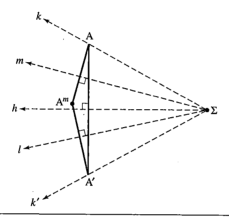

Figure 9.17

(d) If T had another ideal fixed point Ω, then line $\Sigma\Omega$ would be invariant, contradicting part (c).

(e) The proof is the same as the proof of Proposition 9.9, using part (b) instead of Proposition 9.8.

(f) This follows from the classification of motions (Theorem 9.1, later in this chapter), and is inserted here for convenience. ◄

DEFINITION. If A is ordinary and Σ is ideal, the set of all points TA as T runs through the group of parallel displacements about Σ is the *horocycle* through A centered at Σ. (This is the analogue of the circle through A centered at an ordinary point O, which consists of all points TA as T runs through the group of rotations about O.)

One circle can be mapped onto another by a motion iff the two circles have the same radius. We can say that the congruence class of a circle is determined by its radius, where in general two figures are called *congruent* if there is a motion mapping one onto the other. In that sense, *all horocycles are congruent to one another* (Exercise 44).

Glides

We come now to our final type of motion, a *glide* along a line t, defined as a product $T' = R_t T$ of a non-identity translation T along t followed by reflection across t. (If you walk straight through the snow, your consecutive footprints are related by a glide.)

PROPOSITION 9.18. (See Figure 9.18.) Given $l \perp t$ at A, $m \perp t$ at B, $T = R_l R_m$, $T' = R_t T$. Then

- (a) $TR_t = T'$.
- (b) $H_A R_m = T' = R_l H_B$.
- (c) T' maps each side of t onto the opposite side.
- (d) T' has no fixed points.
- (e) The only invariant line of T' is t.
- (f) Conversely, given point B and line l, let t be the perpendicular to l through B. Then $R_l H_B$ is a glide along t if B does not lie on l, and is R_t if B does lie on l.

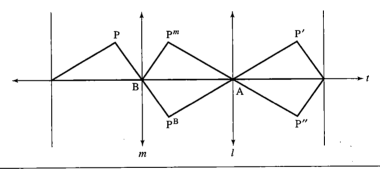

Figure 9.18

PROOF:

Parts (a) and (b) follow from the formulas $H_A = R_t R_l = R_l R_t$ and $H_B = R_t R_m = R_m R_t$. Part (c) is clear, and part (d) follows from it. Part (e) follows from parts (c) and (d). As for part (f), if B lies on l, then $H_B = R_l R_t$, so $R_l H_B = R_l(R_l R_t) = (R_l^2)R_t = R_t$. If B does not lie on l, let $m \neq l$ be the perpendicular to t through B; then we have $T = R_l R_m \neq I$ and $R_l H_B = TR_t$. ◀

NOTE. Glides are characterized in Euclidean geometry by having only one invariant line. In hyperbolic geometry, this characteristic does not distinguish them from translations, so we must add the condition that the two sides of the invariant line are interchanged. The invariant line is called the *axis* of the glide.

HJELMSLEV'S LEMMA. Let G be a glide, l a line not invariant under G, and l' the image of l under G. As point P varies on l and its image P' varies on l', the midpoints of the segments PP' all lie on the axis t of G. Furthermore, those midpoints are all distinct, except in the case of $G = H_M R_l$, where the midpoints all coincide with M; that case occurs if and only if the axis t of G is perpendicular to both l and l'.

The proof will be left for Exercise 21. The lemma gives a method of locating the axis of a glide.

Classification of Motions

Our next objective is to show that every motion is a reflection, a rotation, a translation, a parallel displacement, or a glide. The first step is to describe products of three reflections. Toward that end, we introduce three types of *pencils of lines:*

1. The pencil of all lines through a given point P.
2. The pencil of all lines perpendicular to a given line t.
3. The pencil of all lines through a given ideal point Σ (hyperbolic plane only).

Clearly, two lines l and m determine a unique pencil (if l and m are divergently parallel in the hyperbolic plane, they have a common perpendicular t by Theorem 6.3, Chapter 6). Moreover, if A is any point, there is a line n in that pencil through A. For the three types of pencils, n is:

1. The line $\overleftrightarrow{AP}$ if $A \neq P$.
2. The perpendicular to t through A.
3. The line containing $A\Sigma$.

▩▩ **THE THEOREM ON THREE REFLECTIONS** The first part of the next proposition is called the "theorem on three reflections." F. Bachmann takes it as an axiom for his development of geometry without continuity or betweenness axioms.[9]

---

[9] Bachmann introduces ultra-ideal and ideal points as pencils of the second and third types; using a technique developed by Danish geometer J. Hjelmslev, he is able to prove that the plane so extended is a projective plane coordinatized by a commutative field. See Appendix B.

PROPOSITION 9.19. Let $T = R_l R_m R_n$. (1) If l, m, and n belong to a pencil, then T is a reflection in a unique line of that pencil. (2) If l, m, and n do not belong to a pencil, then T is a glide.

This proposition is wonderful. It leads to the complete classification of motions. You may wonder how the axis of glide T is related to the three given lines; Exercises 55–57 answer this question in the Euclidean case.

PROOF:

Part (1) of Proposition 9.19 follows from Propositions 9.8, 9.11, and 9.17(b), so assume the lines do not belong to a pencil. Choose any point A on l. Let m' be the line through A belonging to the pencil determined by m and n (Figure 9.19). Then line n' exists such that

$$R_{m'} R_m R_n = R_{n'}.$$

Let B be the foot of the perpendicular k from A to n'. Since l, m', and k pass through A, line h exists such that

$$R_l R_{m'} R_k = R_h.$$

Then B does not lie on h (by assumption on l, m, n), so by Proposition 9.18(f), $R_h H_B$ is a glide along the perpendicular to h through B. But

$$R_h H_B = R_h(R_k R_{n'}) = R_l R_{m'} R_k R_k R_{m'} R_m R_n = T. \blacktriangleleft$$

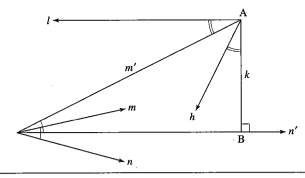

Figure 9.19

COROLLARY. Every product $R_l R_m H_A$ equals a product $R_h R_k$.

PROOF:

Let n be a line through A in the pencil determined by l and m. Let h be the line such that

$$R_l R_m R_n = R_h$$

and let k be the perpendicular to n through A. Then

$$R_l R_m H_A = R_l R_m R_n R_k = R_h R_k. \blacktriangleleft$$

DEFINITION. A motion is called *direct* (or *proper* or *orientation-preserving*) if it is a product of two reflections or else is the identity. It is called *opposite* (or *improper* or *orientation-reversing*) if it is a reflection or a glide.

THEOREM 9.1. Every motion is either direct or opposite and not both. The set of direct motions is a group. The product of two opposite motions is direct, whereas the product of a direct motion and an opposite motion is opposite.

PROOF:

We know that every motion is a product of at most three reflections (corollary to Proposition 9.5), so by Proposition 9.19, every motion is either direct or opposite. *The opposite motions are characterized by having an invariant line whose sides are interchanged.*

Given a product $(R_k R_l)(R_m R_n)$ of direct motions. If l, m, n belong to a pencil, $R_l R_m R_n = R_h$ and the product reduces to $R_k R_h$, which is direct. Otherwise $R_l R_m R_n = R_h H_B$ (Propositions 9.19 and 9.18(b)), and the corollary tells us that $R_k(R_h H_B)$ is direct. It follows that the direct motions form a group.

The product of a reflection and a direct motion is opposite by Proposition 9.19. The product of a glide and a direct motion is a product of five reflections, which reduces to a product of three reflections by the previous paragraph, hence is opposite by Proposition 9.19. Similarly, a product of four to six reflections reduces to a product of two. $\blacktriangleleft$

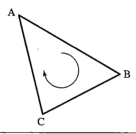

Figure 9.20

NOTE. The intuitive idea behind our classification of motions is that the plane can be given two distinct "orientations," so that, for example, the vertices of $\triangle ABC$ can be ordered in a "clockwise direction" (Figure 9.20). When the triangle is moved by a rotation, translation, or parallel displacement, the orientation of $\triangle A'B'C'$, remains clockwise, whereas under a reflection or a glide, the orientation becomes counterclockwise. (See Exercise 23 for further discussion of orientation.)

NOTE ON ELLIPTIC GEOMETRY. In the elliptic plane, no such invariant exists that is preserved by rotations and reversed by reflections since every reflection is a 180° rotation; there is no distinction between direct and opposite motions in the elliptic plane. The only motions are rotations.

TABLE 9.1

TABLE OF MOTIONS

	Orientation	Fixed points	Invariant lines	Ideal fixed points (hyperbolic plane)
Identity	Direct	All	All	All
Reflection R_l	Opposite	Points on l	l and all $m \perp l$	The two ends of l
Half-turn H_A	Direct	A	All lines through A	None
Rotation about A that is not involutory	Direct	A	None	None
Euclidean translation along t	Direct	None	t and all $u \parallel t$	—
Hyperbolic translation along t	Direct	None	Only t	The two ends of t
Parallel displacement about Ω	Direct	None	None	Ω
Glide along t	Opposite	None	t	The two ends of t

Exercise 71 asks you to construct another column for this table listing invariant cycles.

Automorphisms of the Cartesian Model

Our next objective is to rapidly describe the groups of motions explicitly in terms of coordinates in models of our geometries. We begin with the Cartesian model of the Euclidean plane, and we assume in both this section and the next that the reader has some familiarity with vectors, matrices, and complex numbers.

The easiest transformations to describe are *translations*. As the proof of Proposition 9.10 showed, a translation moves each point a fixed distance and in a fixed direction (in Figure 9.8, it moves the distance $2\overline{AB}$ in the direction $\overrightarrow{BA}$). This can be represented by a vector of length $2\overline{AB}$ emanating from the origin of our coordinate system and pointing in the given direction. If the endpoint coordinates of this vector are (e, f), then, by the definition of vector addition, the translation is given by

$$T(x, y) = (x, y) + (e, f) = (x + e, y + f)$$

(Figure 9.21) or

$$x' = x + e$$
$$y' = y + f.$$

If we apply a second translation T' corresponding to the vector with endpoint (e', f'), then the image (x'', y'') of (x, y) under $T'T$ is given by

$$(x'', y'') = (x', y') + (e', f') = (x + e + e', y + f + f').$$

Thus $T'T$ is the translation by the sum of the vectors determining T and T'. This proves the next proposition.

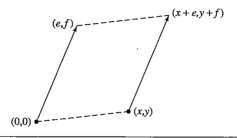

Figure 9.21

PROPOSITION 9.20. In the Cartesian model of the Euclidean plane, the translations form a commutative group isomorphic to the group of vectors under addition.

(According to our general definition of isomorphic models, two groups are *isomorphic* if there is a one-to-one correspondence between them and that correspondence preserves the group laws; here the two groups are considered models of the system of axioms 1 through 4 for a group at the beginning of this chapter.)

We say that the translations form a *two-parameter group* since they depend on two real variables (e, f).[10]

PROPOSITION 9.21. In the Cartesian model of the Euclidean plane, the translations along a fixed line form a one-parameter group isomorphic to the group of real numbers under addition.

PROOF:

Let (e_0, f_0) be a unit vector parallel to the fixed line l (Figure 9.22). Then the vector corresponding to a translation T along l has the form $t(e_0, f_0) = (te_0, tf_0)$, where $|t|$ is the distance translated and t is positive or negative according to whether the direction of translation is the same as (e_0, f_0) or opposite. If T' corresponds to vector $t'(e_0, f_0)$, then $T'T$ corresponds to vector $t(e_0, f_0) + t'(e_0, f_0) =$

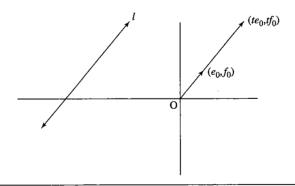

Figure 9.22

---

[10] The theory of groups of transformations that depend continuously on real parameters was first developed by the great Norwegian mathematician Sophus Lie in the late nineteenth century and has become one of the most fruitful ideas in twentieth-century mathematics and physics. (For example, this theory was used to predict the existence of certain subatomic particles—see F. S. Dyson, "Mathematics in the Physical Sciences," *Scientific American*, September 1964.)

$(t + t')(e_0, f_0)$. Thus assigning the parameter t to T gives the isomorphism. ◄

We next discuss rotations about a fixed point A. Our first step is to reduce to the case where A is the origin O: If not, let T be the translation along $\overleftrightarrow{AO}$ taking A to O. Then (by Proposition 9.11)

$$T = R_m R_l = R_{l^*} R_m,$$

where m is the perpendicular bisector of AO and l, l^* are the perpendiculars to $\overleftrightarrow{AO}$ through A, O. The given rotation R about A can be written (by Proposition 9.8) as

$$R = R_l R_k,$$

where k passes through A. Let k^* be the reflection of k across m (see Figure 9.23). Then $R^* = R_{k^*} R_{l^*}$ is a rotation about O and

$$T^{-1} R^* T = (R_l R_m)(R_{k^*} R_{l^*})(R_{l^*} R_m)$$

$$= R_l R_m R_{k^*} (R_{l^*})^2 R_m$$

$$= R_l (R_m R_{k^*} R_m)$$

$$= R_l R_k$$

$$= R.$$

This shows that the rotation R about A is uniquely determined by the rotation R^* about O. Moreover, the mapping $R^* \to T^{-1} R^* T$ is an isomorphism of the group of rotations about O onto the group of rotations about A, as you can easily verify. Thus we may assume A = O.

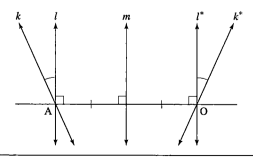

Figure 9.23

By Proposition 9.8, the given rotation about O can be written as $R = R_l R_m$, where m is the x-axis. If $l \perp m$, then R is represented in complex coordinates as

$$z \to -z$$

(Proposition 9.6). Otherwise, if the acute angle from m to l has radian measure $\theta/2$, $0 < \theta/2 < \pi/2$, then R is represented in complex coordinates[11] as

$$z \to e^{i\theta}z \qquad \text{(if } l \text{ has positive slope)}$$
$$z \to e^{-i\theta}z \qquad \text{(if } l \text{ has negative slope)}$$

(see Figure 9.6 and Proposition 9.8). Combining these cases, we see that rotations about O are uniquely represented by the transformations

$$z \to e^{i\theta}z \qquad (-\pi < \theta \leqq \pi).$$

Since $e^{i\phi}(e^{i\theta}z) = (e^{i\phi}e^{i\theta})z$, the product of two rotations about O corresponds to the product $e^{i\phi}e^{i\theta}$ of complex numbers of absolute value 1. This proves the following proposition.

PROPOSITION 9.22. In the Cartesian model of the Euclidean plane, the group of rotations about a fixed point is isomorphic to the one-parameter multiplicative group S^1 of complex numbers $e^{i\theta}$ of absolute value 1 (θ is the real parameter).

Let us combine our results, using complex coordinates. If a point has complex coordinate z, translating it by a vector (e_0, f_0) is the same as adding to z the complex number $z_0 = e_0 + if_0$ since addition of complex numbers is the same as vector addition. Now if T is any *direct* motion and T moves the origin O to the point O' with complex coordinate z_0, follow T by the translation by $-z_0$ to obtain a direct motion fixing O. This motion is a rotation about O by our previous results, hence has the form $z \to e^{i\theta}z$. Therefore our original motion is equal to this rotation followed by translation by z_0. We have proved the following proposition.

PROPOSITION 9.23. The group of direct motions of the Cartesian model of the Euclidean plane is isomorphic to the three-parameter group given in complex coordinates by

$$z \to e^{i\theta}z + z_0.$$

---

[11] Recall that $e^{i\theta} = \cos\theta + i\sin\theta$.

Let us be more explicit on the multiplication law for this group. Let T have complex parameters $(e^{i\theta}, z_0)$ and let T' have complex parameters $(e^{i\theta'}, z_0')$. Then the image of z under $T'T$ is

$$e^{i\theta'}(e^{i\theta}z + z_0) + z_0' = e^{i(\theta+\theta')}z + (e^{i\theta'}z_0 + z_0'),$$

so the complex parameters are $(e^{i(\theta+\theta')}, e^{i\theta'}z_0 + z_0')$. In other words, the rotation parameters multiply, but the translation parameters do not add—there is a "twist" involved in multiplying z_0 by $e^{i\theta'}$. This accounts for the noncommutativity of the group (which technically is a "semidirect product" of S^1 with the additive group $\mathbb{C}$ of complex numbers).

Put another way, let $T = T_1R_1$ and $T' = T_2R_2$, where the T_i are translations and the R_i are rotations about O, $i = 1, 2$. Then

$$T'T = T_2R_2T_1R_1 = T_2R_2T_1(R_2^{-1}R_2)R_1 = T_2(R_2T_1R_2^{-1})(R_2R_1).$$

In this last expression, the factor on the right is the product R_2R_1 of the two rotations about O, and the factor T_2 on the left is the second translation. The middle factor reveals the twist because $R_2T_1R_2^{-1}$ is the translation by $e^{i\theta'}z_0$:

$$R_2T_1R_2^{-1}(z) = R_2T_1(e^{-i\theta'}z)$$

$$= R_2(e^{-i\theta'}z + z_0)$$

$$= e^{i\theta'}(e^{-i\theta'}z + z_0)$$

$$= z + e^{i\theta'}z_0.$$

COROLLARY. Opposite motions of the Cartesian plane have a unique representation in the form

$$z \to e^{i\theta}\bar{z} + z_0.$$

PROOF:

By Theorem 9.1, all opposite motions are obtained by following all the direct motions with one particular opposite motion, which we can choose to be reflection across the x-axis $z \to \bar{z}$. Since $\overline{e^{i\theta}} = e^{-i\theta}$, the complex conjugate of $e^{i\theta}z + z_0$ is $e^{-i\theta}\bar{z} + \bar{z}_0$, and relabeling $-\theta$ for θ, $\bar{z}_0$ for z_0 gives the result. ◄

These results can easily be generalized to similarities.

PROPOSITION 9.24. In the Cartesian model of the Euclidean plane, a similarity is represented in complex coordinates either in the form.

$$z \to w_0 z + z_0 \qquad w_0 \neq 0$$

(in which case it is called *direct*) or in the form

$$z \to w_0 \bar{z} + z_0 \qquad w_0 \neq 0$$

(in which case it is called *opposite*). The direct similarities form a four-parameter group.

Here w_0 ranges through the multiplicative group $\mathbb{C}^*$ of nonzero complex numbers, while z_0 ranges through the additive group $\mathbb{C}$ of all complex numbers; the group of direct similarities is the "semi-direct product" of $\mathbb{C}^*$ with $\mathbb{C}$. The modulus $k = |w_0|$ is the constant of proportionality for the similarity. Geometrically, this representation means that a direct similarity is equal to a dilation centered at the origin followed by a rotation about the origin followed by a translation; an opposite similarity is equal to the reflection in the x-axis followed by a direct similarity. The proof is left for Exercise 24.

Motions in the Poincaré Model

We turn next to the coordinate description of hyperbolic motions, and for this purpose the most convenient representation is the Poincaré upper half-plane model. Recall that the Poincaré lines are either vertical rays emanating from points on the x-axis or semicircles with center on the x-axis. The ideal points are represented in this model by the points on the x-axis and a point at infinity ∞ which is the other end of every vertical ray (Figure 9.24).

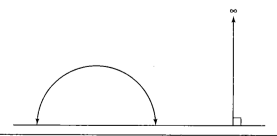

Figure 9.24

We have seen (in the note after Proposition 7.11, p. 326) that hyperbolic reflections are represented in the Poincaré *disk* model either by Euclidean reflections in diameters of the absolute circle γ or by inversions in circles δ orthogonal to γ. Let us show that *in the upper half-plane model, hyperbolic reflections are represented either by Euclidean reflections in the vertical lines or by inversions in circles δ orthogonal to the x-axis*, i.e., circles δ with center on the x-axis.

In Exercise 38, you will show that the mapping

$$E : z \to i\,\frac{i + z}{i - z}$$

sends the unit disk one-to-one onto the upper half-plane, sends i to ∞, and all other points of the unit circle onto the x-axis.

The Poincaré lines of the disk model are mapped onto the Poincaré lines of the upper half-plane model—in fact, all Euclidean circles and lines are mapped into either Euclidean circles or lines by E, and orthogonality is preserved (E is *conformal*); see Figure 9.25. So we can use E as the isomorphism that defines congruence in the upper half-plane interpretation (just as we previously established congruence in the Klein model via the isomorphism F—see Chapter 7).

For simplicity, let us agree to also call the Euclidean reflection in a Euclidean line "inversion"; this will enable us to avoid discussing this special case separately. Figure 9.26 shows a hyperbolic reflection R_l in the disk model represented as inversion. For any point A, drop hyperbolic perpendicular t from A to l and let M be the foot on l of this perpendicular. Let α be the *hyperbolic* circle through A with hyperbolic center M.

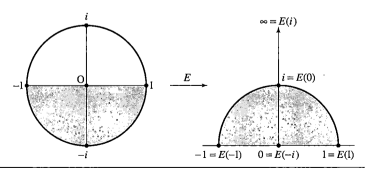

Figure 9.25 Isomorphism E of the disk model to the upper half-plane model.

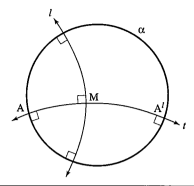

Figure 9.26 Reflection is inversion in the Poincaré models.

Proposition 7.12 showed that α is also a Euclidean circle (with a different Euclidean center). The reflection A^l of A across l is then the other intersection of α with t. Now α is orthogonal to l since l is the extension of a hyperbolic diameter of α. Hence α is mapped onto itself by inversion in l, and so is t (corollary to Proposition 7.6), so A^l must be the inverse of A in l.

If we apply the mapping E, this entire figure is transformed onto an isomorphic figure in the upper half-plane. Thus, the argument just given shows that R_l is also represented in the upper half-plane model by inversion.

We next calculate the formulas for these inversions. For a vertical line $x = k$, the inversion is given by

$$(x, y) \to (2k - x, y).$$

In terms of the single complex coordinate $z = x + iy$, this becomes

$$z \to 2k - \bar{z}.$$

For consistent notation later on, set $b = 2k$ and write this as

$$z \to -\bar{z} + b.$$

For a circle centered at $(k, 0)$ of radius r, make a change of coordinates $x' = x - k$, $y' = y$ (i.e., translate the center to the origin); the complex coordinate change is $z' = z - k$. Then, by the definition of inversion, the image w' of z' is determined by the two equations

$$|z'||w'| = r^2$$

$$\frac{z'}{|z'|} = \frac{w'}{|w'|},$$

whose solution is

$$w' = |w'| \frac{z'}{|z'|} = \frac{r^2}{|z'|} \frac{z'}{|z'|} = \frac{r^2}{\overline{z}'}$$

since $|z'|^2 = z'\overline{z}'$. So in the original coordinate system where we have $w = w' + k$, we get

$$w = \frac{r^2}{\overline{z} - k} + k = \frac{k\overline{z} + r^2 - k^2}{\overline{z} - k}.$$

For convenience, we set $c = 1/r$, $a = kc$, and $b = r(1 - a^2)$. The inversion then takes the form

$$z \to \frac{a\overline{z} + b}{c\overline{z} - a},$$

which includes the previous case when we set $c = 0$ and $a = -1$. We have shown the following.

PROPOSITION 9.25. In the Poincaré upper half-plane model of the hyperbolic plane, reflections are represented in complex coordinates by

$$z \to \frac{a\overline{z} + b}{c\overline{z} - a} \qquad a^2 + bc = 1$$

(where a, b, c are real numbers).

We can next determine the representation of all the *direct* hyperbolic motions since they are products of two reflections. The calculation is simplified by the following general observations.

$\cdot\;$ For any coefficient field K (such as the field $\mathbb{R}$ of all real numbers or the field $\mathbb{C}$ of all complex numbers), we define the *projective line* $\mathscr{P}^1(K)$ over K to be $K \cup \{\infty\}$, where "∞" just means another point not in K. Each point on this "line" will be assigned homogeneous coordinates $[x_1, x_2]$, where $x_1, x_2 \in K$ and are not both zero. These coordinates will be determined only up to multiplication by a nonzero scalar λ; that is,

$$[x_1, x_2] = [\lambda x_1, \lambda x_2] \quad \lambda \neq 0.$$

Specifically, the point $x \in K$ is assigned the homogeneous coordinates

$$[x, 1] = [\lambda x, \lambda] \quad \lambda \neq 0,$$

while the point ∞ is assigned the homogeneous coordinates

$$[1, 0] = [\lambda, 0] \quad \lambda \neq 0.$$

We will operate on the points of $\mathcal{P}^1(K)$ with nonsingular 2×2 matrices with coefficients in K in the usual way that matrices operate on vectors:

$$\begin{bmatrix} a_{11} & a_{12} \\ a_{21} & a_{22} \end{bmatrix}\begin{bmatrix} x_1 \\ x_2 \end{bmatrix} = \begin{bmatrix} a_{11}x_1 + a_{12}x_2 \\ a_{21}x_1 + a_{22}x_2 \end{bmatrix},$$

where the brackets around the matrix again mean that its entries are determined only up to multiplication by a nonzero scalar. These operators are called *projective transformations*, and they form a group under matrix multiplication denoted $PGL(2, K)$. Now a *linear fractional transformation*

$$x \to \frac{ax + b}{cx + d}$$

defined on K can be obtained by operating on the homogeneous coordinates $[x, 1]$ of x with the projective transformation $\begin{bmatrix} a & b \\ c & d \end{bmatrix}$, obtaining $[ax + b, cx + d]$, and then dehomogenizing the coordinates to get $[(ax + b)/(cx + d), 1]$. Viewed thusly it becomes clear that the composite of two such linear fractional transformations can be calculated by multiplying the two matrices, so the composite is again a linear fractional transformation.

Returning to our representation of reflections, we have the added complication of the complex conjugate $\bar{z}$ occurring in the formula of Proposition 9.25; but it is clear that for a product of two reflections, the two conjugations cancel each other out, with the coefficients being unaffected because they are real numbers. Furthermore, the condition $a^2 + bc = 1$ in Proposition 9.25 means that the matrix

$$\begin{bmatrix} a & b \\ c & -a \end{bmatrix}$$

of the transformation has determinant -1. By the formula

$$\det(AB) = (\det A)(\det B)$$

for the determinant of a product of matrices, we see that the product will have determinant $+1$.

We claim that, conversely, every linear fractional transformation with real coefficients and determinant $+1$ is a product of two hyperbolic reflections; i.e., the matrix equation

$$\begin{bmatrix} a' & b' \\ c' & -a' \end{bmatrix}\begin{bmatrix} a & b \\ c & -a \end{bmatrix} = \begin{bmatrix} x & y \\ u & v \end{bmatrix}$$

can be solved for the eight unknowns on the left, given the four real numbers on the right.

CASE 1. $u \neq 0$. Then a solution is $c' = 0$, $a' = -1$, $c = u$, $a = -v$, $b' = (x - v)/u$, $b = vb' - y$.

CASE 2. $u = 0$ and $y = 0$. We may assume $x > 0$. Then a solution is $a = 0 = a'$, $c = \sqrt{x} = b'$, $c^{-1} = b = c'$.

CASE 3. $u = 0$ and $x = v = 1$. Then a solution is $c = 0 = c'$, $a = -1 = a'$, $b' = 0$, and $b = -y$.

CASE 4. $u = 0$. This follows from the preceding cases because

$$\begin{bmatrix} x & y \\ 0 & x^{-1} \end{bmatrix} = \begin{bmatrix} x & 0 \\ 0 & x^{-1} \end{bmatrix} \begin{bmatrix} 1 & y/x \\ 0 & 1 \end{bmatrix}$$

and we know that a product of four reflections reduces to a product of two (Theorem 9.1).

We have proved the next proposition.

PROPOSITION 9.26. In the Poincaré upper half-plane model of the hyperbolic plane, the direct hyperbolic motions are represented by all the linear fractional transformations

$$z \to \frac{az + b}{cz + d} \quad ad - bc = 1$$

$(a, b, c, d, \text{real})$.

This group is denoted $PSL(2, \mathbb{R})$ and is called the *projective special linear group* over the real field. It is a three-parameter group (one of the four parameters being eliminated by the condition that the determinant be $+1$).[12]

We can next obtain all opposite hyperbolic motions by multiplying all the direct motions by one fixed opposite motion. For the latter, let's use the reflection in the y-axis, $z \to -\bar{z}$. The result is

$$z \to \frac{-a\bar{z} + b}{-c\bar{z} + d},$$

which after relabeling gives the next proposition.

---

[12] $SL(2, \mathbb{R})$ is the group of all 2×2 real matrices of determinant $+1$. It is extremely important in analytic number theory—see the book by S. Lang devoted entirely to this group (Reading, Mass.: Addison Wesley, 1975).

PROPOSITION 9.27. In the Poincaré upper half-plane model of the hyperbolic plane, the opposite hyperbolic motions are represented by all the mappings

$$z \to \frac{a\bar{z} + b}{c\bar{z} + d} \quad ad - bc = -1$$

(a, b, c, d real).

We can combine the direct and opposite motions and represent them by all real projective transformations of the real projective line $\mathcal{P}^1(\mathbb{R})$, with the matrices representing direct or opposite motions according to whether the determinant D is positive or negative, since

$$\begin{bmatrix} a & b \\ c & d \end{bmatrix} = \begin{bmatrix} a/\sqrt{|D|} & b/\sqrt{|D|} \\ c/\sqrt{|D|} & d/\sqrt{|D|} \end{bmatrix}$$

and the matrix on the right has determinant ± 1.

COROLLARY (POINCARÉ, 1882). The group of all hyperbolic motions is isomorphic to $PGL(2, \mathbb{R})$.

This isomorphism suggests analogies between one-dimensional real projective geometry and two-dimensional hyperbolic geometry.[13] For example, Proposition 9.16(c) corresponds to the theorem in projective geometry that for any two triples of points on the projective line, there is a unique projective transformation mapping one triple onto the other (see Exercise 65). Also, projective transformations are classified by their fixed points. The equation for a fixed point is

$$x = \frac{ax + b}{cx + d}$$

or

$$cx^2 + (d - a)x - b = 0,$$

showing that the number of finite fixed points is 0, 1, or 2. Since

$$\begin{bmatrix} a & b \\ c & d \end{bmatrix}\begin{bmatrix} 1 \\ 0 \end{bmatrix} = \begin{bmatrix} a \\ c \end{bmatrix},$$

---

[13] Klein's *Erlanger Programme* pointed out many other analogies between geometries whose groups are isomorphic, e.g., the analogies among the *inversive plane* (Exercise P-17, Chapter 7), the one-dimensional complex projective geometry $\mathcal{P}^1(\mathbb{C})$ (which from the real point of view is a geometry on a sphere), and three-dimensional hyperbolic geometry. See the "dictionary" on p. 266 of Coxeter (1998).

we see that ∞ is a fixed point if and only if $c = 0$; in that case, the quadratic equation above becomes linear, and if the transformation is not the identity, it has a finite fixed point only when $a \neq d$.

Our classification of hyperbolic motions (p. 430) showed that the number of ideal fixed points is 0 for rotations, 1 for parallel displacements, and 2 for reflections, translations, and glides.

▦ **EXAMPLE 3.** Let us determine the group of all parallel displacements about ∞. We just showed that these are represented by matrices with $c = 0$ and $a = d$; they form the group of mappings

$$z \to z + b,$$

which is isomorphic to the one-parameter additive group of all real numbers (these are *Euclidean* translations along the x-axis).

▦ **EXAMPLE 4.** Consider next the two ideal fixed points 0 and ∞ and the group of hyperbolic translations along the Poincaré line joining them (the upper half of the y-axis). They are represented by matrices with $c = b = 0$ and $ad = 1$; they form the group of mappings

$$z \to a^2 z,$$

and since a^2 can be any positive number, this group is isomorphic to the multiplicative group of all positive real numbers. By taking the logarithm, this group in turn is isomorphic to the additive group of all real numbers, just as in the Euclidean case of translations along a fixed line—Proposition 9.21. (The mappings are *Euclidean* dilations centered at 0.)

▦ **EXAMPLE 5.** Finally, let us determine the group of all hyperbolic rotations about the point i in the upper half-plane. They are the direct motions that fix i:

$$i = \frac{ai + b}{ci + d},$$

so $a = d$ and $b = -c$, with $1 = ad - bc = a^2 + b^2$. If we set $a = \cos\theta$, $b = -(\sin\theta)$, rotations about i are represented by

$$z \to \frac{(\cos\theta)z - \sin\theta}{(\sin\theta)z + \cos\theta}.$$

But the matrix

$$\begin{bmatrix} \cos\theta & -\sin\theta \\ \sin\theta & \cos\theta \end{bmatrix}$$

is just the matrix of Euclidean rotation through θ about the origin in the Cartesian model of the Euclidean plane. Thus these two groups are isomorphic to each other and to the multiplicative group S^1 of all complex numbers of modulus 1.

NOTE. In the representation of hyperbolic rotations in Example 5, when $\theta = \pi$, z is mapped to z and we get the identity rotation, whereas the Euclidean rotation through π is a half-turn. So in order to have a one-to-one (instead of a two-to-one) mapping of the group of hyperbolic rotations about i onto the group of Euclidean rotations about 0, we must represent the hyperbolic rotation through θ by

$$z \rightarrow \frac{\left(\cos \dfrac{\theta}{2}\right) z - \sin \dfrac{\theta}{2}}{\left(\sin \dfrac{\theta}{2}\right) z + \cos \dfrac{\theta}{2}}.$$

Congruence Described by Motions

In neutral geometry, motions can be used to define congruence of arbitrary figures, namely, S is *congruent* to S' if there is a motion T mapping S onto S'. By Proposition 9.5, this definition gives the same congruence relation for triangles as before (hence the same congruence relation for segments and for angles).

A particularly important figure is a *flag*, which is defined to consist of a point A, a ray $\overrightarrow{AB}$ emanating from A, and a side S of line $\overleftrightarrow{AB}$ (Figure 9.27).

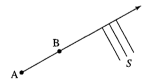

Figure 9.27

LEMMA 9.3. Any two flags are congruent by a unique motion.

PROOF:

Choose B' on the second ray so that $AB \cong A'B'$. Choose any point $C \in S$ and let C' be the unique point in side S' of $\overleftrightarrow{A'B'}$ such that

$\triangle ABC \cong \triangle A'B'C'$ (corollary to SAS, Chapter 3). By Proposition 9.5, there is a unique motion T taking A, B, C, respectively, into A', B', C', and this maps flag $\overrightarrow{AB} \cup S$ onto $\overrightarrow{A'B'} \cup S'$. ◄

We say that the group of motions *operates simply transitively* on the flags. This property expresses the homogeneity of the plane and corresponds to our physical intuition of performing measurements by moving a rigid ruler around. This property is crucial in our next theorem, which answers the converse question: Given a model $\mathcal{M}$ for our incidence and betweenness axioms and given a group $\mathcal{G}$ of betweenness-preserving collineations of $\mathcal{M}$, define congruence by the action of $\mathcal{G}$. For example, *define* AB $\cong$ A'B' to mean that some transformation T in $\mathcal{G}$ maps segment AB onto segment A'B' (similarly for angles). What additional assumptions on $\mathcal{G}$ guarantee that with this definition, our Congruence Axioms C-1 through C-6 hold in $\mathcal{M}$?

THEOREM 9.2. Assume the group $\mathcal{G}$ of betweenness-preserving collineations satisfies the following conditions:

   (i) $\mathcal{G}$ operates simply transitively on the flags.
  (ii) For any two points A, B, there is at least one transformation $T \in \mathcal{G}$ that interchanges A and B.
 (iii) For any two rays r, s emanating from the same vertex, there is at least one transformation $T \in \mathcal{G}$ that interchanges r and s.

   Then, with congruence defined in terms of the action of $\mathcal{G}$, Axioms C-1 through C-6 hold, and $\mathcal{G}$ is the group of motions.

We know that these conditions are necessary for the group of motions—condition (i) by our lemma, condition (ii) using the reflection across the perpendicular bisector of AB, and condition (iii) using the reflection across the bisector of the angle $r \cup s$.

PROOF:

The proof that the conditions are sufficient proceeds in 10 steps.

▓▓▓ **STEP 1.** Congruence is reflexive, symmetric, and transitive (in particular, Axioms C-2 and C-5 hold). $\mathcal{G}$ is a group of automorphisms.

For any figure S, if T maps S onto S', then T^{-1} maps S' onto S; and if T' maps S' onto S'', then $T'T$ maps S onto S''. Obviously, I maps S onto S. (We use here the condition that $\mathcal{G}$ is a group.) If $S \cong S'$, i.e., $S' = TS$, and $U \in \mathcal{G}$, then $US' = (UTU^{-1})US$, so $US \cong US'$; this shows that U is an automorphism.

STEP 2. Existence of reflections.

Let point P lie on line l. Let S_1, S_2 be the two sides of l and let r_1, r_2 be the two rays of l emanating from P. By condition (i), there is a unique $T \in \mathcal{G}$ that interchanges S_1 and S_2 and leaves r_1 invariant. Since T^2 leaves r_1 and S_1 invariant, $T^2 = I$ (by condition (i)). We claim T fixes every point of l: By definition of T, P is fixed. Since T is betweenness-preserving, r_2 is also invariant under T. Suppose point A on l is moved to A′ and, say, P ∗ A ∗ A′. Then $(A')' = A$ since $T^2 = I$, so P′ ∗ (A′)′ ∗ A′ contradicts the assumption that T preserves betweenness. Since the reflection R_l is the involutory automorphism that leaves each point of l fixed, we may write $T = R_l$.

STEP 3. Let r be a ray of line l and let r' be any ray. Then there are exactly two transformations in $\mathcal{G}$ that map r onto r', and they both have the same effect on the points of l.

For by condition (i), given a side S of l, the two transformations are uniquely determined by which side of l' $(r' \subset l')$ S is mapped onto. If T is one such transformation, the other is TR_l, and they both agree on l.

STEP 4. If $AB \cong CD$, then there are exactly two transformations in $\mathcal{G}$ sending A to C and B to D, and they agree on $\overleftrightarrow{AB}$.

By the definition of $\cong$, there is a transformation T in $\mathcal{G}$, mapping segment AB onto segment CD. If T sends A to D and B to C, follow T by a transformation in $\mathcal{G}$ that interchanges C and D (condition (ii)). We can therefore assume A goes to C and B to D. Since betweenness is preserved, ray $\overrightarrow{AB}$ is mapped onto ray $\overrightarrow{CD}$. Hence, step 3 applies.

STEP 5. Congruence Axiom C-1 holds. (This follows from steps 3 and 4.)

STEP 6. Congruence Axiom C-3 holds.

Let T send A to A′ and B to B′. If A ∗ B ∗ C, then T maps ray $\overrightarrow{BC}$ onto ray $\overrightarrow{B'C'}$, where A′ ∗ B′ ∗ C′. If $BC \cong B'C'$, then there is a motion T' sending B to B′ and C to C′. By step 3, T and T' agree at every point on the line through A, B, C. Hence they both send A to A′ and C to C′, so $AC \cong A'C'$.

STEP 7. Congruence Axiom C-4 holds.

Given ∢BAC, ray $\overrightarrow{A'C'}$, and side S' of $\overleftrightarrow{A'C'}$. Let S be the side of $\overleftrightarrow{AC}$ on which B lies. Let T in $\mathcal{G}$ be the unique transformation which, according to condition (i), maps $(\overrightarrow{AC}, S)$ onto $(\overrightarrow{A'C'}, S')$. If B′ is the image of B under T, then ∢BAC $\cong$ ∢B′A′C′. Conversely, if this congruence is effected by a transformation T', where B' lies in S', then T' maps $(\overrightarrow{AC}, S)$ onto $(\overrightarrow{A'C'}, S')$, so by the uniqueness part of condition (i), $T = T'$, and ray $\overrightarrow{A'B'}$ in S' is uniquely determined.

STEP 8. If $\sphericalangle BAC \cong \sphericalangle B'A'C'$, there is a unique transformation in $\mathscr{G}$ sending $\overrightarrow{AB}$ to $\overrightarrow{A'B'}$, and $\overrightarrow{AC}$ to $\overrightarrow{A'C'}$.

By the definition of congruence, there is a $T \in \mathscr{G}$ mapping $\sphericalangle BAC$ onto $\sphericalangle B'A'C'$. If T maps $\overrightarrow{AC}$ onto $\overrightarrow{A'B'}$ and $\overrightarrow{AB}$ onto $\overrightarrow{A'C'}$, then condition (iii) allows us to follow T with a transformation in $\mathscr{G}$ interchanging the two sides of $\sphericalangle B'A'C'$; so we can assume T sends $\overrightarrow{AC}$ to $\overrightarrow{A'C'}$ and $\overrightarrow{AB}$ to $\overrightarrow{A'B'}$. Then uniqueness was shown in the proof of step 7.

STEP 9. Congruence Axiom C-6 (SAS) holds.

Given $AB \cong A'B'$, $\sphericalangle BAC \cong \sphericalangle B'A'C'$, and $AC \cong A'C'$. Let these congruences be effected by transformations $T_1, T_2, T_3 \in \mathscr{G}$, where by steps 4 and 8 we may assume T_1 sends A to A' and B to B', T_2 sends $\overrightarrow{AB}$ to $\overrightarrow{A'B'}$ and $\overrightarrow{AC}$ to $\overrightarrow{A'C'}$, and T_3 sends A to A' and C to C'. By step 3, T_2 agrees with T_1 on $\overleftrightarrow{AB}$, and T_2 agrees with T_3 on $\overleftrightarrow{AC}$. Hence, T_2 sends B to B' and C to C', so that via T_2 we have $BC \cong B'C'$, $\sphericalangle ABC \cong \sphericalangle A'B'C'$, and $\sphericalangle ACB \cong \sphericalangle A'C'B'$.

STEP 10. $\mathscr{G}$ is the group of all motions.

Choose a flag F. A motion T transforms F into F', and by condition (i), an automorphism $T' \in \mathscr{G}$ has the same effect; by Lemma 9.2, we have $T = T'$, so every motion belongs to $\mathscr{G}$. By Lemma 9.3 and the same argument, every member $\mathscr{G}$ is a motion. ◄

Theorem 9.2 is one step in Klein's program to describe the geometry in terms of action of a group. F. Bachmann (1973) carries the program further by describing points, lines, and incidence in terms of involutions in the group—see Exercise 50 and Ewald (1971).

NOTE ON SIMILARITIES. A *pointed flag* is a figure consisting of an ordered pair of distinct points A, B together with a specific side S of line $\overleftrightarrow{AB}$. Every flag with vertex A supports infinitely many pointed flags corresponding to the various choices of point B on the ray of the flag. It is easy to show that for any two pointed flags in the Euclidean plane, there is a unique similarity mapping one onto the other (namely, follow the motion given by Lemma 9.3 with a dilation centered at the vertex). This leads to the following nice characterization.

THEOREM 9.3. In the real Euclidean plane, a transformation is a similarity if and only if it is a collineation mapping circles onto circles.

PROOF:

Let T be such a collineation. In Major Exercise 7, Chapter 5, you showed that given any line l, two points off l are on the same side S of l if and only if they lie on a circle contained in S. Hence T maps S onto a side S' of l'. Also, if T maps circle γ onto circle γ', then T maps the interior of γ onto the interior of γ'—because a point P not on γ lies in the interior of γ if and only if every line through P intersects γ. Next we claim that T maps perpendicular lines onto perpendicular lines. This is because a collineation maps parallelograms onto parallelograms, a parallelogram in the Euclidean plane has a circumscribed circle iff it is a rectangle (Exercise K-20(a), Chapter 7), and since T preserves circles, it must map rectangles onto rectangles. Moreover, since a square is characterized as a rectangle with perpendicular diagonals, T maps squares onto squares. T also maps the center of a square onto the center of the image square (since it is the intersection of the diagonals) and the midpoints of the sides onto the midpoints of the sides (since they form squares with the center and a vertex).

Now let us work with Cartesian coordinates. Consider the basic pointed flag F consisting of the origin $(0, 0)$, the unit point $(1, 0)$ on the x-axis, and the side S of the x-axis containing the unit point $(0, 1)$ of the y-axis. Our transformation T maps F onto some flag F'; let U be the unique similarity that maps F' back onto F. We will show that UT is the identity, so that T is equal to the similarity inverse to U: Namely, UT fixes $(0, 0)$ and $(1, 0)$ and maps each side of the x-axis onto itself. By the remark above about squares, the points $(0, 1)$, $(1, 1)$, $(0, -1)$, and $(1, -1)$ must also be fixed, as must the midpoints and centers of these squares. We can infer successively that all points whose coordinates are integers, half-integers, or dyadic rationals are fixed by UT. Since each point in the plane is interior to an arbitrarily small circle through three fixed points, it must be fixed. (This neat proof is from Werner Fenchel, 1989). ◄

Symmetry

We conclude this chapter with a brief discussion of symmetry, which is one of the main applications of the transformation approach to geometry.

Given a plane figure S, the motions that leave S invariant (i.e., that map S onto itself) are called *symmetries of S*; clearly the symmetries of S form a group. Intuitively, the larger this group, the more symmetric the figure.

For example, a circle γ is highly symmetric. Its symmetry group consists of all rotations about the center O of γ and all reflections across lines through O; this group has the cardinality of the continuum.

A square seems to be fairly symmetric, yet we will show that it has only eight symmetries (see Example 6 later in this section).

The frieze pattern shown in Figure 9.28 has a countably infinite group of symmetries: It consists of all integer powers T^n of a fixed translation T that shifts the pattern one unit to the right.

The first problem is to find a *minimal set of generators* of the group of symmetries. This means finding as small a collection of symmetries as possible with the property that all other symmetries can be expressed as products of the symmetries in this collection and their inverses. For the frieze pattern in Figure 9.28, there is a single generator T (or T^{-1}).

A second problem is to describe the basic *relations* among the generators. For the generator T above, there is no relation: All the powers of T are distinct. Consider, however, Figure 9.29.

The only symmetries of these figures are the identity I, the rotation R about the center O through $120°$ clockwise, and the rotation R^2 about O through $240°$ clockwise. R is a generator of this group and satisfies the relation $R^3 = I$. We say this group is cyclic of order 3.

More generally, a group is called *cyclic of order n* if it has a single generator and n elements; it is called *infinite cyclic* if it has a single generator and infinitely many elements (the group of symmetries of the frieze pattern in Figure 9.28 is infinite cyclic). Let us denote by C_n any cyclic group of order n generated by a rotation through $(360/n)°$ about some point. The constructions in Figure 9.29 can be generalized from 3 to n to obtain a figure having C_n as its symmetry group. The graph in Figure 9.29(b) is called a *triquetrum*; its generalization to $n = 4$ is a swastika. The $2n$-sided convex polygons obtained from generalizing Figure 9.29(c) are called *ratchet polygons*.

Figure 9.28 Infinite cyclic group of symmetries.

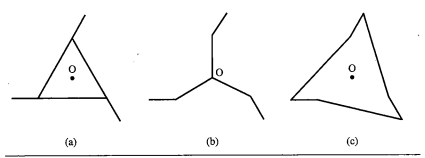

(a)	(b)	(c)

Figure 9.29

A third basic problem is to describe the structure of the symmetry group, showing if possible that it is isomorphic to some familiar group.

▓▓ **EXAMPLE 6.** We will solve these problems for the group of symmetries of a square.

Any symmetry must leave the center O fixed (since, for example, O is the intersection of the diagonals, and each diagonal must be mapped onto itself or the other diagonal). Hence, the symmetries must either be rotations about O or reflections across lines through O. The only rotations about O that are symmetries are I, R, R^2, R^3, where R can be taken to be the counterclockwise rotation through 90°; these form a cyclic subgroup of order 4. There are also four reflections that are symmetries: the two reflections across diagonals c, d and the two reflections across the perpendicular bisectors a, b of the sides (see Figure 9.30). Let T be any one of these reflections—e.g., $T = R_c$. Then $\{R, T\}$ is a minimal set of generators of the group.

For R can be written in four ways as a product of reflections,

$$R = R_bR_d = R_cR_b = R_aR_c = R_dR_a,$$

so that

$$TR = R_c(R_cR_b) = (R_c)^2R_b = R_b$$

$$TR^2 = (TR)R = R_b(R_bR_d) = R_d$$

$$TR^3 = (TR^2)R = R_d(R_dR_a) = R_a.$$

The basic relations between these generators are

$$R^4 = I$$

$$T^2 = I$$

$$RT = TR^3$$

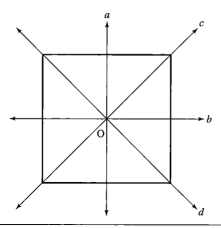

Figure 9.30 Symmetries of a square.

(the last because $RT = (R_aR_c)R_c = R_a = TR^3$). This last relation shows that the group is non-commutative. It is denoted D_4 and is called the *dihedral group* of order 8.

More generally, if $n \geqq 3$, D_n denotes the group of symmetries of a regular n-gon. It has $2n$ symmetries and is generated by two elements $\{R, T\}$, where R is rotation about the center of the n-gon through $(360/n)°$, and T is reflection across a line joining a vertex to the center. For $n = 2$, D_2 denotes the group generated by a half-turn H_P and a reflection across a line through P, while for $n = 1$, D_1 denotes a cyclic group of order 2 generated by a reflection.

The following remarkable theorem has been attributed to Leonardo da Vinci.

LEONARDO'S THEOREM. In both the Euclidean and hyperbolic planes, the only *finite* groups of motions are the groups C_n and D_n ($n \geqq 1$).

The proof will be based on Lemmas 9.4–9.9.

LEMMA 9.4. A finite group of motions cannot contain non-identity translations, parallel displacements, or glides.

PROOF:

The point is to show that if T is any of these three types of motions, no power T^n, $n \neq 0$, of T is equal to the identity I. We will show this for the case where $T = R_lR_m$, $l \parallel m$, leaving the case of a glide for Exercise 18. By Exercise 7, we can also write $T = R_k R_l$, where k is the reflection of m across l, so that

$$T^2 = (R_k R_l)(R_l R_m) = R_k R_m.$$

Repeating this argument, we can show by induction that for any positive integer n, we can write

$$T^n = R_h R_m,$$

where h lies in the half-plane bounded by m and containing l—in particular, $h \parallel m$, so $T^n \neq I$. Applying this result to T^{-1}, we have that $T^n \neq I$ for negative n as well.

Now if T belonged to a finite group, we would have $T^n = T^m$ for distinct integers n, m; hence $T^{n-m} = I$, a contradiction. ◄

LEMMA 9.5. If a finite group of motions contains rotations, all those rotations have the same center.

PROOF:

Let T be a rotation about A and U a rotation about B $\neq$ A and let $l = \overleftrightarrow{AB}$. By Proposition 9.8, there is a unique line m through A (respectively, n through B) such that $T = R_l R_m$ (respectively, $U = R_l R_n$). Then $U^{-1} T^{-1} U T$, which belongs to the finite group, is equal to $(R_n R_l R_m)^2$, which is a translation (see Exercise 15). This contradicts Lemma 9.4 unless the translation is the identity, but in that case $UT = TU$, which can happen only when at least one of U, T is the identity (see warning after Proposition 9.9). ◄

LEMMA 9.6. If a finite group of motions contains reflections, the axes of those reflections are concurrent.

PROOF:

Otherwise we would obtain a contradiction of the previous lemmas since the group must contain the products $R_l R_m$. ◄

LEMMA 9.7. If a finite group of motions contains a rotation and a reflection, then the center of the rotation lies on the axis of the reflection.

PROOF:

Otherwise the product of the rotation and the reflection would be a glide (Proposition 9.19), contradicting Lemma 9.4. ◄

COROLLARY. A finite group of motions of order >2 has a unique fixed point.

PROOF:

Since the order is >2, the group must contain a non-identity rotation (by Lemmas 9.4 and 9.6), and every symmetry in the group fixes the center of that rotation, which is uniquely determined (by Lemmas 9.4, 9.5, and 9.7). ◄

LEMMA 9.8. If a finite group of motions of order n contains only rotations, then it is cyclic or order n.

PROOF:

The cases $n = 1$ or 2 are trivial, so assume $n > 2$. Let O be the center of all the rotations (Lemma 9.5). Choose any point $P_1 \neq O$. The images $P_1, \ldots, P_n$ of P_1 under the rotations in our group are all distinct since the rotations are distinct (Exercise 19). We assume these images numbered so that $(\sphericalangle P_1OP_2)°$ is the minimum of all the degrees $(\sphericalangle P_iOP_j)°$. Let R be the rotation taking P_1 to P_2. We claim that R generates the group.

Let Q_i be the image of P_1 under R^{i-1}, $i = 1, 2, \ldots$ (so that we get $Q_1 = P_1$, $Q_2 = P_2$). Then

$$(\sphericalangle Q_iOQ_{i+1})° = (\sphericalangle P_1OP_2)°$$

for each i. If some P_j were not among the Q_i's, ray $\overrightarrow{OP_j}$ would lie between some $\overrightarrow{OQ_i}$ and $\overrightarrow{OQ_{i+1}}$, hence $(\sphericalangle P_jOQ_i)°$ would be smaller than $(\sphericalangle P_1OP_2)°$, contradicting our choice of P_2. Therefore, every rotation in our group is equal to a power of R. ◄

LEMMA 9.9. If a finite group of motions contains a reflection, then it is a dihedral group D_n.

PROOF:

Partition the group $\mathscr{G}$ into its set $\mathscr{D}$ of direct motions and its set $\mathscr{E}$ of opposite motions. Let n be the number of elements in $\mathscr{D}$. By Lemma 9.8, $\mathscr{D}$ is a cyclic subgroup of order n generated by a rotation R. Let T be any reflection. Then the opposite motions in $\mathscr{E}$ are also reflections and since the product of two of them is a direct motion, they can be written uniquely in the form

$$TR^i, \quad i = 0, 1, \ldots, n.$$

This is the group D_n. ◄

Thus, Leonardo's theorem is proved. ◄

Much is known about particular types of infinite groups of motions. For instance, a *frieze group* is a group of motions that has an invariant line t and whose translations form an infinite cyclic group $\langle T \rangle$ generated by one particular translation T along t. It is not difficult to prove that *there are exactly seven frieze groups* in Euclidean or hyperbolic planes:

1. $\langle T \rangle$.
2. The group $\langle T, R_t \rangle$ generated by T and the reflection across t.
3. The group $\langle T, R_u \rangle$ generated by T and a reflection across a perpendicular u to t.
4. The infinite cyclic group $\langle G \rangle$ generated by the unique glide G such that $G^2 = T$.
5. The group $\langle T, H_A \rangle$ generated by T and a half-turn about a point A on t.
6. The group $\langle T, H_A, R_t \rangle$.
7. The group $\langle G, H_A \rangle$.

(For the proof, see Martin, 1982, and Exercise 60.)

Another type of infinite group is called a *wallpaper group*, whose subgroup of translations is generated by two translations along distinct intersecting lines. In the Euclidean plane, *there are exactly 17 wallpaper groups*. The ornamental patterns designed on the walls of the Alhambra in Spain by the Moors illustrate these 17 groups.

The classic treatise on symmetry is Hermann Weyl's *Symmetry*.[14] You will find therein a discussion of the 17 wallpaper groups, plus an analysis of three-dimensional symmetry, including the generalization of Leonardo's theorem to three dimensions. Most important, you will find a fascinating treatment in words and pictures of how these purely mathematical abstractions relate to the physical universe in the form of crystals, biological specimens, and works of art throughout the ages.[15]

[14] Hermann Weyl, *Symmetry* (Princeton, N.J.: Princeton University Press, 1952).

[15] Other excellent references on this subject are: J. N. Kapur, *Transformation Geometry* (Affiliated East-West Press, 1976); Joe Rosen, *Symmetry Discovered* (Cambridge: Cambridge University Press, 1975); E. H. Lockwood and R. H. Macmillan, *Geometric Symmetry* (Cambridge: Cambridge University Press, 1978); and I. Stewart and M. Golubitsky, *Fearful Symmetry: Is God a Geometer?* (Oxford: Blackwell, 1992). Ian Stewart, *Why Beauty is Truth: A History of Symmetry* (New York: Perseus, 2007). You can find pictures of the Alhambra designs on the web.

Review Exercise

Which of the following statements are correct?

(1) In the Cartesian model, the equations for reflection across the y-axis are $y' = y$ and $x' = -x$.

(2) In the Cartesian model, the equations for the 90° clockwise rotation about the origin are $y' = x$ and $x' = -y$.

(3) In the Euclidean plane, a similarity that is not a motion must be a dilation.

(4) In the Cartesian model, the equations for the translation moving the origin to (1, 1) are $y' = y + 1$ and $x' = x + 1$.

(5) An involution is a transformation equal to its own inverse but not equal to the identity.

(6) If a motion leaves a circle invariant, it must be a rotation about the center of the circle.

(7) In the Cartesian model, if k is the x-axis, l the line $y = x$, m the y-axis, and n the line $y = -x$, then $R_k R_l R_m = R_n$.

(8) In the Cartesian model, the equations for the half-turn about the point (1, 0) are $x' = -x + 2$ and $y' = -y$.

(9) In the Cartesian model, the glide along the x-axis mapping of (0, 1) to (1, −1) is given by the equations $x' = x + 1$ and $y' = -y$.

(10) If A, A' are distinct points in the Euclidean plane, there is a unique translation T such that $T(A) = A'$; i.e., the group of translations operates simply transitively on the points.

(11) If A, A' are distinct points in the hyperbolic plane, there are infinitely many translations T such that $T(A) = A'$.

(12) In neutral geometry, if A, A' are distinct points and O is any point on the perpendicular bisector of AA', then there is a unique rotation T with center O such that $T(A) = A'$.

(13) In neutral geometry, the set of all translations is a group.

(14) In neutral geometry, if the product of two rotations is a non-identity translation, then the rotations are half-turns about distinct points.

(15) In hyperbolic geometry, for any two distinct points A, A', there are exactly two parallel displacements T such that $T(A) = A'$.

(16) In Euclidean geometry, the product of a non-identity rotation and a translation is a rotation.

(17) Reflections in distinct lines never commute.

(18) In hyperbolic geometry, the product of two rotations about the same point could be any of the three types of direct motion.

(19) In Euclidean geometry, the set of all half-turns and all translations is a group.

(20) If $R_l R_m R_n$ is a glide, then lines l, m, and n do not lie in a pencil.

(21) In the Cartesian model, the equations for a direct similarity have the form $x' = ax - by + e$, $y' = bx + ay + f$, where we have $a^2 + b^2 \neq 0$, and all lengths are multiplied by $k = \sqrt{a^2 + b^2}$ under this transformation.

(22) A rotation through angle θ can be written as a product of reflections in two lines that meet and form an angle θ.

(23) The group of motions operates simply transitively on the set of all equilateral triangles whose sides have length 1.

(24) In hyperbolic geometry, the group of motions operates simply transitively on the set of all ordered triples of distinct ideal points.

(25) A finite group of motions cannot contain more than one half-turn.

(26) In the Poincaré upper half-plane model, the linear fractional transformation

$$z \to \frac{3z + 4}{z + 1}$$

represents a direct hyperbolic motion.

(27) In the Poincaré upper half-plane model, the transformation $z \to z - 1$ represents a parallel displacement about ∞.

(28) In the Cartesian model, the product of a translation along the x-axis with the reflection across the y-axis is a reflection.

(29) The group of symmetries of a regular pentagon is cyclic of order 5.

(30) No opposite motion commutes with a non-identity direct motion.

(31) In Euclidean geometry, two figures are congruent if and only if there is an automorphism mapping one onto the other.

(32) If there exists a reflection leaving a triangle invariant, then that triangle is isosceles.

(33) If A and A' are any two points on opposite sides of line t, then there exists a glide T along t such that $T(A) = A'$.

(34) In the Cartesian model, the equations for rotation through angle θ about the origin are $x' = x \cos \theta + y \sin \theta$ and, equivalently, $y' = x \sin \theta - y \cos \theta$.

(35) In neutral geometry, if a motion has a unique invariant line, then it is a glide.

(36) A motion that is a product of an odd number of reflections is opposite.

(37) In the Cartesian model, the equations for reflection across the line $y = x$ are $y' = x$ and $x' = y$.

(38) In Euclidean geometry, the group of symmetries of a quadrilateral has order $\geqq 4$ if and only if the quadrilateral is a rectangle.

(39) In hyperbolic geometry, the group of symmetries of a quadrilateral must have order < 4.

(40) In Euclidean geometry, if the group of symmetries of a convex quadrilateral has order 2, then the quadrilateral must be an isosceles trapezoid.

(41) In Euclidean geometry, the group of symmetries of every triangle has order $\geqq 3$.

(42) In neutral geometry, if there exists an automorphism that is not a motion, then the geometry is Euclidean.

(43) In neutral geometry, the product of reflections in two parallel lines is a translation.

(44) In neutral geometry, an automorphism with exactly one fixed point must be a rotation.

(45) In the Cartesian model, an opposite similarity that fixes the origin O is equal to a product DR, where D is a dilation centered at O, R is a reflection in a line through O, and D and R commute.

(46) In the Poincaré upper half-plane model, the transformation

$$z \to \frac{6\bar{z} - 4}{8\bar{z} - 6}$$

represents a reflection.

(47) In the Cartesian model, for any complex number z_0, the transformation $z \to \bar{z} + z_0$ represents a reflection.

(48) In Euclidean geometry, a betweenness-preserving transformation that doubles the length of every segment must be a similarity.

(49) In hyperbolic geometry, there is no collineation that doubles the length of every segment.

(50) In neutral geometry, every motion has an invariant line or a fixed point or both.

Exercises

Outline: Exercises 2–22 consist of supplementary results and proofs left to the reader on the classification of motions; Exercise 23 is an essay question on orientation; Exercises 24–33 give more information on similarities; Exercises 34–35 are about linear and affine transformations; Exercises 36–38 discuss Möbius transformations; Exercises 39–49 deal with orbits of groups of transformations (particularly with horocycles and equidistant curves in the Poincaré model); Exercise 50 exhibits algebraic equations in the group of motions and their geometric meaning; Exercise 51 refers to invariant sets of transformations; Exercise 52 presents another attempt to prove the parallel postulate using rotations, and Exercise 53 shows what happens when translations are used (*parallel transport*); Exercises 54–57 determine the axis of a glide in the Euclidean plane; Exercise 58 presents a natural hyperbolic transformation that is not a collineation; Exercises 59–61 are about symmetry; Exercises 62–63 give some unusual models; Exercise 64 raises the question of two-dimensionality in incidence planes; Exercises 65–68 are fundamental for one-dimensional projective geometry; Exercise 69 furnishes an example of an automorphism of a non-Archimedean Pythagorean plane that is not a similarity; Exercises 70–71 augment the classification of motions; Exercise 72 treats the circular points at infinity; Exercises 73–77 are about the nine-point circle and other topics in higher Euclidean plane geometry.

1. Show that there are 168 collineations of the seven-point projective plane (Figure 9.1). (Hint: A collineation is uniquely determined by its effect on four points, no three of which are collinear.)
2. Prove that the set of all automorphisms of a model of neutral geometry is a group, and that the set of motions is a subgroup.
3. Finish the proof of Proposition 9.2.
4. Prove that a transformation of the plane that multiplies all lengths by a constant $k > 0$ preserves betweenness and congruence of segments.
5. Prove that a reflection is a motion.
6. In a Euclidean plane, prove that $\triangle ABC \sim \triangle A'B'C'$ if and only if there is a similarity sending A, B, C, respectively, onto A', B', C' and that similarity is unique. (Hint: Use Lemma 9.2, Proposition 9.5, and Exercise 12 of Chapter 5.)
7. Prove that the invariant lines of a reflection R_m are m and all lines perpendicular to m. Prove also that $R_m R_k R_m = R_{k^*}$, where k^* is the reflection of k across m.

8. Prove the corollary to Proposition 9.6.

9. Prove that if an automorphism is an involution, then it is either a reflection or a half-turn. (Hint: If A and A' are interchanged, show that the midpoint of AA' is fixed and apply Propositions 9.4 and 9.6.)

10. Show that the product $T'T$ of rotations about distinct points can be any of the three types of direct motions. (Hint: Apply Proposition 9.8 to the line joining the centers of rotation.)

11. Prove that a non-identity rotation that is not a half-turn has no invariant lines. (Hint: Apply Proposition 9.8.)

12. Let T be a translation along t, l a line that is not invariant under T. If l meets t (automatic in the case where the plane is Euclidean, by Proposition 9.13), prove that $l \parallel l'$. Suppose now the plane is hyperbolic. If l is asymptotically parallel to t in direction Ω, prove that l' is also (in particular, $l \parallel l'$). If l is divergently parallel to t, show by diagrams from a Poincaré model that l' could meet l, be divergently parallel to l, or be asymptotically parallel to l.

13. Prove the corollary to Proposition 9.15.

14. Show that in the hyperbolic plane, the product of translations along distinct lines could be any of the three types of direct motions and that two such translations do not commute unless at least one of them is the identity. (Hint: Apply the warning after Proposition 9.9 to the invariant lines.)

15. If R_1, R_2, R_3 are reflections, prove that $(R_1 R_2 R_3)^2$ is a translation. (Hint: Use Propositions 9.18 and 9.19.)

16. Let T, T' be glides along perpendicular lines. Prove that TT' is a half-turn if and only if the plane is Euclidean. (Hint: Use Propositions 9.15 and 9.18.)

17. In the hyperbolic plane, prove that every direct motion can be expressed as a product of three half-turns. (Hint: Refer back to Figure 9.9. Show that for any two lines l, u, there is a perpendicular m to u that is divergently parallel to l.)

18. If T is a glide and $n \neq 0$, prove that $T^n \neq I$. (Hint: This has already been proved for translations.)

19. If rotations T, T' about O have the same effect on a point $P \neq O$, prove that $T = T'$.

20. If T is any motion, A any point, and l any line, prove that $TH_A T^{-1}$ is a half-turn and that $TR_l T^{-1}$ is a reflection.

21. Prove Hjelmslev's lemma. (Hint: See Figure 9.31 on the next page.)

22. (*Hjelmslev's theorem*) Let l and l' be distinct lines and let T be a motion transforming l onto l'. As point P varies on l and its image

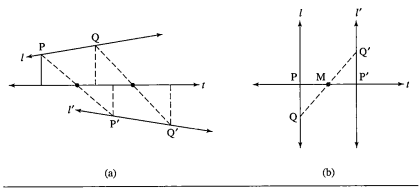

Figure 9.31 Hjelmslev's lemma.

P' under T varies on l', the midpoints of the segments PP' either are distinct and collinear or else they all coincide. (Hint: Show that you can assume that T is an opposite motion and apply Exercise 21 if T is a glide.)

23. How would you go about precisely defining the notion of an "orientation" of the (neutral) plane? The requirements are that you must show that there are exactly two "orientations," which are interchanged by opposite motions and preserved by direct motions. (If your definition uses words such as "clockwise" and "counterclockwise," you must define them precisely.) Making this notion precise is surprisingly tricky. It can be done in several ways, all of which appear artificial—see Ewald (1971), p. 65, for one. Perhaps the reason for the artificiality is, as Hermann Weyl says, that "to the scientific mind there is no difference, no polarity between left and right. . . . It requires an arbitrary act of choice to determine what is left and what is right; . . . in all physics nothing has shown up indicating an intrinsic difference of left and right." (See Weyl's *Symmetry*, pp. 16–38, for a fascinating discussion of this problem, illustrated with examples from physics, biology, and art. The Nobel Prize–winning work of C. N. Yang and T. D. Lee, done after Weyl's death, does indicate a physical difference between left and right at the subatomic level.)

24. Prove Proposition 9.24. (Hint: Follow the given similarity by a translation, a rotation about 0, and a dilation centered at 0—if necessary—to obtain a similarity fixing 0 and 1; then apply Lemma 9.1.)

25. Prove that every similarity that is not a motion has a unique fixed point. (Hint: Use Proposition 9.24. For a synthetic proof of this, see Coxeter, 2001, Section 5.4.)

26. Prove that the reflections in the Cartesian model are characterized among all opposite motions $z \rightarrow e^{i\theta}\bar{z} + z_0$ by the equation

$$e^{i\theta}\bar{z}_0 + z_0 = 0$$

and that the axis of such a reflection is the line

$$2(yy_0 + xx_0) = x_0^2 + y_0^2$$

if $z_0 = x_0 + iy_0 \neq 0$, whereas if $z_0 = 0$, it is the line

$$y \cos(\theta/2) - x \sin(\theta/2) = 0.$$

(Hint: If $z_0 \neq 0$, the axis must pass through $z_0/2$ and be perpendicular to the line joining 0 to z_0; if $z_0 = 0$, the axis passes through $e^{i(\theta/2)}$.) Conversely, given a line $ax + by = c$, find z_0 and θ for the reflection aross this line.

27. What is the representation in complex coordinates of a dilation of ratio k whose center has complex coordinate z_1?

28. The definition of direct and opposite similarities given in Proposition 9.24 depended on the representation in complex coordinates. Prove that a coordinate-free description is given as follows: A similarity is direct (respectively, opposite) if and only if it is the product of a dilation with a direct (respectively, opposite) motion. Prove also that if AB and A′B′ are any two segments, there is a unique *direct* similarity taking A to A′, and B to B′.

29. Let γ_1 and γ_2 be circles in the Euclidean plane with distinct centers O_1, O_2 and distinct radii r_1, r_2. Prove that there are two dilations D_1, D_2 which transform γ_1 onto γ_2. (Hint: Choose any point A on γ_1 and let A_1, A_2 be the ends of the diameter of γ_2 that is parallel to $\overleftrightarrow{O_1A}$; then O_2A_1 (respectively, O_2A_2) will be the image of O_1A under D_1 (respectively, D_2).) The centers of D_1 and D_2 are called the *centers of similitude* of the two circles. (See Coxeter, 2001, Section 5.3, for an application to the famous nine-point circle.)

30. Given any $\triangle ABC$ in the Euclidean plane, let A′, B′, C′ be the midpoints of its sides, labeled so that the medians are AA′, BB′, CC′. Show that there is a unique dilation of ratio $-\frac{1}{2}$ taking $\triangle ABC$ onto $\triangle A'B'C'$. (Hint: See Problem 8, and Coxeter, 2001, Section 1.4.)

31. Show that the set of all dilations (with all possible centers and ratios) is not a group, whereas the set of all translations and dilations (of a Euclidean plane) is a group and that this group is noncommutative.

32. Prove that dilations are geometrically characterized among all similarities by the two properties of (i) mapping each line l onto a line equal to or parallel to l; and (ii) having a fixed point.

33. A *point at infinity* for the Cartesian plane is an equivalence class of all lines equal to or parallel to some given line, and the *line at infinity* is the set of all points at infinity (compare Chapter 2). Since automorphisms (in fact, collineations) of the Cartesian plane preserve parallelism, they induce transformations of the line at infinity onto itself, and we can investigate the fixed points at infinity of these transformations. Show that an automorphism which (i) fixes every point at infinity is either a dilation or a translation; (ii) fixes two points at infinity is an opposite similarity (if l, m determine the two fixed points, then $l \perp m$); (iii) has no fixed points at infinity is either a rotation that is not the identity and not a half-turn, or the product of such a rotation with a dilation.

   (Hint: In the representation Proposition 9.24 gives for similarities, let $w_0 = \alpha + \beta i$; if the similarity is direct, it takes lines of slope m onto lines of slope $(\beta + \alpha m)/(\alpha - \beta m)$, whereas if the similarity is opposite, it takes lines of slope m onto lines of slope $(\beta - \alpha m)/(\alpha + \beta m)$.)

34. A *linear transformation* of the Cartesian plane is a transformation T given in coordinates by

$$x' = a_{11}x + a_{12}y$$
$$y' = a_{21}x + a_{22}y,$$

where the matrix

$$A = \begin{bmatrix} a_{11} & a_{12} \\ a_{21} & a_{22} \end{bmatrix}$$

has a nonzero determinant. In vector notation, the transformation has the form

$$z' = Az.$$

An *affine transformation* is a linear transformation followed by a translation:

$$z \rightarrow Az + z_0.$$

Prove that an affine transformation is a collineation and that the set of all affine transformations is a group (called the *affine group*). It can be shown that, conversely, every collineation of the *real* Cartesian plane is an affine transformation—see Artzy (1965), p. 155.

35. If F is a field, we have shown (in Example 8, Chapter 2) how to extend the structure of an affine plane on F^2 by adding points and a line at infinity to obtain the projective plane $P^2(F)$ coordinatized by F. Show that a collineation T of F^2 transforms parallel lines into parallel lines, hence it can be extended naturally to a collineation T^* of $P^2(F)$, leaving the line at infinity l_∞ invariant by defining $T^*([l]) = [l']$, where as usual $l' = T(l)$ and $[l]$ is the point at infinity at the end of l (defined to be the equivalence class of l, consisting of l and all lines parallel to l—see Chapter 2). Conversely, every collineation of $P^2(F)$ that leaves l_∞ invariant induces a collineation of F^2 by restricting it to the points not at infinity.

   A *projective transformation* of $P^2(F)$ is determined by a 3×3 matrix B of elements of F whose determinant is $\neq 0$. If $[x, y, z]$ is any point of $P^2(F)$, writing its homogeneous coordinates as a column vector and multiplying on the left by B is the way the projective transformation operates on points. If $[a, b, c]$ is any line of $P^2(F)$, multiplying this row vector on the right by the inverse matrix B^{-1} is the way the projective transformation operates on lines. Show that with these definitions the projective transformation preserves incidence, hence is a collineation. Recall that l_∞ is the line $[0, 0, 1]$ (i.e., the line whose equation is $z = 0$). Determine which matrices B determine transformations that leave l_∞ invariant and for such a transformation determine the 2×2 matrix A and the 2-vector z_0 of the affine transformation T of which the projective transformation is the extension T^*.

36. Linear fractional transformations

$$T : z \to \frac{az + b}{cz + d}$$

   with complex coefficients and a nonzero determinant

$$\delta = ad - bc$$

   are called *Möbius transformations* or *homographies*. Show that such a transformation with $c \neq 0$ can be factored into a composite

$$T = T_4 T_3 T_2 T_1,$$

   where T_1 is the translation $z \to z + (a/c)$, T_2 is the similarity $z \to (-\delta/c^2)z$, T_3 is the mapping $z \to z^{-1}$, and T_4 is the translation $z \to z + (d/c)$.

37. Show that a Möbius transformation maps the set of all circles and lines in the Cartesian model onto itself, preserving orthogonality.

(Hint: In the case where $c \neq 0$, use the factorization in the previous exercise, observing that T_3 is the composite of inversion in the unit circle with reflection across the x-axis and using Exercise P-17, Chapter 7. In the case where $c = 0$, use the fact that T is an automorphism of the Cartesian model.)

38. Show that the mapping

$$E : z \rightarrow i \, \frac{i + z}{i - z}$$

maps the open unit disk one-to-one onto the upper half-plane, sends i to ∞, and maps all the other points of the unit circle onto the x-axis. (Hint: Calculate the real and imaginary parts of $E(z)$ and use the previous exercise.)

39. Let $\mathscr{G}$ be a subgroup of the group of motions. For any point P, the *orbit* $\mathscr{G}P$ of P under $\mathscr{G}$ is defined to be the set of all images of P under motions in $\mathscr{G}$. For example, if $\mathscr{G}$ is the entire group of motions, then $\mathscr{G}P$ is the entire plane. Let $\mathscr{G}$ be the group of all rotations about a point O; if $P \neq O$, show that the orbit $\mathscr{G}P$ is the circle through P centered at O.

40. Let $\mathscr{G}$ be the group of all translations along a line t. If P lies on t, prove that $\mathscr{G}P = t$. Suppose P doesn't lie on t. Show that $\mathscr{G}P$ is the unique parallel to t through P in the case where the plane is Euclidean and show that $\mathscr{G}P$ is the *equidistant curve* to t through P if the plane is hyperbolic.

41. Let Ω be an ideal point and let A be an ordinary point in the hyperbolic plane. We defined the *horocycle* through A about Ω to be the orbit of A under the group $\mathscr{G}$ of parallel displacements about Ω. Show that an equivalent description is that it consists of A and all points A′ such that the perpendicular bisector of AA′ has Ω as an end. (Hint: Use Proposition 9.17.)

42. Let $l \mid m$, "meeting" at ideal point Ω, let A ɪ l, and let B be the foot of the perpendicular from A to m. By the method of Major Exercise 11, Chapter 6, there is a unique point P on m such that PQ $\cong$ AB, where Q is the foot of the perpendicular from P to l. Let t be the perpendicular bisector of AP. Prove that t is the *symmetry axis* of l and m—i.e., that $R_t(l) = m$—and hence, if we fix A and Ω and let m vary through all lines through Ω, the locus of such *corresponding points* P (as Gauss called them) fills out the horocycle through A centered at Ω. (Hint: Show that AB meets PQ in a point C and use Major Exercise 5, Chapter 6, and AAA to show that $\triangle$CPB $\cong$ $\triangle$CAQ and hence C ɪ t and $\measuredangle$PAQ $\cong$ $\measuredangle$APB. Deduce that t lies between l and m and so passes through Ω. See Figure 9.32.)

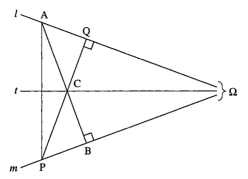

Figure 9.32 Corresponding point P on the horocycle through A.

43. Prove that the symmetry axes t, u, v of the pairs of sides of a trebly asymptotic triangle are concurrent in a point G that has the properties described in Exercise K-13 of Chapter 7 (see Figure 7.48). (Hint: Show first that t meets u at some point G and then that $R_t(u) = v$, so that $G = R_t(G)$ lies on v also.) Show that $R_v R_u R_t = R_u$. (Hint: Use the theorem on three reflections.) We have succeeded in proving the result of Exercise K-13 without resorting to Euclidean geometry.

44. Prove that all horocycles are congruent to one another. (Hint: Use the fact that all rays are congruent to one another—Lemma 9.3.)

45. In the Poincaré upper half-plane model, let t be the upper half of the y-axis. Show that the equidistant curves of t are the nonvertical rays in the upper half-plane emanating from 0. (Hint: See Example 4.)

46. In the Poincaré upper half-plane model, show that the horocycles about ∞ are the horizontal lines in the upper half-plane. (Hint: See Example 3.) Show that the horocycles about x_0 are the circles in the upper half-plane tangent to the x-axis at x_0. (Hint: Use the inversion $z \to \bar{z}^{-1}$ in the unit circle to map ∞ to 0 and the horocycles about ∞ to the horocycles about 0; then use the parallel displacement $z \to z + x_0$.)

47. In the Poincaré upper half-plane model, prove that the equidistant curves are either (1) nonvertical rays emanating from a point on the x-axis or (2) intersections with the upper half-plane of circles cutting the x-axis in two points with the centers of the circles not lying on the x-axis. (Hint: Use a real linear fractional transformation to map the upper half of the y-axis onto any other Poincaré line and apply Exercises 37 and 45.)

48. Find the Euclidean radius and Euclidean center of the hyperbolic circle of hyperbolic center i passing through $2i$. (Hint: Apply Proposition 7.5 to the unit circle.)

49. Use the Poincaré upper half-plane model to demonstrate that in the hyperbolic plane, three points lie on a line, a circle, an equidistant curve, or a horocycle.

50. Verify the following translation of geometric statements into algebraic equations in the group of motions (which leads to new proofs of geometric theorems—see Bachmann, 1973):

    (1) P lies on $l \Leftrightarrow (H_P R_l)^2 = I$.

    (2) $l \perp m \Leftrightarrow (R_l R_m)^2 = I$.

    (3) l, m, n belong to a pencil $\Leftrightarrow (R_l R_m R_n)^2 = I$.

    (4) l is the perpendicular bisector of AB $\Leftrightarrow R_l H_B R_l H_A = I$.

    (5) Let m and n be intersecting lines. Then l is a bisector of one of the angles formed by m and $n \Leftrightarrow R_l R_m R_l R_n = I$. (Describe l geometrically for the case where this equation holds and $m \parallel n$.)

    (6) M is the midpoint of AB $\Leftrightarrow H_M H_B H_M H_A = I$.

    (7) $l \perp \overleftrightarrow{AB} \Leftrightarrow (R_l H_A H_B)^2 = I$.

    (8) $(H_A R_l R_m)^2 = I \Leftrightarrow$ A lies on a common perpendicular to l and m.

    (9) Given A $\neq$ B. In Euclidean geometry, $H_A R_l H_A = H_B R_l H_B \Leftrightarrow l = \overleftrightarrow{AB}$ or $l \parallel \overleftrightarrow{AB}$. In hyperbolic geometry, the equation holds $\Leftrightarrow l = \overleftrightarrow{AB}$.

    (10) In hyperbolic geometry, A, B, C are collinear $\Leftrightarrow (H_A H_B H_C)^2 = I$.

    (11) Assume A, B, C not collinear and that G is the centroid of △ABC (the point of intersection of the medians). Then in Euclidean geometry, $H_P H_C H_P H_B H_P H_A = I \Leftrightarrow P = G$. (Hint: Recall that $H_P H_Q$ is the translation by a vector of length $2\overline{PQ}$ in the direction of ray $\overrightarrow{QP}$.)

    (12) Given □ABCD in the Euclidean plane. Then $H_A H_B H_C H_D = I \Leftrightarrow$ □ABCD is a parallelogram.

51. A set $\mathscr{S}$ of transformations is called *invariant* under a group $\mathscr{G}$ of transformations if for every $S \in \mathscr{S}$ and $T \in \mathscr{G}$, TST^{-1} belongs to $\mathscr{S}$. (In the case where $\mathscr{S}$ is a subgroup of $\mathscr{G}$ and is invariant under $\mathscr{G}$, then $\mathscr{S}$ is called a *normal subgroup*.) For instance, you showed in Exercise 20 that the sets of half-turns and reflections are each invariant under the group of motions. Determine whether each of the following sets is invariant under the indicated groups:

    (i) $\mathscr{S} =$ all rotations about one given point, $\mathscr{G} =$ all motions.

    (ii) $\mathscr{S} =$ all rotations about all points, $\mathscr{G} =$ all motions.

    (iii) $\mathscr{S} =$ all translations along one given line, $\mathscr{G} =$ all motions.

    (iv) $\mathscr{S} =$ all translations along all lines, $\mathscr{G} =$ all motions.

    (v) (Euclidean geometry) $\mathscr{S} =$ all motions, $\mathscr{G} =$ all similarities.

(vi) (Euclidean geometry) $\mathscr{S}$ = all dilations, $\mathscr{G}$ = all similarities.

(vii) (Hyperbolic geometry) $\mathscr{S}$ = all parallel displacements, $\mathscr{G}$ = all motions.

(viii) $\mathscr{S}$ = all glides, $\mathscr{G}$ = all motions.

(ix) (Euclidean geometry) $\mathscr{S}$ = all direct motions, $\mathscr{G}$ = all similarities.

(x) (Euclidean geometry) $\mathscr{S}$ = all rotations about O, $\mathscr{G}$ = all similarities having O as a fixed point.

(xi) (Euclidean geometry) $\mathscr{S}$ = all translations along t, $\mathscr{G}$ = all similarities having t as an invariant line.

52. In 1809, B. F. Thibaut attempted to prove Euclid's parallel postulate with the following argument using rotation; find the flaw. Given any triangle $\triangle ABC$. Let $D * A * B$, $A * C * E$, $C * B * F$ (see Figure 9.33). Rotate $\overleftrightarrow{AB}$ about A through $\sphericalangle DAC$ to $\overleftrightarrow{AC}$; then at C, rotate $\overleftrightarrow{AC}$ to $\overleftrightarrow{BC}$ through $\sphericalangle ECB$; finally, at B, rotate $\overleftrightarrow{BC}$ to $\overleftrightarrow{AB}$ through $\sphericalangle FBA$. After these three rotations, $\overleftrightarrow{AB}$ has returned to itself and has thus been rotated through 360°. Adding up the individual angles of rotation gives

$$[180° = (\sphericalangle A)°] + [180° = (\sphericalangle C)°] + [180° - (\sphericalangle B)°] = 360°,$$

so that the angle sum of $\triangle ABC$ is 180° and Euclid's parallel postulate follows (assuming Archimedes' axiom).

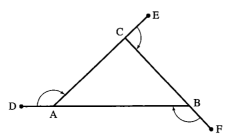

Figure 9.33

53. Given $\triangle ABC$ in the hyperbolic plane, let T_1 be the translation along $\overleftrightarrow{AB}$ taking A to B, T_2 the translation along $\overleftrightarrow{BC}$ taking B to C, and T_3 the translation along $\overleftrightarrow{AC}$ taking C to A. Show that $T_3 T_2 T_1$ is a rotation about A through $d°$, where $d°$ is the defect of $\triangle ABC$.

54. Given $\triangle ABC$ that is not a right triangle. Let $a = \overleftrightarrow{BC}$, $b = \overleftrightarrow{AC}$, $c = \overleftrightarrow{AB}$ and let D, E, F be the feet of the perpendiculars from A, B, C to a, b, c. We know from Proposition 9.19 that the product

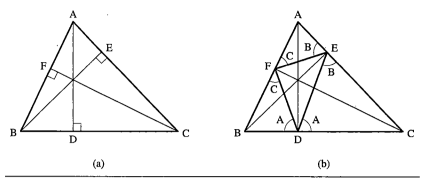

(a)   (b)

Figure 9.34 Axis of a Euclidean glide around a triangle.

$G = R_a R_b R_c$ of the reflections in the sides of $\triangle ABC$ is a glide. Assume the geometry is Euclidean. Prove that the axis of G is the line $t = \overleftrightarrow{DF}$. (Hint: Show that right triangles $\triangle ABE$ and $\triangle ACF$ are similar and then that $\triangle AEF \sim \triangle ABC$; hence $\measuredangle AEF \cong \measuredangle B$, $\measuredangle AFE \cong \measuredangle C$. Applying the same argument to $\triangle BFD$ and $\triangle CED$, deduce that $\measuredangle AFE \cong \measuredangle BFD$, $\measuredangle AEF \cong \measuredangle CED$, $\measuredangle BDF \cong \measuredangle CDE$. From these congruences deduce that line t is invariant under G. See Figure 9.34.)

55. With the same notation as in the previous exercise, assume that $\triangle ABC$ is acute-angled and let $G = TR_t$, where T is a translation along t. Show that this translation is in the direction $\overrightarrow{FD}$ through a distance equal to the perimeter of $\triangle DEF$. (Hint: Determine the image of F under G.)

56. With the same notation as above, determine the axis of G in the case where $\triangle ABC$ is a right triangle. (Hint: It depends on whether or not $\measuredangle B$ is a right triangle; use Hjelmslev's lemma applied to line $\overleftrightarrow{AB}$.)

57. Given three lines a, b, c in the Euclidean plane such that $a \parallel c$, b meets c at A, and b meets a at C (see Figure 9.35).
    (i) If b is perpendicular to both a and c, prove that b is the axis of the glide $R_a R_b R_c$ and $R_a R_b R_c = R_a R_c R_b = R_b R_a R_c$.
    (ii) If b is not perpendicular to a and c, let D, F be the feet of the perpendiculars from A, C to a, c. Prove that $\overleftrightarrow{DF}$ is the axis of glide $R_a R_b R_c$. (Hint: Use Hjelmslev's lemma to show that F and D lie on the axis.) Prove that the axis of glide $R_a R_c R_b$ (respectively, $R_c R_a R_b$) is the line through D (respectively, F) parallel to b. (Hint: Find the midpoint of C and its image and the midpoint of A and its image.)

58. Given ideal point Ω in the hyperbolic plane and distance d, define a transformation T that sends point P onto the unique point P' on

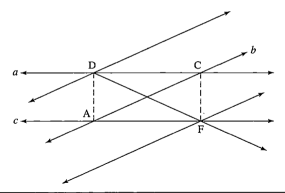

Figure 9.35 Axis of a Euclidean glide when one vertex is at infinity.

ray $P\Omega$ such that $\overline{PP'} = d$. (The analogous transformation in a Euclidean plane would be a translation through distance d.) Describe T explicitly in the Poincaré upper half-plane model when $\Omega = \infty$. (Hint: The Poincaré distance from $x + iy$ to $x + iy'$ is $|\log y/y'|$.) Show that T is not a collineation by showing that the image under T of a line not through ∞ is not a line.

59. Instead of generating the dihedral group D_n by one reflection and one rotation, show that it can be generated by two reflections and give the three basic relations for these generators $(n > 1)$.

60. Find the symmetry group of each of the following infinite patterns (describe the group by generators and relations):
    (i)   . . . LLLLL . . .
    (ii) . . . LΓLΓL . . .
    (iii) . . . VVVVV . . .
    (iv) . . . NNNNN . . .
    (v)   . . . VΛVΛV . . .
    (vi) . . . DDDDD . . .
    (vii) . . . HHHHH . . .

61. Which pairs of groups in the previous exercise are isomorphic? (Hint: See Coxeter, 2001, Section 3.7.)

62. Refer to the model in Exercise 35, Chapter 3, in which length has been changed along the x-axis, causing the SAS criterion to fail. Show that the only similarities that are automorphisms of this model are the ones that leave the x-axis invariant.

63. Suppose the model in the previous exercise is modified so that length along three non-concurrent lines is converted to three different units of measurement from the unit on all other lines. Show that the identity is the only automorphism of this model (hence "nothing can be

moved" in this model, everything is invariant, and Klein's *Erlanger Programme* gives no insight into the "geometry").

64. Define an *incidence plane* to be a model of incidence geometry that is *two-dimensional* in the following sense: If l, m, n are three lines forming a triangle and P any point, then there exists a line t through P such that t meets $l \cup m \cup n$ in at least two points (see Figure 9.36). Show that a model of both the incidence and betweenness axioms is automatically two-dimensional. Let T be a one-to-one mapping of the set of points of an incidence plane onto itself such that if O, P, Q are collinear, then their images O', P', Q' are collinear. Prove that, conversely, if O', P', Q' are collinear, then so are O, P, Q. Is there a model of incidence geometry that is not two-dimensional in which this converse fails? (I don't know the answer.)

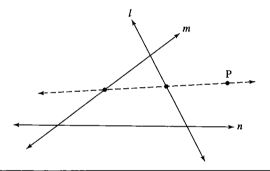

Figure 9.36

65. Prove the following analogue of Proposition 9.16(c) in one-dimensional projective geometry over an arbitrary field K: For any two triples of points on the projective line $\mathcal{P}^1(K)$, there is a unique projective transformation mapping one triple onto the other. (Hint: This is an exercise in two-dimensional linear algebra based on the fact that there is a unique non-singular 2×2 matrix mapping one pair of linearly independent vectors onto another—see Ewald, 1971, p. 215.)

66. Given four points P_1, P_2, P_3, P_4 on the projective line $\mathcal{P}^1(K)$, let the homogeneous coordinates of P_i be $[s_i, t_i]$ for $i = 1, 2, 3, 4$. Define the *cross-ratio* $(P_1 P_2, P_3 P_4)$ by

$$\frac{\begin{vmatrix} s_1 & t_1 \\ s_3 & t_3 \end{vmatrix} \begin{vmatrix} s_2 & t_2 \\ s_4 & t_4 \end{vmatrix}}{\begin{vmatrix} s_2 & t_2 \\ s_3 & t_3 \end{vmatrix} \begin{vmatrix} s_1 & t_1 \\ s_4 & t_4 \end{vmatrix}} = (P_1 P_2, P_3 P_4),$$

where the four terms in this ratio are 2×2 determinants (which are not zero because the points are distinct). If a projective transformation maps point P_i onto P'_i, $i = 1, 2, 3, 4$, prove that the cross-ratio is preserved: $(P_1P_2, P_3P_4) = (P'_1P'_2, P'_3P'_4)$. (Hint: If M is a matrix of the projective transformation, each determinant occurring in the formula for $(P'_1P'_2, P'_3P'_4)$ is the product of det M with the corresponding determinant in the formula for (P_1P_2, P_3P_4).)

67. Since the determinants in the formula for the cross-ratio of Exercise 66 may be negative (when K is an ordered field), this cross-ratio is not the same as the positive cross-ratio defined on p. 319. Show that for points $\neq \infty$, this cross-ratio is the same as the *signed* cross-ratio defined in Exercises H-4 and H-7, Chapter 7. (Hint: Use inhomogeneous coordinates $[s_i, 1]$.)

68. Deduce from Exercises 65 and 66 that if (P_1, P_2, P_3) and (P'_1, P'_2, P'_3) are any two triplets of points on $\mathcal{P}^1(K)$, if T is the unique projective transformation carrying the first triple onto the second, and if P_4 is any fourth point, then the image P'_4 of P_4 under T is uniquely determined by the equation

$$(P_1P_2, P_3P_4) = (P'_1,P'_2, P'_3P'_4).$$

This shows that the cross-ratio is the fundamental invariant of the one-dimensional projective group.

69. Here is an example of an automorphism T of a non-Archimedean Pythagorean plane such that T is not a similarity. Let $F = \mathbb{R}((t))$ be the ordered Pythagorean field of formal Laurent series with real coefficients (see Project 2, Chapter 4) and consider the Cartesian plane F^2. The field F has nontrivial order-preserving automorphisms such as the σ that fixes $\mathbb{R}$ and sends the indeterminate t to $t + 1$. Show that $T(x, y) = (\sigma x, \sigma y)$ defines an automorphism of the Pythagorean plane F^2 and that if $A = (0, 0)$, $B = (1, 0)$, and $C = (1, t)$, then ∢CAB is not mapped onto a congruent angle by T.

70. Justify the assertions made about $H_AH_BH_C$ in the note after the corollary to Proposition 9.15.

71. Add another column to Table 9.1 listing the invariant cycles (if any) for the different types of motion. A *cycle* in a hyperbolic plane is a circle, an equidistant curve, or a horocycle. For example, the only invariant cycles for a rotation about A are the circles with center A.

72. The real projective plane can be extended to the projective plane coordinatized by the complex numbers. Within the latter are two special points at infinity, sometimes referred to as *Isaac and Jacob*: $I = [1, i, 0]$ and $J = [1, -i, 0]$. They are the two points at infinity

that satisfy the equation $x^2 + y^2 = 0$. Show that every circle in $\mathbb{R}^2$ passes through I and J (first explain what that means). It can be shown that, conversely, if a conic in $\mathbb{R}^2$ passes through I and J, then it is a circle, hence I and J are called the *circular points at infinity*. Show that an affine transformation of $\mathbb{R}^2$ is a similarity if and only if, when extended to the complex projective plane in the natural way, it leaves the set {I, J} invariant. (Hint: Theorem 9.3.)

Note: The remaining exercises are in Euclidean geometry.

73. Show that the center U of the nine-point circle is the harmonic conjugate of the circumcenter O with respect to the orthocenter H and the centroid G (see Problem 9, p. 407).

74. Show that the dilation T' with the centroid G as center and ratio -2 also maps the nine-point circle onto the circumcircle (hence, in the terminology of Exercise 29, G and H are the two *centers of similitude* of these circles).

75. Show that the distance from the circumcenter O of a triangle to a side is half the distance from the orthocenter H to the opposite vertex. (Hint: See Problem 8, p. 406.)

76. Justify all the assertions in the solutions to Problems 1–9 that have not been justified there.

77. Report on Feuerbach's theorem, which states that the nine-point circle is tangent to the inscribed circle and the three escribed circles (see H. Eves, 1972, or Coxeter and Greitzer, 1967).

78. Let $\triangle ABC$ be any triangle whose largest angle is $<120°$. For any point P, let $d(P) = \overline{PA} + \overline{PB} + \overline{PC}$. Prove that there is a unique point P_0 at which the function $d(P)$ achieves its minimum value, that P_0 lies in the interior of $\triangle ABC$, that it can be constructed with straightedge and compass, that

$$120° = (\sphericalangle AP_0B)° = (\sphericalangle BP_0C)° = (\sphericalangle CP_0A)°,$$

and that

$$d(P_0)^2 = a^2 + b^2 - 2ab \cos\left(C + \frac{\pi}{3}\right),$$

where $a = \overline{BC}$, $b = \overline{AC}$. (Hint: Rotate $\triangle ABC$ about each of its vertices through 60°. See Kay (1969), p. 271, where P_0 is called the *Fermat point*, or H. Rademacher and O. Toeplitz, *The Enjoyment of Mathematics*, Princeton, N.J.: Princeton University Press, 1957, p. 33.)

Advanced Project on Klein's Erlanger Programme

The *Erlanger Programme* of Klein was not stated precisely. We know that it focuses on the group of automorphisms of a geometry. It gives us much useful information and a new point of view—e.g., we understand the undefined notion of *congruence* of figures in terms of transformations. We also have a precise approach to *symmetry* by way of transformations. We appreciate the importance of *invariants*. We saw that certain geometries can be organized in a hierarchy of groups of their automorphisms (see p. 401).

But this description presumes that we have geometries to begin with. Later on, the process was reversed: One started with a group operating on a set, and with certain restrictions, a "geometry" was said to be given. For a precise definition of these *Klein geometries*, see http://en.wikipedia.org/wiki/Klein_geometry. It is the study of certain *homogeneous spaces*: http://en.wikipedia.org/wiki/Homogeneous_space. Riemannian and certain other geometries are not in general homogeneous.

There is an interesting discussion of the many ramifications of Klein's ideas at http://lists.meer.net/pipermail/junk-l/2005-October/000253.html. You will have to learn much advanced material in order to understand it fully, but you can get the flavor. A good survey article is "What Is Geometry?" by S-s. Chern (*American Mathematical Monthly*, **97**(8) (1990): 679–686).

Read these and other relevant sources that you can find and write a report.

10

Further Results in Real Hyperbolic Geometry

The theorems of this geometry appear to be paradoxical and, to the uninitiated, absurd; but calm, steady reflection reveals that they contain nothing at all impossible.

C. F. Gauss

In this chapter, we will penetrate further into real plane hyperbolic geometry, using the ability to measure with real numbers to calculate basic formulas for length, area, and trigonometry. We will reason from the axioms when that is not overly burdensome; otherwise we will work within the (isomorphic) real Klein or Poincaré models. For certain results, we will refer elsewhere for proofs that are too lengthy to include.[1]

---

[1] The logical development of real hyperbolic geometry strictly from its axioms is carried out in full detail in Borsuk and Szmielew (1960). Hartshorne develops an algebraic analogue of hyperbolic trigonometry, using Hilbert's arithmetic of ends, which is valid in hyperbolic planes over any Euclidean field; a real hyperbolic plane is isomorphic to the real Klein or Poincaré models, so whatever we prove in those models also has a proof from the axioms (see Appendix B, Part I).

Throughout this chapter, we will measure angles in *radians* because experience has shown that formulas come out simpler using radians. The measure of a right angle is then $\pi/2$, and the number of radians in any angle is $\pi/180$ times the number of degrees. As we stated in Chapter 1, the number $\pi = 3.14159 \ldots$ must be defined analytically by some infinite series or infinite product. In a real hyperbolic plane, it is *not* the ratio of circumference to diameter for any circle in the hyperbolic plane, as is proved in Theorem 10.5.

Area and Defect

In 1799, in answer to a letter from Farkas Bolyai in which Bolyai claimed to have proved Euclid's fifth postulate, Gauss wrote:

> . . . the way in which I have proceeded does not lead to the desired goal, the goal that you declare you have reached, but instead to a doubt of the validity of [Euclidean] geometry. I have certainly achieved results which most people would look upon as proof, but which in my eyes prove almost nothing; if, for example, one can prove that there exists a right triangle whose area is greater than any given number, then I am able to establish the entire system of [Euclidean] geometry with complete rigor. Most people would certainly set forth this theorem as an axiom; I do not do so, though certainly it may be possible that, no matter how far apart one chooses the vertices of a triangle, the triangle's area still stays within a finite bound. I am in possession of several theorems of this sort, but none of them satisfy me.[2]

One of the most surprising facts in hyperbolic geometry is that there is an upper limit to the possible area a triangle can have, even though there is *not* an upper limit to the lengths of the sides of the triangle.

To see how this can be, we have to review the way in which area is calculated in Euclidean geometry. The simplest figure is a rectangle, whose area we calculate as the length of the base times the length of the side (Figure 10.1). This formula is arrived at by noticing that exactly bh unit squares fill up the interior of the rectangle, where a unit square is a square whose side has length 1. Keep in mind that the unit of length is arbitrary, so that if we measure area in square inches, we

[2] R. Bonola (1955). Charles Dodgson, on p. 14 of his *Curiosa Mathematica* (1890), rejected hyperbolic geometry because he found it unthinkable that a triangle could retain a finite area when its sides were indefinitely lengthened. An in-depth study of areas of surfaces using calculus shows that area is a rather subtle notion.

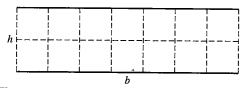

Figure 10.1 Area $= bh$.

get a different number than if we measure in square feet (but the latter number is always proportional to the former, the proportionality factor being $144 = 12^2$).

From the area of a rectangle we can calculate the area of a right triangle. A diagonal of a rectangle divides it into two congruent right triangles, and since we want congruent triangles to have the same area, the area of the right triangle must be half the area of the rectangle (see Figure 10.2).

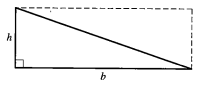

Figure 10.2 Area $= \frac{1}{2}bh$.

We can decompose the interior of an arbitrary triangle into the union of the interiors of two right triangles by dropping an appropriate altitude. Since we want the area of the whole to be the sum of the areas of its parts, we find the area of the triangle to be $\frac{1}{2}b_1h + \frac{1}{2}b_2h$, and since $b = b_1 + b_2$, we again get half the base times the height for the area of a general triangle (Figure 10.3).

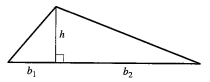

Figure 10.3 Area $= \frac{1}{2}bh$ where $b = b_1 + b_2$.

You can see that by taking b and h to be sufficiently large, the area $\frac{1}{2}bh$ can be made as large as you like.

So why doesn't this work equally well in hyperbolic geometry? Because the whole system of measuring area is based on square units, and as we have seen (Theorem 6.1), rectangles (in particular, squares) do not exist in hyperbolic geometry.

What, then, does "area" mean in hyperbolic geometry? We can certainly say intuitively that it is a way of assigning to every triangle a certain positive number called its *area*, and we want this area function to have the following properties:

1. *Invariance under congruence.* Congruent triangles have the same area.
2. *Additivity.* If a triangle T is split into two triangles T_1 and T_2 by a segment joining a vertex to a point of the opposite side, then the area of T is the sum of the areas of T_1 and T_2 (Figure 10.4).

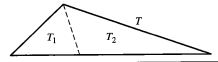

Figure 10.4 Area T = area T_1 + area T_2.

Having described area, we then ask how it is calculated. Here we find one of the most beautiful aspects of mathematics, a direct relationship between two concepts that at first seem totally unrelated. You may have recognized this relationship in reading area properties 1 and 2, for in Proposition 6.1, we proved that the *defect* also has these properties. Recall that in hyperbolic geometry the angle sum of any triangle is always less than π (Theorem 6.1) so that if we define the defect to be π minus the angle sum, we get a positive number. When a mathematician sees two functions with the same properties, he suspects they are related. Gauss discovered this relationship as early as 1794 (he was only 17 years old) and called it the first theorem on the subject.[3]

---

[3] Theorem 10.1 seems to imply that Euclidean and hyperbolic area theories have little in common. Yet F. Bolyai, P. Gerwien, and W. Wallace discovered a wonderful theorem on area that is valid in *both* geometries. Define the area of a polygon to be the sum of the areas of the triangles used to triangulate the polygon. (It is not difficult to show that this definition does not depend on the choice of triangulation.) The theorem states that two polygons S and S' have the same area if and only if for some n, polygon S (respectively, S') has a triangulation $\{T_1, \ldots, T_n\}$ (respectively, $\{T'_1, \ldots, T'_n\}$) such that $T_j \cong T'_j$ for all $j = 1, \ldots, n$. (For a proof, see E. E. Moise (1990, pp. 394–410). To appreciate the depth of this theorem, try to prove it for two rectangles in the Euclidean plane.) Archimedes' axiom is needed for this theorem.

THEOREM 10.1. In hyperbolic geometry, there is a positive constant k such that for any $\triangle ABC$,

$$\text{Area}(\triangle ABC) = k^2 \times \text{defect}(\triangle ABC).$$

For the proof, which is not difficult although it is somewhat lengthy, see Moise (1990, p. 413). The theorem says that the area of any triangle is proportional to its defect, with proportionality constant k^2. This constant depends on the unit of measurement, i.e., on whichever triangle is taken to have area equal to 1.

We can now see why there is an upper limit to the area of all triangles. Namely, the defect measures how much the angle sum is less than π. Since the angle sum can never get below 0, the defect can never get above π. Thus we have the following corollary.

COROLLARY. In hyperbolic geometry, the area of any triangle is at most πk^2.

Of course, there is no finite triangle whose area equals the maximal value πk^2, although we can approach this area as closely as we wish (and achieve it with an infinite trebly asymptotic triangle). However, J. Bolyai showed how to construct a *circle* of area πk^2 and a regular four-sided polygon with a 45° angle that also has this area (see Exercise 29 and the last section of this chapter).

If we were to use hyperbolic geometry to model the universe, astronomical measurement with a fixed star as one vertex of a triangle would give the parallax of the star as an upper bound for the defect (Figure 10.5). But cosmologists do not use such a simple hyperbolic model (see Weeks, 2002, *The Shape of Space*).

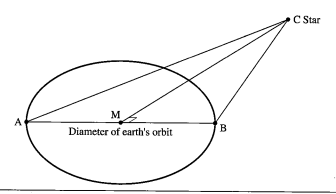

Figure 10.5 Defect$(\triangle ACM)° = 90° - (\sphericalangle CAM)° - (\sphericalangle ACM)° < 90° - (\sphericalangle CAM)° =$ parallax in degrees.

The Angle of Parallelism

Recall that given any line l and any point P not on l, there exist limit-ing parallel rays $\overrightarrow{PX}$ and $\overrightarrow{PY}$ to l that are situated symmetrically about the perpendicular PQ from P to l (Figure 10.6; see Chapter 6). We proved that $\angle XPQ \cong \angle YPQ$ (Proposition 6.6), so either of these angles can be called the *angle of parallelism* for P with respect to l.

It is not difficult to show that the number α of radians in the angle of parallelism depends only on the distance d from P to Q, not on the particular line l or the particular point P (see Major Exercise 5, Chap-ter 6). Lobachevsky denoted α as $\Pi(d)$.

The following formula relating α and d was discovered by J. Bolyai and Lobachevsky.

THEOREM 10.2. Formula of Bolyai–Lobachevsky:

$$\tan \frac{\Pi(d)}{2} = e^{-d/k}.$$

This is certainly one of the most remarkable formulas in all of math-ematics, and it is astonishing how few mathematicians know it. In this formula, the number e is the base for the natural logarithms (e is ap-proximately 2.718 . . .), and $\tan \alpha/2$ is the trigonometric tangent (de-fined analytically, not by opposite over adjacent in a hyperbolic right triangle) of half of α.

J. Bolyai wondered about the meaning of the constant k; we say more about it in the section on curvature. In Theorem 7.2, Chapter 7, we proved this formula for the Poincaré unit disk model and obtained $k = 1$. For a proof of the formula directly from the axioms for the real hyperbolic plane, using corresponding arcs of concentric horocycles, see Borsuk and Szmielew (1960), Chapter 6, Section 26. Hartshorne has a purely algebraic version in terms of the multiplicative length μ

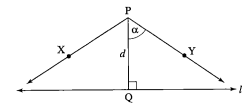

Figure 10.6

of a segment and his algebraically defined tangent half-angle—see Appendix B, Part I.

Cycles

The older proofs of the Bolyai–Lobachevsky formula (Theorem 10.2) all make use of a curve that is peculiar to hyperbolic geometry, called either a *limiting curve* or a *horocycle* in the literature. It is obtained as follows (see also Exercises 41–44, Chapter 9).

Start with a line l and a point Q on l and erect perpendicular $\overleftrightarrow{PQ}$ to l through Q. Then consider the circle δ with center P and radius PQ, which is tangent to l at Q (see Figure 10.7). Now let P recede from l along the perpendicular. The circle δ will increase in size, remaining tangent to l, and will approach a limiting position as P recedes arbitrarily far from Q. In Euclidean geometry, the limiting position of δ would just be the line l, but in hyperbolic geometry the limiting position of δ is a new curve h called a *limiting curve* or *horocycle*.

We can visualize this in the Poincaré model as follows. Let l be a diameter of the Euclidean circle γ whose interior represents the hyperbolic plane and let Q be the center of γ. We proved in Chapter 7 that the hyperbolic circle with hyperbolic center P is represented by a Euclidean circle whose Euclidean center R lies between P and Q (see Figure 10.8).

As P recedes from Q toward the ideal point represented by S, R is pulled up to the Euclidean midpoint of SQ, so that the horocycle h is a Euclidean circle tangent to γ at S and tangent to l at Q. It can be shown in general that all horocycles are represented in the Poincaré model by Euclidean circles inside γ and tangent to γ. Moreover, all the

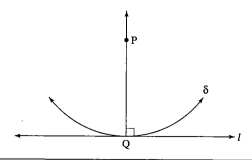

Figure 10.7

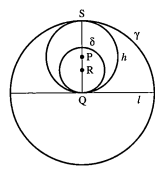

Figure 10.8 Horocycle h in Poincaré model.

Poincaré lines passing through the ideal point S are orthogonal to h; a hyperbolic ray from a point of h out to the ideal point S is called a *diameter* of h (see Exercises 41, 44, and 46, Chapter 9).

In the Poincaré model, two horocycles tangent to γ at S are said to be *concentric* with center S (see Figure 10.9). There is an analogous construction in hyperbolic space called the *horosphere*. Instead of taking the limit of circles to get the horocycle, one takes the limit of spheres to get the horosphere or *limiting surface*.

Another important curve in hyperbolic geometry that has no Euclidean counterpart is the *equidistant curve* (Figure 10.10)—we discussed it in Chapter 5. Start with a line l and a point P not on l. Consider the locus of all points on the same side of l as P and at the same perpendicular distance from l as P. In Euclidean geometry, this locus would just be the unique line through P parallel to l, but in hyperbolic geometry it is not a line—it is the *hypercycle*, or *equidistant curve*, through P.

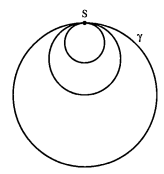

Figure 10.9 Concentric horocycles.

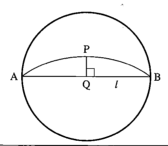

Figure 10.10 Equidistant curve.

In the Poincaré model, let A and B be the ideal endpoints of l. It turns out that the equidistant curve to l through P is represented by the arc of the Euclidean circle passing through A, B, and P (Exercises 40, 45, and 47, Chapter 9). This curve is orthogonal to all Poincaré lines perpendicular to the line l.

In the Euclidean plane, three points lie either on a uniquely determined line or on a uniquely determined circle. Not so in the hyperbolic plane—look at the Poincaré model. A Euclidean circle represents:

1. A *hyperbolic circle* if it is entirely inside γ;
2. A *horocycle* if it is inside γ except for one point where it is tangent to γ;
3. An *equidistant curve* if it cuts γ nonorthogonally in two points;
4. A *hyperbolic line* if it cuts γ orthogonally.

It follows that in the hyperbolic plane, three noncollinear points lie on a circle, a horocycle, or a hypercycle accordingly as the perpendicular bisectors of the triangle are "concurrent" in an ordinary, ideal, or ultra-ideal point—see the penultimate section of this chapter.

The Curvature of the Hyperbolic Plane

One of the difficulties with the Poincaré model is that, although it faithfully represents angles of the hyperbolic plane (i.e., it is a *conformal model*), it distorts distances. So it is natural to ask whether another model exists that also represents hyperbolic lengths faithfully by Euclidean lengths. If there is such a model, it would be called *isometric*. An equally natural idea is to seek as a model some surface in Euclidean three-dimensional space. The lines of the hyperbolic plane would

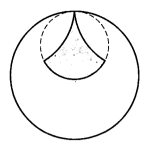

Figure 10.11 Horocyclic sector in Poincaré model.

then be represented by *geodesics* on the surface, and we would expect the surface to be curved so as to mirror our expectation that hyperbolic lines are "really curved." (See Appendix A for the definition of "geodesic." If a shortest path exists on a surface between two given points, it must be an arc of a geodesic; but conversely, an arc of a geodesic need not be the shortest path, for on a sphere, there are two arcs of one great circle joining two points, and if the points are not antipodal, one arc is shorter.)

A difficult theorem of Hilbert (see Do Carmo, 1976, p. 446) states that it is impossible to embed the entire hyperbolic plane isometrically as a surface in Euclidean three-space. On the contrary, it *is* possible to embed the Euclidean plane isometrically in hyperbolic space, as the surface of the horosphere. This result, proved by both J. Bolyai and Lobachevsky, was already recognized by Wachter in 1816.[4]

But all is not lost. Consider a *horocyclic sector*, bounded by an arc of a horocycle and the two diameters cutting off this arc. Such a sector in the Poincaré model is shown in Figure 10.11.

By identifying the bounding diameters of the sector, a surface in Euclidean three-space called the *pseudosphere* is obtained. Gauss called it the "opposite of a sphere." We have already described it and discussed how Beltrami used it in the section about him in Chapter 7. The most beautiful surface of constant negative curvature in Euclidean three-

[4] The smallest n for which it is known at this time that the hyperbolic plane can be isometrically C^∞-embedded in $\mathbb{R}^n$ is $n = 6$ (D. Blanuša, 1955). For the elliptic plane, it is $n = 4$ (Do Carmo, 1976, pp. 436–438). For those who understand this language, strangely enough there does exist a continuously differentiable embedding of the hyperbolic plane into Euclidean three-space. This was proved in 1955 by N. Kuiper, using analytic methods (see *Indagationes Mathematicae*, **17**: 683). T. Milnor proved in 1972 that no "nice" (e.g., C^2) embeddings exist.

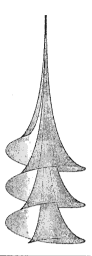

Figure 10.12 Dini's surface.

space is *Dini's surface*, a twisted version of the pseudosphere that opens like a flower growing up to infinity. You will find many pictures of it online; see Figure 10.12 for an example.

F. Minding was the first to publish an article about the pseudosphere, in 1839, but Gauss had written an unpublished note about it around 1827. Curiously, neither of them recognized that it could be used to connect differential geometry to hyperbolic geometry, as Beltrami did in 1868. The pseudosphere is related to the hyperbolic plane as a cylinder is to the Euclidean plane.

The representation on the pseudosphere enables us to give some geometric meaning to the fundamental constant k that appears in Theorems 10.1 and 10.2. The point is that there is a way (discovered by Gauss) of measuring the *curvature* of any surface. We cannot give the precise definition since it involves a knowledge of differential geometry (see Appendix A). In general, the curvature K varies from point to point, being close to zero at points where the surface is rather flat, and large at points where the surface bends sharply. For some surfaces the curvature is the same at all points, so naturally these are called *surfaces of constant curvature K*. An important property of such surfaces is that figures can be moved around on them without changing size or shape.

In 1827, Gauss proved (for constant K) a fundamental formula relating the curvature, area, and angular measure. He took a geodesic triangle $\triangle ABC$ with vertices A, B, and C and sides geodesic segments. By integration, he calculated the area of the triangle. He determined that, if $(\sphericalangle A)^r$ denotes the radian measure of angle A, then

$$K \times \text{area } \triangle ABC = (\sphericalangle A)^r + (\sphericalangle B)^r + (\sphericalangle C)^r - \pi.$$

He then showed what this meant by considering the three possible cases.

CASE 1. K is positive, hence both sides of the equation are positive. In this case, Gauss' formula shows that the angle sum in radians of a geodesic triangle is greater than π and that the area is proportional to the *excess* (the number on the right side of the equation), with a proportionality factor of $1/K$. An example is the surface of a sphere of radius r whose curvature is $K = 1/r^2$. The larger the radius, the smaller the curvature, and the more the surface resembles a plane. (Gauss' formula in the special case of a sphere had already been discovered by Girard in the seventeenth century—Appendix A.) According to a theorem of H. Liebmann, H. Hopf, and W. Rinow, spheres are the only complete surfaces of constant positive curvature in Euclidean three-space, so the elliptic plane cannot be embedded in Euclidean three-space either.

CASE 2. $K = 0$. In this case, Gauss' formula shows that the angle sum in radians is equal to π. An example is the Euclidean plane; another example is an infinitely long cylinder.

CASE 3. K is negative. In this case, Gauss' formula shows that the angle sum in radians is less than π and that the area is proportional to the *defect*. An example of such a surface is the pseudosphere. We can compare Gauss' formula with the formula in Theorem 10.1 relating area to defect. The comparison gives $K = -1/k^2$. Thus $-1/k^2$ is the *curvature of a hyperbolic plane*.

Here we can recapture the analogy with case 1 by setting $r = ik$, where $i = \sqrt{-1}$. Then $K = 1/r^2$, so the hyperbolic plane can be described as a "sphere of imaginary radius $r = ik$," as Lambert speculated (see Chapter 5).

Notice that as k gets very large, the curvature K approaches zero, and the geometry of the surface resembles more and more the geometry of the Euclidean plane. It is in this sense that Euclidean geometry is a "limiting case" of hyperbolic geometry.

We will also see in the next sections that the geometry of an "infinitesimal region" in the hyperbolic plane is Euclidean.

Finally, here is the definition of a real hyperbolic plane from Riemann's viewpoint: It is *a complete, simply connected, two-dimensional Riemannian manifold of constant negative curvature.* (See Appendix A for an explanation of all these terms.) Two such manifolds are isomorphic if and only if they have the same curvature. Similarly, the geometry on different-sized spheres is the same spherical geometry, but different-sized spheres have different curvatures and are not isomorphic as Riemannian manifolds. Curvature explains the constant that puzzled J. Bolyai. No such constant appears in the purely synthetic axiomatic development because it depends on the measurement scale chosen.

We see here the breakthrough Riemann made in expanding our understanding of what is "real" in geometry. When, in 1865, Cayley demanded "a *real* geometric interpretation of Lobachevsky's system of equations," Beltrami in 1868 provided a partial solution with his demonstration that the geometry on a pseudosphere in $\mathbb{R}^3$ is hyperbolic. But a pseudosphere is not a complete hyperbolic plane, and Hilbert proved in 1901 that it is impossible to isometrically embed the hyperbolic plane in $\mathbb{R}^3$. So the hyperbolic plane just lives as an abstract surface in "conceptual space," and mathematicians certainly consider it to be "real." The Poincaré and Klein models are distorted versions of it.

Hyperbolic Trigonometry

Trigonometry is the study of the relationships among the sides and angles of a triangle. We reviewed a few formulas of Euclidean trigonometry in Exercises 14 and 15 of Chapter 5. There, the theory of similar triangles, which is valid only in Euclidean geometry, was used to establish definitions. It defined, for example, the *sine* of an acute angle to be the ratio of the opposite side to the hypotenuse in any right triangle having that acute angle as one of its angles (similarly for the *cosine, tangent,* and other trigonometric functions). We will need to use these functions for hyperbolic trigonometry. What can this mean when these functions have been defined in Euclidean geometry?

An evasive answer is to do hyperbolic trigonometry only in a conformal Euclidean model such as a Poincaré model; since angles are measured in the Euclidean way, the sine, cosine, and tangent of an angle have their usual Euclidean meanings. This answer will have to suffice for those readers who have not yet studied infinite series.

The rigorous answer is to redefine these trigonometric functions purely analytically, without reference to geometry, and then apply them to the different geometries. The definition is in terms of the Taylor series expansions:

$$\sin x = \sum_{n=0}^{\infty} (-1)^n \frac{x^{2n+1}}{(2n+1)!} \qquad \cos x = \sum_{n=0}^{\infty} (-1)^n \frac{x^{2n}}{(2n)!}$$

(and $\tan x = \sin x/\cos x$, etc.). In good calculus texts, one can find proof that these series converge for all x (by the ratio test). Good calculus texts also show how to develop all the familiar formulas for these *circular functions* (so called because $x = \cos \theta$ and $y = \sin \theta$ are parametric equations for the Cartesian unit circle $x^2 + y^2 = 1$; in particular, this equation and the Pythagorean theorem imply that a right triangle in the Euclidean plane with legs of length $\sin \theta$ and $\cos \theta$ has a hypotenuse of length 1, so that $\sin \theta$ is indeed the ratio of opposite side to hypotenuse for the appropriate acute angle in that triangle).

Hyperbolic trigonometry involves, in addition to the circular functions, the *hyperbolic functions,* defined analytically by

$$(1) \qquad \sinh x = \frac{e^x - e^{-x}}{2} \qquad \cosh x = \frac{e^x + e^{-x}}{2}$$

(and $\tanh x = \sinh x/\cosh x$, etc.). These functions were studied by Lambert in 1770. (V. Riccati had written about these functions a few years earlier.) Their graphs are shown in Figure 10.13. The hyperbolic sine and cosine have the Taylor series expansions

$$(2) \qquad \sinh x = \sum_{n=0}^{\infty} \frac{x^{2n+1}}{(2n+1)!} \qquad \cosh x = \sum_{n=0}^{\infty} \frac{x^{2n}}{(2n)!},$$

which are obtained from expansion of the circular sine and cosine by omitting the coefficients $(-1)^n$; this can be seen by recalling that the Taylor series for the exponential function is

$$e^x = \sum_{n=0}^{\infty} \frac{x^n}{n!}.$$

In fact, by introducing the imaginary number $i = \sqrt{-1}$, we see that

$$\sinh x = i \sin \frac{x}{i} \qquad \cosh x = \cos \frac{x}{i}.$$

The name "hyperbolic functions" stems from the identity

$$(3) \quad \cosh^2 x - \sinh^2 x = \frac{e^{2x} + 2 + e^{-2x}}{4} - \frac{e^{2x} - 2 + e^{-2x}}{4} = 1,$$

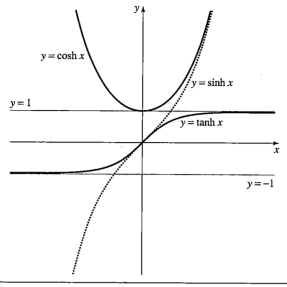

Figure 10.13

from which the parametric equations $x = \cosh\theta$ and $y = \sinh\theta$ give one branch of the hyperbola $x^2 - y^2 = 1$ in the Cartesian plane. (Here the number θ has the geometric interpretation of twice the area bounded by the hyperbola, the x-axis, and the line joining the origin to the point $(\cosh\theta, \sinh\theta)$. There is an analogous interpretation for θ for the circular functions when we replace the hyperbola with the circle.)

(You may well wonder what this hyperbola in the Cartesian model for Euclidean geometry has to do with hyperbolic geometry! Nothing, so far as. I know. Felix Klein coined the names "hyperbolic" and "elliptic" geometries because lines in these geometries have two and zero ideal points at infinity, respectively; this is analogous to affine hyperbolas and ellipses, which have two and zero points at infinity, respectively. A Euclidean line has only one ideal point, and this is analogous to an affine parabola, which has one point at infinity.)

On the following page is a list of identities for hyperbolic and circular functions that will be used in the sequel.

We are going to state the formulas of hyperbolic trigonometry under the simplifying assumption that $k = 1$ (where k is the constant in the Bolyai–Lobachevsky formula); this can be shown to mean that we have chosen our unit of length so that the ratio of the length of

Hyperbolic	Circular
$\cosh^2 x - \sinh^2 x = 1$	$\cos^2 x + \sin^2 x = 1$
$1 - \tanh^2 x = \operatorname{sech}^2 x$	$1 + \tan^2 x = \sec^2 x$
$\sinh(x \pm y) = \sinh x \cosh y$	$\sin(x \pm y) = \sin x \cos y$
$\pm \cosh x \sinh y$	$\pm \cos x \sin y$
$\cosh(x \pm y) = \cosh x \cosh y$	$\cos(x \pm y) = \cos x \cos y$
$\pm \sinh x \sinh y$	$\mp \sin x \sin y$
$\tanh(x + y) = \dfrac{\tanh x + \tanh y}{1 + \tanh x \tanh y}$	$\tan(x + y) = \dfrac{\tan x + \tan y}{1 - \tan x \tan y}$
$\sinh^2 \dfrac{x}{2} = \dfrac{\cosh x - 1}{2}$	$\sin^2 \dfrac{x}{2} = \dfrac{1 - \cos x}{2}$
$\cosh^2 \dfrac{x}{2} = \dfrac{\cosh x + 1}{2}$	$\cos^2 \dfrac{x}{2} = \dfrac{1 + \cos x}{2}$
$\tanh^2 \dfrac{x}{2} = \dfrac{\cosh x - 1}{\cosh x + 1}$	$\tan^2 \dfrac{x}{2} = \dfrac{1 - \cos x}{1 + \cos x}$
$\tanh \dfrac{x}{2} = \dfrac{\sinh x}{\cosh x + 1}$	$\tan \dfrac{x}{2} = \dfrac{\sin x}{1 + \cos x}$
$= \dfrac{\cosh x - 1}{\sinh x}$	$= \dfrac{1 - \cos x}{\sin x}$
$\sinh x = 2 \sinh \dfrac{x}{2} \cosh \dfrac{x}{2}$	$\sin x = 2 \sin \dfrac{x}{2} \cos \dfrac{x}{2}$
$\sinh x \pm \sinh y$	$\sin x \pm \sin y$
$= 2 \sinh \dfrac{1}{2}(x \pm y) \cosh \dfrac{1}{2}(x \mp y)$	$= 2 \sin \dfrac{1}{2}(x \pm y) \cos \dfrac{1}{2}(x \mp y)$
$\cosh x + \cosh y$	$\cos x + \cos y$
$= 2 \cosh \dfrac{1}{2}(x + y) \cosh \dfrac{1}{2}(x - y)$	$= 2 \cos \dfrac{1}{2}(x + y) \cos \dfrac{1}{2}(x - y)$
$\cosh x - \cosh y$	$\cos x - \cos y$
$= 2 \sinh \dfrac{1}{2}(x + y) \sinh \dfrac{1}{2}(x - y)$	$= 2 \sin \dfrac{1}{2}(x + y) \sin \dfrac{1}{2}(x - y)$

corresponding arcs on concentric horocycles is equal to e when the distance between the horocycles is 1 (see Borsuk and Szmielew, 1960, Sections 18–25, Chapter VI). This choice is entirely analogous to the choice of unit of angle measure such that a right angle has (radian) measure $\pi/2$—it makes the formulas come out nicely. Furthermore, the fundamental formula of Bolyai–Lobachevsky for the radian measure of the angle of parallelism becomes

(4) $\Pi(x) = 2 \arctan e^{-x}.$

Straightforward calculation using double angle formulas for the circular functions then yields the following formulas:

(5)	$\sin \Pi(x) = \operatorname{sech} x = 1/\cosh x,$
(6)	$\cos \Pi(x) = \tanh x,$
(7)	$\tan \Pi(x) = \operatorname{csch} x = 1/\sinh x.$

Thus the function Π provides a link between the hyperbolic and the circular functions.

Given $\triangle ABC$, we will use the standard notation $a = \overline{BC}$, $b = \overline{AC}$, $c = \overline{AB}$ for the lengths of the sides. We will write expressions such as "cos A" to abbreviate "cosine of the number of radians in $\sphericalangle A$," and we repeat that this does *not* mean the ratio of adjacent side to hypotenuse in a hyperbolic triangle. We will develop the first formula (10) of hyperbolic trigonometry from the Poincaré model, i.e., using Euclidean trigonometry; after that we can deduce the remaining formulas without referring to the Poincaré model. For segments that are part of diameters of the absolute circle κ, there is ambiguity in the notation for length; as in Chapter 7, we will write $\overline{AB}$ for the Euclidean length and $d(AB)$ for the Poincaré length. We will take κ to have radius 1. For a segment OB with one endpoint at the center O of κ, the proof of Lemma 7.4 showed that

$$e^{d(OB)} = \frac{1 + \overline{OB}}{1 - \overline{OB}}.$$

Writing, for brevity, $x = d(OB)$ and $t = \overline{OB}$ in this formula, a little algebra gives the basic relations

(8) $\sinh x = \dfrac{2t}{1 - t^2} \qquad \cosh x = \dfrac{1 + t^2}{1 - t^2},$

so that

(9) $\tanh x = \dfrac{2t}{1 + t^2} = F(t),$

where F is the isomorphism of the Poincaré model onto the Klein model defined in Chapter 7.

THEOREM 10.3. Given any *right* triangle $\triangle ABC$, with $\sphericalangle C$ right, in the hyperbolic plane (with $k = 1$). Then

(10)
$$\sin A = \frac{\sinh a}{\sinh c} \qquad \cos A = \frac{\tanh b}{\tanh c},$$

(11)
$$\cosh c = \cosh a \cosh b = \cot A \cot B,$$

(12)
$$\cosh a = \frac{\cos A}{\sin B}.$$

PROOF:

Before indicating a proof of this theorem, let us compare these formulas to the formulas for a Euclidean right triangle. The first equality in formula (11) is the *hyperbolic analogue of the Pythagorean theorem;* for if we expand both sides in Taylor series using formula (2), the formula becomes

$$1 + \frac{1}{2}c^2 + \cdots = 1 + \frac{1}{2}(a^2 + b^2) + \cdots.$$

And if we neglect the higher-order terms (when $\triangle ABC$ is sufficiently small), this reduces to

$$c^2 \approx a^2 + b^2.$$

Similarly (when $\triangle ABC$ is sufficiently small), formula (10) becomes approximately

$$\sin A \approx \frac{a}{c} \qquad \cos A \approx \frac{b}{c}.$$

Let us be more precise: Consider right triangles with fixed $\sphericalangle A$ and with $c \to 0$. Then by Proposition 4.5, $a \to 0$ (since $a < c$). By formula (2) and the geometric series formula,

$$\frac{1}{\sinh c} = \frac{1}{c(1 + u)} = \frac{1}{c}(1 - u + u^2 - u^3 + \cdots),$$

where $\lim\limits_{c \to 0} u = 0$. Thus

$$\frac{\sinh a}{\sinh c} = \frac{a}{c}\left(1 + \frac{a^2}{3!} + \frac{a^4}{5!} + \cdots\right)(1 - u + u^2 - u^3 + \cdots),$$

and we see that

$$\lim_{c \to 0} \frac{a}{c} = \lim_{c \to 0} \frac{\sinh a}{\sinh c} = \sin A.$$

A similar argument applies to cos A. So *it is appropriate to say that the hyperbolic trigonometry of "infinitesimal" triangles is Euclidean.*

Formula (12) and the second equality in formula (11) have no counterparts in Euclidean geometry because in Euclidean geometry the angles do not determine the lengths of the sides.

All the geometry of a right triangle is incorporated in formula (10), for all the other formulas follow from (10) by pure algebra and identities. Namely, the identity $\sin^2 A + \cos^2 A = 1$ and (10) give

$$1 = \frac{\tanh^2 b}{\tanh^2 c} + \frac{\sinh^2 a}{\sinh^2 c},$$

$$\sinh^2 c = \cosh^2 c \tanh^2 b + \sinh^2 a,$$

$$1 + \sinh^2 c = \cosh^2 c \left(\frac{\sinh^2 b}{\cosh^2 b}\right) + 1 + \sinh^2 a,$$

$$\cosh^2 c \cosh^2 b = \cosh^2 c \sinh^2 b + \cosh^2 a \cosh^2 b,$$

$$\cosh^2 c(\cosh^2 b - \sinh^2 b) = \cosh^2 a \cosh^2 b,$$

$$\cosh^2 c = \cosh^2 a \cosh^2 b,$$

which gives the first equality in formula (11). Applying formula (10) to B instead of A gives

$$\sin B = \frac{\sinh b}{\sinh c},$$

so that we get formula (12):

$$\frac{\cos A}{\sin B} = \frac{\tanh b}{\tanh c} \frac{\sinh c}{\sinh b} = \frac{\cosh c}{\cosh b} = \cosh a.$$

Multiplying this by the analogous formula for cosh b yields the second equality in formula (11).

Finally, to prove formula (10), we do the geometry under the assumption that vertex A of the right triangle coincides with the center O of the absolute (this can always be achieved by a suitable inversion as in the verification of SAS, Chapter 7). The points B′, C′ in Figure 10.14 are the images of B, C under the isomorphism F (see also Figure 7.35). From Euclidean right triangle $\triangle B'C'O$ and formula (9) we get

$$\cos A = \frac{\overline{OC'}}{\overline{OB'}} = \frac{\tanh b}{\tanh c}.$$

Let B″ be the other intersection of $\overrightarrow{OB}$ with the orthogonal circle κ_1 containing Poincaré line $\overleftrightarrow{BC}$. By Proposition 7.5, B″ is the inverse of B in κ, so that

$$\overline{BB''} = \overline{OB''} - \overline{OB} = \frac{1}{t} - t = \frac{1 - t^2}{t} = \frac{2}{\sinh x}$$

in the notation of formula (8). In the standard notation,

$$\overline{BB''} = \frac{2}{\sinh c} \quad \text{and} \quad \overline{CC''} = \frac{2}{\sinh b}.$$

Let B_1 be the foot of the perpendicular from the center O_1 of κ_1 to BB″, so that B_1 is the midpoint of BB″ (base of an isoceles triangle). Then $\sphericalangle BO_1B_1 \cong \sphericalangle B (= \sphericalangle GBB_1$ in Figure 10.14, where $\overleftrightarrow{GB}$ is the tangent to κ_1 at B), because both these angles are complements of $\sphericalangle B_1BO_1$. Hence

$$\sin B = \frac{\overline{BB_1}}{\overline{O_1B}} = \frac{\overline{BB''}}{2\overline{O_1C}} = \frac{\overline{BB''}}{\overline{CC''}} = \frac{\sinh b}{\sinh c}.$$

Since $\sphericalangle B$ is an arbitrary acute angle in a right triangle, we can relabel, interchanging A and B, to get the second formula in (10). ◄

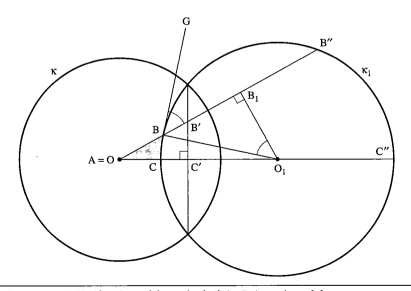

Figure 10.14 Verification of formula (10) in Poincaré model.

THEOREM 10.4. For any $\triangle ABC$ in the hyperbolic plane (with $k = 1$ and standard notation for the sides),

(13) $\qquad \cosh c = \cosh a \cosh b - \sinh a \sinh b \cos C$

(14) $\qquad \dfrac{\sin A}{\sinh a} = \dfrac{\sin B}{\sinh b} = \dfrac{\sin C}{\sinh c},$

(15) $\qquad \cosh c = \dfrac{\cos A \cos B + \cos C}{\sin A \sin B}.$

Formula (13) is the *hyperbolic law of cosines*, and formula (14) is the *hyperbolic law of sines*; they are analogous to the Euclidean laws and reduce to the latter for "infinitesimal" triangles as before. Formula (15) has no Euclidean analogue.

This theorem can be proved by dropping an altitude to create two right triangles and by applying the preceding theorem, some algebra, and identities such as

$$\cosh(x \pm y) = \cosh x \cosh y \pm \sinh x \sinh y.$$

We leave the details as an exercise.

NOTE ON ELLIPTIC GEOMETRY. Analogously, elliptic geometry with $k = 1$ can be developed from its model on a sphere of radius 1 with antipodal points identified (see Kay, 1969, Chapter 10). The elliptic law of cosines is

$$\cos c = \cos a \cos b + \sin a \sin b \cos C,$$

and the elliptic law of sines is

$$\frac{\sin A}{\sin a} = \frac{\sin B}{\sin b} = \frac{\sin C}{\sin c}.$$

In a triangle with right angle at C, the elliptic analogue of formula (10) is

$$\sin A = \frac{\sin a}{\sin c} \qquad \cos A = \frac{\tan b}{\tan c}.$$

In each of these formulas, if we replace the ordinary trigonometric functions by the corresponding hyperbolic trigonometric functions whenever the argument is the length of a side, we get the formulas above for hyperbolic geometry with $k = 1$. These replacements were first noticed by Lambert and Taurinus with regard to spherical trigonometry (see Chapter 5).

Circumference and Area of a Circle

THEOREM 10.5 (GAUSS). The circumference C of a circle of radius r is given by $C = 2\pi \sinh r$.

PROOF:

Of course C is defined as the limit $\lim\limits_{n\to\infty} p_n$ of the perimeter p_n of the regular n-gon inscribed in the circle (Figure 10.15). Recall first how the formula $C = 2\pi r$ is derived in Euclidean geometry. From Figure 10.15 and Euclidean trigonometry, we see that

$$p_n = r2n \sin \frac{\pi}{n} = r2n\left[\frac{\pi}{n} - \frac{1}{3!}\left(\frac{\pi}{n}\right)^3 + \frac{1}{5!}\left(\frac{\pi}{n}\right)^5 - \cdots\right]$$

$$= 2\pi r - \frac{2r\pi^2}{n^2}\left[\frac{\pi}{3!} - \frac{1}{5!}\frac{\pi^3}{n^2} + \cdots\right]$$

$$\lim_{n\to\infty} p_n = 2\pi r.$$

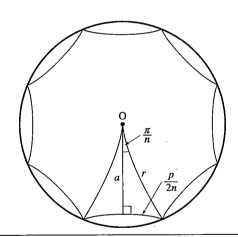

Figure 10.15

In the hyperbolic case, we use instead formula (10) of Theorem 10.3 to get

$$\sinh(p/2n) = \sinh r \sin(\pi/n),$$

which when expanded in series becomes

$$\frac{p}{2n}\left[1 + \frac{1}{3!}\left(\frac{p}{2n}\right)^2 + \frac{1}{5!}\left(\frac{p}{2n}\right)^4 + \cdots\right]$$

$$= \frac{\pi}{n}\sinh r\left[1 - \frac{1}{3!}\left(\frac{\pi}{n}\right)^2 + \frac{1}{5!}\left(\frac{\pi}{n}\right)^4 - \cdots\right]$$

(where $p = p_n$ for typographical simplicity). Multiplying both sides by $2n$ and taking $\lim\limits_{n \to \infty}$ gives the formula we seek. (Note once more that for a circle of "infinitesimal radius," the hyperbolic formula reduces to the Euclidean formula.) ◄

This theorem enables us to rewrite the *law of sines* (14) *in a form that is valid in neutral geometry.*

COROLLARY (J. BOLYAI). The sines of the angles of a triangle are to one another as the circumference of the circles whose radii are equal to the opposite sides.

Bolyai denoted the circumference of a circle of radius r by Or and wrote this result in the form

$$Oa : Ob : Oc = \sin A : \sin B : \sin C.$$

Next, consider formulas for *area*. By Theorem 10.1 and our convention $k = 1$, the area K of a triangle is equal to its defect in radians; i.e.,

$$K = \pi - A - B - C.$$

Let us calculate this defect for a right triangle with right angle at C, so that $K = \pi/2 - (A + B)$.

THEOREM 10.6. $\tan K/2 = \tanh a/2 \tanh b/2$. (For Euclidean geometry, the formula for area K is $K/2 = a/2 \cdot b/2$.)

PROOF:

Here are the main steps in the proof:

$$\tanh^2 \frac{a}{2} \tanh^2 \frac{b}{2} = \frac{(\cosh a) - 1}{(\cosh a) + 1} \frac{(\cosh b) - 1}{(\cosh b) + 1}$$

$$= \frac{1 - \sin(A + B)}{1 + \sin(A + B)} \frac{\cos (A - B)}{\cos (A - B)}$$

$$= \frac{1 - \cos K}{1 + \cos K}$$

$$= \tan^2 \frac{K}{2}.$$

Steps 1 and 4 are just identities for $\tanh^2(x/2)$ and $\tan^2(x/2)$, respectively. Step 2 follows from substituting formula (12) for $\cosh a$

and cosh b and doing a considerable amount of algebra using trigonometric identities.[5] Step 3 just uses the identity $\cos(\pi/2 - x) = \sin x$. ◄

THEOREM 10.7. The area of a circle of radius r is $4\pi \sinh^2(r/2) = 2\pi(\cosh r - 1)$.

PROOF:

Here again we define the area A of a circle to be the limit $\lim\limits_{n\to\infty} K_n$ of the area K_n of the inscribed regular n-gon. Referring to Figure 10.15 again, using the previous theorem, and writing a, K, p, for a_n, K_n, p_n, we have

$$\tan \frac{K}{4n} = \tanh \frac{p}{4n} \tanh \frac{a}{2}.$$

If we multiply both sides by $4n$ and pass to the limit as $n \to \infty$, we obtain

(16) $$A = C \tanh \frac{r}{2},$$

using $\lim\limits_{n\to\infty} p_n = C$, $\lim\limits_{n\to\infty} a_n = r$, continuity of tan and tanh, and the series

$$4n \tan \frac{K}{4n} = K + \frac{K}{3}\left(\frac{K}{4n}\right)^2 + \cdots,$$

$$4n \tanh \frac{p}{4n} = p - \frac{p}{3}\left(\frac{p}{4n}\right)^2 + \cdots.$$

Then we substitute in formula (16) the formula for C from Theorem 10.5 and use the identities

$$\tanh \frac{r}{2} = \frac{\sinh r}{\cosh r + 1},$$

$$\sinh^2 r = \cosh^2 r - 1,$$

$$2 \sinh^2 \frac{r}{2} = \cosh r - 1,$$

to obtain Theorem 10.7. ◄

[5] See Exercise 5. From now on, you will be offered the opportunity to exercise your algebraic technique to fill in such gaps.

By expanding this formula in a series, we can show how much larger the area of a hyperbolic circle is than that of a Euclidean circle with the same radius:

$$A = \pi\left(r^2 + \frac{r^4}{12} + \cdots\right).$$

EXAMPLE 1. Consider any trebly asymptotic triangle. In Exercise K-13, Chapter 7, you showed (in the Klein model) that its altitudes are concurrent in a point G equidistant from its sides, and the distance r of G from the sides satisfies $\Pi(r) = \pi/3$. Applying formulas (5)–(7) and Theorems 10.5 and 10.7, we get $\tanh r = \frac{1}{2}$, and we calculate that the inscribed circle δ of a trebly asymptotic triangle has circumference $2\pi/\sqrt{3}$ and area $\frac{2}{3}\pi(2\sqrt{3} - 3)$.

EXAMPLE 2. Consider the *pedal triangle* formed by the feet of the altitudes of a trebly asymptotic triangle. It is an equilateral triangle (by Proposition 9.16(c)), δ is its circumscribed circle with center G, and by the hyperbolic law of cosines applied to the isosceles triangle formed by G and two of the vertices of the pedal triangle, its side c satisfies $\cosh c = \frac{3}{2}$. Hence the circle with radius c has area π (like the Euclidean circle of radius 1) and circumference $\pi\sqrt{5}$.

EXAMPLE 3. Let p be Schweikart's segment, determined by $\Pi(p) = \pi/4$ (equivalently, by formula (7), $\sinh p = 1$). Then $\cosh p = \sqrt{2}$, so the circle with radius p has circumference 2π (like the Euclidean circle of radius 1) and area $2\pi(\sqrt{2} - 1)$.

EXAMPLE 4. Let q be the hypotenuse of the isosceles right triangle with legs p. By the hyperbolic Pythagorean theorem (formula (11) in Theorem 10.3), $\cosh q = 2$. Then the circle with radius q has area 2π. Since $\sinh q = \sqrt{3}$, the circumference is $2\pi\sqrt{3}$.

NOTE ON ELLIPTIC GEOMETRY. The formulas for circumference and area of a circle of radius r are

$$C = 2\pi \sin r,$$

$$A = 4\pi \sin^2(r/2).$$

Bolyai's formula is valid in elliptic geometry (so it is indeed a theorem in *absolute geometry*).

Saccheri and Lambert Quadrilaterals

We next consider a Saccheri quadrilateral with base b, legs of length a, and summit of length c. We saw in Basic Theorem 6.1 that $c > b$. We now make this more precise.

THEOREM 10.8. For a Saccheri quadrilateral,

$$\sinh \frac{c}{2} = \cosh a \sinh \frac{b}{2}.$$

(Since $\cosh^2 a = 1 + \sinh^2 a > 1$, we have $\sinh(c/2) > \sinh(b/2)$, hence $c > b$.)

PROOF:

Theorem 10.8 is proved by letting $d = \overline{AB'}$ and $\theta = (\sphericalangle A'AB')^r$ in Figure 10.16, applying formula (13) from Theorem 10.4 to get

$$\cosh c = \cosh a \cosh d - \sinh a \sinh d \cos \theta,$$

using formulas (10) and (11) from Theorem 10.3 to get

$$\cos \theta = \sin\left(\frac{\pi}{2} - \theta\right) = \frac{\sinh a}{\sinh d},$$

$$\cosh d = \cosh a \cosh b,$$

and eliminating d to obtain

$$\cosh c = \cosh^2 a \cosh b - \sinh^2 a$$

$$= \cosh^2 a(\cosh b - 1) + 1.$$

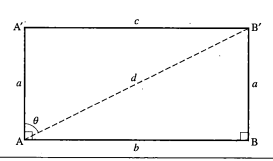

Figure 10.16

Finally, the identity

$$2 \sinh^2 \frac{x}{2} = \cosh x - 1$$

gives the result. ◄

Corollary. Given a Lambert quadrilateral, if c is the length of a side adjacent to the acute angle and b is the length of the opposite side, then

$$\sinh c = \cosh a \sinh b,$$

where a is the length of the other side adjacent to the acute angle (in particular, $c > b$).

The corollary follows from representing the Lambert quadrilateral as half of a Saccheri quadrilateral (see Figure 10.17). There are additional remarkable formulas for Lambert quadrilaterals that we will derive next. They are based on the concept of *complementary segments:* These are segments whose lengths x, x^* are related by

$$(17) \qquad\qquad \Pi(x) + \Pi(x^*) = \frac{\pi}{2}.$$

The geometric meaning of this equation is shown in Figure 10.18, where the "fourth vertex" of the Lambert quadrilateral is the ideal point Ω.

If we apply our earlier formulas (4)–(7) for the angle of parallelism, we obtain

$$(18) \qquad\qquad \sinh x^* = \operatorname{csch} x,$$

$$(19) \qquad\qquad \cosh x^* = \coth x,$$

$$(20) \qquad\qquad \tanh x^* = \operatorname{sech} x,$$

$$(21) \qquad\qquad \tanh \frac{x^*}{2} = e^{-x}.$$

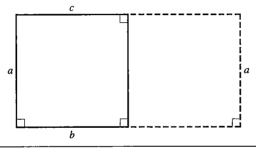

Figure 10.17 Lambert quadrilateral is half a Saccheri quadrilateral.

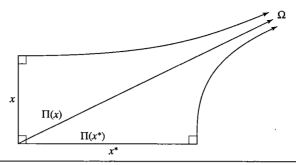

Figure 10.18 Complementary segments.

For example: $\sinh x^* = \cot \Pi(x^*) = \tan \Pi(x) = \operatorname{csch} x$ by formula (7); formula (21) follows from formulas (18), (19), and the identity

$$\tanh(t/2) = (\sinh t)/(1 + \cosh t).$$

THEOREM 10.9 (ENGEL'S THEOREM).   There exists a right triangle with the parameters shown in Figure 10.19 if and only if there exists a Lambert quadrilateral with the parameters shown in Figure 10.20. Note that PQ is a complementary segment to the segment (not shown) whose angle of parallelism is ∢A.

The geometric meaning of Engel's theorem is shown in Figure 10.21. It includes J. Bolyai's parallel construction (Figure 6.11), for if B = X is the point between R and S such that PX ≅ QR, Engel's theorem says $(∢BAC)^r = \Pi(\overline{PQ}^*)$, and since $(∢QPX)^r = \pi/2 - (∢BAC)^r$, $(∢QPX)^r = \Pi(\overline{PQ})$; i.e., $\overrightarrow{PX}$ is limiting parallel to $\overrightarrow{QR}$.

Engel's theorem also says that the ray emanating from R limiting parallel to $\overrightarrow{SP}$ makes an angle with $\overrightarrow{RS}$ that is congruent to ∢ABC, and that the ray emanating from X limiting parallel to $\overrightarrow{SP}$ makes an angle with $\overrightarrow{XS}$ that is congruent to the acute ∢R of the Lambert quadrilateral.

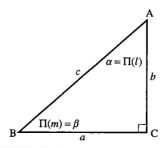

Figure 10.19

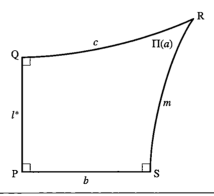

Figure 10.20

PROOF:

For the proof, start with a Lambert quadrilateral labeled as in Figure 10.22. We've already shown that

(i) $\quad\quad\quad\quad\quad\quad\quad \sinh w = \cosh z \sinh v,$

(i') $\quad\quad\quad\quad\quad\quad\quad \sinh z = \cosh w \sinh u.$

Let $\theta = (\sphericalangle SPR)^r$, $d = \overline{PR}$. By Theorem 10.3, $\sinh w = \sin \theta \sinh d = \cos(\pi/2 - \theta) \sinh d = \tanh v \cosh d = \tanh v(\cosh u \cosh w)$, so that

(ii) $\quad\quad\quad\quad\quad\quad\quad \tanh w = \tanh v \cosh u,$

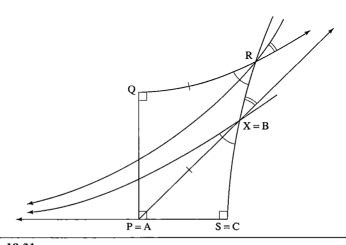

Figure 10.21

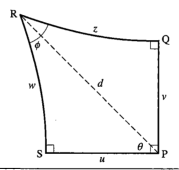

Figure 10.22

and by symmetry,

(ii′) $\tanh z = \tanh u \cosh v.$

Let $\phi = (\sphericalangle R)^r$. By the law of sines and Theorem 10.3,

$$\frac{\sin \phi}{\sin \overline{QS}} = \frac{\sin(\sphericalangle QSR)^r}{\sinh z} = \frac{\cos(\sphericalangle PSQ)^r}{\sinh z} = \frac{\tanh u}{\tanh \overline{QS} \sinh z},$$

so by formula (i′) and Theorem 10.3 we have

(iii) $\sin \phi = \dfrac{\tanh u \cosh \overline{QS}}{\sinh z} = \dfrac{\tanh u(\cosh u \cosh v)}{\sinh u \cosh w} = \dfrac{\cosh v}{\cosh w},$

and by symmetry,

(iii′) $\sin \phi = \dfrac{\cosh u}{\cosh z}.$

Now let X be the point between R and S such that $\overline{PX} = z$ and consider right triangle $\triangle PSX$ (Figure 10.21). By formulas (i′), (ii′), and (iii′), respectively, we get (using Theorem 10.3)

$$\sin(\sphericalangle PXS)^r = \frac{\sinh u}{\sinh z} = \operatorname{sech} w,$$

$$\cos(\sphericalangle XPS)^r = \frac{\tanh u}{\tanh z} = \operatorname{sech} v = \tanh v^*,$$

$$\cosh \overline{XS} = \frac{\cosh z}{\cosh u} = \csc \phi,$$

so that

$$(\sphericalangle PXS)^r = \Pi(w),$$
$$(\sphericalangle XPS)^r = \Pi(v^*),$$
$$\Pi(\overline{XS}) = \phi,$$

by formulas (5), (6), and (5), respectively. Thus if we relabeled P as A, X as B, and S as C, we would obtain right triangle ∢ABC corresponding to our given Lambert quadrilateral as asserted.

Conversely, given right triangle △PSX, we can recover □PQRS by setting R equal to the unique point on $\overrightarrow{SX}$ such that $\Pi(\overline{RS}) = (\angle PXS)^r$ and setting Q equal to the foot of the perpendicular from R to the line through P perpendicular to $\overleftrightarrow{PS}$. ◄

The correspondence in Theorem 10.9 provides for a whole series of existence theorems. For example, it says that from the existence of a right triangle with parameters $(a, \Pi(m), c, \Pi(l), b)$ we can deduce the existence of a Lambert quadrilateral with parameters $(l^*, c, \Pi(a), m, b)$, as in Figure 10.20, ordering the parameters by a clockwise progression in the figure. Now read the parameters backward! This gives Figure 10.23, from which we deduce the existence of a second right triangle in Figure 10.24 having parameters $(a, \Pi(c), m, \Pi(b^*), l^*)$.

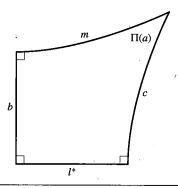

Figure 10.23

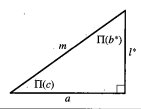

Figure 10.24

We can continue the process of reading these parameters backward, obtaining a second Lambert quadrilateral, etc. We then end up with the existence of five Lambert quadrilaterals and four other right triangles that are implied by the existence of the first right triangle. The results are summarized in the following tabulation.

| $\triangle ABC$, $\sphericalangle C$ right | | | | | Lambert $\square SPQR$, $\sphericalangle R$ acute | | | | |
BC	$\sphericalangle B$	AB	$\sphericalangle A$	AC	PQ	QR	$\sphericalangle R$	RS	SP
a	$\Pi(m)$	c	$\Pi(l)$	b	l^*	c	$\Pi(a)$	m	b
a	$\Pi(c)$	m	$\Pi(b^*)$	l^*	c^*	m	$\Pi(l^*)$	b^*	a
l^*	$\Pi(m)$	b^*	$\Pi(a^*)$	c^*	m^*	b^*	$\Pi(c^*)$	a^*	l^*
c^*	$\Pi(b^*)$	a^*	$\Pi(l)$	m^*	b	a^*	$\Pi(m^*)$	l	c^*
m^*	$\Pi(a^*)$	l	$\Pi(c)$	b	a	l	$\Pi(b)$	c	m^*

Note also that since Theorem 10.3 gave us formulas showing how a right triangle is uniquely determined by any two of its five parameters, Theorem 10.9 gives us the same result for a Lambert quadrilateral (e.g., starting with u and v, w is given by formula (ii), z by formula (ii'), and ϕ by formula (iii) in the proof of Theorem 10.9).

RIGHT TRIANGLE CONSTRUCTION THEOREM. Given positive numbers λ, μ such that $\mu + \lambda < \pi/2$. Then a right triangle can be constructed having λ, μ as the measures of its acute angles (that triangle is unique up to congruence by AAA).

PROOF:

We will use the same letters for the angles as for their angle measures. Let μ^* be the complementary angle to μ and construct the length m^* such that $\Pi(m^*) = \mu^*$. On one side of angle λ, lay off a segment of length m^*. Since $\mu + \lambda < \pi/2$, we have $\lambda < \mu^*$. It follows from the definition of the angle of parallelism that the perpendicular ray from the end of that segment intersects the other side of angle λ to form a right triangle. Let the other leg have length c^* and the hypotenuse length a^*. Construct the angles $\Pi(c^*)$ and $\Pi(a^*)$, their complements γ and α, and then the lengths c, a such that $\Pi(c) = \gamma$ and $\Pi(a) = \alpha$. Now on the sides of angle μ, lay off segments of lengths c, a. Join the endpoints of those segments. The result is a right triangle in which a is a leg, c is the hypotenuse, and the other acute angle is λ: We have gone from the right triangle in line 4 of the table above to the right triangle in line 1 ($\lambda = \Pi(l)$ and $\mu = \Pi(m)$). ◀

COROLLARY. If $\lambda < \pi/3$, an equilateral triangle can be constructed with angle measure λ. For any segment, an equilateral triangle can be constructed having that segment as a side.

PROOF:

Take $\mu = \frac{1}{2}\lambda$. Then the right triangle so constructed can be doubled to form an equilateral triangle with angle λ. Given a segment a, use our trigonometric formulas for the right triangle with legs a, $a/2$ to determine the angle for its equilateral triangle. (Note that Euclid's proof of his I.1 cannot·be used until it is proved that circle-circle continuity holds in any hyperbolic plane—see ·Appendix B.) ◄

Coordinates in the Real Hyperbolic Plane

Choose perpendicular lines through an origin O and fix coordinate systems on each of them so that they can be called the u-axis and v-axis. For any point P, let U and V be the perpendicular projections of P on these axes and let u and v be the respective coordinates of U and V. We then have a Lambert quadrilateral $\square$UOVP. We label the remaining sides with coordinates w, z such that

(22) $\tanh w = \tanh v \cosh u,$

                  $\tanh z = \tanh u \cosh v$

(see Figure 10.25). Formulas (ii) and (ii′) in the proof of Theorem 10.9 showed that if P is in the first quadrant (i.e., $u > 0$ and $v > 0$), then $w = \overline{PU}$ and $z = \overline{PV}$. We·also set

(23) $x = \tanh u, \quad y = \tanh v,$

(24) $T = \cosh u \cosh w, \quad X = xT, \quad Y = yT.$

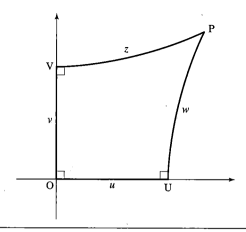

Figure 10.25

Then we call (u, v) the *axial coordinates*, (u, w) the *Lobachevsky co-ordinates*, (x, y) the *Beltrami coordinates*, and (T, X, Y) the *Weierstrass coordinates* of point P. Lobachevsky coordinates are the main tool for calculus in the real hyperbolic plane (see Martin, 1982, or Wolfe, 1945). Beltrami coordinates are used to prove that the abstract real hyperbolic plane is isomorphic to the Klein model (see Theorem 69, Chapter VI, of Borsuk–Szmielew, 1960).

THEOREM 10.10 (still assuming $k = 1$). Assigning to each point P its pair (x, y) of Beltrami coordinates gives an isomorphism of the hyperbolic plane onto the Beltrami–Klein model. In particular, we see that $Ax + By + C = 0$ is an equation of a line in Beltrami coordinates if and only if $A^2 + B^2 > C^2$ and every line has such an equation. The distance $\overline{P_1 P_2}$ between two points is given in terms of Beltrami coordinates by

$$(25) \qquad \cosh \overline{P_1 P_2} = \frac{p_1 \cdot p_2}{\|p_1\| \, \|p_2\|},$$

where $p_i = (1, x_i, y_i)$, the inner product $p_1 \cdot p_2$ is defined by

$$p_1 \cdot p_2 = 1 - x_1 x_2 - y_1 y_2,$$

and $\|p_i\| = \sqrt{p_i \cdot p_i}$. Similarly, if $A_i x + B_i y + C_i = 0$ are the equations of two lines l_i, $i = 1, 2$, intersecting in a nonobtuse angle of radian measure θ, then

$$(26) \qquad \cos \theta = \frac{l_1 \cdot l_2}{\|l_1\| \, \|l_2\|},$$

where now the inner product is defined by

$$l_1 \cdot l_2 = A_1 A_2 + B_1 B_2 - C_1 C_2$$

and $\|l_i\| = \sqrt{l_i \cdot l_i}$ (in particular, $0 = l_1 \cdot l_2$ is the necessary and sufficient condition for the lines to be perpendicular).

Assigning to each point P its triple (T, X, Y) of Weierstrass coordinates maps the hyperbolic plane onto the locus

$$T^2 - X^2 - Y^2 = 1, \qquad T \geqq 1,$$

which is one of the two sheets of a hyperboloid in Cartesian three-space. The equation of a line in Weierstrass coordinates is linear homogeneous (i.e., of the form $AX + BY + CT = 0$).

Before giving the proof, note that the Weierstrass representation gives one interpretation of the hyperbolic plane as a "sphere of imaginary radius i." Namely, if we replace the usual positive definite quadratic form $X^2 + Y^2 + T^2$ (that measures distance squared from the origin) with the indefinite quadratic form $X^2 + Y^2 - T^2$, then the sphere of radius i with respect to this "distance" has the equation

$$X^2 + Y^2 - T^2 = i^2 = -1,$$

which is the equation of a hyperboloid. This indefinite metric is the three-dimensional analogue of the metric determined by the form $X^2 + Y^2 + Z^2 - T^2$ in four-dimensional space-time, which is used for special relativity (see Taylor and Wheeler, 1992). Note that the "lines" in the Weierstrass model are intersections with the sheet of the hyperboloid of Euclidean planes through the origin. To picture this model, just imagine one branch of the hyperbola $T^2 - X^2 = 1$ in the (T, X) plane rotated around the T-axis (see Figure 7.19, p. 312).

PROOF:

The proof of Theorem 10.10 is based on the trigonometry of Lambert quadrilaterals obtained in the preceding theorem.

As the graph in Figure 10.13 showed, $u \to \tanh u$ is a one-to-one mapping of the entire real line onto the open interval $(-1, 1)$. That the pairs (x, y) of Beltrami coordinates satisfy the relation $x + y^2 < 1$ follows from the fact that the perpendiculars to the axes at U and V intersect if and only if $|u| < |v|^*$ (see Figure 10.18); i.e.,

$$\tanh^2 u < \tanh^2 |v|^* = \text{sech}^2 v = 1 - \tanh^2 v$$

(using formula (20)).

To derive the distance formula, introduce the *polar coordinates* (r, θ) for point P in Figure 10.25 defined by

$$r = \overline{OP},$$

$$\theta = \begin{cases} \sphericalangle(XOP)^r & \text{if } v \geq 0, \\ -\sphericalangle(XOP)^r & \text{if } v \leq 0. \end{cases}$$

The relations with axial coordinates are then

(27) $\tanh r \cos \theta = \tanh u = x,$

          $\tanh r \sin \theta = \tanh v = y,$

by formula (10) for the cosine of an angle in a right triangle and the identity $\sin \theta = \cos(\pi/2 - \theta)$. Hence

$$\tanh^2 r = \tanh^2 u + \tanh^2 v = x^2 + y^2.$$

From the identity $\text{sech}^2 r = 1 - \tanh^2 r$, we get

$$\cosh r - (1 - x^2 - y^2)^{-1/2} = \|p\|^{-1}$$

if $p = (1, x, y)$, which is the distance formula when we have $P_1 = P$ and $P_2 = O$. For general P_1 and P_2, formula (27) gives

$$\cos(\theta_2 - \theta_1) = \cos\theta_1 \cos\theta_2 + \sin\theta_1 \sin\theta_2$$

$$= \frac{x_1 x_2 + y_1 y_2}{\tanh r_1 \tanh r_2}.$$

Suppose first that O, P_1, P_2 are collinear, so we have $\cosh \overline{P_1 P_2} = \cosh(r_1 \pm r_2)$. Since $\cos(\theta_2 - \theta_1) = \pm 1$,

$$\cosh \overline{P_1 P_2} = \cosh r_1 \cosh r_2 - \sinh r_1 \sinh r_2 \cos(\theta_2 - \theta_1)$$

$$= \cosh r_1 \cosh r_2[1 - \tanh r_1 \tanh r_2 \cos(\theta_2 - \theta_1)].$$

But this formula also holds when O, P_1, P_2, are not collinear by the law of cosines (13). Substituting the two preceding formulas gives the desired formula (25).

To show that the mapping $P \to (x, y)$ sends hyperbolic lengths onto Klein lengths means reconciling formula (25) with the formula in Exercise K-14, Chapter 7. This follows from a calculation based on the formula

$$(28) \qquad \tanh \overline{P_1 P_2} = \frac{[(x_1 - x_2)^2 + (y_1 - y_2)^2 - (x_1 y_2 - x_2 y_1)^2]^{1/2}}{p_1 \cdot p_2}$$

and the identity

$$(29) \qquad\qquad \text{arctanh } t = \frac{1}{2} \ln \frac{1 + t}{1 - t}.$$

Formula (28) is obtained from formula (25) by means of the identity $\tanh^2 t = 1 - \cosh^{-2} t$. (The term in brackets on the right side of formula (28) could be written as $(p_1 \cdot p_2)^2 - \|p_1\|^2\|p_2\|^2$. Incidentally, the $\frac{1}{2}$ occurring in formula (29) explains why the factor $\frac{1}{2}$ appeared in the formula for Klein length in Theorem 7.4.)

Because $P \to (x, y)$ is an isometry, it is a collineation; so lines in the hyperbolic plane are mapped onto chords of the absolute in the Klein model, which have linear equations as described in the theorem.

The formula (26) for $\cos\theta$ is an assertion about angle measure in the Klein model once we pass to that model by means of the isomorphism $P \to (x, y)$. Suppose the two lines meet at point P_0 with coordinates (x_0, y_0) and suppose we write the ith line as $\overleftrightarrow{P_0P_i}$, where P_i has coordinates (x_i, y_i), $i = 1, 2$. Then the coefficients in the equation for the ith line are given by $A_i = y_i - y_0$, $B_i = x_0 - x_i$, $C_i = x_iy_0 - y_ix_0$. Suppose $P_0 = O$, the center of the absolute. Then formula (26) reduces to

$$\cos\theta = \frac{x_1x_2 + y_1y_2}{(x_1^2 + y_1^2)^{1/2}(x_2^2 + y_2^2)^{1/2}},$$

which is the Euclidean formula for the cosine of the angle $\sphericalangle P_1OP_2$. But the Klein model is conformal at the special point O, so we have verified formula (26) in this case.

If $P_0 \neq O$, let us find a hyperbolic motion T such that $T(O) = P_0$ and let $Q_i = T^{-1}(P_i)$. Since T preserves angle measure, all we then have to do is show that formula (26) is equal to the cosine of $\sphericalangle Q_1OQ_2$. The natural candidate for T is the reflection across the perpendicular bisector of OP_0. We need two lemmas (which are generalized in Exercise 9).

LEMMA 10.1 The coordinates of the Klein midpoint M of O and P are

$$\left(\frac{x}{1 + \|p\|}, \frac{y}{1 + \|p\|}\right),$$

where $\|p\| = \sqrt{1 - x^2 - y^2}$ and P has coordinates (x, y).

PROOF:

Let $r = \overline{OP}$; we've seen that $\cosh r = \|p\|^{-1}$, $x = \tanh r \cos\theta$, $y = \tanh r \sin\theta$. The coordinates (x', y') of M are then given by $x' = \tanh(r/2) \cos\theta$, $y' = \tanh(r/2) \sin\theta$, i.e., $x' = x \tanh(r/2)/\tanh r$, $y' = y \tanh(r/2)/\tanh r$. But

$$\frac{\tanh(r/2)}{\tanh r} = \frac{\sinh r}{\cosh r + 1} \cdot \frac{\cosh r}{\sinh r}$$

$$= \left(1 + \frac{1}{\cosh r}\right)^{-1} = (1 + \|p\|)^{-1}. \blacktriangleleft$$

LEMMA 10.2. The perpendicular bisector of OP_0 has the equation $x_0x + y_0y + \|p_0\| - 1 = 0$, where $\|p_0\| = \sqrt{1 - (x_0^2 + y_0^2)}$ and P_0 has coordinates (x_0, y_0).

PROOF:

The perpendicular bisector of OP_0 passes through the midpoint and has slope $-x_0/y_0$ (since Klein perpendicularity is the same as

Euclidean perpendicularity when one chord is a diameter of the absolute). ◄

If we now apply the general formula for reflection in the Klein model that you checked in Exercise K-16, Chapter 7, then Lemma 10.2 implies that reflection across the perpendicular bisector of OP_0 is given by

$$x' = \frac{x[\|p_0\|^2 - \|p_0\|] - x_0(x_0x + y_0y + \|p_0\| - 1)}{\|p_0\|^2 - \|p_0\| + [\|p_0\| - 1](x_0x + y_0y + \|p_0\| - 1)},$$

$$y' = \frac{y[\|p_0\|^2 - \|p_0\|] - y_0(x_0x + y_0y + \|p_0\| - 1)}{\|p_0\|^2 - \|p_0\| + [\|p_0\| - 1](x_0x + y_0y + \|p_0\| - 1)}.$$

Using these formulas, another long calculation shows that formula (26) is equal to the cosine of $\sphericalangle Q_1OQ_2$.

As a check on the formula, note that $\cos\theta = 0$ if and only if we have that $A_1A_2 + B_1B_2 + C_1(-C_2) = 0$, which equation says line l_1 passes through the pole $(A_2, B_2, -C_2)$ of line l_2.

We leave the assertions about the Weierstrass coordinates as an exercise. ◄

Polar coordinates (r, θ) of a point P with respect to an origin O and the positive x-axis are defined in the same way as in Euclidean geometry: r is the hyperbolic distance of the point P from O, and θ is the radian measure of the angle between the positive x-axis and the ray $\overrightarrow{OP}$, taken to be positive in the first two quadrants and negative in the third and fourth quadrants, $-\pi < \theta \leq \pi$. The equation of a ray emanating from O in polar coordinates is $\theta = $ a constant. If we denote the constant by θ_0, then the equation of the opposite ray is $\theta = \theta_0 - \pi$ if $\theta_0 > 0$, and $\theta = \theta_0 + \pi$ if $\theta_0 \leq 0$. From the identity $\tan(x \pm \pi) = \tan x$, we see that the equation of a line l through O in polar coordinates is $\tan\theta = \tan\theta_0$, where θ_0 (a *slope angle* for l) is the angle either one of the rays of l emanating from O makes with the positive x-axis.

Let us use polar coordinates to determine the equation of an equidistant curve κ to a line l through O for some distance d. If point P varies on κ, r is the hypotenuse of a right triangle whose side opposite angle $\theta - \theta_0$ has length d (Figure 10.26). Equation (10) of Theorem 10.3 for a right triangle yields

(30-equidistant) $\qquad \sinh r = \dfrac{\sinh d}{\sin(\theta - \theta_0)}$

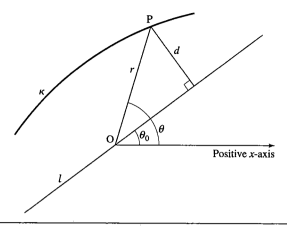

Figure 10.26 Polar coordinate equation for equidistant curve.

for the polar coordinate equation of κ, with the following proviso: For each distance d, there is an equidistant curve at that distance from l on each side of l. We can distinguish them by using *signed distances*, i.e., allowing d to have negative values (the sign being determined by an orientation of the plane). For the equidistant having d positive, then $\theta - \theta_0$ will take on only positive values; for the other equidistant, where d is negative, $\theta - \theta_0$ will take on only negative values. In the negative case, both the numerator and the denominator of the right side of (30-equidistant) will be negative numbers, so their quotient can equal the number on the left, which is always positive.

Next consider a line l not through O. Let Q be the foot of the perpendicular from O to l and let b be the signed distance from O to l (plus or minus the length of segment OQ). Let m be the perpendicular to OQ through O and let θ_0 be a slope angle for m. Then for a variable point $P \neq Q$ on l, consideration of the right triangle $\triangle OQP$, use of formula (10), of Theorem 10.3, and of the trigonometric identity $\cos[(\pi/2) - x] = \sin x$ yields the polar equation for l,

(30-h-line)
$$\tanh r = \frac{\tanh b}{\sin(\theta - \theta_0)}.$$

This equation is also valid for point Q because its polar coordinates are $(|b|, (\pi/2) + \theta_0)$. From Figure 10.27 you can see that the second polar coordinate θ of a point P on l varies strictly between $(\pi/2) - \Pi(|b|) + \theta_0$ and $(\pi/2) + \Pi(|b|) + \theta_0$.

The analogous formula in a Euclidean plane is

(30-e-line)
$$r = \frac{b}{\sin(\theta - \theta_0)}.$$

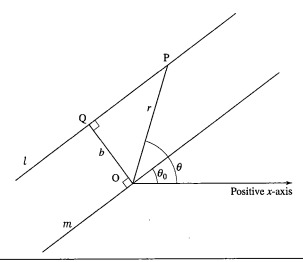

Figure 10.27 Polar coordinate equation for line not through O.

▨▨▨ **APPLICATION: A MODEL OF THE EUCLIDEAN PLANE WITHIN THE HYPERBOLIC PLANE.** We gave an incomplete description of this model in Project 1, Chapter 7. Recall that the "points" of the model are all the points of the hyperbolic plane. The "lines" are all the hyperbolic lines through a fixed hyperbolic point O plus all the equidistant curves having those lines as axes. "Betweenness" is induced by the betweenness in the hyperbolic plane.

With the polar coordinate description given above, we can now define a continuous isomorphism Φ of this interpretation onto the Cartesian model $\mathbb{R}^2$. The origin O will be mapped to $(0, 0) \in \mathbb{R}^2$, and the x- and y-axes will be mapped to the x- and y-axes in $\mathbb{R}^2$. The equidistant curve for the x-axis with signed distance u has the equation

$$(31) \qquad \sinh r = \frac{\sinh u}{\sin \theta},$$

while the equidistant curve for the y-axis with signed distance v has the equation

$$(32) \qquad \sinh r = \frac{\sinh(-v)}{\sin(\theta - (\pi/2))} = \frac{\sinh v}{\cos \theta},$$

using the trigonometric identities $\sin(\theta - (\pi/2)) = -\cos \theta$ and $\sinh(-v) = -\sinh v$. The point P in the hyperbolic plane with polar coordinates (r, θ) will then be mapped by Φ to the point $\Phi(P) = (x, y)$, where $x = \sinh u$ and $y = \sinh v$. The Euclidean distance of $\Phi(P)$ from the origin in $\mathbb{R}^2$ is then sinh r—compute $\sqrt{(x^2 + y^2)}$ using equations (31) and (32). Hyperbolic distances from O are stretched out via the hyperbolic sine function to get the Euclidean distances from $(0, 0)$. Let $\rho = \sinh r$. Equations (31) and (32) show that $\cos \theta$ and $\sin \theta$ are the same in both planes, so angles with a vertex at the origin corresponding to one another via Φ have the same radian measure θ. So from the definitions of x and y, we see that the polar coordinates of $\Phi(P)$ in $\mathbb{R}^2$ are (ρ, θ). The equidistant curve whose equation is (30-equidistant) is then mapped by Φ to the line in $\mathbb{R}^2$ whose equation in polar coordinates is

$$(30') \qquad\qquad \rho = \frac{d'}{\sin(\theta - \theta_0)},$$

where $d' = \sinh d$ is the Euclidean distance from the origin in $\mathbb{R}^2$ to that line and θ_0 is a slope angle for the Euclidean parallel line through the origin. Since angles with a vertex at the origin are measured the same in both planes, the line through O, $\tan \theta = \tan \theta_0$, in the hyperbolic plane is mapped by Φ to the line through the origin in $\mathbb{R}^2$ having that same equation, where for a point P having polar coordinates (r, θ_0) on such a line, $\Phi(P) = (\rho \cos \theta_0, \rho \sin \theta_0)$. We leave it as an exercise to show that betweenness on the equidistant curves, induced in a natural way from betweenness in the hyperbolic plane, corresponds by Φ to the betweenness in $\mathbb{R}^2$.

If we then transport the congruence relations in $\mathbb{R}^2$ back to our interpretation in the hyperbolic plane via Φ^{-1}, in the same way that we transported congruence from the Poincaré model to the Klein model via the isomorphism F in Chapter 7, we see that we do indeed have a model of the Euclidean plane within the hyperbolic plane. (See also Project 1.)

The coordinates $x = \sinh u$ and $y = \sinh v$ for a point P in the hyperbolic plane may be called *Ramsay–Richtmyer coordinates*, referring to their 1995 text, Section 7.10. *This model shows that plane hyperbolic geometry and plane Euclidean geometry are equally consistent—we don't have to go to the horosphere in hyperbolic three-space to prove that.*

The Circumscribed Cycle of a Triangle

You learned in Exercise 5, Chapter 5, that the existence of a circumscribed circle for every triangle is equivalent to the Euclidean parallel postulate. The circumscribed circle exists if and only if the perpendicular bisectors of the sides are concurrent in an ordinary point (Exercise 10, Chapter 6). In Exercise 11, Chapter 6, and Major Exercise 7, Chapter 6, you showed that the perpendicular bisectors are always "concurrent" in an ideal or ultra-ideal point if the circumscribed circle does not exist.

In the ultra-ideal case, you showed (see Figure 6.21) that the vertices A, B, C of the given triangle are all equidistant from the common perpendicular t to the perpendicular bisectors. This implies that they lie on an equidistant curve having t as an axis. According to our definition of "equidistant curve," it is required that A, B, C all lie on the same side of t.

Some authors (e.g., Coxeter, Sommerville) define "equidistant curve" differently; i.e., they define it to be the locus of all points at the same distance from an axis t, no matter which side of t. These authors would designate our "equidistant curve" one of the two "branches" of theirs. Let us call the equidistant curve of Coxeter and Sommerville a *doubly equidistant curve*, indicating the union of two equidistant curves having the same axis, each being the reflection of the other across the axis. In Exercise 2(a), Chapter 6, you showed that every triangle is circumscribed by three doubly equidistant curves whose axes are the medial lines that join pairs of midpoints of the sides (Figure 6.15).

Refer to the Poincaré upper half-plane model: The Euclidean circle through A, B, and C is a hyperbolic circle if it lies entirely in the upper half-plane (compare Proposition 7.12); it is a horocycle with an ideal center Ω if it is tangent to the x-axis at Ω (Exercise 46, Chapter 9), and its arc in the upper half-plane is an equidistant curve otherwise (Exercise 47, Chapter 9).

Figure 10.28 shows the three doubly equidistant curves and a hyperbolic circle circumscribing $\triangle ABC$ in this model.

The next theorem gives trigonometric criteria to decide which type of cycle circumscribes $\triangle ABC$.

THEOREM 10.11. With standard notation for $\triangle ABC$, let a be the length of a longest side, so that $\sphericalangle A$ is a largest angle. The cycle circumscribing $\triangle ABC$ is a

$$\left.\begin{array}{r}\text{Circle}\\\text{Horocycle}\\\text{Equidistant curve}\end{array}\right] \Leftrightarrow \sinh\frac{a}{2}\left\{\begin{array}{c}<\\=\\>\end{array}\right\} \sinh\frac{b}{2} + \sinh\frac{c}{2}$$

$$\Leftrightarrow (\sphericalangle A)^r \left\{\begin{array}{c}<\\=\\>\end{array}\right\} \Pi\left(\frac{b}{2}\right) + \Pi\left(\frac{c}{2}\right).$$

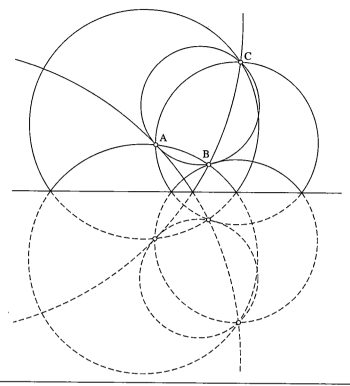

Figure 10.28 Cycles circumscribing triangle ABC.

PROOF:

Consider first the case where the perpendicular bisectors are asymptotically parallel through an ideal point Ω. According to Lemma 6.1 in Major Exercise 7, Chapter 6, Figure 10.29 holds, where A', B', C' are the midpoints. This shows that $(\angle A)^r = (\angle C'A\Omega)^r + (\angle B'A\Omega)^r = \Pi(c/2) + \Pi(b/2)$.

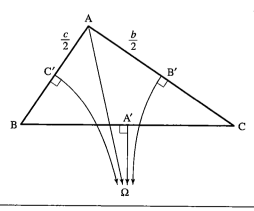

Figure 10.29

In the case where the perpendicular bisectors have a common perpendicular t, Figure 10.30 holds.

Since $\angle C'A\Omega > \angle C'A\Lambda$ and $\angle B'A\Omega > \angle B'A\Sigma$, we see that

$$(\angle A)^r > (\angle C'A\Lambda)^r + (\angle B'A\Sigma)^r = \Pi(c/2) + \Pi(b/2).$$

In the case where the perpendicular bisectors meet, we must have

$$(\angle A)^r < \Pi(c/2) + \Pi(b/2)$$

since this is the only other possibility. Thus the second criterion is established.

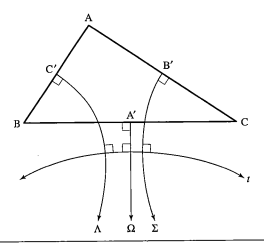

Figure 10.30

The derivation of the first criterion in terms of hyperbolic sines from the second criterion involves a calculation using identities and our earlier formulas. First, by the hyperbolic law of cosines (13),

$$\cos A = \frac{\cosh b \cosh c - \cosh a}{\sinh b \sinh c}$$

$$= \frac{\left(2 \sinh^2 \dfrac{b}{2} + 1\right)\left(2 \sinh^2 \dfrac{c}{2} + 1\right) - \left(2 \sinh^2 \dfrac{a}{2} + 1\right)}{4 \sinh \dfrac{b}{2} \cosh \dfrac{b}{2} \sinh \dfrac{c}{2} \cosh \dfrac{c}{2}}$$

$$= \frac{2 \sinh^2 \dfrac{b}{2} \sinh^2 \dfrac{c}{2} + \sinh^2 \dfrac{b}{2} + \sinh^2 \dfrac{c}{2} - \sinh^2 \dfrac{a}{2}}{2 \sinh \dfrac{b}{2} \sinh \dfrac{c}{2} \cosh \dfrac{b}{2} \cosh \dfrac{c}{2}}$$

Second, by the identity for $\cos(x + y)$ and formulas (5) and (6),

$$\cos \left[\Pi\left(\frac{b}{2}\right) + \Pi\left(\frac{c}{2}\right) \right]$$

$$= \cos \Pi\left(\frac{b}{2}\right) \cos \Pi\left(\frac{c}{2}\right) - \sin \Pi\left(\frac{b}{2}\right) \sin \Pi\left(\frac{c}{2}\right)$$

$$= \tanh \frac{b}{2} \tanh \frac{c}{2} - \frac{1}{\cosh \dfrac{b}{2} \cosh \dfrac{c}{2}}$$

$$= \frac{\sinh \dfrac{b}{2} \sinh \dfrac{c}{2} - 1}{\cosh \dfrac{b}{2} \cosh \dfrac{c}{2}}.$$

The first criterion then follows from these equations after some straightforward algebra. ◄

COROLLARY. An isosceles triangle whose base is not longer than its sides (in particular, an equilateral triangle) has a circumscribed circle. If the base is longer than the sides, then the circumscribed cycle is a

$$\left.\begin{array}{c}\text{Circle} \\ \text{Horocycle} \\ \text{Equidistant curve}\end{array}\right\} \Leftrightarrow \cosh a \left\{\begin{array}{c}< \\ = \\ >\end{array}\right\} 4 \cosh b - 3,$$

where a is the length of the base and b is the length of a side. We leave the proof for Exercise 10.

Our final theorem gives a lovely formula relating the radius of the circumscribed circle to the area (which equals the defect) of a triangle.

THEOREM 10.12. If $\triangle ABC$ has a circumscribed circle of radius R, then with standard notation, the area K of $\triangle ABC$ is given by

(33)
$$\sin \frac{K}{2} = \frac{\tanh \dfrac{a}{2} \tanh \dfrac{b}{2} \tanh \dfrac{c}{2}}{\tanh R}.$$

NOTE. If we look at only the leading terms in the power series expansion of sin and tanh (i.e., we look at only an "infinitesimal" hyperbolic triangle), this formula reduces to the Euclidean formula

$$K = \frac{abc}{4R}.$$

In Euclidean geometry, we could replace K by $\frac{1}{2}bc \sin A$ and solve for R; in hyperbolic geometry, Exercise 28 provides a formula for R purely in terms of the sides of the triangle.

Here is a proof of the Euclidean formula. Choose B to be a vertex such that the diameter BD of the circumscribed circle κ intersects side AC. Then $\sphericalangle$D of $\triangle$BDC and $\sphericalangle$A subtend the same arc BC of κ, so we have $\sin A = \sin D = a/2R$ (since $\sphericalangle$BCD is right, being inscribed in a semicircle). Substitute for $\sin A$ in $K = \frac{1}{2}bc \sin A$ to get the formula.

Note that in Euclidean geometry, the common ratio $S = (\sin X)/x$ in the law of sines is equal to $2R$. See Exercise 20 for the hyperbolic case.

The proof of Theorem 10.12 will be indicated in Exercises 20–28.

Bolyai's Constructions in the Hyperbolic Plane

János Bolyai is alleged (by otherwise reliable writers) to have "squared the circle" using only a straightedge and compass in the hyperbolic plane.[6] Since squares do not exist in a hyperbolic plane, what is meant by "squaring" is to construct a regular 4-gon (a quadrilateral with all sides and angles congruent) having the same area as the circle. In the Euclidean plane, circle squaring was considered impossible with those tools, although a rigorous proof of that impossibility was not obtained until 1882, when Lindemann proved the much stronger result that π is transcendental. Even as early as 450 B.C., the Greek playwright Aristophanes poked fun at circle squarers in his comedy *The Birds*. So it was with some caution that Bolyai published his result as part of the appendix to his father's *Tentamen*. He wrote, in Section 33, as his last point in a five-point summary: "V. Finally, to friendly readers it will not be unacceptable, that for the case wherein not Σ but S is reality, a rectilinear figure is constructed equivalent to a circle." Bolyai denoted Euclidean geometry by Σ and what we now call hyperbolic geometry by S. Point I of his summary stated: "Whether Σ or some one S is reality remains to be decided." (The words "some one" refer to the different possible values of the distance scale k.)

Throughout this section, the word "construct" or "construction" refers only to "straightedge-and-compass construction," and k will still be taken to be 1. At the end of this section, we will discuss a new re-

---

[6] For one example, see J. Gray (2004, pp. 67, 69).

lationship between constructions in the hyperbolic plane and constructible real numbers.

The problem of circle squaring, like the problem of angle trisection, is to find a *general method* of constructing the desired figure from the given figure. In the case of angle trisection, many particular angles can be constructively trisected, such as a 135° angle (bisect a right angle). But there is no general method of angle trisection as there is for bisecting an arbitrary angle. Similarly, some Euclidean circles can be squared, for if you start with a circle whose radius is $a/\sqrt{\pi}$, where a is some constructible length, then that circle can obviously be Euclidean-squared. If the only radii considered are those whose length is a constructible real number, then *no* Euclidean circle can be squared because $\sqrt{\pi}$, being transcendental, is not a constructible real number.

Bolyai did not provide a general method for squaring circles in the hyperbolic plane because *no such method exists*—for one thing, because areas of circles are unbounded (by the formula in Theorem 10.7 for their area in terms of the radius), whereas areas (equal to their defects) of regular 4-gons are bounded by 2π. What he did was construct both a circle having area π and a regular 4-gon whose angles measure $\pi/4$; this regular 4-gon has area π as well. He also saw the general answer to when *both* such constructions are possible, as we shall describe below.

Consider the latter construction first. The regular 4-gon with angles measuring $\pi/4$ is obtained by first constructing a right triangle with acute angles $\pi/4$ and $\pi/8$ and then reflecting it seven times (Figure 10.31). Such a right triangle can be constructed by the method in the proof of the right triangle construction theorem (p. 506).

The radius r of a circle of area π is the side of the equilateral pedal triangle of a trebly asymptotic triangle (example 2, p. 499); it satisfies $\cosh r = 3/2$. Thus there are constructions in the hyperbolic plane of both a circle and a regular 4-gon having an area equal to π. *That is another key difference between Euclidean and hyperbolic geometries.*

We know from Theorem 10.7 that the area of a circle of radius r is $4\pi \sinh^2(r/2)$. Bolyai showed that the key to construction of the radius r of a desired circle is the following general theorem.

BOLYAI'S CIRCLE-ANGLE THEOREM. Given either an acute angle of radian measure θ or a segment of length r (distance scale $k = 1$), then there is a straightedge-and-compass construction of one from the other satisfying $\tan \theta = 2 \sinh(r/2)$.

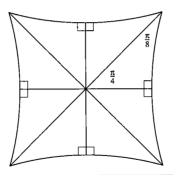

Figure 10.31 Regular 4-gon of area π.

Figure 10.32 shows the geometric relationship between θ and r. See Project 2 for indications of the proof (an excellent exercise). Since the angle $\theta = \pi/4$ is constructible, this construction also provides the radius of a circle of area π.

COROLLARY. There is a constructive correspondence between circles of radius r and acute angles θ such that the area of the circle is $\pi \tan^2 \theta$.

PROOF:

Substitute in the formula of Theorem 10.7 for the area of a circle. ◄

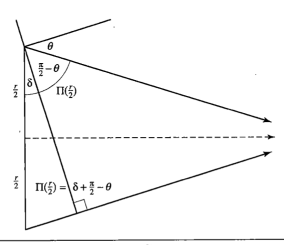

Figure 10.32 Angle θ associated to radius r.

REMARK. If we let R = tan θ, the area of the circle is πR^2. This correspondence associates with each circle in the hyperbolic plane of radius r a circle in the Euclidean plane of radius R having the same area. But it is the angle θ that is of interest now.

As a key step in his proof of Bolyai's circle-angle theorem, George Martin exhibits a simple construction for the *segment of parallelism*.

GEORGE MARTIN'S THEOREM. Given an acute angle ⊀BAC labeled so that C is the foot of the perpendicular from B to $\overleftrightarrow{AC}$. Let CΩ be the opposite ray to $\overrightarrow{CB}$. On the opposite side of $\overleftrightarrow{AC}$ from B, construct the ray s emanating from A for which the angle formed by ray s and $\overrightarrow{AC}$ is the angle of parallelism for segment AB (using Bolyai's parallel construction, Chapter 6). Then ray s intersects ray CΩ at some point D, and AD is a segment of parallelism for ⊀BAC (see Figure 10.33).

PROOF:

Since AB > AC (hypotenuse greater than leg), $\Pi(AB) < \Pi(AC)$ (exterior angle theorem, Major Exercise 4, Chapter 6). Hence point D exists (by definition of a limiting parallel ray). By formulas (6) and (10) of hyperbolic trigonometry,

$$\cos \sphericalangle BAC = \frac{\tanh AC}{\tanh AB} = \frac{\tanh AC}{\cos \Pi(AB)} = \tanh AD = \cos \Pi(AD),$$

so ⊀BAC = $\Pi(AD)$. ◄

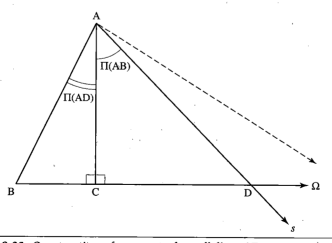

Figure 10.33 Construction of segment of parallelism AD.

. The general question of when it is possible to construct *both* a circle and a regular 4-gon having the same area was answered by Bolyai, though he did not justify a crucial step. A justification was given in a 1995 article by William Jagy.[7] The answer is very surprising. It comes down to the famous theorem of Gauss and Wantzel determining for which n the regular n-gon can be constructed in a Euclidean plane.

We sketch what Jagy did. He showed that it is generally impossible, starting with an arbitrary circle of area $<2\pi$, to construct a regular 4-gon having the same area (Theorem B); conversely, beginning with an arbitrary regular 4-gon, it is impossible in general to construct a circle having the same area (Theorem C). For Theorem B he gave N. M. Nestorovich's example of the circle with radius r—constructible by the Mordukhai–Boltovski (M-B) theorem below—such that

$$\sinh \frac{r}{2} = \frac{\sqrt{2 - \sqrt{2}}}{2}.$$

The regular 4-gon having the same area has angle $\sigma = \pi\sqrt{2}/4$, which is not constructible by Theorem A below. For Theorem C, Jagy used two theorems about irrational numbers[8] to produce a whole family of counterexamples—a remarkable insight. So we will focus on the problem of simultaneously constructing *both* a circle and a regular 4-gon having the same area.

As Bolyai's circle-angle theorem showed, the problem comes down to constructing two angles—the auxiliary angle θ for which the circle has area $\pi \tan^2 \theta$ and the acute angle σ of the regular 4-gon. The equation for equal areas is then

$$\pi \tan^2 \theta = 2\pi - 4\sigma.$$

A key result needed to solve this problem is the following.

---

[7] W. C. Jagy, "Squaring Circles in the Hyperbolic Plane," *Mathematical Intelligencer*, 1995, **17**(2), 31–36. The result is also discussed in N. M. Nesterovich, "On the Quadrature of the Circle and Circulature of a Square in Lobachevskiĭ Space (Russian), *Doklady Academiya SSR* (N.S.), **63** (1948), 613–614; see http://zakuski.math.utsa.edu/~gokhman/ftp/translations/quadratic.pdf. Bolyai restricted the case to where the cosh of the radius is rational; Jagy proved that was not really a restriction.

[8] The *Gelfond–Schneider theorem* states that a^b is transcendental if a is algebraic, $\neq 0$, or 1, and b is algebraic and irrational. *Olmstead's theorem* states that if τ is a rational multiple of π, the only possible values of $\tan \tau$ that are rational are 0, 1, and -1. Proofs of these theorems are in I. Niven's *Irrational Numbers* (Washington, DC: Mathematical Association of America, 1956).

Angle Construction Theorem (ACT). An angle can be constructed in the hyperbolic plane if and only if it can be constructed in the Euclidean plane.

Plan for a proof:
Use the theorem that the abstract hyperbolic plane is isomorphic to its *conformal* Poincaré disk model (Hartshorne, Corollary 43.3). More details are given below.

Corollary. There is no general construction for trisecting all angles in the hyperbolic plane.

Proof:
By the theorem of Pierre Wantzel that there is no such construction in the Euclidean plane.[9] ◄

Continuing with Jagy's argument: If θ and σ are both constructible, then $\omega = 2\pi - 4\sigma$ is a constructible angle and we have $x = \tan^2 \theta$ as a constructible length. Using the theorem of Gelfond–Schneider on transcendental numbers, Jagy proves x *must be rational*. Writing $x = m/n$ in lowest terms, he shows that $2\pi/n$ *is a constructible angle*. Juxtaposing n copies of this angle, a circle centered at their common vertex is cut by the rays of these angles to give n points that can be joined by segments to form a constructible regular n-gon. Gauss determined all integers n for which the regular n-gon in the Euclidean plane is constructible, and Wantzel completed his argument.

Theorem of Gauss–Wantzel. A Euclidean regular n-gon is constructible if and only if any odd primes in the prime factorization of n occur to the first power and are *Fermat numbers,* numbers of the form

$$F_k = 2^{2^k} + 1.$$

For a proof, see Hartshorne, Theorem 29.4. Stated differently, the result applies to *both* the Euclidean and hyperbolic planes, by the ACT, because it's a criterion for constructing the angle $2\pi/n$. For $k = 0, 1, 2, 3, 4$, these Fermat numbers are 3, 5, 17, 257, and 65,537, all of which

---

[9] See Hartshorne, Section 28. Wantzel's 1837 proof has a gap that was filled by J. Petersen in 1871 in an article in Danish. See Hartshorne's note, p. 490.

are prime. For $5 \leqq k \leqq 23$ these Fermat numbers are not prime, and that is all that is presently known about Fermat numbers.

Solving for σ in terms of the common area ω, which is an angle in the hyperbolic plane (it is a defect), we see that one of them is a rational multiple of π, with the denominator n of the rational number being of the above form if and only if the other is; in that case, $\tan \theta$, being the square root of a rational number, is constructible, whence θ is easily constructible in the Euclidean plane. Hence θ is constructible in the hyperbolic plane by the ACT. So the final result is as follows.

Bolyai's Construction Theorem (Jagy's Theorem A). Suppose that a regular 4-gon with acute angle σ and a circle in the hyperbolic plane have the same area $\omega < 2\pi$. Then both are constructible if and only if σ is an integer multiple of $2\pi/n$, where n is a number for which the angle $2\pi/n$ is constructible (there are infinitely many such n since the power of 2 in the factorization of n can be arbitrary, for example, when $\omega = \pi$, $n = 8$).

On the final page of Bolyai's great appendix, he referred admiringly to "the theory of polygons of the illustrious Gauss (remarkable invention of our, nay of every age)." We can applaud the invention/discovery/development of hyperbolic geometry by him and by Lobachevsky[10] as being at least as remarkable.

Relation between Constructions and Constructible Real Numbers. In the Euclidean case, this topic has been treated carefully in many texts. Recall from Example 4, Chapter 3, that a *constructible real number* is one that can be obtained from rational numbers by finitely many applications of addition, subtraction, multiplication, division, and square roots of positive numbers.

If we ask, starting with the origin and the unit points, which other points (a, b) in $\mathbb{R}^2$ can be constructed, the classical answer is exactly those points whose coordinates are constructible real numbers—

---

[10] We have exhibited a deep result of J. Bolyai that Lobachevsky didn't consider. Gauss, in his infamous reply to F. Bolyai after receiving his *Tentamen*, posed to J. Bolyai the difficult problem of determining the volume V of a tetrahedron in hyperbolic space, saying "As the area of a triangle can be obtained so simply, one would expect a simple expression for that volume also. This expectation appears to be treacherous." Lobachevsky solved it: See Coxeter (1998, p. 289). The answer is in terms of what has come to be called *the Lobachevsky function*:

$$L(x) = \int_0^x \ln(\sec \theta) \, d\theta.$$

essentially because lines and circles have linear and quadratic equations, respectively (see Hartshorne, Theorem 13.2, or Moise, Chapter 19, for the lengthy details). From that result, the impossibility of in general trisecting an angle, of duplicating a cube, and of squaring a circle can be proved, and the values of n for which the regular n-gon can be constructed can be determined (see Hartshorne, Chapter 6).

For the hyperbolic plane, the result quoted in some texts and articles is the following.

MORDUKHAI–BOLTOVSKI THEOREM (1927). In the real hyperbolic plane with distance scale $k = 1$, a segment of length r is constructible if and only if sinh r (equivalently, cosh r or tanh r or e^r) is a constructible real number.

The equivalence follows from algebraic relations among e^r, sinh r, cosh r, and tanh r. Note that according to this M-B theorem, *for a distance scale $k = 1$, the unit segment cannot be constructed* because that would imply that the number e is constructible; but Hermite proved in 1873 that e is transcendental.

COROLLARY. A non-right angle θ in the real hyperbolic plane is constructible if and only if tan θ (equivalently, sin θ or cos θ) is a constructible real number.

PROOF:

We may restrict our attention to acute angles θ (by bisecting an obtuse angle). Then we know that $\theta = \Pi(x)$ for a uniquely determined length x and that θ is constructible if and only if x is (Bolyai's construction in Chapter 6 of the limiting parallel ray provides θ given x, and George Martin's construction provides x given θ). The corollary now follows from formulas (5), (6), and (7) relating the trigonometric functions of θ to the hyperbolic trigonometric functions of x. ◄

The corollary also holds in the Euclidean plane for a more elementary reason: We can go back and forth between θ and any of its trigonometric values by constructing a suitable right triangle having θ as one of its acute angles: If tan θ is constructible, construct the right triangle having tan θ as the length of one leg and 1 as the length of the other; then θ is the angle opposite the leg of length tan θ. Conversely, if acute angle θ is constructible, lay off a segment of length 1 on one side of the angle. At the end of that segment, erect a perpendicular; it must intersect the other side of the angle (by Euclid V) to form a right triangle in which the side opposite θ has length tan θ.

Thus the ACT also follows from the M-B theorem.

Knowing that the theory of constructions in Euclidean planes does not require the full power of the real numbers, I saw that for hyperbolic planes, the discussion about distance scale $k = 1$ and the use of hyperbolic functions are not necessary for the characterization of constructible segments in a hyperbolic plane. The key to avoiding them is Hartshorne's concept of the *multiplicative length* μ of a segment (Chapter 7, p. 320), recovered in a real plane from the additive length by exponentiating. That led me to conjecture this generalization of the M-B theorem to arbitrary (not necessarily real) hyperbolic planes.

Hyperbolic Constructible Segments Theorem. In a hyperbolic plane, given two perpendicular lines and a choice of one end labeled 1 on one of those lines and a choice of ends labeled ∞ and 0 on the other. Then with respect to Hilbert's field of ends based on those data, *a segment is constructible if and only if its multiplicative length μ is a constructible number.* For any constructible number $a > 1$, there exists a constructible segment having a as its multiplicative length.

Since the additive length d in a real hyperbolic plane is given by $d = \log \mu$, the M-B theorem follows from this theorem. In Appendix B, Part I, we will describe the field of ends and give Hartshorne's proof.

Review Exercise

All statements in this exercise refer to hyperbolic geometry (unless explicit mention of other geometries is made). Which of the following statements are correct?

(1) The area of a triangle is proportional to its defect.

(2) The angle of parallelism $\Pi(x)$ in radians relates the circular and hyperbolic functions by means of an equation such as tanh $x =$ cos $\Pi(x)$.

(3) In all right triangles having a fixed number of radians for $\angle A$ (and standard notation, right angle at C), the ratio a/c is the same and is called the sine of $\angle A$.

(4) J. Bolyai discovered a formulation of the law of sines that is valid in neutral geometry.

(5) The segment length x^* complementary to x is uniquely determined by the formula $\Pi(x^*) = \pi/2 - \Pi(x)$.

(6) The equations relating Beltrami coordinates to Lobachevsky coordinates are $x = \tanh u$ and $y = \tanh v$.

(7) With standard notation, if a is the largest side of $\triangle ABC$, then the cycle circumscribing $\triangle ABC$ is a circle if and only if we have $\sinh (a/2) > \sinh (b/2) + \sinh (c/2)$.

(8) The representation by Weierstrass coordinates helps make sense of Lambert's description of the hyperbolic plane as a "sphere of imaginary radius i."

(9) The curvature of the hyperbolic plane is $1/k^2$, where k^2 times the defect in radians equals the area of a triangle.

(10) The analogue of the Pythagorean theorem is $\cosh c = \cosh a \cosh b$ (for a right triangle with right angle at C, standard notation, $k = 1$).

Exercises

1. Verify all the identities for hyperbolic functions listed in the table on p. 490.

2. Verify formulas (5), (6), (7) in which Π provides a link between hyperbolic and circular functions. Graph the function $\Pi(x)$.

3. The proof of Theorem 10.3 required a complicated argument using the Poincaré model. Give a shorter proof using the Klein model. (Hint: According to the note at the end of Chapter 7, you can assume $A = O$, the center of the absolute. Show that $\cos A = \overline{AC}/\overline{AB}$ and $\sin A = \overline{BC}/\overline{AB}$ (Euclidean lengths), $\overline{AB} = \tanh c$ (where $c = d'(AB)$ is the Klein length), $\overline{AC} = \tanh b$, and $\overline{BC} = (1 - \overline{AC}^2)^{1/2}$ $\tanh a = \tanh a/\cosh b$ (use Theorem 7.4). Conclude by deducing the formula $\cosh c = \cosh a \cosh b$ from the Pythagorean theorem. See Figure 10.34).

4. Prove Theorem 10.4.

5. Verify step 2 in the proof of Theorem 10.6.

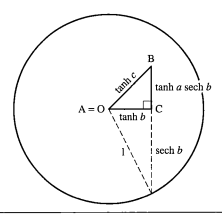

Figure 10.34

6. Verify formulas (18) through (21) for complementary lengths. Graph the function

$$f(x) = x^* = \ln \frac{e^x + 1}{e^x - 1}.$$

7. Prove the assertions about Weierstrass coordinates in Theorem 10.10. (Hint: Derive the equation of a line in Weierstrass coordinates from the equation of a line in Beltrami coordinates.)

8. Verify formulas (28) and (29) in the proof of Theorem 10.10 and use them to reconcile with the distance formula in Exercise K-14, Chapter 7.

9. Generalize Lemmas 10.1 and 10.2 by showing that if (x_1, y_1) and (x_2, y_2) are distinct points in the Klein model, then the midpoint and perpendicular bisector of the segment they determine are given, respectively, by

$$\left(\frac{x_1 s_2 + x_2 s_1}{s_1 + s_2}, \frac{y_1 s_2 + y_2 s_1}{s_1 + s_2} \right)$$

$$(x_1 s_2 - x_2 s_1)x + (y_1 s_2 - y_2 s_1)y + (s_1 - s_2) = 0,$$

where $s_i = \sqrt{1 - x_i^2 - y_i^2}$, $i = 1, 2$. (Hint: Use Lemma 10.2 to find the point Q in the Cartesian plane where the perpendicular bisectors of OP_1 and OP_2 meet. Then joining Q to the pole of $\overleftrightarrow{P_1 P_2}$ gives the perpendicular bisector of $P_1 P_2$, and intersecting it with $P_1 P_2$ gives the midpoint.)

10. Prove the corollary to Theorem 10.11.

11. In a right triangle with right angle at C, prove that the circumscribed cycle is a

$$\left.\begin{array}{r}\text{Circle}\\ \text{Horocycle}\\ \text{Equidistant curve}\end{array}\right\} \Leftrightarrow \frac{a}{2} \left\{\begin{array}{c}<\\ =\\ >\end{array}\right\} \left(\frac{b}{2}\right)^*.$$

(Hint: Apply the second criterion of Theorem 10.11 with $\angle C$ the largest angle, using the fact that Π is a decreasing function. Or else argue directly from the definition of complementary lengths.)

12. Verify A. P. Kotelnikov's rule for remembering the relations among the parts of a right triangle with the right angle at C, standard notation (x^* denoting the complementary length to x): In Figure 10.35, *the sine of each angle is equal to the product of the tangents of the two adjacent angles and is equal to the product of the cosines of the two opposite angles.* For example,

$$\sin A = \tan \Pi(a^*) \tan \Pi(c) = \cos \Pi(b^*) \cos B.$$

(This rule is the hyperbolic analogue of the rule John Napier published in 1614 for the trigonometry of a right triangle on a unit sphere in Euclidean space. For Napier's rule, use a, b, A', c', B' in cyclic order, where A' denotes the complementary angle to $\angle A$ and $c' = \pi/2 - c$. J. Bolyai and Lobachevsky discovered that spherical trigonometry in hyperbolic space is the same as in Euclidean space.)

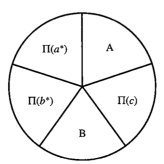

Figure 10.35 Kotelnikov's rule.

13. In Euclidean geometry, every circle can be inscribed in a triangle (the tangents at three appropriate points on the circle meet to form a triangle). Show that in hyperbolic geometry (with $k = 1$), an inscribed circle of a triangle must have a diameter less than $\ln 3$. (Hint: In Figure 10.37, show that $(\angle AIB')^r + (\angle BIC')^r + (\angle CIA')^r = \pi$ and that each of these three angles is less than $\Pi(r)$; then apply the Bolyai–Lobachevsky formula to find x such that $\Pi(x) = \pi/3$.)

14. (a) Show that (with $k = 1$) a trebly asymptotic triangle has an inscribed circle of diameter ln 3 (see Exercise K-13, Chapter 7).
   (b) Show that the ratio of the area to the circumference of a circle of radius r is tanh $r/2$.

15. Show that for any three positive numbers α, β, γ such that we have $\pi > \alpha + \beta + \gamma$, there exists a triangle having these numbers as the radian measures of its angles. (Hint: Use Theorem 10.4.)

16. In a singly asymptotic triangle ABΩ, if $c = \overline{AB}$, then

$$\cosh c = \frac{\cos A \cos B + 1}{\sin A \sin B}.$$

(Hint: Let C approach Ω in formula (15), Theorem 10.4. For a proof without using continuity, note that when ∢A and ∢B are acute,

$$c = \Pi^{-1}(\alpha) + \Pi^{-1}(\beta),$$

where $\alpha = (∢A)^r$ and $\beta = (∢B)^r$—see Figure 10.36.) Show that the generalization of the Bolyai–Lobachevsky formula to the case where $\beta < \pi/2$ is

$$\tan \frac{\alpha}{2} = e^{-c} \cot \frac{\beta}{2}.$$

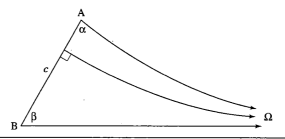

Figure 10.36

17. Write down equations that show how the side and the angle of an equilateral triangle determine each other.

18. (a) In a right triangle with standard notation and right angle at C, show that tan $A = $ tanh $a/$sinh b.
   (b) Deduce that in an isosceles triangle with base b and side a, summit at B, and one base angle at A,

$$\tanh a \, \cos \frac{B}{2} = \tan A \, \sinh \frac{b}{2},$$

$$\sin A \cosh \frac{b}{2} = \cos \frac{B}{2}.$$

(Hint: Drop the altitude to the base.)

19. In a right triangle $\triangle ABC$ with right angle at C (and standard notation), show that

$$\sin K = \frac{\sinh a \sinh b}{1 + \cosh a \cosh b},$$

where $K =$ the area = the defect of $\triangle ABC$. (Hint: Use Theorem 10.3 and trigonometric identities.)

20. Given $\triangle ABC$, if h is the length of the altitude from vertex B, show that (in standard notation) the product $\sinh b \sinh h$ is independent of the choice of which vertex is labeled B; this is the hyperbolic analogue of the Euclidean theorem that bh is constant. (Hint: Show that $\sinh b \sinh h = S \sinh a \sinh b \sinh c$, where S is the constant ratio occurring in the law of sines.) The next exercises will shed light on the geometric significance of the constant $\frac{1}{2} \sinh b \sinh h$, which we will denote by H (for Heron); by the hint, we have $2H = \sin C \sinh a \sinh b = \sin B \sinh c \sinh a = \sin A \sinh b \sinh c$.

In Exercises 21–28, s will denote the *semiperimeter* $\frac{1}{2}(a + b + c)$ of $\triangle ABC$.

21. Show that (in standard notation)

$$\sin \frac{A}{2} = \sqrt{\frac{\sinh(s - b) \sinh(s - c)}{\sinh b \sinh c}},$$

$$\cos \frac{A}{2} = \sqrt{\frac{\sinh s \sinh(s - a)}{\sinh b \sinh c}}.$$

(Hint: Square both sides; use identities and Theorem 10.4.)

22. Show that $H = \sqrt{\sinh s \sinh(s - a) \sinh(s - b) \sinh(s - c)}$. (Hint: Use the identity $\sin A = 2 \sin(A/2) \cos(A/2)$.)

23. "Infinitesimally," the Heron is equal to $\sqrt{s(s - a)(s - b)(s - c)}$. Show that in Euclidean geometry, this quantity is equal to the area of $\triangle ABC$. (Hint: See Coxeter, 2001, p. 12.)

24. Suppose the inscribed circle of $\triangle ABC$ has radius r and touches BC at A', CA at B', and AB at C'. Show that in neutral geometry, $\overline{AB'} = s - a = \overline{AC'}$, $\overline{BC'} = s - b = \overline{BA'}$, $\overline{CA'} = s - c = \overline{CB'}$ (see Figure 10.37). (Hint: Review the construction of the inscribed circle in Exercise 18, Chapter 4.)

25. Deduce from Exercise 24 that in hyperbolic geometry,

$$\tanh r \sinh s = H,$$

whereas in Euclidean geometry, rs = the area of $\triangle ABC$. (Hint: In hyperbolic geometry, use Exercises 18, 21, 22, and 24 to compute $\tan (A/2) \sinh (s - a)$; in Euclidean geometry, add up the areas of triangles IAB, IBC, and ICA in Figure 10.37.)

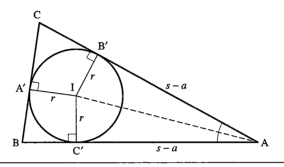

Figure 10.37

26. Prove Gauss' equations:

$$\cosh \frac{1}{2} c \sin \frac{1}{2}(A + B) = \cosh \frac{1}{2}(a - b) \cos \frac{1}{2} C,$$

$$\cosh \frac{1}{2} c \cos \frac{1}{2}(A + B) = \cosh \frac{1}{2}(a + b) \sin \frac{1}{2} C,$$

$$\sinh \frac{1}{2} c \sin \frac{1}{2}(A - B) = \sinh \frac{1}{2}(a - b) \cos \frac{1}{2} C,$$

$$\sinh \frac{1}{2} c \cos \frac{1}{2}(A - B) = \sinh \frac{1}{2}(a + b) \sin \frac{1}{2} C.$$

(Hint: Use identities such as

$$\sinh x + \sinh y = 2 \sinh \frac{1}{2}(x + y) \cosh \frac{1}{2}(x - y)$$

and analogous identities for the circular functions; then apply the half-angle formulas of Exercise 21.)

27. Show that a hyperbolic analogue of Heron's Euclidean area formula in Exercise 23 is the formula

$$\sin \frac{K}{2} = \frac{H}{2 \cosh \dfrac{a}{2} \cosh \dfrac{b}{2} \cosh \dfrac{c}{2}},$$

where K = area = defect of $\triangle ABC$. (Hint: Use Gauss' equations, the identity $\sin K/2 = \cos \frac{1}{2} (A + B + C)$, other trigonometric identities, and the formula $H = \frac{1}{2} \sin C \sinh a \sinh b$.)

28. If $\triangle ABC$ has a circumscribed circle of radius R, show that

$$\tanh R = \frac{2 \sinh \dfrac{a}{2} \sinh \dfrac{b}{2} \sinh \dfrac{c}{2}}{H},$$

which by Exercise 27 is equivalent to formula (30) of Theorem 10.12. (Hint: In Figure 10.38, we have that $\sin A = \sin(\beta' + \gamma')$ and $H = \frac{1}{2} \sin A \sinh b \sinh c$; use Theorem 10.3 to determine $\cos \gamma'$ and $\cos \beta'$ and use Exercise 18 to determine $\sin \gamma'$ and $\sin \beta'$, obtaining with the help of identities the formula

$$H \tanh R = \left[\cos \frac{\gamma}{2} \sinh \frac{b}{2} + \cos \frac{\beta}{2} \sinh \frac{c}{2} \right] 2 \sinh \frac{b}{2} \sinh \frac{c}{2}.$$

Show finally that the term in brackets equals $\sinh a/2$ by using Theorem 10.3 to derive expressions for $\sinh a/2$, $\sinh b/2$, and $\sinh c/2$ and plugging in $\sin(\gamma + \beta)/2 = \sin \alpha/2$.)

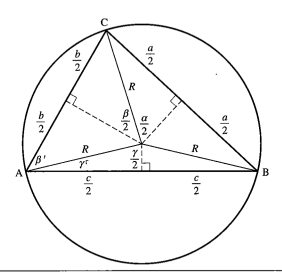

Figure 10.38

29. Let $\Omega\Sigma\Lambda$ be a trebly asymptotic triangle and $\triangle ABC$ its *pedal triangle* formed by the feet of the perpendiculars from each ideal vertex to the opposite side. Since all trebly asymptotic triangles are congruent to one another by Proposition 9.16(c), $\triangle ABC$ is equilateral.

Show that in the Poincaré upper half-plane model, the radian measure θ of an angle of $\triangle ABC$ is given by the relation $\tan \frac{1}{2}\theta = \frac{1}{2}$ or $\tan \theta = \frac{4}{3}$, $\sin \theta = \frac{4}{5}$, $\cos \theta = \frac{3}{5}$ and that the length c of a side is given by $\cosh c = \frac{3}{2}$. Deduce that *a circle whose radius is a side of* $\triangle ABC$ *has an area equal to* π. Show further that the Heron H, the circumradius R, and the inradius r of $\triangle ABC$ are given by

$$H = \frac{1}{2},$$

$$\tanh R = \frac{1}{2},$$

$$\tanh r = \frac{1}{4}.$$

(Hint: There are many ways to obtain these results using the previous exercises and the models. In the Poincaré upper half-plane model, taking $\Omega = -1$, $\Sigma = \infty$, $\Lambda = 1$ gives $A = i$, $B = 1 + 2i$, $C = -1 + 2i$. Show that $\overleftrightarrow{BC}$ is the upper semicircle of $x^2 + y^2 = 5$, that $\overleftrightarrow{AB}$ is the upper semicircle of $(x - 2)^2 + y^2 = 5$, and that the tangents to these circles at B have slopes $-\frac{1}{2}$, $\frac{1}{2}$, respectively.

This and the double angle formulas yield the assertions about θ. Exercises 18(a), 20, 25, and 28 can then be applied. Or use the Klein model, choosing triangle $\Omega \Sigma \Lambda$ so that the origin is the incenter and circumcenter of $\triangle ABC$.)

30. In any Hilbert plane, if two altitudes of a triangle meet (which is automatic if no angle is obtuse), then the third altitude is concurrent with them (Hartshorne, Theorem 43.15). For $\triangle ABC$ with an obtuse angle at C in a hyperbolic plane, the lines containing the altitudes are concurrent in a point H that may be ordinary, ideal, or ultra-ideal (see Exercise K-18, Chapter 7, for the Klein model). Using hyperbolic trigonometry, this result can be made more precise. Let h be the length of the altitude dropped from vertex C.

(a) Show that H is ordinary, ideal, or ultra-ideal according to whether $-\cos C \tanh a \tanh b$ is less than, equal to, or greater than $\tanh h$, respectively.

(b) Suppose the triangle is isosceles $(a = b)$. Then H is ordinary, ideal, or ultra-ideal according to whether $\coth a$ is greater than, equal to, or less than $\sec \frac{1}{2}C - 2 \cos \frac{1}{2}C$. In particular, if we have $(\sphericalangle C)^r \le 2\pi/3$, then H must be ordinary.

(c) This geometric argument shows that for any triangle with obtuse angle at C, if H is ideal, then $\theta = (\sphericalangle C)^r > 2\pi/3$. Draw the diagram to illustrate this argument and justify the assertions. Let F be the foot of the altitude from C to AB. Let θ_1, θ_2 measure $\sphericalangle$FCA, $\sphericalangle$FCB, respectively, so that $\theta = \theta_1 + \theta_2$. Let Ω, Λ be the respective ends of rays $\overrightarrow{CB}$, $\overrightarrow{FC}$. By hypothesis, $\Lambda\Lambda$ is perpendicular to the ray opposite to $C\Omega$ at some point D, so that $\theta_2 = \Pi(CD)$. Since $\overrightarrow{FB}$ intersects $C\Omega$ at B, CF < CD. Compare right triangles ADC and AFC, which have the common hypotenuse AC. We see that $\theta_1 > \pi - \theta$ (e.g., by the second formula in (10), Theorem 10.4). Since $\overrightarrow{CA}$ intersects the ray opposite to $D\Lambda$ at A, we also see that $\theta_2 > \pi - \theta$. Hence we have $\theta/2 > \pi - \theta$ as claimed.

31. In a Euclidean plane, one method of trisecting right angles is to construct an equilateral triangle by the method of Euclid's proof of his Proposition 1 and then bisect its angle. This method does not work in the hyperbolic plane because it gives an angle with measure $<\pi/6$. Yet the ACT asserts that an angle of measure $\pi/6$ can also be constructed in the hyperbolic plane. Find an explicit construction. (Hint: See Exercise K-13, Chapter 7, for one method.)

32. Show that in the hyperbolic plane, the maximum radius r of a circle which can be inscribed in an ordinary or asymptotic triangle satisfies tanh $r = 1/2$. (Hint: See the examples on p. 499.) Prove that in a Euclidean plane, every circle can be inscribed in a triangle. *This is another significant difference between Euclidean and hyperbolic geometry.*

33. An *asymptotic quadrilateral* has a *symmetry point* or *center* S obtained by intersecting its diagonal lines (see Figure 6.39, p. 285). Construct the *regular asymptotic quadrilateral R* in which S, by definition, is equidistant from the sides. Find the circumference and area of the inscribed circle δ of R with center S. The feet of the perpendiculars dropped from S to the sides of R form a regular 4-gon; find the length of its side and its area.

Projects

1. We described congruence in the hyperbolic plane model of the Euclidean plane indirectly using the isomorphism $\Phi^{-1}(\rho, \theta) = $ (arcsinh ρ, θ) from $\mathbb{R}^2$. In Chapter 9, we saw that in any H-plane, congruence is determined by reflections, which generate the group of motions. If we pull back Euclidean reflections via Φ^{-1}, we obtain new mappings of the hyperbolic plane that we can call

e-reflections. Show that the e-reflection in a hyperbolic line l through our origin O is the same as the hyperbolic reflection in l (hence e-rotations about O are the same as hyperbolic rotations about O). But if κ is an equidistant curve for l, can the e-reflection in κ also be described geometrically within the hyperbolic plane? Think of the Poincaré disk model, where h-reflections are described geometrically as inversions in circles orthogonal to the rim of the disk. Is there an analogous hyperbolic geometric theory of e-reflections in curves equidistant from lines through O? In suitable coordinates, express the e-reflection in κ of any point P in terms of data from P, κ, and l.

In the Poincaré disk model, e-reflections are *not* inversions in the circles which cut out those equidistant curves in the disk. The *conformal* disk model, in which lines are diameters of the unit disk or arcs of circles cutting the rim at the ends of a diameter, is *Klein's model of the elliptic plane with one line removed* (the rim with antipodal points identified is the missing line)—see Coxeter (1998, Section 14.6 and Figure 14.8A). This model is obtained from the lower hemisphere model in $\mathbb{R}^3$ of the elliptic plane, in which lines are great semicircles, by stereographic projection from the north pole onto the equatorial disk. Although the interpretation of points and lines is the same, determine the different formulas for distance that distinguish the elliptic geometry disk model from the Euclidean geometry disk model.

2. From Figure 10.32, use hyperbolic trigonometry and the formula for $\sinh 2x$ to prove that $\tan \theta = 2 \sinh(r/2)$. As for constructibility, it is easy to see from that figure how to construct θ given r. The reverse construction is more subtle—see Martin (1982, Construction 7, Chapter 34), or Gray (2004, pp. 69–74). It uses theorems that the three altitudes and the three angle bisectors of a singly asymptotic triangle—suitably interpreted—are each concurrent. Explain and report on those.

3. The right triangle construction theorem is a special case of the following more general theorem: In a hyperbolic plane, given any angles λ, μ, ν such that $\lambda + \mu + \nu < \pi$, a triangle can be constructed (unique up to congruence) having λ, μ, ν as its angles. If you cannot figure out how to construct this triangle, report on the construction using a pentagon in Martin (1982), p. 486.

4. Explain how Jagy shows that if θ and σ are constructible, then we see that $x = \tan^2 \theta$ is a constructible length, and by the theorem of Gelfond–Schneider on transcendental numbers, *x must be rational*. Explain how Jagy gets the counterexamples for his Theorem C.

5. In the hyperbolic plane with $k = 1$, it is a theorem that *Schweikart's segment p cannot be trisected with a straightedge and compass*. This can be proved via the M-B theorem and the identity

$$\sinh 3t = 4 \sinh^3 t + 3 \sinh t,$$

provided you know that certain cubic equations have no solution in the field K of constructible numbers. Report on a proof of that impossibility (e.g., in Ramsay and Richtmyer, 1995, Section 11.3). Show how to trisect, with a straightedge and compass, an arbitrary segment in a Euclidean plane. *This is another significant difference between Euclidean and hyperbolic geometry.*

6. It is a theorem that the tangent at one endpoint of a horocyclic arc is asymptotically parallel to the line extending the diameter of the horocycle from the other endpoint when the length of the arc is 1 (distance scale $k = 1$). The *tangent* is defined to be the line that is perpendicular to the diameter at that point on the horocycle. Report on a proof of this theorem—e.g., disentangling it from Martin (1982), Theorem 32.2.

7. If you know calculus, report on what calculus is like in the real hyperbolic plane (see Martin, 1982, Section 34.2, or Wolfe, 1945). Mention the differential formulas for arc length and area, both in Lobachevsky coordinates and in polar coordinates, and apply those formulas to circles, equidistant curves, and horocycles. Alternatively, do this report for the Poincaré upper half-plane model (there are many references, such as Stahl, 1993).

8. In the Poincaré upper half-plane model, the ideal points are the points on the x-axis together with a new point ∞; geometrically, they form a circle, as can be seen from the isomorphism with the disk model. Analogously, in the upper half-space model of hyperbolic three-space, the ideal points are the points in the (x, y)-plane Π_0 together with a new point ∞; geometrically, they form a sphere, as can be seen by an analogous isomorphism with another model, the interior of a sphere. The geometry of hyperbolic three-space can be described in terms of the Euclidean geometry of circles on a sphere (*inversive geometry*). From the point of view of Klein's *Erlanger Programme*, these geometries are equivalent because they have the same group of direct automorphisms, *the group of Möbius transformations PGL(2, $\mathbb{C}$)*. Report on this (see Coxeter, 1998, Section 14.9; Stillwell, 1992, Section 4.9; and search the web for other references). Recall from Chapter 9 that the group of direct automorphisms of the real hyperbolic plane is *PGL*(2, $\mathbb{R}$), so going from

two hyperbolic dimensions to three corresponds to going from $\mathbb{R}$ to $\mathbb{C}$ in the group of Möbius transformations.

9. Learn about three-dimensional hyperbolic geometry from the literature and the web and write a report about it. Explain what a *horosphere* is and how, with "lines" interpreted as horocycles on the horosphere and with a suitable notion of congruence, a model of a Euclidean plane is obtained, as F. L. Wachter first observed in an 1816 letter to Gauss. One reference is Ramsay and Richtmyer (1995, Sections 4.7–4.9, 9.4). This model shows that the Euclidean plane can be embedded as a surface in hyperbolic three-space, but Hilbert showed that the hyperbolic plane cannot be smoothly embedded as a surface in Euclidean three-space. Report on that result if you know differential geometry.

10. A major project is to learn advanced Euclidean geometry and see how much of it carries over or has analogues in hyperbolic geometry. As some examples, is there an interesting theory of cyclic quadrilaterals? Is there an Euler line? What can be said about regular pentagons and hexagons (including asymptotic ones) and their inscribed circles? About pentagons and hexagons in which all angles are right angles? In the Euclidean plane, every angle can be trisected using a *marked* straightedge and compass (Hartshorne, Chapter 6). Is that also the case in the hyperbolic plane?

Once hyperbolic geometry was accepted toward the end of the nineteenth century and early twentieth century, there were many articles and books published about it. D. M. Y. Sommerville compiled a bibliography of those articles in 1911, which has been republished and somewhat updated to 1970. Some of this work may have been described in those articles.

A

Elliptic and Other Riemannian Geometries

The dissertation submitted by Herr Riemann offers convincing evidence . . . of a creative, active, truly mathematical mind, and of a gloriously fertile imagination.

C. F. Gauss

Elliptic Geometry

In Euclidean geometry there is exactly one parallel to a line l through a point P not on l; in hyperbolic geometry there is more than one parallel. A third geometry could be studied, one in which there is *no* parallel to l through P, i.e., a geometry in which parallel lines do not exist.

However, if we simply add the latter as a new parallel axiom to replace the other parallel axioms, the system we get is inconsistent. In Corollary 2 to Theorem 4.1 we proved that parallel lines do exist in neutral geometry, so that we would get a contradiction by adding such a parallel axiom.

To avoid this, we have to modify some of our other axioms. We can see what modifications need to be made by thinking of the surface of a sphere and interpreting "line" as "great circle." Then, indeed, there

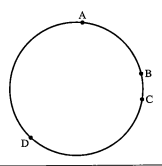

Figure A.1 (A, C | B, D).

are no parallel lines. But other things change as well. It is impossible to talk about one point B being "between" two other points A and C on a circle. So all the axioms of betweenness have to be scrapped. They are replaced instead by seven axioms of *separation*. In Figure A.1, A and C separate B and D on the circle since you can't get from B to D without crossing either A or C.

Let us designate the undefined relation "A and C separate B and D" by the symbol (A, C | B, D). The separation axioms are then:

SEPARATION AXIOM 1. If (A, B | C, D), then points A, B, C, and D are collinear and distinct.

SEPARATION AXIOM 2. If (A, B | C, D), then we have (C, D | A, B) and (B, A | C, D).

SEPARATION AXIOM 3. If (A, B | C, D), then not (A, C | B, D).

SEPARATION AXIOM 4. If points A, B, C, and D are collinear and distinct, then (A, B | C, D) or (A, C | B, D) or (A, D | B, C).

SEPARATION AXIOM 5. If points A, B, and C are collinear and distinct, then there exists a point D such that (A, B | C, D).

SEPARATION AXIOM 6. For any five distinct collinear points A, B, C, D, and E, if (A, B | D, E), then either (A, B | C, D) or (A, B | C, E).

To state the last axiom, we recall the notion of a *perspectivity* from one line onto another (from Chapter 7). Let l and m be any two lines and let O be a point not on either of them. For each point A on l, the line $\overleftrightarrow{OA}$ intersects m in a unique point A' (Figure A.2; remember the ellip-

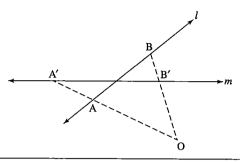

Figure A.2 Perspectivity with center O.

tic parallel property); the one-to-one correspondence that assigns A' to A for each A on l is called the *perspectivity* from l to m with center O.

SEPARATION AXIOM 7. Perspectivities preserve separation; i.e., if (A, B | C, D), with l the line through A, B, C, and D, and if A', B', C', and D' are the corresponding points on line m under a perspectivity, then (A', B' | C', D').

Without the notion of betweenness we have to carefully reformulate all the geometry using that relation. For example, the *segment* AB consists of the points A and B and all the points between them. Yet this doesn't make sense on a circle. We can only talk about the segment ABC determined by *three* collinear points: It consists of the points A, B, and C and all the points not separated from B by A and C.

Similarly, we have to redefine the notion of a triangle since its sides are no longer determined by the three vertices (see Figure A.3).

Once these notions have been redefined, the axioms of congruence and continuity all make sense when rephrased and can be left intact.

There is still a difficulty with Incidence Axiom 1, which asserts that two points do not lie on more than one line. This is false for great

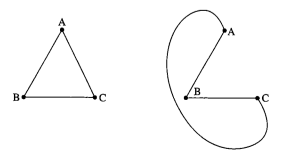

Figure A.3 Two different "triangles" with the same vertices.

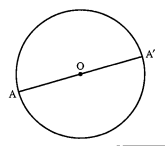

Figure A.4 A and A′ are identified.

circles on the sphere since antipodal points (such as the poles) lie on infinitely many lines.

Klein saw that the way to remedy this is to *identify* antipodal points; i.e., just as we interpret "line" to mean "great circle" in this model, we interpret "point" to mean "pair of antipodal points" (Figure A.4). This means that *in our imagination* we have pasted together two antipodal points so that they coalesce into a single point. It can be proved, as you might guess, that such pasting cannot actually be carried out in Euclidean three-dimensional space. But we can still identify antipodal points in our minds—every time we move from one to the other, we think of ourselves as being back at the original point.

In making these identifications, we discover another surprising property: A line no longer divides the plane into two sides, for you can "jump across" a great circle by passing from a given point to its now equal antipodal point that used to be on the other side. If we cut out a strip from this plane, it will look like a Möbius strip, which has only one side (see Figure A.5). The technical name for this property of "onesidedness" is *non-orientability*.

To sum up, the axioms of plane elliptic geometry are the same as the incidence, congruence, and continuity axioms of neutral geometry (with the new definitions of segment, triangle, etc.). The betweenness axioms are replaced by separation axioms, and the parallel postulate is replaced by an axiom stating that no two lines are parallel. A model, which shows that elliptic geometry is just as consistent as Euclidean geometry, consists of the great circles on the sphere with antipodal points identified.[1]

---

[1] The geometry of the sphere itself is sometimes misleadingly called "double elliptic geometry." References for spherical geometry and trigonometry are Brannan, Esplen, and Gray (1999, Chapter 7) and McCleary (1994, Chapter 1). For more on the history of spherical geometry, consult Rosenfeld (1988, Chapter 1) and Katz (1998, Chapter 4).

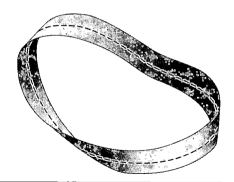

Figure A.5 Möbius strip.

As you might expect from this model, it is a theorem in elliptic geometry that *lines have finite length*. Moreover, all the lines perpendicular to a line l are not parallel to each other but are concurrent; i.e., all the perpendiculars to l have a point in common called the *pole of l*. In the model, for instance, the pole of the equator is the north (or, what is the same, the south) pole.

Another model for plane elliptic geometry (due to Klein) is *conformal*—as in the Poincaré model for hyperbolic geometry, angles are accurately represented by Euclidean angles. In this model, "points" are the Euclidean points inside the unit circle in the Euclidean plane as well as pairs of antipodal points on the circle; "lines" are either diameters of the unit circle or arcs of Euclidean circles that meet the unit circle at the ends of a diameter (see Coxeter, 1998, Section 14.6). This representation shows that the angle sum of a triangle is greater than 180° in elliptic geometry (see Figure A.6 and the proof on p. 546).

These two models of the elliptic plane are isomorphic by stereographic projection, using as an intermediary model the lower hemisphere L, with great semicircles C as lines plus the equator, which is

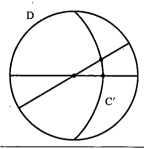

Figure A.6 Klein conformal elliptic model.

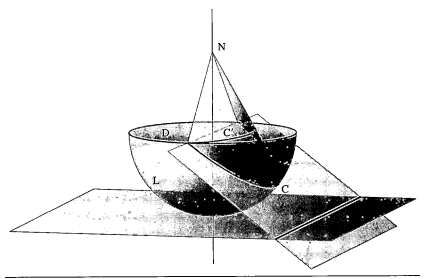

Figure A.7 Isomorphism of elliptic models.

also a line after its antipodal points are identified: Just project L onto the equatorial disk D from the north pole N (see Figure A.7).

Here is an intuitive proof (Weeks, 2002, pp. 139–142) that for the unit sphere S and its associated model E of the elliptic plane, **the angle sum of any triangle in E is greater than 2π and the *excess* of that sum over 2π is equal to the area of the triangle.**

Given a triangle $\triangle$ in E, its preimage on S consists of two disjoint congruent spherical triangles ABC and A′B′C′, the points of one being the antipodes of the other. The angles of ABC have the same measure as the corresponding angles of $\triangle$. The area Σ of $\triangle$ equals the area of ABC. Extend the sides of ABC into three great circles on S (Figure A.8).

A *lune* is one of the regions bounded by two great circles meeting at two antipodal points, making, say, an angle θ at their meeting points. We accept that the area of the lune depends linearly on θ. Since the extreme case $\theta = \pi$ gives a hemisphere, which has surface area 2π (Archimedes), the area of that lune is 2θ by linearity. If α, β, γ are the angles of our triangle, the sphere is covered by three lunes having those angles and their three antipodal lunes (which have the same area). In adding up those lune areas to get the total area of the sphere, we see that the areas of our congruent triangles ABC and A′B′C′ are counted three times. Hence

$$4\alpha + 4\beta + 4\gamma = 4\pi + 4\Sigma,$$

which gives the formula for Σ that we claimed.

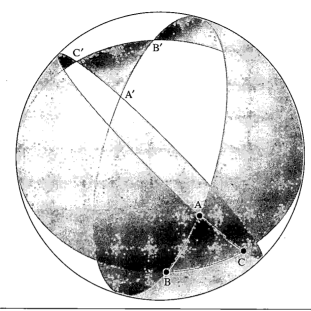

Figure A.8 Excess of elliptic triangle equals its area.

Elliptic geometry becomes even more interesting when you pass from two to three dimensions. In three dimensions, orientability is restored and a new kind of parallelism occurs. Two lines are called *Clifford-parallel* if they are equidistant from each other; the lines are joined to each other by a continuous family of common perpendicular segments of the same length. Such lines cannot lie in a plane (in an elliptic plane, two lines must intersect), so they are skew lines. Moreover, in general, in elliptic space for any point P not on a line *l* there exist exactly two lines through P that are Clifford-parallel to *l*, called the *right and left Clifford parallels* to *l* through P. We say "in general" because there is a special line *l**, called the *absolute polar* of *l*: If P lies on *l**, there is only one Clifford parallel to *l* through P, which is *l**. (Naturally, this is difficult to visualize! See Coxeter, 1998, Chapter 7.)

Elliptic space is finite but unbounded—finite because all lines have finite length and look like circles, and unbounded because there is no boundary, just as on the surface of a sphere there is no boundary. In a universe having this geometry, with light rays traveling along elliptic lines, you could conceivably look through a very powerful telescope and see the back of your own head! (Although you might have to wait a few billion years for the light to travel all the way around.)

Riemannian Geometry

It is impossible to rigorously explain the ideas of Riemannian geometry without using the language and results of differential and integral calculus, as Gauss, Riemann, and their many successors did. So in this section, an attempt will be made to roughly describe the intuitive ideas, with a modicum of precision and a few calculus formulas for those in the know. The main references I recommend for follow-up study of this subject by novices are McCleary (1994), O'Neill (2006), and Weeks (2002).

░░░░ CURVATURE

As indicated in Chapter 10, the basic notion is *curvature*, and the basic reason a real non-Euclidean plane differs from the real Euclidean plane is that it is somehow "curved," whereas the Euclidean plane is not curved or is "flat." Let us define this notion.

Consider first dimension one. The two simplest one-dimensional figures in a Euclidean plane are a line and a circle. We think of a line as not being curved, so we assign curvature $k = 0$ to a Euclidean line. A circle's curvature intuitively depends on its radius r. The larger r is, the more gradually the circle curves as we move around it, and in the limit as r approaches ∞, if we fix one point P, the Euclidean circle approaches its tangent line at P. As r approaches zero, the circle curves more and more sharply. So it is natural to define the curvature k of a circle of radius r by $k = 1/r$. In these definitions, the curvature is *constant*—it is the same at every point on the line or the circle.

Now consider an arbitrary smooth curve γ in $\mathbb{R}^2$—i.e., a curve that has a continuously turning tangent line at every point P on γ. The tangent line at P is the limiting position of a line joining P to a second point Q on γ as Q approaches P along γ (see Figure A.9) The limiting line must be the same whether you approach P from the left or from the right—e.g., the graph of $y = |x|$ is not smooth at (0, 0) because those limits are different, so there is a corner singularity at the origin. (For those who know calculus, a *parametrized smooth plane curve* is the image of an open interval in $\mathbb{R}$ under a sufficiently differentiable mapping $f(t) = (x(t), y(t))$ into $\mathbb{R}^2$ having a nowhere zero derivative $f'(t) = (x'(t), y'(t))$; this provides a nonzero tangent "velocity" vector at each point and its second derivative provides an "acceleration" vector field along the curve.)

Besides the fixed point P, we can also consider two other points P_1 and P_2 on γ. These three points generally determine a circle δ. Fix P

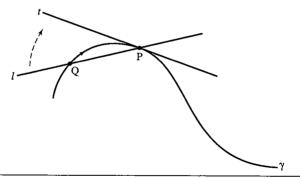

Figure A.9 Tangent line is limit of secants.

and let P_1 and P_2 both approach P along γ. The limiting position of circle δ as P_1 and P_2 approach P is the circle that "best fits" the curve γ at P and is called the *osculating circle* to γ at P (from the Latin *osculare*, "to kiss"); see Figure A.10. It is reasonable to define the *curvature* of γ at P as the curvature of its osculating circle at P, i.e., the reciprocal $k = 1/r$ of the radius r of the osculating circle (r is also called the *radius of curvature* of γ at P).

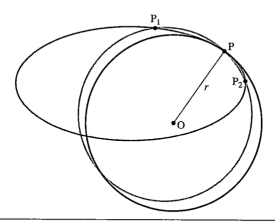

Figure A.10 Osculating circle.

It is clear from Figure A.11 that the osculating circle will vary in size as we move along the curve γ, so that the curvature k varies from point to point along γ. Notice also that the tangent to γ at a point P is also the tangent to the osculating circle at P.

Notice further in Figure A.11 that the osculating circle may be on different sides of the curve γ. It is convenient to *redefine curvature* so that it is positive on one side of γ and negative on the other. Once this is done, it becomes clear that in Figure A.11 there must be a point I

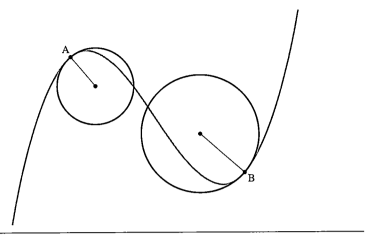

Figure A.11 Varying osculating circles.

between A and B on γ at which the curvature is zero if we assume γ is smooth enough for the curvature to vary continuously. This point I is called a *point of inflection*, and at such a point the osculating "circle" degenerates into a line, the tangent line at I (see Figure A.12).

What we have said about plane curves applies equally well to curves in Euclidean space, with the following modifications. The osculating circle lies in a unique plane through P called the *osculating plane* Π of γ at P (except in the degenerate case of a point P at which γ has curvature zero). Since Π varies with P if γ is not a plane curve, we no longer have a smooth way to assign positive and negative values to the curvature. We assign a *curvature vector* **k** lying in the osculating plane, emanating from P and perpendicular to the tangent line to γ at P (which also lies in the osculating plane), of length $1/r$ and pointing toward the center of the osculating circle (called the *center of curvature*). If γ has

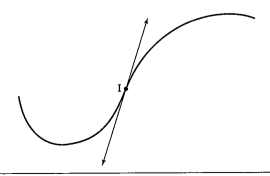

Figure A.12 Point of inflection $k = 0$.

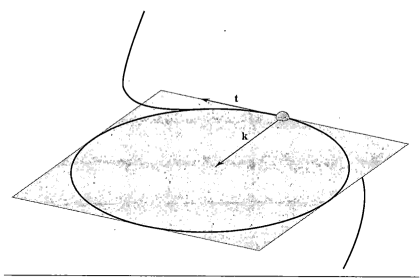

Figure A.13 Curvature vector **k**.

curvature zero at P, we take **k** to be the zero vector. If the curve is parametrized by its arc length s starting from some initial point (or any other *unit speed* parameter), then its "velocity" tangent vector has length 1 at every point, and its *acceleration vector equals the curvature vector* at every point (Figure A.13).

We next pass to a surface S in $\mathbb{R}^3$. Locally (i.e., in the neighborhood of each point on it), S looks like the graph of a sufficiently well-behaved function of two real variables. Again, we are interested in a *smooth surface* that has a continuously turning tangent plane T (defined analogously to the tangent line to a curve) at all of its points. Locally, there are two continuous fields of unit *normal* vectors perpendicular to the tangent planes at points in that neighborhood, the vectors in one field being the negatives of the other. Choosing one of those vector fields orients that neighborhood, determining a positive side of the surface—the side to which those normal vectors "point." If such a unit normal vector field exists globally on S, this is one criterion for S to be *orientable*, and the choice of a global unit normal vector field is called an *orientation* of S. For example, a Möbius strip (first presented in 1865) minus its edge is a smooth surface that is not orientable.

Suppose S is oriented. If γ is a smooth curve on S, then at each point P on γ we can again define a *signed curvature* k that is positive or negative according to whether the center of curvature of γ at P lies on the positive or negative side of S.

Consider the line through P that is perpendicular to T, called the *normal line* at P. A plane that contains the normal line intersects the surface in a plane curve. We can imagine this plane rotating around the normal line, and as it does so, it cuts out different curves on the surface passing through P. We have already explained how to define the curvature at P for each of these normal sections. In general, these curvatures vary as we rotate around the normal line. (In the special case of a sphere, these curvatures are constant and equal to the reciprocal of the radius of the sphere since the curves cut out are all great circles on the sphere.) Euler proved in 1760 that these curvatures achieve a maximum value k_1 and a minimum value k_2 as we rotate, and the corresponding normal sections (called *principal curves*) are perpendicular to each other if distinct. The product $K = k_1 k_2$ of these maximum and minimum curvatures is now called the *Gaussian curvature*—after Gauss, who defined it differently (O'Neill, 2006, Theorem 8.4, Chapter 6)—or simply "the curvature" of the surface at the point P. Once again, K in general changes as P varies over the surface; if K happens to stay constant, we obtain as examples the three geometries discussed in Chapter 10, according to whether K is negative (pseudosphere, for example), zero (plane), or positive (sphere).

In Figure A.14, the tangent plane, normal line, and principal curves for a saddle-shaped surface are shown. For point P on this surface, the Gaussian curvature will be a negative number, according to our convention, since the osculating circles for the two principal curves lie on

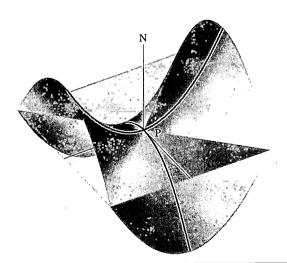

Figure A.14 Point where curvature is negative.

Figure A.15 Point where curvature is positive.

different sides of the tangent plane. On the other hand, for the surface in Figure A.15, the two principal curves lie on the same side of the tangent plane, so the Gaussian curvature is positive.

Refer back to the pseudosphere shown in Figure 7.2, which is obtained by revolving a tractrix around its asymptote. At any point on this surface, it can be shown that the two principal curves are the (horizontal) circle of revolution through that point and the (vertical) tractrix through that point. Since these curves lie on opposite sides of the tangent plane, we see that the surface curvature is negative. As we move up the surface, the circle shrinks and its curvature increases indefinitely, while the tractrix flattens and its curvature decreases to zero (the curvature of its asymptote); this makes it plausible that the product K of the principal curvatures could stay constant (see O'Neill, 2006, Example 7.6, Chapter 5 for the calculation).

Similarly, in the case of a cylinder obtained, say, by rotating a vertical line around a parallel vertical line, the principal curves at any point are the horizontal circle and the vertical line, which has curvature $k_2 = 0$; hence, the Gaussian curvature $K = k_1 k_2$ at any point on a cylinder

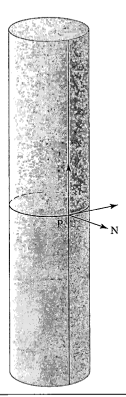

Figure A.16 Gaussian curvature of a cylinder is zero.

will also be zero (see Figure A.16). We can better grasp this surprising result if we think of a cylinder as a "rolled-up plane." Surely, in any sensible definition of surface curvature, a flat plane should be assigned zero curvature. In the process of "rolling up" a rectangular plane strip, the local arc lengths and angles between curves on the strip are not changed, and in this sense the "intrinsic geometry" is not changed.

Gauss was looking for a definition of surface curvature that depended only on the intrinsic geometry of the surface, not on the particular way the surface was embedded in Euclidean three-space. He was able to prove that the curvature K does not change if the surface is subjected to a "bending" in which local arc lengths and angles of all curves on the surface are left invariant. Thus, K describes the intrinsic curvature of the surface independent of the way it is bent to fit into Euclidean three-space. This is all the more remarkable because the principal curvatures k_1 and k_2 and their average (the *mean curvature*) may change under such a "bending"; nevertheless, their product $K = k_1 k_2$

stays the same.[2] Gauss was so excited about this result that he named it the *theorema egregium*, "the extraordinary theorem." In a letter to the astronomer Hansen, he wrote, prophetically: "These investigations deeply affect many other things; I would go so far as to say they are involved in the metaphysics of the geometry of space."

Gauss also solved the problem of determining this intrinsic curvature K without reference to the ambient three-space. Imagine a two-dimensional creature living on a surface and having no conception of a third dimension, being unable to conceive of the normal lines we used to define the curvature K. How could this creature calculate K? We will have to use the language of differential calculus to give Gauss' answer.

In the Euclidean plane, a point is determined by its x- and y-coordinates. If these coordinates are subjected to infinitesimal changes denoted dx and dy, then the point moves an infinitesimal distance ds whose square is given by the Pythagorean formula $ds^2 = dx^2 + dy^2$. Now on a smooth surface, a point will also be determined locally by two coordinates x and y. If these coordinates are subjected to infinitesimal changes dx and dy, then the point moves a distance ds on the surface whose square is given by the more complicated expression

$$ds^2 = E\ dx^2 + 2F\ dx\ dy + G\ dy^2,$$

where E, F, and G may vary as the point varies. The functions E, F, and G could in principle be determined by the two-dimensional creature making measurements on this surface. In 1828, Gauss found a complicated formula for his curvature K in terms of E, F, and G (see McCleary, 1994, pp. 148–150, or O'Neill, 2006, Proposition 6.3, Chapter 6). Thus, the creature could also calculate K from this formula and discover that his world was curved, although he would have difficulty visualizing what that might mean. (The creature could also estimate the curvature of his world by measuring the circumference of a small *polar circle*—see O'Neill, 2006, Corollary 3.9, Chapter 8.)

Although this talk about a two-dimensional creature may seem bizarre, it is not, as Riemann demonstrated. Riemann reasoned that we are in an entirely analogous situation, living in a three-dimensional

---

[2] Nowadays the vague term "bending" is replaced by the more precise notion of *local isometry*—a smooth mapping that maps a neighborhood of every point isometrically onto a neighborhood of the image point. We will soon explain what "isometrically" means in this context. Another interpretation of "bending" is a smooth isometric deformation of one surface onto another, as for example the deformation of a catenoid to a helicoid and back—see the animation at http://mathmuse.sci.ibaraki.ac.jp/deform/DeformationE.html.

universe in which an infinitesimal change in distance ds is given by an analogous formula involving the three infinitesimals dx, dy, and dz:

$$ds^2 = g_{11}\ dx^2 + g_{22}\ dy^2 + g_{33}\ dz^2 + 2g_{23}\ dy\ dz + 2g_{31}\ dz\ dx + 2g_{12}\ dx\ dy.$$

From this formula Riemann was able to define a "curvature tensor" analogous to the Gaussian curvature for a surface, only more complicated: Gauss' curvature involved only a single number K, whereas Riemann's depended on six different numbers. Riemann discovered this curvature almost accidentally in his research on heat transfer. In fact, he developed such a curvature tensor for abstract geometries of any dimension n, and Einstein was able to apply Riemann's ideas to his four-dimensional space-time continuum.

So we are in the same position as that two-dimensional creature. We can make measurements to determine the Riemannian curvature of our universe. Astronomers have been performing such measurements. If we find that the Riemannian curvature is not zero, we know that the geometry is not Euclidean. However, this does not mean that our space is embedded in some higher-dimensional physical space in which it is somehow "curved." When we say, loosely, that "space is curved," we mean only that its geometric properties differ from the properties of Euclidean space in a very specific way given by Riemann's formulas.

RIEMANNIAN MANIFOLDS

It was in his 1854 inaugural lecture, "Über die Hypothesen welche der Geometrie zugrunde liegen" (On the hypotheses that form the foundation for geometry), that Riemann introduced the idea of an n-dimensional space whose intrinsic geometry is determined by a quadratic formula for the infinitesimal change in distance ds. Such a structure is now called a *Riemannian manifold*. Here is an idea of what that structure is.

On the first level, the structure is a *topological space*. This means that certain subsets, which together cover the whole set, are designated as *open*, and a *neighborhood* of a point is defined to be an open set containing that point. In $\mathbb{R}^3$, an open set is just a union of small open balls around each of its points. Finite intersections and all unions of open sets are open. This structure enables one to define a *continuous mapping* as one for which the inverse image of an open set is open. A one-to-one mapping of a topological space onto another that is continuous and whose inverse mapping is also continuous is called a *homeomorphism*. Homeomorphic spaces are "topologically the same."

The second level is that of a *topological n-manifold*. This means that everywhere locally the space looks like an open set in $\mathbb{R}^n$ by means of given homeomorphisms (called *charts* or *patches*) from open sets in $\mathbb{R}^n$. Often other "niceness" conditions are assumed in the definition, such as the Hausdorff property that distinct points can be separated by disjoint open sets, or the property that the family of open sets is generated by a countable number of them. It is possible to define what an *orientation* of a topological manifold is using homology theory (see Greenberg and Harper, 1982).

The third level is that of a *differentiable n-manifold*. This means that the patches can be chosen so that on the intersection of the images of two patches, one patch followed by the inverse of the other patch is a differentiable or analytic function of n real variables, a *diffeomorphism* of those open sets in $\mathbb{R}^n$. If these changes of coordinates can be chosen to have the property that they preserve a given orientation on $\mathbb{R}^n$, then the manifold is *oriented*. This level enables one to do calculus on the manifold M. Most importantly, it enables the definition of an n-dimensional tangent space for each point (McCleary, 1994, Definition 16.3); those tangent spaces bundled together form a $2n$-dimensional differentiable manifold, the *tangent bundle* of M, which projects down onto M. Sections of the tangent bundle (right inverses of the projection—smooth assignments of tangent vectors) over all or part of M are *vector fields* whose study (initiated by Heinz Hopf) is fundamental in differential topology.

The final level of a Riemannian structure is the imposition of a positive-definite quadratic form on each of the tangent spaces that varies smoothly. Called a *Riemann metric*, it is an n-dimensional dot product that enables the measurement of lengths of tangent vectors and the angles between them. The classical notation for the square of the length of a tangent vector is the ds^2 mentioned above. By integrating ds along an arc of a curve, one calculates the arc length induced by the manifold's Riemannian structure.

The theory becomes quite abstract, and you have to learn to think abstractly to understand it!

We will mainly focus on two-dimensional Riemannian manifolds, called *geometric surfaces*. Smooth surfaces in $\mathbb{R}^3$ automatically inherit a Riemannian structure from $\mathbb{R}^3$. All the *intrinsic* geometric notions for surfaces in $\mathbb{R}^3$ have definitions for abstract geometric surfaces; in particular, *curvature* can be defined (e.g., by using Élie Cartan's method of differential forms from moving frames—O'Neill, 2006, Theorem 2.1, Chapter 7).

GEODESICS

Let us return to a surface Σ in $\mathbb{R}^3$. For each point P on Σ and each curve γ lying on Σ and passing through P, the curvature vector $\mathbf{k}$ decomposes naturally into a vector sum

$$\mathbf{k} = \mathbf{k}_n + \mathbf{k}_g$$

of its projection $\mathbf{k}_n$ on the normal line and its projection $\mathbf{k}_g$ on the tangent plane T to Σ at P; these projections are called, respectively, the *normal* and *tangential curvature vectors*. If γ is again parametrized with unit speed, then $\mathbf{k}$ is the acceleration vector for that parametrization, and its tangential component is considered the *acceleration of γ along the surface* (Figure A.17). The length of $\mathbf{k}_g$ is called the *geodesic curvature k_g* of the curve γ at P relative to Σ. We call γ a *geodesic* if $k_g = 0$. In that case, γ has *zero curvature relative to the surface Σ*; it may have nonzero curvature relative to Euclidean three-space, but then its curvature vector points along the normal line to the surface at P. Another way to describe a geodesic γ that has nonzero $\mathbf{k} = \mathbf{k}_n$ is that its osculating plane Π at P is perpendicular to the tangent plane T (since Π contains the normal line at P). From this description we see imme-

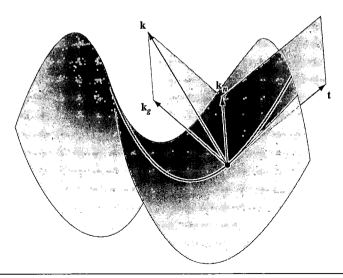

Figure A.17 Normal and tangential components of curvature vector.

diately that the geodesics on a sphere are its great circles ($\Pi \perp T$ iff Π passes through the center of the sphere).

This description of geodesic curvature refers to the ambient Euclidean three-space. But F. Minding in 1830 showed that it too is an intrinsic quantity for Σ: It depends only on the functions E, F, and G and the curve γ (McCleary, 1994, Theorem 11.3). Hence the notion of "geodesic" can be defined more generally on a Riemannian manifold. And it gives us the correct interpretation of the heretofore confusing term *straight line* on such a manifold. For a geometric surface, when γ is parametrized with unit speed, we want to say that it is a geodesic when its acceleration along the surface is zero, as for surfaces in $\mathbb{R}^3$. Acceleration is defined as the derivative (rate of change) of velocity, but now that our surface is not necessarily embedded in $\mathbb{R}^3$, a new kind of derivative is needed; called the *covariant derivative*, it is defined in terms of intrinsic data on the surface (O'Neill, 2006, Section 7.3, Chapter 7). The covariant derivative operation generalizes to Riemannian manifolds M of any dimension and operates on any vector field on a part or all of M. A curve on M whose velocity tangent vector field has a covariant derivative equal to zero at every point is defined to be a *geodesic* (McCleary, 1994, Definition 16.21, or O'Neill, 2006, Section 7.4, Chapter 7).[3]

A proposed alternative interpretation of the term "straight line segment" is "the shortest path on the surface joining two points on the surface." It can be proved that *if such a shortest path exists, it must be an arc of a geodesic*. But a shortest path may not exist: Let Σ be a punctured plane or a punctured sphere; two points on opposite sides of the puncture cannot be joined by a shortest path on Σ. On *complete* manifolds M (meaning that every geodesic segment can be extended indefinitely—Euclid's Postulate II for geodesics), it can be proved that shortest paths always exist when M is connected (theorem of Hopf-Rinow). However, an arc of a geodesic need not be the shortest path:

---

[3] The existence of a covariant differentiation operator with the required properties is equivalent to the existence of a *parallel transport* operation on tangent vectors along a curve in the manifold. The velocity tangent vectors to a geodesic are all "parallel" in this sense. This operation was first revealed by T. Levi-Civita in 1917; Einstein independently discovered this operation later. The structure underlying it is called the *Levi-Civita connection* on M; its existence and uniqueness is called the *fundamental theorem of Riemannian geometry*. S.-S. Chern, dubbed "the father of modern differential geometry" by S.-T. Yau, wrote that "most properties of Riemannian geometry derive from its Levi-Civita parallelism, an infinitesimal transport of the tangent spaces" (Chern, 1990, p. 682). Parallel transport on a torus is discussed and illustrated at http://www.rdrop.com/~half/math/torus/torus.geodesics.pdf.

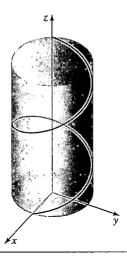

Figure A.18 A helix is a geodesic on a cylinder.

Consider the longer great-circular arc joining two non-antipodal points on a sphere, or consider the helical arc joining two points on a vertical line on the cylinder in Figure A.18. So the definition of "geodesic" (found in several books by nonmathematicians) as "the shortest path" is inadequate because it excludes such arcs.

What is the case, even for incomplete manifolds, is that a geodesic arc is the shortest path between two sufficiently close points on it—*a geodesic minimizes arc length locally*, and conversely, a curve with this property must be a geodesic. At each point P on a Riemannian manifold *M*, there exists exactly one geodesic issuing in each direction, and the entire route of the geodesic is determined by its point P and its tangent vector at P. In fact, there is a fundamental smooth one-to-one map *exp* (the *exponential map*) from a neighborhood of the origin O in the tangent space *T* at P to a neighborhood of P in *M* that takes Euclidean line segments through O in that neighborhood to geodesic arcs through P on *M*; exp is defined on all of *T* for all P if and only if *M* is *complete* (O'Neill, 2006, Chapter 8). When *M* is the one-dimensional manifold of positive real numbers, exp is the usual exponential function.

IMPORTANT PROPERTIES MANIFOLDS CAN HAVE
A differentiable manifold is called *connected* if any two points can be joined by a piecewise smooth curve. On a connected manifold, the *distance* between two points can be defined as the greatest lower bound of the lengths of such curves joining the points. All examples we dis-

cuss are connected. On a *complete* connected manifold, any two points lie on a geodesic, and the distance between two points is realized as the arc length of a geodesic joining them (for dimension 2, see O'Neill, 2006, Chapter 8).

A smooth mapping of one connected manifold onto another is an *isometry* if it preserves the distance between any two points. Hence it is a one-to-one mapping. It also preserves the sizes of angles but may reverse the direction of oriented angles. The motions (reflections, rotations, translations, glides, and parallel displacements in the hyperbolic plane) studied in Chapter 9 are the simplest examples of isometries, and we showed there how they can be used to define the congruence of figures in Euclidean and hyperbolic planes. If M has enough self-isometries to map any point onto any other point, it is called *homogeneous*. This is a very strong restriction on a geometric surface, for it guarantees that M has constant curvature and is complete (O'Neill, 2006, Theorem 5.5, Chapter 8). Only the hyperbolic plane is homogeneous among complete geometric surfaces of constant negative curvature.

A smooth mapping of one geometric surface onto another is a *local isometry* if it maps a neighborhood of each point isometrically onto a neighborhood of the image point. Such a mapping is locally one-to-one but not necessarily globally. For example, the natural mapping that wraps the plane around the cylinder is a local but not a global isometry. A local isometry preserves the curvature at each point of a geometric surface since the curvature at a point is defined by local data. A theorem of Minding states that for any two points on surfaces of the same constant curvature, there exists an isometry of a neighborhood of one point onto a neighborhood of the other point (McCleary, 1994, Theorem 13.5). An important method of endowing smooth surfaces with a geometry is the following: Suppose M is a geometric surface and F is a local diffeomorphism of M onto a smooth surface N; if M satisfies the necessary compatibility condition with respect to F, then N can be given a unique geometric structure such that F becomes a local isometry (O'Neill, 2006, Proposition 2.6, Chapter 7). We will use this construction below.

Compactness is an important topological finiteness notion. For subspaces of $\mathbb{R}^n$, it means "closed and bounded." For example, the sphere, the closed disk, and the torus are compact. Any continuous image of a compact space is also compact, so the elliptic plane, being a continuous image of the sphere, is compact. The Euclidean and hyperbolic planes and the pseudosphere are not compact. A compact Riemannian manifold is always complete (O'Neill, 2006, Corollary 2.4, Chapter 8). A complete connected Riemannian manifold is compact if and only if the distance between any two of its points is bounded. For a compact surface S in $\mathbb{R}^3$, at a point of S of maximum distance from the origin, the curvature must be positive (O'Neill, 2006, Theorem 3.5, Chapter 6). A theorem of O. Bonnet states that a complete connected geometric surface of positive curvature bounded away from zero must be compact (O'Neill, 2006, Theorem 7.2, Chapter 8). A compact geometric surface of constant negative curvature has only finitely many self-isometries (theorem of H. A. Schwarz).

A manifold is *simply connected* if it is connected and every closed curve on it can be continuously shrunk to a point. For example, the plane, the disk, and the sphere are simply connected; the cylinder, the torus, the pseudosphere, and the elliptic plane (see Stillwell, 1992, p. 70) are not. When a connected manifold M is not simply connected, it can be "unwound" to a simply connected manifold U that "evenly covers it" by a local homeomorphism $U \to M$, with the pre-image of any point of M being a discrete set; U is called *the universal covering space* of M. For example, the sphere is the universal cover of the elliptic plane E by the two-to-one covering map that identifies antipodal points; that map conveys the local spherical geometric structure to E, the map becoming a local isometry. The plane is the universal covering space of a torus ($M = \mathbb{R}^2/\mathbb{Z}^2$); by means of that covering, the torus can be similarly endowed with a geometric structure having constant curvature zero—the *flat torus*. (For an illustration, see http://www.geom.uiuc.edu/~banchoff/script/b3d/hypertorus.html.)

The only *complete, simply connected* geometric surfaces of *constant curvature* are the sphere, the Euclidean plane, and the hyperbolic plane (O'Neill, 2006, Corollary 6.3, Chapter 8). If we drop only the hypothesis that the curvature is constant but require it to be everywhere ≤ 0, a theorem of Hadamard guarantees that there are no closed geodesics on the surface, that the geodesic joining two points is *unique* (which is our Axiom I-1), and that such a surface is diffeomorphic to $\mathbb{R}^2$ (O'Neill, 2006, Theorem 7.7, Chapter 8).

For a geometric surface S, S is *orientable* if it does *not* contain a Möbius strip. If it is non-orientable, it has a canonical orientable surface which maps onto it by a two-to-one covering map, as we have seen for the elliptic plane E. For another example, the non-orientable *Klein bottle* has a two-to-one covering map from the torus. By means of that map, the Klein bottle can be endowed with a *flat* geometric structure; see http://www.geom.uiuc.edu/~banchoff/Klein4D/Klein4D.html. In fact, there are only five types of complete flat geometric surfaces: Euclidean plane, cylinder, twisted cylinder, torus, and Klein bottle (O'Neill, 2006, p. 420; Stillwell, 1992, Chapter 2).

EMBEDDINGS AND IMMERSIONS IN $\mathbb{R}^3$

Although we have highlighted Riemann's generalization of Gauss' ideas from two dimensions to higher dimensions, Riemann's formulation gives new information about geometric surfaces that cannot be embedded in $\mathbb{R}^3$. First, it enables us to **define what a real hyperbolic plane is from the point of view of Riemannian geometry**: It is *a complete simply connected geometric surface of constant negative curvature*. Second, **a real elliptic plane is a complete connected geometric surface of constant positive curvature such that any two points lie on a unique geodesic. Neither of these geometric surfaces can be embedded in $\mathbb{R}^3$**—i.e., there is no isometry of either of them onto a surface in $\mathbb{R}^3$.

That the elliptic plane cannot be embedded in $\mathbb{R}^3$ follows from two facts: (1) Every compact $(n-1)$-dimensional topological manifold in $\mathbb{R}^n$ is orientable (a consequence of Alexander duality—see Greenberg and Harper, 1982, Theorem 27.11; or, for differentiable manifolds, see H. Samelson, "Orientability of Hypersurfaces in $\mathbb{R}^n$," *Proceedings of the American Mathematical Society*, 22(1) (July 1969), 301–302)[4]; (2) the elliptic plane is compact and non-orientable. There is an even stronger theorem of H. Liebmann that *a compact connected surface in $\mathbb{R}^3$ of constant curvature must have positive curvature and must be a sphere* (Mc-Cleary, 1994, Theorem 13.6, or O'Neill, 2006, Theorem 3.7, Chapter 6).

We can find models of the elliptic plane E in $\mathbb{R}^3$ if we relax the embedding requirement and instead just seek an *immersion*. There are three types of immersions: continuous, smooth, and isometric. They are all locally one-to-one. A smooth mapping is an immersion if, at every point, the linear mapping of the tangent plane it induces is one-to-one; the smooth immersion is isometric if that linear mapping preserves the Riemannian metric—hence the curvature. Three famous continuous immersions of E in $\mathbb{R}^3$ are the closed cross-cap, Steiner's Roman surface, and Boy's surface (Figure A.19). Only Boy's surface is a smooth immersion, without pinch points. (Interestingly, Hilbert conjectured that there was no smooth immersion of E in $\mathbb{R}^3$ and assigned his student Werner Boy to prove that conjecture; instead, Boy proved Hilbert was wrong!)

---

[4] Tom Banchoff told me he has a simpler proof in dimension 2, using a parity lemma that states that a closed curve in $\mathbb{R}^3$ transversely intersecting a compact surface intersects it an even number of times.

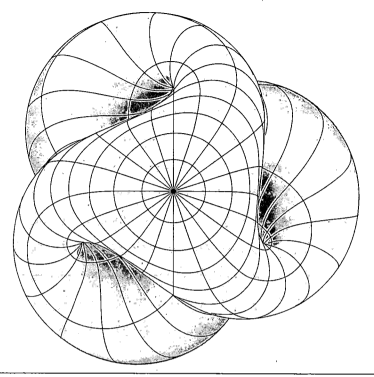

Figure A.19 Boy's surface. It has a continuous double-point curve, which meets itself in a triple point.

There are marvelous computer drawings online of these immersed surfaces, which you can rotate to view from all angles at http://xahlee. org/surface/gallery_o.html. See also http://en.wikipedia.org/wiki/Cross-cap, where the term "cross-cap" is often used for the surface obtained by removing a small open disk from the closed cross-cap, the result being homeomorphic to a Möbius strip. There is a detailed discussion of Boy's surface with excellent illustrations at http://en.wikipedia.org/wiki/Boy's_surface. See also Gray et al., 2006, Chapter 11.

As those models of E show, the image of a smooth surface under an immersion that is not an embedding can have self-intersections and other singularities—i.e., not be a smooth surface. Going up one dimension, there is a smooth (but not isometric) embedding of E in $\mathbb{R}^4$ induced by the Veronese map of the unit sphere:

$$f(x, y, z) = (x^2 - y^2, xy, xz, yz)$$

Similarly, the Klein bottle can be smoothly embedded in $\mathbb{R}^4$ (Do Carmo, 1976, p. 437).

Pretty as those models of E are, they severely distort the geometry of E, which is homogeneous with positive constant curvature. There is no isometric immersion of E in $\mathbb{R}^3$ or $\mathbb{R}^4$. Current knowledge about isometric immersions is summarized at http://eom.springer.de/I/i052800.htm.

We have to rely on the models studied in Chapter 7 to view the entire hyperbolic plane because Hilbert proved that no complete geometric surface of constant negative curvature can be isometrically immersed in $\mathbb{R}^3$ (Do Carmo, 1976, Section 5-11).[5] The upper half-plane model has the simplest Riemann metric among those models, as follows:

$$ds = \frac{\sqrt{dx^2 + dy^2}}{y}.$$

Or, using the complex infinitesimal $dz = dx + i\, dy$,

$$ds = \frac{|dz|}{y}.$$

For the Poincaré unit disk model, the Riemann metric is

$$ds = \frac{|dz|}{1 - |z|^2}.$$

(For the Riemann metric on the Beltrami–Klein disk model, see McCleary, 1994, p. 221.) By contrast, the Riemann metric on Klein's conformal disk model of the elliptic plane is

$$ds = \frac{2|dz|}{|z|^2 + 1}$$

(Coxeter, 1998, Section 14.64). By integrating ds along a smooth arc, one calculates its arc length in these models.

From the quadratic expression for ds^2 in terms of the functions E, F, and G in local coordinates given previously, the following formula for *infinitesimal area* dA holds:

$$dA = \sqrt{EG - F^2}\, dx\, dy$$

---

[5] Hilbert's theorem was strengthened in 1954 by N. V. Efimov, who proved that a complete geometric surface of negative curvature bounded away from 0 cannot be isometrically immersed in $\mathbb{R}^3$. For $\mathbb{R}^4$, see I. Kh. Sabitov, "Isometric Immersions of the Lobachevsky Plane in E.4," translated in *Siberian Mathematical Journal* **30**(5) (1989), 805–811.

John Nash (see *A Beautiful Mind* by Sylvia Nasar) proved that every Riemannian manifold of dimension m can be embedded in $\mathbb{R}^n$ for n sufficiently large compared to m. This is a much deeper result than the one for which he was awarded the Nobel Prize in economics!

(see Gray et al., 2006, Section 12.4). The (double) integral of this area form over a suitable region of the surface gives the area of that region. In the Poincaré upper half-plane hyperbolic model, $E = G = y^{-2}$, $F = 0$, yielding the global area form

$$dA = \frac{dx\,dy}{y^2}.$$

An abstract surface is *orientable* if and only if it has a global area form (O'Neill, 2006, Lemma 7.5, Chapter 6 and Exercises for Section 7.1, Chapter 7—but see his list of errata on his website).[6]

GAUSS–BONNET THEOREM

In the section on curvature in Chapter 10, the formula of Gauss relating curvature K, area, and the angle sums of geodesic triangles was discussed, but only for surfaces of constant K. Gauss and O. Bonnet proved a much stronger theorem than that for any orientable geometric surface. Let's state it for the special case of an "embedded triangle" T on the surface S. Using calculus notation, it says

$$\iint_T K\,dA = 2\pi - \sum \alpha_i - \int_{\partial T} \kappa_g\,ds.$$

The double integral on the left is the total surface curvature of the triangular region. The single integral on the right is the total geodesic curvature of the boundary ∂T of the region. The summation on the right is the sum of the (radian measures of the) three "jump" angles formed at each of the vertices of the triangle by the two tangent vectors to the arcs meeting at those points. In the case where the arcs of the boundary are *geodesics*, their geodesic curvature is zero, so the single integral vanishes. The jump angles are then exterior angles of the geodesic triangle, so replacing each α by $\pi - \beta$, where β is the adjacent interior angle, we find that *the total surface curvature of the geodesic triangular region equals the angle sum of the triangle minus* π. When the surface has constant curvature K, this reduces to the result stated in

---

[6] The arc length for a smooth arc in $\mathbb{R}^3$ is the limit of the lengths of polygonal paths inscribed in the arc. One may be tempted, analogously, to define the *area* of a smooth piece of surface in $\mathbb{R}^3$ to be the limit of the surface areas of inscribed polyhedra. This method does not work, not even for a section of a cylinder—see Section 37 of *Differential Geometry* by E. Kreyszig, Dover Publications, 1991, or "Surface Area and the Cylinder Area Paradox" by F. Zames, *The Two Year College Mathematics Journal*, 8(4), September 1977, pp. 207–211. There have been many different and unequal definitions of surface area given in the literature. The one via area forms is the natural one for abstract orientable geometric surfaces.

Chapter 10 that K times the area of the geodesic triangle equals the angle sum of the triangle minus π.

When the orientable surface is *compact*, a theorem of T. Rado states that the surface has a triangulation with the edges of the triangles being smooth arcs. Applying the Gauss–Bonnet formula and taking into account the cancellations that take place from the orientations of the triangles, one obtains the Gauss–Bonnet theorem: For *compact* orientable surfaces S, the *total curvature* of S equals 2π times a topological invariant called the *Euler characteristic* $\chi(S)$ of S. In calculus notation, the formula is

$$\iint_S K \, dA = 2\pi\chi(S)$$

(see McCleary, 1994, pp. 175–178 or O'Neill, 2006, Section 7.6).

The total curvature is the number on the left obtained by integrating the Gaussian curvature function K over the entire compact surface; when K is constant, the total curvature is just K times the area of the surface. The Euler characteristic of S is defined to be $V - E + F$, where for any triangulation of S, V is the number of vertices, E the number of edges, and F the number of faces (so, in particular, $V - E + F$ is a *constant*, independent of the triangulation, as Descartes first noticed for polyhedra). H. Hopf proved that the Euler characteristic is also equal to the sum of the indices of any global vector field on S. A fundamental theorem in topology states that two compact orientable topological surfaces are homeomorphic if and only if they have the same Euler characteristic. More specifically, a compact orientable surface is homeomorphic to a sphere with $g \geq 0$ handles attached, and

$$\chi(S) = 2 - 2g.$$

"Attaching" is in terms of the *connected sum* of two surfaces, obtained by removing a disk from each of them and "gluing" them along the boundary curves that result. A "handle" is a torus with a small disk removed (see Weeks, 2002, Chapter 5 and Appendix C).

The Gauss–Bonnet theorem[7] shows that the curvature determines the topology of a compact orientable geometric surface S, and conversely

---

[7] For proofs of Gauss–Bonnet, see O'Neill, 2006, Section 7.6, or McCleary, 1994, Chapter 12. It has been vastly generalized to higher dimensions—see "The Many Faces of the Gauss–Bonnet Theorem" by L. I. Nicolaescu, http://www.nd.edu/%7Elnicolae/GradStudSemFall2003.pdf, and "All the Way with Gauss–Bonnet," by D. H. Gottlieb, http://hopf.math.purdue.edu//Gottlieb/bonnet.pdf. It is certainly one of the greatest theorems in all of mathematics!

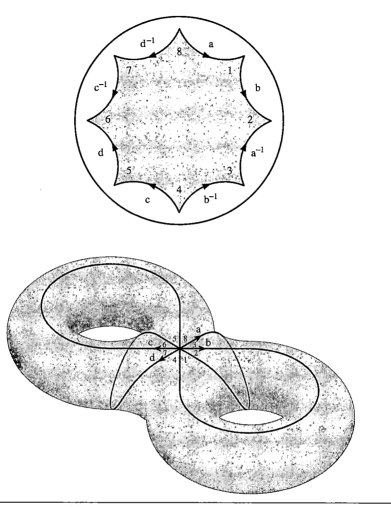

Figure A.20 Double torus constructed by identifying sides of a hyperbolic octagon having 45° angles. The figure is accurate topologically but is necessarily distorted geometrically because no hyperbolic compact surface, with its constant negative curvature, can be embedded in $\mathbb{R}^3$.

the topology limits the possible curvature for geometric structures on S. For example, compact topological surfaces of genus $g > 1$ cannot have a geometric structure with curvature everywhere ≥ 0 since their Euler characteristic is negative; they can be endowed with *hyperbolic structures*—geometric structures of constant negative curvature (O'Neill, 2006, Theorem 6.5, Chapter 8, or Stillwell, 1992, Chapters 5–8). For given g, all those hyperbolic structures have the same area, by Gauss–Bonnet. For $g = 2$, a hyperbolic structure can be constructed, for example, by suitably identifying sides of a hyperbolic octagon having $45°$ angles (see Figure A.20) (Stillwell, 1992, Section 6.6).

Consider a compact connected orientable surface S of positive curvature everywhere. Its total curvature must then be positive, so by Gauss–Bonnet, its Euler characteristic is positive. By the classification of compact surfaces, S must be diffeomorphic to a sphere. If in addition the curvature is constant, then S must be *isometric* to a sphere. Consider an abstract elliptic plane E as we defined it above. Since it is complete with constant positive curvature, it is compact. Since two points lie on a unique geodesic, it can't be a sphere; if it were orientable, it would be a sphere by the classification, hence it can't be orientable. But its orientable double-covering S is compact with constant positive curvature, hence is a sphere. Thus E must be isometric to the model obtained from S by identifying antipodal points.

By Gauss–Bonnet, for $g = 1$, the total curvature of any geometric structure on a torus is zero; if the curvature is constant, it must be zero, so a torus cannot have a hyperbolic structure. We've seen that it has a flat structure. This example illustrates the fact that a given smooth surface can be endowed with different geometric structures. $\mathbb{R}^2$ can be given the geometric structure of the hyperbolic plane since it is diffeomorphic to the disk. $\mathbb{R}^2$ can also be given a geometric structure of constant positive curvature since it is diffeomorphic (by stereographic projection) to a sphere minus a point.

In higher dimensions, M. Kervaire discovered in 1960 that there exist topological manifolds which have no differentiable structure. In 1956, John Milnor discovered exotic differentiable structures on the seven-sphere. Most bizarrely, with major implications for physics, Simon Donaldson discovered in 1986 that $\mathbb{R}^4$ has uncountably many different differentiable structures, whereas $\mathbb{R}^n$ for $n \neq 4$ has only one!

To carry out precisely all that has been mentioned about Riemannian geometry in this appendix, the tools of analysis (linear and multilinear algebra, advanced calculus, differential equations, calculus of variations) and topology are needed.

A *complex manifold* is an even-dimensional topological manifold whose charts are complex analytic (holomorphic) mappings. A *Riemann surface* is a *complex manifold of complex dimension one*. It is not the same as a geometric surface! But a connected *orientable* geometric surface can be given two structures of a Riemann surface, one for each of its orientations. Gauss proved that result by showing the existence of local isothermal coordinates for which the Riemann metric takes the form

$$ds^2 = \lambda(x, y)(dx^2 + dy^2).$$

The mappings studied in Riemann surface theory are the *conformal mappings* (see G. Jones, and D. Singerman, 1987, *Complex Functions: An Algebraic and Geometric Viewpoint*, New York: Cambridge University Press). Hurwitz's great theorem about a compact Riemann surface of genus $g > 1$ is that it has at most 84 $(g - 1)$ automorphisms. As mentioned in Chapter 8, the *Klein quartic* has genus 3 and has the maximum 168 automorphisms.

Some idea of Riemann's influence on modern mathematics can be gleaned from the following list of concepts, methods, and theorems that have been named after him: Riemannian curvature of Riemannian manifolds, Riemann integral, Riemann–Lebesgue lemma, Riemann surfaces, Riemann–Roch theorem, Riemann matrices, Riemann hypothesis about the Riemann zeta function, Riemann's method in the theory of trigonometric series, Riemann's method for hyperbolic partial differential equations, Riemann mapping theorem, and Cauchy–Riemann equations.[8]

---

[8] For the story of Riemann's difficult life, see Bell (1961); the biography by Laugwitz (2004) goes more deeply into Riemann's mathematical work. For the study of geometric surfaces, O'Neill (2006) avoids the "debauch of indices" that plagues old-fashioned presentations of the subject and provides excellent examples. McCleary (1994) is useful because it focuses on the history of hyperbolic geometry and its relationships to differential geometry. The two books by Do Carmo are clear and classical. Gray et al. (2006) has excellent computer graphics. A comprehensive presentation of Riemannian geometry can be found in the five volumes of Spivak (1999). The more general *Cartan Geometry*, which synthesizes the geometries of Klein and Riemann, is developed in Sharpe (1997) and is highly recommended to advanced students.

Einstein used a modified version of Riemannian geometry for his geometry of space-time in general relativity. The modification replaces the Riemann metric with one that is not positive-definite, the *Lorentz metric*. Einstein used it to show that gravity can be explained by the curvature of space-time. Introductory explanations of his theory can be found in Lanczos (1970) and Taylor and Wheeler (1992). A rigorous treatment is O'Neill (1983), *Semi-Riemannian Geometry with Applications to Relativity*, New York: Academic Press.

For the classic effort to explain the differential geometry and topology of curves and surfaces "in its visual and intuitive aspects," see Hilbert and Cohn-Vossen (1990, Chapters 4 and 6). For a delightful and more elementary recent effort that includes the eight fundamental Riemannian geometries on three-dimensional manifolds, see Weeks (2002).

Gauss proved the existence of a local conformal mapping of a curved surface to the plane (isothermal coordinates) only in the case of a real analytic surface; for a general geometric surface, this was not proved until a century later—see Spivak (1999). Isothermal coordinates are also used in the study of *minimal surfaces* (e.g., soap films—see Hildebrandt, S. and A. Tromba, 1986, *Mathematics and Optimal Form*, New York: W. H. Freeman).

B

Hilbert's Geometry Without Real Numbers

> With this approach there is no need for the real numbers, no appeal to continuity. In this way the true essence of geometry can develop most naturally and economically.
>
> Robin Hartshorne

Hilbert's treatment of hyperbolic planes appeared as a journal article soon after the 1899 publication of his *Grundlagen der Geometrie*, and it was subsequently included as Appendix III in later editions. It begins with the following statement (italics were in the original):

In the following investigation I replace the [Euclidean] axiom of parallels by a corresponding requirement of Bolyai–Lobachevskian geometry, and then show *that it is possible to develop Bolyai–Lobachevskian geometry in the plane exclusively with the plane axioms without the use of the continuity axioms.* This new development of Bolyai–Lobachevskian geometry, as it appears to me, is superior, because of its simplicity, to the hitherto well-known development schemes, namely those of Bolyai and Lobachevsky, who both use the limiting sphere, and that of F. Klein by means of the projective method. Those developments essentially use space as well as continuity.

The "limiting sphere" (or horosphere) referred to by Hilbert is the limit of a sphere in real hyperbolic three-space held tangent to a plane as its radius increases indefinitely (in Euclidean three-space, that limit is the plane). When Klein used the projective method, he also needed three-dimensional projective space in order to apply the theorems of Desargues and Pappus (see Projects 1 and 3, Chapter 2). Instead, Hilbert assumed his hyperbolic axiom of parallelism (the existence of limiting parallel rays—see Chapter 6) for a purely two-dimensional development of plane hyperbolic geometry. He dispensed with Archimedes' axiom and with his completeness axiom equivalent to Dedekind's axiom and coordinatization by the real numbers. Hilbert's presentation is very concise, omitting many details, which have been carefully worked out and expanded upon by later authors (e.g., F. Enriques, H. Liebmann, J. Gerretsen, P. Szasz). But it contains many of the key ideas of the subsequent axiomatic treatment of plane hyperbolic geometry, which renders the methods of J. Bolyai and Lobachevsky mainly of historic interest. Hartshorne's book meticulously carries out Hilbert's program.

It was not only hyperbolic geometry whose development was worked out without real numbers. Geometers also developed as much Euclidean geometry as possible without even assuming the circle-circle or line-circle continuity principles. Neutral geometry without real numbers is the study of Hilbert planes (which we abbreviate as *H-planes* in this appendix). The H-planes satisfying the Euclidean parallel postulate are isomorphic to Cartesian planes F^2, where F is an ordered Pythagorean field (Hartshorne, Theorem 21.1); I suggest they be called *Pythagorean planes* since the Pythagorean equation can be proved for them and F is Pythagorean. If the study of Euclidean planes (Pythagorean planes satisfying the circle-circle continuity principle) is the geometry of constructions with a straightedge and compass, the study of Pythagorean planes is the study of constructions with a straightedge and *dividers* or *gauge* (a tool that permits the execution of Axiom C-1, transporting segments). You might expect that a tool permitting the execution of Axiom C-4, transporting angles, would also be needed, but that turns out to follow in Pythagorean planes (Hartshorne, Exercise 20.21). You will find more information about constructions with these *Hilbert tools* in Hartshorne's treatise.

Where does the Pythagorean field F come from? You saw in Major Exercise 9, Chapter 4, that addition and order can be defined for segment congruence classes in any H-plane so as to have the usual properties. When the plane is Pythagorean, Hilbert showed how to also define a multiplication for segment congruence classes with the usual algebraic properties (once a unit segment is chosen); this is accomplished by developing the *theory of similar triangles* in such a plane—again without using real numbers. That development is a major accomplishment, for the theory of similar triangles presented in Euclid is based on Eudoxus' treatment of irrational proportions, and Hilbert completely avoids those complications. Then the segment congruence classes form the set of positive elements of an ordered Pythagorean field F, the *field of segment arithmetic*. See Hartshorne, Chapter 4. (A key step in this development is his Proposition 5.8 about *cyclic quadrilaterals*—quadrilaterals that have a circumscribed circle.)

Returning to hyperbolic planes, Hilbert showed that they too have a field hidden in their geometry, the *field of ends*, which turns out to be a Euclidean field (see Part I below). The hyperbolic plane is then isomorphic to a Poincaré or Klein model coordinatized by that field of ends F. One nice application of this theorem is a proof that the circle-circle continuity principle holds in a hyperbolic plane (because that principle holds in F^2 when F is a Euclidean field—see Moise (1990, Section 16.5)—hence in the Poincaré models over F, where circles are Euclidean circles). Another pretty application is a proof that a segment is constructible if and only if its multiplicative length is a constructible number (see Part I), which implies the M-B theorem of Chapter 10.

Geometry without continuity assumptions was developed by European researchers for many years after Hilbert; Victor Pambuccian is writing a history of that work. Some of the prominent contributors were F. Schur, M. Dehn, G. Hessenberg, and particularly the Danish geometer J. Hjelmslev. The work of the latter two mathematicians showed that every H-plane also has a field hidden in its geometry (see Part II below). Using that field and a projective embedding, W. Pejas was able to classify all H-planes.

Part I. Hilbert's Field of Ends in a Hyperbolic Plane[1]

Recall from Chapter 6, Note 2 after Major Exercise 4, that in a hyperbolic plane, an *end* (*ideal point*) is an equivalence class of limiting parallel rays. We arbitrarily choose one line as our "x-axis," and we arbitrarily label its ends 0 and ∞. We choose a perpendicular to this axis,

---

[1] We follow Hartshorne's development, with some simplifications and clarifications. Our work in Chapter 9 plays a crucial role, particularly the theorem on three reflections and its constructive proof.

labeling the intersection point as O, one end as 1, the other as -1. As a set, our field F consists of all ends except ∞; those ends will be called "finite."

We know from Major Exercise 8, Chapter 6, that any two distinct ends α, β are the ends of a unique constructed line in the hyperbolic plane, which we will now denote $\alpha \wedge \beta$. If α is a finite end, we denote *reflection in the line* $\alpha \wedge \infty$ by σ_α.

DEFINITION OF ADDITION. If α, β are finite ends, their sum $\alpha + \beta$ is defined by the equation (in the group of motions) $\sigma_{\alpha+\beta} = \sigma_\beta \sigma_0 \sigma_\alpha$.

How do we know the product is equal to the reflection in a unique line having ∞ as one end and another end which we have denoted $\alpha + \beta$? This is the special case of the theorem on three reflections in Proposition 9.17(b) when $\alpha \neq \beta$, both $\neq 0$. Namely, apply that result to the parallel displacement $T = \sigma_\beta \sigma_0$ and to the line $k = \alpha \wedge \infty$.

Here is another *description of addition* we need: Let $A = \sigma_\alpha O$ and $B = \sigma_\beta O$. By the asymptotically parallel case of the perpendicular bisector theorem (Major Exercise 7, Chapter 6), the perpendicular bisector of AB has ∞ as one end. We claim its other end is $\alpha + \beta$: Namely, $\sigma_{\alpha + \beta}A = \sigma_\beta \sigma_0 O = \sigma_\beta O = B$, so the reflection $\sigma_{\alpha+\beta}$ must be the reflection in the perpendicular bisector of AB (Figure B.1). Since AB and BA are the same segment, it follows that $\alpha + \beta = \beta + \alpha$, so the addition we have defined is commutative. From the definition, $\alpha + 0 = \alpha = 0 + \alpha$ for any finite end α.

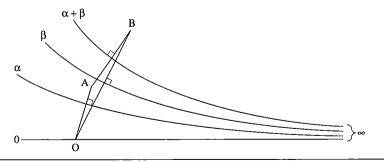

Figure B.1

When $\alpha = \beta$, let $D = \sigma_\alpha \sigma_0 \sigma_\alpha O$. Then $(\alpha + \alpha) \wedge \infty$ is the perpendicular bisector of OD, which has ∞ as one end by Major Exercise 7, Chapter 6, applied to triangle DOC, with $C = \sigma_{-\alpha} O = \sigma_0 A$. We use here the fact that if $-\alpha$ denotes the reflection of α in our x-axis $0 \wedge \infty$, then $-\alpha$ so defined is the additive inverse of α: $\alpha + (-\alpha) = 0$. This is because, using the above description with $\beta = -\alpha$, our x-axis is the perpendicular bisector of AB.

For the associative law $(\alpha + \beta) + \gamma = \alpha + (\beta + \gamma)$, a calculation shows that the reflection determined by both sides of this equation is equal to $\sigma_\gamma \sigma_0 \sigma_\beta \sigma_0 \sigma_\alpha$. In fact, the mapping $\alpha \to \sigma_\alpha \sigma_0$ is *an isomorphism of the additive group of finite ends onto the multiplicative group of parallel displacements about* ∞.

To define order and multiplication of finite ends, consider our "y-axis" with one chosen end denoted 1. According to our description of additive inverses, the other end is -1; in general, the ends of any line perpendicular to the x-axis have the form α and $-\alpha$.

DEFINITION OF ORDER. An end on the same side of the x-axis as 1 is called *positive*, and on the opposite side, negative. Then $\alpha < \beta$ means $\beta + (-\alpha) = \beta - \alpha$ is positive.

From the way we have drawn our diagrams, $\alpha < \beta$, when α, β are positive, is shown by α drawn to the left of β. For any finite $\alpha \neq 0$, denote *reflection in the line whose ends are α and $-\alpha$* by τ_α.

DEFINITION OF MULTIPLICATION. If α, β are finite ends $\neq 0$, their product $\alpha\beta$ is defined by the equation (in the group of motions) $\tau_{\alpha\beta} = \tau_\beta \tau_1 \tau_\alpha$ plus the order condition below. Multiplication by 0 is defined to be 0.

Once again the theorem on three reflections in the specific form of Proposition 9.11 guarantees that $\tau_\beta \tau_1 \tau_\alpha$ is a reflection in a line h perpendicular to our x-axis (when α, β, 1 are distinct). To specify which of the ends of h is $\alpha\beta$ and which is $-\alpha\beta$ is decided by the usual algebraic rules: pos $\times$ pos = pos, pos $\times$ neg = neg, and neg $\times$ neg = pos.

Let A, B be the points on our x-axis that are the feet of the perpendiculars from α, β, respectively. Let C be the point on the x-axis such that $\underline{OC} = \underline{OA} + \underline{OB}$, where O is the origin, addition of segments is the usual juxtaposition, and the underlining denotes *signed segments* as in Exercise H-4, Chapter 7, with the positive direction chosen to go from 0 to ∞. We claim that the perpendicular to the x-axis through C has as its ends $\alpha\beta$ and $-\alpha\beta$, so that *multiplication of finite ends corresponds to addition of signed segments*. To verify this claim, just reread the proof of Proposition 9.11, taking $T = \tau_\beta \tau_1$ and line m to be the perpendicular from α to the x-axis, in the notation of that proof.

Since addition of signed segments is commutative and associative and has "the zero segment" as its identity element, we see that multiplication is commutative and associative and has 1 as its identity element. We also see that *reflection in the y-axis sends α to α^{-1}* (Figure B.2).

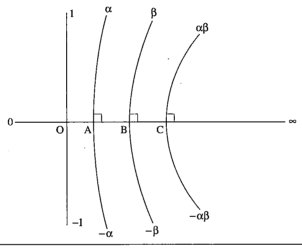

Figure B.2 $\underline{OC} = \underline{OA} + \underline{OB}$.

The correspondence above restricted to the multiplicative subgroup of positive ends is an isomorphism with the additive group of all signed segments on the directed x-axis. Equivalently, the correspondence $\alpha \rightarrow \tau_\alpha \tau_1$ is an isomorphism with the multiplicative group of translations along the x-axis. That isomorphism also takes care of the definition of multiplication when α, β, 1 are not distinct.

Now we see *why every positive end has a square root*: It is because every segment constructively has a midpoint (Proposition 4.3, Chapter 4), so that every signed segment can be halved.

From our definitions and descriptions of addition and multiplication, it is clear that sums and products of positive ends are positive.

Here is another way to view multiplication of finite ends: Let γ be nonzero. Let the perpendicular from γ to the x-axis cut it at C. Multiplication by γ is an operation on the set of finite ends. If γ is positive, this operation is the *translation T* along the x-axis which takes O to C; if γ is negative, it is the inverse translation, except in the case where $\gamma = -1$, when it is the reflection across the x-axis. From that observation we can verify the *distributive law* of multiplication with respect to addition: $\gamma(\alpha + \beta) = \gamma\alpha + \gamma\beta$. Namely, since T fixes 0 and ∞ and is a motion, it maps the lines used to describe $\alpha + \beta$ onto the lines used to describe $\gamma\alpha + \gamma\beta$.

THEOREM. The set of finite ends, with the operations of addition, multiplication, and order defined above (given ends 0, 1, and ∞) is a Euclidean field F.

NOTE. Our definitions of the structure on F depended on our initial choice of three ends, 0, 1, and ∞. If we had chosen a different triple of ends, the resulting structure of the ordered field would be isomorphic to the original one because of Proposition 9.16(c): There is a unique motion of the plane sending the first triple into the second, and that motion preserves all the geometry used in defining our structure.

In Hartshorne's treatise, he develops an analytic geometry and a trigonometry for hyperbolic planes without using real numbers. We will review his main results—see his Sections 41 and 42. Of particular interest is his *geometric definition of the multiplicative length* $\mu(AB)$ of a segment AB as follows.

DEFINITION. By Axiom C-1, there is a unique segment OC, starting from our origin O, on the positive ray $O\infty$ of our x-axis such that we have $OC \cong AB$. Let γ be the positive end of the perpendicular to our x-axis through C. Then $\mu(AB) = \gamma$.

PROPOSITION B-1. Multiplicative length has the following properties:
    (a) $\mu(AB) > 1$.
    (b) $AB \cong A'B'$ if and only if $\mu(AB) = \mu(A'B')$.
    (c) $AB < A'B'$ if and only if $\mu(AB) < \mu(A'B')$.
    (d) $\mu(AB + A'B') = \mu(AB)\,\mu(A'B')$.

PROOF:
Easy exercise. ◄

Thus the function μ determines segment congruence. The next function determines angle congruence.

DEFINITION. Given angle θ, by Axiom C-4 there is a unique congruent angle having vertex at the origin O, one side being the negative ray

O0 of our x-axis and the other side lying on the positive side of our x-axis (Figure B.3). The end of that other side is denoted *tan θ/2*.

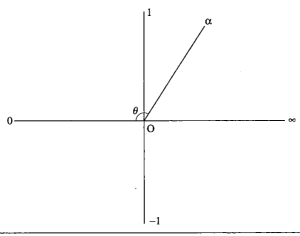

Figure B.3 $\alpha = tan\ (\theta/2)$.

When we examine the Poincaré half-plane model later, we will see that this function is the usual tangent of a half-angle function in trigonometry.

PROPOSITION B-2.

    (a) *tan θ/2 > 0.*

    (b) *θ ≅ ψ* if and only if *tan θ/2 = tan ψ/2.*

    (c) If *θ < ψ*, then *tan θ/2 < tan ψ/2.*

    (d) *θ* is a right angle if and only if *tan θ/2 = 1.*

    (e) Addition formula: If the angle *θ + ψ* obtained by juxtaposition is defined, then

$$tan\ (\theta + \psi)/2 = \frac{tan\ \dfrac{\theta}{2} + tan\ \dfrac{\psi}{2}}{1 - tan\ \dfrac{\theta}{2}\ tan\ \dfrac{\psi}{2}}$$

The first four parts of this proposition are immediate from our definition. For the proof of the addition formula, see Hartshorne, Proposition 41.8.

Here is the analytic geometry result we will need: Given any point P. It is uniquely the intersection of a line $b \wedge \infty$ having ∞ as one end, and a "vertical" line with ends a and $-a$, where $a > 0$. Since b is on the opposite side of that vertical line from ∞, it satisfies $|b| < a$ (Figure B.4). Example: If P = O, $a = 1$ and $b = 0$.

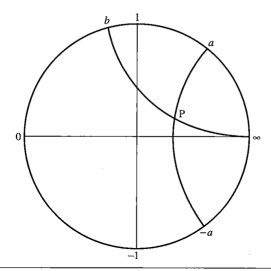

Figure B.4 Point P is determined by ends a, b.

PROPOSITION B-3. With the above notation, a line with finite ends u, v passes through P if and only if its ends satisfy the equation

$$uv - b(u + v) + a^2 = 0,$$

where $|b| < a$.

For the proof, see Hartshorne, Section 41.6. Example: The lines through the origin O (other than our x-axis) are given by the equation $uv = -1$. (Hilbert introduced the field elements x, y given by $y = uv$, $2x = u + v$; in terms of them, point P is determined by the *linear* equation $y = 2bx - a^2$. Hilbert used linear equations for points to indicate how to prove Pappus' and Desargues' theorems algebraically for a hyperbolic plane. In contrast to the analytic geometry for Euclidean planes, in hyperbolic planes directed lines are determined by ordered pairs of ends, and points are determined by certain linear equations in finite ends. Hartshorne, Proposition 41.5, also provides algebraic formulas for certain motions of the plane.)

Here is our first main result.

CHARACTERIZATION OF THE HYPERBOLIC PLANES THEOREM.
- (a) Two hyperbolic planes are isomorphic if and only if their fields of ends are isomorphic.
- (b) If Π is a hyperbolic plane with a field of ends F, then Π is isomorphic to the Poincaré upper half-plane model coordinatized by F.

SKETCH OF PROOF OF (a):

Given an isomorphism of hyperbolic planes, since the algebraic operations and order on their fields of ends were defined geometrically, it is an easy exercise to show that the induced mapping of ends is an isomorphism of fields.

Suppose, conversely, we are given an isomorphism $f\colon F_1 \to F_2$ of the fields of finite ends of hyperbolic planes Π_1, Π_2. We extend f to the missing ends by setting $f(\infty_1) = \infty_2$. We must construct from f an isomorphism φ from Π_1 onto Π_2. By definition, φ is composed of a map of lines and another map of points (both of which will be denoted φ, *par abus*) which preserve incidence, betweenness, and congruence. Now a line is uniquely determined by its ends, so we naturally define $\varphi(\alpha \wedge \beta)$ to be the line $f(\alpha) \wedge f(\beta)$. In order to define φ on points, we use the fact that a point P is uniquely determined by the family of all lines passing through it. We have given the equation of all lines through P in Proposition B-3 above, in terms of constants α, β in the field. So we define $\varphi(P)$ to be the point determined by the constants $f(\alpha)$, $f(\beta)$. Since f is a field isomorphism, the equation for lines through P will be mapped to the equation for lines through $\varphi(P)$, so the one-to-one mapping φ preserves incidence.

Betweenness of points can be expressed in terms of the ordering of the field of ends, and since f preserves order, φ preserves betweenness.

That φ preserves congruence follows from the facts that congruence of segments is determined by the multiplicative length function and congruence of angles by the tangent function, both having values in the field of ends (Propositions B-1 and B-2). Explicitly, what must be checked is that $\mu(\varphi(A)\varphi(B)) = f(\mu(AB))$ for any segment AB and for any angle θ, $\tan(\varphi(\theta)/2) = f(\tan(\theta/2))$. ◀

PROOF OF (b):

By the result for part (a) and the previous result that the field of ends is a Euclidean field, what this comes down to is the following: Let Π be the Poincaré upper half-plane model coordinatized by a Euclidean field F. Let F_1 be the field of ends of Π. *Then F and F_1 are isomorphic ordered fields.*

The ordered field structure on F_1 depended on the choice of three special ends. The most convenient choices for our purposes are to take ∞ to have its usual meaning in the Poincaré upper half-plane model (the second point at infinity on the upper imaginary axis), to take the end **0** to be the point $(0, 0)$, and to take the end **1** to be the point $(1, 0)$. (Here 0 and 1 are the elements of the field F; if we use "complex number" notation for points in the plane F^2 coordinatized by F, then indeed these special ends are the "complex numbers" **0** and **1** in the field $F(i)$ obtained from F by adjoining $i = \sqrt{-1}$.) With these choices, what we previously called the "x-axis" is now the upper half $\mathbf{0} \wedge \infty$ of the imaginary axis, the "y-axis" is the upper semicircle of the unit circle centered at **0**, the origin O is $i = (0, 1)$, and the other end of the "y-axis" is $-\mathbf{1} = (-1, 0)$. The remaining ends in the Poincaré upper half-plane model are the points $\alpha = (a, 0)$ of the horizontal axis in F^2. Since the P-line from $\alpha = (a, 0)$ perpendicular to $\mathbf{0} \wedge \infty$ is the upper semicircle of the circle centered at **0** of radius $|a|$, we see that $-\alpha = (-a, 0)$ is its other end. Since the positive ends are by definition the ones on the same side of P-line $\mathbf{0} \wedge \infty$ as **1**, we see that $\alpha > 0$ if and only if $a > 0$.

The mapping φ from F_1 to F is defined very simply in this model by $\varphi(\alpha) = a$, the first Cartesian coordinate of the end α as a point on the horizontal axis in F^2. We must show that φ is an isomorphism of ordered fields. We just showed that φ sends positive ends to positive elements of F.

Recall that σ_α is the P-reflection in the P-line $\alpha \wedge \infty$. In the upper half-plane model, this P-line is just the open vertical ray whose vertex is $(a, 0)$, and the P-reflection is the same as the Cartesian reflection. Using "complex number" notation as in the Chapter 9 section on motions in this model, the explicit formula is

$$\sigma_\alpha(z) = -\bar{z} + 2a,$$

where $\bar{z}$ as usual denotes the "complex conjugate." Addition of ends was defined by

$$\sigma_{\alpha + \beta} = \sigma_\beta \sigma_0 \sigma_\alpha.$$

If $\beta = (b, 0)$, we calculate $\sigma_{\alpha+\beta}(z) = -\bar{z} + 2(a + b)$, which is reflection in the open vertical ray whose vertex is $(a + b, 0)$. This shows that the mapping φ preserves addition:

$$\varphi(\alpha + \beta) = a + b = \varphi(\alpha) + \varphi(\beta).$$

Consider next multiplication of finite ends, which for positive ends was defined by the formula

$$\tau_{\alpha\beta} = \tau_\beta \tau_1 \tau_\alpha.$$

Recall that τ_α is reflection in the P-line with ends α, $-\alpha$, which is the upper semicircle of the circle centered at $\mathbf{0}$ of radius a. So τ_α is inversion in that circle, which is given in "complex number" polar notation by

$$\tau_\alpha(re^{i\theta}) = \frac{a^2}{r}\, e^{i\theta}$$

If $\beta = (b, 0)$, we calculate $\tau_{\alpha\beta}(re^{i\theta}) = \dfrac{a^2b^2}{r}\, e^{i\theta}$, which is inversion in the circle centered at $\mathbf{0}$ of radius ab. This shows that the mapping φ preserves multiplication of positives:

$$\varphi(\alpha\beta) = ab = \varphi(\alpha)\varphi(\beta).$$

Since the case of one or both negatives is an easy exercise, we have proved φ is an isomorphism of ordered fields. ◄

NOTE. In Hartshorne's treatise, he proves part (b) for the Poincaré disk model, but his direct proof of that is more cumbersome. We obtain that result as a corollary because we know the Poincaré and Klein disk models are both isomorphic to the upper half-plane model by the isomorphisms given explicitly in Chapters 7 and 9.

Let us examine the interpretation in this Poincaré upper half-plane model of the mysterious function *tan* $\theta/2$ previously defined for abstract hyperbolic planes. Given any angle θ, we laid off a congruent copy of the angle with vertex at the origin (which is i in the model), with one side being the ray from the origin to the end $\mathbf{0}$, and the positive end α of the other side being, by definition, *tan* $\theta/2$. Now apply our isomorphism φ. **We claim**

$$\varphi(tan\ \theta/2) = \tan\ \theta/2,$$

where tan (without italics) is the usual tangent function in the Cartesian plane. For example, we know that if θ is a right angle, *tan* $\theta/2 = 1$, and $\varphi(1) = 1 = \tan \theta/2$.

Figure B.5 illustrates the case where θ is an acute angle; we leave the case of an obtuse angle as a similar exercise.

In the model, the ray of θ having $\alpha = (a, 0)$ as end is the arc from i to α of a Cartesian semicircle Γ with center $C = (-c, 0)$ on the horizontal axis, for some $c > 0$. Since by definition $\varphi(\alpha) = a$, we must show that $a = \tan \theta/2$.

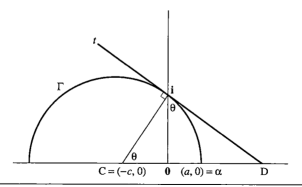

Figure B.5

Let t be the Cartesian tangent line to Γ at i and let t intersect the horizontal axis at D. By the definition of angle congruence in the conformal Poincaré upper half-plane model, $\theta = \sphericalangle Di0$ in the Cartesian plane. Since the angle sum of a triangle is 180° in the latter plane, and $\sphericalangle DiC$ is right, we have $\theta = \sphericalangle iC0$. Since the Cartesian length of $i0$ is 1, we see from right triangle C0i that $c = \cot \theta$ and the radius of Γ is $\sqrt{1 + c^2}$. The equation for Γ is then $(x + c)^2 + y^2 = 1 + c^2$. Setting $y = 0$ gives us the intersections of Γ with the horizontal axis, so we have $a = -c + \sqrt{1 + c^2}$. Now invoke the trigonometric identity $1 + \cot^2 \theta = \csc^2 \theta$, so that

$$a = \csc \theta - \cot \theta = \frac{1 - \cos \theta}{\sin \theta} = \tan \frac{\theta}{2}$$

by other trigonometric identities. ◄

Now let us calculate the interpretation in the Poincaré upper half-plane model of the multiplicative length function μ on segments. By definition, a segment iC congruent to the given segment is laid off on the ray $i\infty$; then a perpendicular through C is constructed to that ray, and the positive end α of that perpendicular is the value of μ. Let $C = (0, c)$ in Cartesian coordinates, with $c > 1$. The perpendicular through C is then the upper semicircle of the Cartesian circle centered at 0 passing through C, which is a circle of radius c, intersecting the positive ray of the horizontal axis at $(c, 0) = \alpha$. Applying our isomorphism φ, we see that $\varphi(\alpha) = c$ is the interpretation of the multiplicative length. This is just the reciprocal of the cross-ratio $(iC, 0\infty)$ suitably understood, and that is the multiplicative length for the Poincaré model (see p. 320, Important Remark).

Multiplicative Bolyai–Lobachevsky Formula:

$$\tan \frac{\Pi(AB)}{2} = \mu(AB)^{-1}$$

(This becomes the usual formula for a real hyperbolic plane with distance scale $k = 1$ when we substitute $\mu = e^d$, using the calculations above which relate *tan* defined in terms of ends to the usual tan in the Cartesian plane, and μ defined in terms of ends to the cross-ratio.)

Proof:

By definition, the left side of the formula is the positive end a of one side of the angle $\theta = \angle OOa$ congruent to $\Pi(AB)$, the other side being the negative ray of the x-axis. The line $a \wedge -a$ is perpendicular to the x-axis at a point C such that $CO \cong AB$. Now reflect across the y-axis (Figure B.6). The reflection of C is a point D on the positive ray $O\infty$, and the reflection of the perpendicular through C is the perpendicular to the x-axis through D. By definition of multiplicative inverse and of μ, the positive end of that perpendicular is $a^{-1} = \mu(AB)$. ◄

We previously proved the additive distance version of this formula for the Poincaré disk model (Theorem 7.2) and the Klein model (p. 346). Note that for the Klein model, the cross-ratio (AB, PQ) is equal to the *square* of the reciprocal of $\mu(AB)$.

One consequence of this formula is that *Bolyai's parallel construction works in any hyperbolic plane* (not just the real one): Hartshorne,

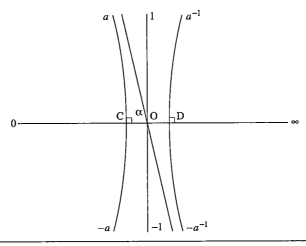

Figure B.6

Proposition 41.10, presents a proof by a lengthy calculation using his hyperbolic analytic geometry. We have justified that construction geometrically in the Klein model on pp. 344–345, Chapter 7, using a Euclidean perspectivity, and since we now know that the abstract hyperbolic plane is isomorphic to the Klein model, that proof is generally valid. ◄

However, Bolyai's construction does not prove the existence of limiting parallel rays! By the definition of "hyperbolic plane," that existence is Hilbert's axiom, and from the consequences of that axiom one can then prove that Bolyai's parallel construction works.

Another vital consequence is that a version of hyperbolic trigonometry can be developed in any hyperbolic plane without using real numbers. The key is the tangent half-angle function $t = \tan \theta/2$ defined above. We can use the familiar formulas in the Cartesian plane as *definitions* for the hyperbolic plane:

$$\sin \theta = \frac{2t}{1 + t^2}, \qquad \cos \theta = \frac{1 - t^2}{1 + t^2}, \qquad \tan \theta = \frac{\sin \theta}{\cos \theta}.$$

Then it is a straightforward exercise to verify that all the usual formulas for these trigonometric functions hold. In particular, the definition of $\tan \theta$ is consistent with the definition of t.

A generalized hyperbolic trigonometry is based on the multiplicative Bolyai–Lobachevsky formula above. Using it, we avoid the introduction of hyperbolic trigonometric functions (or we could, if we like, define them formally via formulas (5)–(7) of Chapter 10—e.g., define $\tanh x$ to be $\cos \Pi(x)$). For a right triangle $\triangle ABC$ with right angle at C, here are analogues of formulas (10)–(12) from Chapter 10, using the standard notation (e.g., $\Pi(a)$ is the angle of parallelism for the segment a opposite vertex A).

PROPOSITION B-4. With standard notation for $\triangle ABC$ with right angle at C,
    (a) $\tan A = \cos \Pi(a) \tan \Pi(b)$
    (b) $\cos \Pi(b) = \cos A \cos \Pi(c)$
    (c) $\sin A = \cos B \sin \Pi(b)$

with similar equations for B in place of A.
    (d) $\tan \Pi(c) = \sin A \tan \Pi(a) = \sin B \tan \Pi(b)$
    (e) $\sin \Pi(c) = \tan A \tan B$
    (f) $\sin \Pi(c) = \sin \Pi(a) \sin \Pi(b)$

PROOF:

Routine algebra, or see Hartshorne, Propositions 42.2 and 42.3. Formula (f) is the hyperbolic analogue of the Pythagorean theorem. ◄

Using this trigonometry, we can verify, without real numbers, George Martin's construction of the segment of parallelism. Refer to Figure 10.33 and the statement of his construction on p. 523. Apply formula (b) first to ∢BAC and then to ∢DAC = Π(AB); then deduce that cos ∢BAC = cos Π(AD), as in Chapter 10. (Good exercise: Check this construction in the Klein model by taking A to be the center of the unit disk.)

From Martin's construction we obtain a construction of the line of enclosure of any angle: First, bisect the angle. Then, starting from the vertex of the angle, construct the segment of parallelism on that bisector for half of the angle; then erect the perpendicular at the other end of that segment: *Voila,* the line of enclosure of the original angle! This is a much simpler construction than Hilbert's original one (Major Exercise 8, Chapter 6), but the existence of the line of enclosure was used in the development of the field of ends.

Also, using this construction, one can give a simpler construction than Hilbert's original one (Theorem 6.3) of the common perpendicular to two divergently parallel lines: Just use Major Exercise 12, Chapter 6, to locate the symmetry point S of those two lines and drop the perpendicular from S. Once again, Hilbert's construction was used to construct the line of enclosure, so we cannot dispense with it in the logical development until someone finds a better one from the axioms.

Our final and main result is the following theorem and corollary about constructions with a straightedge and compass in a hyperbolic plane. Its proof uses some of the constructions already proved for Hilbert planes, plus the constructions special to hyperbolic planes such as the unique common perpendicular to two divergently parallel lines, the line of enclosure of any angle, the line joining two ends, and the perpendicular from an end to a line.

HYPERBOLIC CONSTRUCTIBLE SEGMENTS THEOREM. In a hyperbolic plane, given two perpendicular lines (intersecting at a point O), a choice of one end labeled 1 on one of those lines and a choice of ends labeled ∞ and 0 on the other, then with respect to Hilbert's field of ends based on those data, *a segment is constructible if and only if its multiplicative length μ is a constructible number.* For any constructible number $a > 1$, a constructible segment exists having a as its multiplicative length.

PROOF (HARTSHORNE):

Observe first of all that segment AB is constructible if and only if angle Π(AB) is constructible, by the constructions of Bolyai and Martin.

Next, since the finite ends are also the numbers in our field F, we distinguish two types of constructibility for them: We say a is *geometrically constructible* if line $a \wedge \infty$ is constructible. We say a is *algebraically constructible* if it is obtained from 0 and 1 by finitely many operations of addition, subtraction, multiplication, division, and taking square roots of positive numbers.

Let $\alpha = \Pi(AB)$. Then $\mu(AB)$ is algebraically constructible if and only if $tan(\alpha/2)$ is algebraically constructible, by the multiplicative Bolyai–Lobachevsky formula.

Proof of "if" direction: If a is algebraically constructible, then a is geometrically constructible. That is because the ends ∞, 0, and 1 are given, and the five arithmetic operations on ends are all constructed geometrically. (That is the beauty of Hilbert's end-arithmetic!) Similarly, $-a$ is geometrically constructible.

Suppose $a > 1$. Then the line of enclosure $a \wedge -a$ for angle $aO(-a)$ is constructible, as is the point C where it intersects the positive ray of the x-axis $0 \wedge \infty$, as is segment OC. By the definition of multiplicative length, $\mu(OC) = a$. If $a = \mu(AB)$, then $AB \cong OC$, and AB can be constructed from OC by transport of segment.

Proof of "only if" direction: Here we use the isomorphism of the given plane with the Poincaré unit disk model over F. There, if P-segment AB is constructible with h-straightedge and h-compass, our work in Chapter 7 shows that it is also constructible with Euclidean straightedge and compass. By another Euclidean construction, we perform an inversion sending A to O (the center of the disk) and B to some point C. If $\mu(OC) = a$ and x is the Euclidean length $\overline{OC}$, then $a = (1 + x)/(1 - x)$ (see Hartshorne's proof of Lemma 39.7). Since OC is Euclidean constructible, the classical theorem for F^2 tells us that x is a constructible number, hence a is. ◄

COROLLARY TO THE PROOF (ACT THEOREM). An angle in a hyperbolic plane is constructible if and only if one of the trigonometric functions of that angle is a constructible number (hence all trigonometric functions of that angle are constructible numbers).

This result, together with the classical result on constructions in Euclidean planes, shows that for Euclidean and hyperbolic planes, if "elementary geometry" is defined as the geometry of straightedge-and-compass constructions starting with the minimal possible data, the correct field on which to coordinatize and do analytic geometry is the field K of constructible numbers.

For a lovely tie-in of the multiplicative length to hyperbolic Pythagorean triples, to an old number theory problem of Euler, and to research in algebraic geometry on surfaces, see Hartshorne and Van Luijk (2006).

Part II. Pejas' Classification of H-Planes

The most comprehensive treatise on geometry without real numbers is F. Bachmann's (1973) *Aufbau der Geometrie aus dem Spiegelungsbegriff* (Construction of Geometry Based on the Concept of Reflection), in which plane geometries based only on axioms of incidence, perpendicularity, and reflections are studied.[2] Bachmann is justified in calling this study *plane absolute geometry* since it includes elliptic, hyperbolic, and Euclidean geometries as special cases (as well as other unusual geometries). Bachmann has succeeded in dispensing with betweenness axioms as well as continuity axioms. It is unfortunate that he calls models of his axioms "metric" planes—since "metric spaces" has a quite different meaning in which a "metric" measures distance by real numbers; I propose calling them *absolute planes*.

The key ideas for this study are:

1. Embed the absolute plane in a projective plane.
2. Show that the points and lines in this projective plane have homogeneous coordinates from some field K (as in Project 3, Chapter 2).
3. Show that two lines are perpendicular if and only if their coordinates satisfy a particular homogeneous quadratic equation.
4. Use algebra to get further information about the absolute plane.

This program has been completely successful in classifying H-planes (the problem of determining all absolute planes remains unsolved). Let us describe the embedding briefly. The intuitive idea comes from our discussion of the Klein model, in which new "ideal" and "ultra-ideal"

---

[2] For a presentation of Bachmann's axioms in English, see Ewald (1971). I. M. Yaglom, in the foreword to the Russian translation of Bachmann's book, calls it "indisputably the most significant development in the foundations of geometry in decades." You were introduced to Bachmann's methods in Proposition 9.19 and Exercise 50 of Chapter 9.

points are introduced so that lines that were previously parallel now meet when extended through these new points (see Major Exercise 13, Chapter 6). Abstractly, the new points are simply pencils of lines. The old points are in one-to-one correspondence with the pencils of the first kind; i.e., A corresponds to the pencil $p(A)$ of all lines through A.

An ultra-ideal point is a pencil of the second kind; it is a pole $P(t)$, which for a fixed line t consists of all the perpendiculars to t. But the pencils of the third kind must be described carefully to avoid circular reasoning. We described such a pencil in Chapter 9 (p. 427) as consisting of all lines through a fixed ideal point, the ideal point being defined as an equivalence class of limiting parallel rays. Since we don't know that limiting parallel rays exist now, we instead use the theorem on three reflections (Proposition 9.19) as our definition: The pencil $p(lm)$ determined by parallel lines l and m which do not have a common perpendicular consists of all lines n such that the product

$$R_l R_m R_n$$

is a reflection. Certain properties that were previously obvious now require a considerable amount of ingenuity to prove—for example, if h, $k \in p(lm)$, then $p(hk) = p(lm)$. Hjelmslev[3] discovered how to prove that a pencil of the first kind $p(D)$ and a pencil of the third kind $p(a'c')$ have a unique line b in common (see Figure B.7). He dropped perpendiculars a, c from D to a', c' at points A, C, then dropped perpendicular d from D to $\overleftrightarrow{AC}$. Then the line b is uniquely determined by the equation

$$R_a R_b R_c = R_d.$$

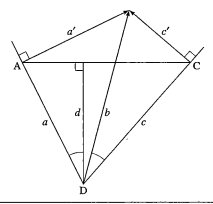

Figure B.7 Hjelmsler's construction of b.

[3] J. Hjelmslev, "Neue Begründung der ebenen Geometrie," *Mathematische Annalen,* **64** (1907):449–474.

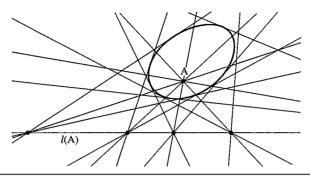

Figure B.8 The polar of a point.

We now know the new points of our projective plane. Our old lines
a can be extended to new lines *l(a)* so that any pencil containing *a* is
by definition incident with *l(a)*. But we need more new lines to fill out
our projective plane. For example, for any old point A, the polar *l(A)*,
consisting of all poles of lines through A, should be a new line; in the
Klein model, this line lies entirely outside the absolute circle (in the
Cartesian model, it's the line at infinity). See Figure B.8.

But how are we to describe the lines in the Klein model that are
tangent to the absolute circle, and how are we to verify the axioms for
a projective plane? To accomplish this, Hjelmslev discovered a remarkable device: He fixed a point O and fixed a pair *u, v* of non-perpendicu-
lar lines through O—think of these lines as determining an acute angle
θ. He then defined a transformation that fixed O and sent any A $\neq$ O
to the midpoint A* of the segment joining A to its image under rotation $R_u R_v$ about O through 2θ (Figure B.9); this transformation is called
the *half-rotation* (or *snail map*) about O corresponding to *u, v*.

Hjelmslev observed that in the Klein model, half-rotations extend
to collineations of the projective plane, and that any projective line except *l(O)* could be mapped onto an extended Klein line *l(a)* by a suit-

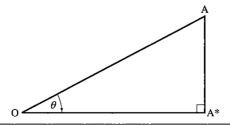

Figure B.9 A* = image of A under half-rotation through θ.

able half-rotation about O. So he proposed to call a set of pencils a "line" if it is mapped onto some $l(a)$ by some half-rotation about O, or if it is $l(O)$. With this definition, he was then able to verify the axioms for a projective plane.

The execution of idea 2—construction of the field K of coordinates— requires even more technique. The key tool is a complicated theorem of Hessenberg, which generalizes the Euclidean theorem that tells when a quadrilateral can be circumscribed by a circle. The method of constructing K is the standard method (due to Emil Artin) for any affine plane in which Pappus' theorem holds (Ewald, 1971, Chapter 3); here the affine plane is obtained by removing the polar $l(O)$ of O, and a special case of Pappus' theorem states: Let P, Q, R lie on a, and let $\overline{P}$, $\overline{Q}$, $\overline{R}$ lie on b, such that a does not meet b in any of these six points (Figure B.10). If $P\overline{Q} \parallel \overline{P}Q$ and $Q\overline{R} \parallel \overline{Q}R$, then $P\overline{R} \parallel \overline{P}R$. Bachmann's approach is to verify Pappus' theorem by brute force; there is another approach due to Lingenberg that I believe gives more insight (see Lenz, 1967, p. 206ff., on the Euclidean "pseudoplane"). Pappus' theorem implies that the definition given for "lines" in our projective plane does not depend on the choice of O.

If we started with an H-plane, it is then possible to define an order on K in terms of the betweenness relation in the H-plane. We need only specify the set P of positive scalars (since $x < y \Leftrightarrow y - x > 0$).

Choose any two points A, B from the H-plane. For any third point X collinear with A, B, the *ratio*

$$AX:BX$$

is defined to be that unique scalar in K which, multiplying the vector from B to X, gives the vector from A to X. We then define P to consist of 1 and all ratios AX:BX as X runs over all third points on the affine line $\overleftrightarrow{AB}$ that do *not* lie between A and B (this includes all ideal and ultra-ideal points on $\overleftrightarrow{AB}$). It can be shown from invariance of the ratio

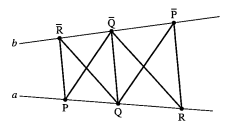

Figure B.10 A special case of Pappus' theorem.

under parallel projection that this definition of P does not depend on the choice of A, B.

Moreover, a theorem of Menelaus (compare Exercise H-5, Chapter 7) can be proved for these ratios and can be used to demonstrate that P has all the properties required for a set of positive numbers that makes K an *ordered field* (see Lenz, 1967, p. 223ff.). In turn, this enables us to extend the betweenness relation to all triples of collinear points in the affine plane and to show that the H-plane is embedded as a convex open subset of the affine plane, containing the origin O. The embedding is locally Euclidean at O in that perpendicularity for lines through O has the familiar meaning from Cartesian analytic geometry.

It can be proved (from free mobility—Lemma 9.3, p. 444) that the field K is Pythagorean (the sum of two squares is a square), but K is not Euclidean unless the line-circle continuity principle holds (a neat algebraic criterion for this geometric property).

As for idea 3 on our list, further argument shows that there is a constant $k \in K$ such that lines having homogeneous coordinates $[a_1, a_2, a_3]$, $[b_1, b_2, b_3]$ are perpendicular if and only if

$$a_1b_1 + a_2b_2 + ka_3b_3 = 0.$$

Here "perpendicularity" has been extended to all pairs of lines in the projective plane, and there may exist certain lines (called *isotropic*) that are perpendicular to themselves (e.g., when $k = -1$, the line $[0, 1, -1]$; i.e., $y = 1$). To each line $[a_1, a_2, a_3]$ is associated its *pole* $[a_1, a_2, ka_3]$ (except for the line at infinity $[0, 0, 1]$ when $k = 0$), and we see that the perpendiculars to this line are precisely the lines passing through its pole; isotropic lines pass through their own poles, and for $k \neq 0$, the locus of poles of isotropic lines is given by the affine equation

$$x^2 + y^2 = -k^{-1}$$

and may be called *the absolute*. The points (x, y) in the H-plane satisfy the inequality $|k| (x^2 + y^2) < 1$. They form a convex open neighborhood of the origin.

The fourth angle of a Lambert quadrilateral is acute, right, or obtuse accordingly as $k < 0$, $k = 0$, or $k > 0$. The constant k is only determined up to multiplication by a nonzero square, but whether it is negative, zero, or positive is uniquely determined. Thus k is a purely algebraic indicator of the "curvature" of the H-plane. In 1900, Dehn gave an example of an H-plane with $k = 1$, an example he called *non-Legendrean*.

Hartshorne (note for Section 43) wrote:

To understand the development of ideas leading to Pejas' theorem is tantamount to reviewing the entire history of the role of projective geometry in the foundations of elementary geometry.

Pejas' classification of H-planes is used in my article "Aristotle's Axiom in the Foundations of Geometry," *Journal of Geometry*, **33** (1988), 53–57. An important observation in this article is that if Aristotle's axiom holds, the H-plane is *maximal*; Bachmann showed earlier that if Archimedes' axiom holds, the H-plane is *minimal* (Hartshorne, Corollary 43.6), and conversely (using Pejas' classification). These sizings refer to the notion of one H-plane being isomorphic to a *full subplane* Π of another H-plane Π', meaning that the points of Π are a subset of the points of Π', the lines of Π are all the nonempty intersections of lines of Π' with the set of points of Π, and betweenness and congruence are induced from Π'. For example, the constructible Cartesian plane K^2 is *not* a full subplane of the real Cartesian plane $\mathbb{R}^2$ because, for example, the line $y = \pi x$ intersects K^2 in just the origin, so if its intersection were a line of K^2, Axiom I-2 for K^2 would be violated.

Here is why *Aristotle's axiom implies maximality*: Assume Π is not maximal. Pejas tells us that Π can be taken to be a convex open neighborhood of an origin O inside Π'. If point C' is in Π' and not in Π, let B be a point $\neq$ O in Π on the perpendicular through O to $\overrightarrow{OC'}$; B exists, and $\sphericalangle BC'O$ is congruent to an angle in Π because Π is a full subplane. But every point C of Π on ray $\overrightarrow{OC'}$ lies between O and C' by convexity, so by the exterior angle theorem, $\sphericalangle BCO > \sphericalangle BC'O$. This contradicts the important corollary to Aristotle's axiom in Chapter 3.

Since Archimedes' axiom implies Aristotle's (Exercise 2, Chapter 5), *a minimal H-plane must also be maximal*. Dehn's non-Legendrean H-plane is maximal but does not satisfy Aristotle's axiom because it satisfies the obtuse angle hypothesis (non-obtuse-angle theorem, Chapter 4).

Using those notions, we can describe Schur's model: Let F be a non-Archimedean Euclidean field (Project 2, Chapter 4) and let Π' be the Klein model inside F^2, which is a hyperbolic plane. Since Aristotle's axiom holds in any hyperbolic plane (Exercise 13, Chapter 6), it holds in Π'. Define Π to be the full subplane whose points are the ones in Π' that have *infinitesimal* coordinates. Then Π is an H-plane in which the acute angle hypothesis is satisfied, but it is not maximal, so Hilbert's hyperbolic axiom of parallelism fails (that can easily be seen directly). Here $k = -1$.

Dehn's examples are similar. For his semi-Euclidean plane Π for which Hilbert's Euclidean axiom of parallelism fails, $\Pi' = F^2$ with F as above, and Π is the full subplane whose points are the ones in Π' that have infinitesimal coordinates $(k = 0)$. For his non-Legendrean plane, the points of Π are again the points of F^2 with infinitesimal coordinates, but now Π is given the structure of H-plane for which $k = 1$ (it is not an H-subplane of F^2).

Another unusual class of H-planes are the ones in which *any two parallel lines have exactly one common perpendicular*—call them *HE-planes* (*halb-elliptisch*). They arise only over Pythagorean fields F that are not Euclidean fields (so line-circle continuity fails for them), and they can satisfy either the acute or the obtuse angle hypothesis. Schur's example is not an HE-plane since its lines, which are asymptotically parallel in the larger plane Π', do not have a common perpendicular.

If, in addition, F is Archimedean (e.g., the smallest Pythagorean subfield of $\mathbb{R}$), the HE-plane must satisfy the acute angle hypothesis (by the Saccheri–Legendre theorem). An example of an Archimedean HE-plane satisfying the acute angle hypothesis can be found in Pejas (1961); there Π is the interior of a *virtual conic*—the affine equation above for the absolute has no solutions in F. Pejas takes F to be the intersection of the real closures of $\mathbb{Q}(\sqrt{2})$ under two different orderings. The constant k is positive in one ordering but negative in the other, which is why the absolute is empty. (For a similar example, see Hartshorne, Exercises 39.25–39.30. Of course, Aristotle's axiom holds in these Archimedean planes.)

The main result in my article is the advanced theorem mentioned in Chapter 6: *A non-Euclidean H-plane is hyperbolic if and only if it satisfies the line-circle continuity principle and Aristotle's axiom.* Here is the argument: If the plane Π is hyperbolic, we showed in Part I that its field F of ends is Euclidean and Π is isomorphic to a Poincaré model over F, in which the line-circle and circle-circle continuity principles hold. Exercise 13, Chapter 6, shows that Aristotle's axiom also holds. Conversely, line-circle continuity implies that the field F in Pejas' classification is Euclidean, and since the plane Π is non-Euclidean, its metric constant k can be taken to be ± 1. Aristotle's axiom eliminates the case $k = 1$ (by the non-obtuse-angle theorem, Chapter 4, or by another argument in my article using the important corollary to Aristotle's axiom and the uniformity theorem of Chapter 4). Pejas' classification then tells us that Π is isomorphic to a full subplane of the Klein model over

F. But Aristotle's axiom ensures that Π is maximal, so it is isomorphic to that Klein model, which is hyperbolic. ◄

Conclusion

We have come to the end of our long journey toward understanding the role of the parallel postulate in elementary geometry. Hilbert planes are our models for such geometry without a parallel postulate. They have all been determined by Pejas by means of an embedding in a projective plane coordinatized by an ordered Pythagorean field F.

In order to prove all the results in Euclid that are independent of his parallel postulate, one needs to assume the line-circle and circle-circle continuity principles as well. Line-circle continuity guarantees that the field F is Euclidean and eliminates the HE-planes (among others) from Pejas' list. But unusual H-planes such as Schur's and Dehn's remain on the list, including H-planes satisfying the obtuse angle hypothesis. They are eliminated by assuming Aristotle's axiom, which reduces us to the classical Euclidean and hyperbolic planes coordinatized by F but which allows the non-Archimedean planes first studied by Veronese, since F might be a non-Archimedean Euclidean field.[4]

Real numbers have been shown to be unnecessary for this study of elementary geometry. The methods needed to fully develop the subject without them are admittedly sophisticated. The counterexamples exhibited in that development revealed the impossibility of proving various geometric statements. Here is what Hilbert had to say, in the conclusion of his *Grundlagen*, about such a study:

> The present treatment is a critical investigation of the principles of geometry. In this investigation, the ground rule was to discuss every question that arises in such a way as to find out at the same time whether it can be answered in a specified way with some limited means. This ground rule seems to me to contain a general and natural guide-line. . . .

---

[4] Philip Ehrlich has written a fascinating history called "The Rise of Non-Archimedean Mathematics and the Roots of a Misconception. I: The Emergence of Non-Archimedean Systems of Magnitudes"—see http://www.springerlink.com/content/bp78g22780212561/fulltext.pdf or the printed version in *Archives of the History of Exact Sciences*, **60** (2006), 1–121. Cantor, whose infinite cardinal and ordinal numbers were so revolutionary, resisted the idea that infinitesimals could be mathematically acceptable.

The *impossibility* of certain solutions and problems thus plays a prominent role in modern mathematics and the drive to answer questions of this type was oftentimes the cause for the discovery of new and fruitful areas of investigation. Recall only Abel's proof of the impossibility of solving the fifth degree equation by radicals, the realization of the impossibility of proving the axiom of parallels, and Hermite and Lindemann's theorems on the impossibility of constructing the numbers e and π algebraically.

The ground rule according to which the principles of the possibility of a proof should be discussed at all is very intimately connected with the requirement for the "purity" of the methods of proof which has been championed by many mathematicians with great emphasis. This requirement is basically none other than a subjective form of the ground rule followed here. In fact, the present investigation seeks to uncover which axioms, hypotheses, or aids are necessary for a proof of a fact in elementary geometry. . . .

Axioms

Incidence Axioms

AXIOM I-1. For every point P and for every point Q not equal to P there exists a unique line l incident with P and Q.

AXIOM I-2. For every line l there exist at least two distinct points that are incident with l.

AXIOM I-3. There exist three distinct points with the property that no line is incident with all three of them.

Betweenness Axioms

AXIOM B-1. If A * B * C, then A, B, and C are three distinct points all lying on the same line, and C * B * A.

AXIOM B-2. Given any two distinct points B and D, there exist points A, C, and E lying on $\overleftrightarrow{BD}$ such that A * B * D, B * C * D, and B * D * E.

AXIOM B-3. If A, B, and C are three distinct points lying on the same line, then one and only one of the points is between the other two.

AXIOM B-4. For every line l and for any three points A, B, and C not lying on l:

(i) If A and B are on the same side of l and if B and C are on the same side of l, then A and C are on the same side of l.

(ii) If A and B are on opposite sides of l and if B and C are on opposite sides of l, then A and C are on the same side of l.

Congruence Axioms

AXIOM C-1. If A and B are distinct points and if A' is any point, then for each ray r emanating from A' there is a *unique* point B' on r such that B' ≠ A' and AB ≅ A'B'.

AXIOM C-2. If AB ≅ CD and AB ≅ EF, then CD ≅ EF. Moreover, every segment is congruent to itself.

AXIOM C-3. If A * B * C, A' * B' * C', AB ≅ A'B', and BC ≅ B'C', then AC ≅ A'C'.

AXIOM C-4. Given any angle ⊀BAC (where by the definition of "angle" $\overrightarrow{AB}$ is not opposite to $\overrightarrow{AC}$), and given any ray $\overrightarrow{A'B'}$ emanating from a point A', then there is a *unique* ray $\overrightarrow{A'C'}$ on a given side of line $\overleftrightarrow{A'B'}$ such that ⊀B'A'C' ≅ ⊀BAC.

AXIOM C-5. If ⊀A ≅ ⊀B and ⊀A ≅ ⊀C, then ⊀B ≅ ⊀C. Moreover, every angle is congruent to itself.

AXIOM C-6 (SAS). If two sides and the included angle of one triangle are congruent, respectively, to two sides and the included angle of another triangle, then the two triangles are congruent.

Hilbert Plane Axioms

The incidence, betweenness, and congruence axioms above.

Continuity Axioms

DEDEKIND'S AXIOM. Suppose that the set $\{l\}$ of all points on a line l is the disjoint union $\Sigma_1 \cup \Sigma_2$ of two non-empty subsets such that no point of either subset is between two points of the other. Then there exists a unique point O on l such that one of the subsets is equal to a ray of l with vertex O and the other subset is equal to the complement.

ARCHIMEDES' AXIOM. If CD is any segment, A any point, and r any ray with vertex A, then for every point B $\neq$ A on r there is a number n such that when CD is laid off n times on r starting at A, a point E is reached such that $n \cdot$ CD $\cong$ AE and either B = E or B is between A and E.

ARISTOTLE'S AXIOM. Given any side of an acute angle and any segment AB, there exists a point Y on the given side of the angle such that if X is the foot of the perpendicular from Y to the other side of the angle, XY > AB.

CIRCLE-CIRCLE OR CIRCULAR CONTINUITY PRINCIPLE. If a circle γ has one point inside and one point outside another circle γ', then the two circles intersect in two points.

LINE-CIRCLE CONTINUITY PRINCIPLE. If a line passes through a point inside a circle, then the line intersects the circle in two points.

Parallelism Axioms

HILBERT'S EUCLIDEAN AXIOM OF PARALLELISM. For every line l and every point P not lying on l there is at most one line m through P such that m is parallel to l.

EUCLID'S FIFTH POSTULATE. If two lines are intersected by a transversal in such a way that the sum of the degree measures of the two interior angles on one side of the transversal is less than 180°, then the two lines meet on that side of the transversal.

HILBERT'S HYPERBOLIC AXIOM OF PARALLELISM. For every line l and every point P not on l, a limiting parallel ray $\overrightarrow{PX}$ emanating from P exists and it does *not* make a right angle with $\overrightarrow{PQ}$, where Q is the foot of the perpendicular from P to l.

Pythagorean Plane Axioms

Hilbert plane axioms plus Hilbert's Euclidean axiom of parallelism.

Euclidean Plane Axioms

Pythagorean plane axioms plus circle-circle continuity principle (line-circle continuity principle follows and for Pythagorean planes implies the circle-circle continuity principle).

Hyperbolic Plane Axioms

Hilbert plane axioms plus Hilbert's hyperbolic axiom of parallelism (equivalently, by the advanced theorem, plus the line-circle principle, Aristotle's axiom, and the negation of Hilbert's Euclidean axiom of parallelism).

Field Axioms/Definitions

A *field* is a set with two binary operations—multiplication and addition—and two distinguished elements 1 and 0 satisfying the following axioms:

$1 \neq 0$
$\forall x \ (1x = x \ \& \ 0 + x = x)$
$\forall x \forall y \ (xy = yx \ \& \ x + y = y + x)$
$\forall x \forall y \forall z \ (x(yz) = (xy)z \ \& \ x + (y + z) = (x + y) + z)$
$\forall x \ (x \neq 0 \Rightarrow \exists! y(xy = 1))$
$\forall x \ \exists! y \ (x + y = 0)$
$\forall x \forall y \forall z \ (x(y + z) = xy + xz).$

The field is *ordered* if it is also provided with a binary relation $<$ satisfying the axioms:

$\forall x \forall y \ (x < y \lor x = y \lor y < x)$ and exactly one of these three relations holds.

$0 < 1$

$\forall x \, \forall y \, \forall z \; (x < y \Rightarrow x + z < y + z)$

$\forall x \, \forall y \, \forall z \; (x < y \; \& \; 0 < z \Rightarrow xz < yz)$

An ordered field is *Euclidean* if it also satisfies the axiom

$\forall x \; (x > 0 \Rightarrow \exists y (x = y^2))$.

A field is *Pythagorean* if it also satisfies the axiom

$\forall x \, \exists y (y^2 = 1 + x^2)$.

The *constructible field* K is the field generated from 0 and 1 by the operations of addition, subtraction, multiplication, division by a nonzero number, and taking the square root of a positive number. In terms of the field of real numbers $\mathbb{R}$, it is the smallest Euclidean subfield of $\mathbb{R}$.

A field is *Archimedean* if for every $x > 0$, there is a natural number n such that $n \cdot 1 > x$.

Bibliography

Introductory and General

Chern, S.-S. 1990. "What Is Geometry?" *American Mathematical Monthly*, **97**(8):679–686.

Courant, R., and H. Robbins. 1941. *What Is Mathematics?* New York: Oxford University Press.

Gamow, G. 1956. "The Evolutionary Universe," *Scientific American*, **195**(September):136–154 (Offprint no. 211).

Hartshorne, R. 2000. "Teaching Geometry According to Euclid," *Notices of the American Mathematical Society*, **47**(4):460–465.

Hilbert, D., and S. Cohn-Vossen. 1952. *Geometry and the Imagination*, New York: Chelsea.

Kac, M., and S. M. Ulam. 1992. *Mathematics and Logic*, New York: Dover.

Kline, M. 1979. *Mathematics: An Introduction to Its Spirit and Use: Readings from Scientific American*, New York: W. H. Freeman and Company. (See especially the articles on geometry by Kline.)

Osserman, R. 1996. *Poetry of the Universe: A Mathematical Exploration of the Cosmos*, New York: Anchor.

History and Biography

Bell, E. T. 1969. "Father and Son, Wolfgang and Johann Bolyai," in *Memorable Personalities in Mathematics: Nineteenth Century*, Stanford, Calif.: School Mathematics Study Group.

———. 1986. *Men of Mathematics*, Carmichael, Calif.: Touchstone.

———. 1992. *The Development of Mathematics*, New York: Dover.

Bonola, R. 1955. *Non-Euclidean Geometry*, New York: Dover.

Bos, H. J. M. 1993. "'The Bond with Reality Is Cut'—Freudenthal on the Foundations of Geometry around 1900," *Educational Studies in Mathematics*, **25(1–2)**:51–58.

Boyer, C. B., and U. Merzbach. 1991. *A History of Mathematics*, 2nd ed., New York: Wiley.

Dantzig, T., J. Mazur, and B. Mazur. 2005. *Number: The Language of Science*, Masterpiece Science Edition, New York: Penguin.

Derbyshire, J. 2006. *Unknown Quantity: A Real and Imaginary History of Algebra*, Washington, D.C.: Joseph Henry Press.

Dodgson, C. L. (Lewis Carroll). 1890. *Curiosa Mathematica: A New Theory of Parallels*, London: Macmillan.

Dunnington, G. W. 1955. *Carl Friedrich Gauss: Titan of Science*, New York: Hafner.

Ehrlich, P. 2006. "The Rise of Non-Archimedean Mathematics and the Roots of a Misconception I: The Emergence of Non-Archimedean Systems of Magnitudes," *Archives of the History of Exact Sciences*, **60**:1–121.

Engel, F., and P. Stäckel. 1895. *Theorie der Parallellinien von Euklid bis auf Gauss*, Leipzig: Teubner.

Freudenthal, H. 1962. "The Main Trends in the Foundations of Geometry in the Nineteenth Century," in *Logic, Methodology and Philosophy of Science*, E. Nagel, P. Suppes, and A. Tarski, eds., Stanford, Calif.: Stanford University Press, pp. 613–621.

Gray, J. J. 1989. *Ideas of Space: Euclidean, Non-Euclidean, and Relativistic*, 2nd ed., New York: Oxford University Press.

———. 2000. *The Hilbert Challenge*, New York: Oxford University Press.

Hall, T. 1970. *C. F. Gauss: A Biography*, Cambridge, Mass: MIT Press.

Heath, T. L. 2003. *A Manual of Greek Mathematics*, New York: Dover.

Holme, A. 2002. *Geometry: Our Cultural Heritage*, New York: Springer.

Katz, V. J. 1998. *A History of Mathematics: An Introduction*, 2nd ed., Reading, Mass.: Addison-Wesley Longman.

Lanczos, C. 1970. *Space through the Ages*, New York: Academic Press.

Laugwitz, D. 2004. *Bernhard Riemann 1826–1866: The Turning Points in the Conception of Mathematics*, tr. A. Shenitzer, New York: Springer.

Livio, M. 2002. *The Golden Ratio: The Story of Phi*, New York: Broadway Books.

Maor, E. 1994. *e: The Story of a Number*, Princeton, N.J.: Princeton University Press.

———. 2007. *The Pythagorean Theorem: A 4,000-Year History*, Princeton, N.J.: Princeton University Press.

Nagel, E. 1939. "Formation of Modern Conceptions of Formal Logic in the Development of Geometry," *Osiris* (**7**):142–224.

Nahin, P. J. 1998. *An Imaginary Tale: The Story of $\sqrt{-1}$*, Princeton, N.J.: Princeton University Press.

O'Shea, D. 2007. *The Poincaré Conjecture*, New York: Walker.

Posamentier, A. S., and I. Lehmann 2004. *π: A Biography of the World's Most Mysterious Number*, Amherst, N.Y.: Prometheus Books.

Reid, C. 1970. *Hilbert*, New York: Springer-Verlag.

Roberts, S. 2006. *King of Infinite Space: Donald Coxeter, the Man Who Saved Geometry*, New York: Walker.

Rosenfeld, B. A. 1988. *A History of Non-Euclidean Geometry*, tr. A. Shenitzer, New York: Springer-Verlag.

Schmidt, F., and P. Stäckel. 1972. *Briefwechsel zwischen C. F. Gauss und W. Bolyai*, New York: Johnson Reprint Corporation.

Stäckel, P. 1913. *Wolfgang und Johann Bolyai, Geometrische Untersuchungen*, 2 vols., Leipzig: Teubner.

Stein, S. 1999. *Archimedes: What Did He Do Besides Cry Eureka?* Washington, D.C.: Mathematical Association of America.

Stillwell, J. 1996. *Sources of Hyperbolic Geometry*, Providence, R.I.: American Mathematical Society.

——. 2002. *Mathematics and Its History*, 2nd ed., New York: Springer.

Van der Waerden, B. L. 1961. *Science Awakening*, New York: Oxford University Press.

Voelke, J.-D. 2005. *Renaissance de la géométrie non euclidienne entre 1860 et 1900*, Bern, Switzerland: Peter Lang.

Yaglom, I. M. 1988. *Felix Klein and Sophus Lie: Evolution of the Idea of Symmetry in the Nineteenth Century*, Boston: Birkhäuser.

Zebrowski, E. 2000. *A History of the Circle*, New Brunswick, N.J.: Rutgers University Press.

Philosophical

DeLong, H. 2004. *A Profile of Mathematical Logic*, New York: Dover.

Grünbaum, A. 1968. *Geometry and Chronometry in Philosophical Perspective*, Bloomington: University of Minnesota Press.

Hadamard, J. 1945. *The Psychology of Invention in the Mathematical Field*, Princeton, N.J.: Princeton University Press.

Hardy, G. H. 1940. *A Mathematician's Apology*, New York: Cambridge University Press.

Hempel, C. G. 1945. "Geometry and Empirical Science," *American Mathematical Monthly,* **52**:7–17; also in vol. 3, *World of Mathematics,* J. R. Newman, ed., pp. 1619–1646.

Meschkowski, H. 1965. *Evolution of Mathematical Thought,* San Francisco: Holden-Day.

Poincaré, H. 1952. *Science and Hypothesis,* New York: Dover. Originally published in French, 1902.

Polanyi, M. 1964. *Personal Knowledge,* New York: Harper and Row.

Renyi, A. 1967. *Dialogues on Mathematics,* San Francisco: Holden-Day.

Stein, H. 1970. "On the Paradoxical Time-Structures of Gödel," *Journal of the Philosophy of Science,* **37**:589ff.

Torretti, R. 2001. *Philosophy of Geometry from Riemann to Poincaré,* New York: Springer.

Zeeman, E. C. 1974. "Research, Ancient and Modern," *Bulletin of the Institute of Mathematics and Its Applications,* Warwick University, England, **10**:272–281.

Mathematically Elementary to Moderate

Artzy, R. 1965. *Linear Geometry,* Reading, Mass.: Addison-Wesley.

Brannan, D. A., M. F. Esplen, and J. J. Gray. 1999. *Geometry,* New York: Cambridge University Press.

Bos, Henk J. M. 2001. *Redefining Geometrical Exactness: Descartes' Transformation of the Early Modern Concept of Construction,* New York: Springer.

Coxeter, H. S. M. 1992. *The Real Projective Plane,* 3rd ed., New York: Springer.

——. 2001. *Introduction to Geometry,* 2nd ed. (paperback), New York: Wiley.

——. 2003. *Projective Geometry,* 2nd ed., New York: Springer.

Coxeter, H. S. M., and S. L. Greitzer. 1967. *Geometry Revisited,* New York: Random House.

Descartes, R. 1954. *The Geometry of René Descartes,* in French and English, New York: Dover.

Euclid. 1956. *Thirteen Books of the Elements,* 3 vols., tr. T. L. Heath, with annotations, New York: Dover.

Eves, H. 1972. *A Survey of Geometry,* revised ed., Boston: Allyn and Bacon.

Fetisov, A. I., and Y. S. Dubnov. 2006. *Proof in Geometry and Mistakes in Geometric Proofs,* New York: Dover.

Gray, J. J. 2004. *János Bolyai: Non-Euclidean Geometry and the Nature of Space*, Cambridge, Mass.: MIT Press. Informative review by Robert Osserman, *Notices of the American Mathematical Society*, **52**(9)October 2005:1030–1034.

Hallett, M., and U. Majer, eds. 2004. *David Hilbert's Lectures on the Foundations of Geometry 1891–1902*, New York: Springer.

Hartshorne, R. 1967. *Foundations of Projective Geometry*, Menlo Park, Calif.: Benjamin Cummings.

——. 2003. "Non-Euclidean III.36," *American Mathematical Monthly*, **110**(6):495–502.

Heilbron, J. L. 1998. *Geometry Civilized: History, Culture and Technique*, Oxford: Clarendon.

Henderson, D., and D. Taimina. 2005. *Experiencing Geometry*, 3rd ed., Upper Saddle River, N.J.: Pearson Prentice Hall.

Hessenberg, G., and J. Diller. 1967. *Grundlagen der Geometrie*, Berlin: de Gruyter.

Jagy, W. C. 1995. "Squaring Circles in the Hyperbolic Plane," *Mathematical Intelligencer*, **17**(2):31–36.

Jones, A., A. Morris, and K. Pearson. 1991. *Abstract Algebra and Famous Impossibilities*, New York: Springer-Verlag.

Kay, D. C. 1969. *College Geometry*, New York: Holt, Rinehart and Winston.

Klein, F. 2004. *Geometry*, part 2 of *Elementary Mathematics from an Advanced Standpoint*, tr. C. Noble, E. Hedrick, New York: Dover.

——. 2007. *Famous Problems of Elementary Geometry*, tr. D. Smith, W. Berman, New York: Dover.

Kleven, D. J. 1978. "Morley's Theorem and a Converse," *American Mathematical Monthly*, **85**:100–105.

Luneburg, R. K. 1947. *Mathematical Analysis of Binocular Vision*, Princeton, N.J.: Princeton University Press.

Martin, G. E. 1982. *The Foundations of Geometry and the Non-Euclidean Plane*, New York: Springer-Verlag.

——. 1997. *Geometric Constructions*, New York: Springer.

Meschkowski, H. 1964. *Noneuclidean Geometry*, New York: Academic Press.

Mumford, D., C. Series, and D. Wright. 2002. *Indra's Pearls: The Vision of Felix Klein*, New York: Cambridge University Press.

Pedoe, D. 1988. *Geometry: A Comprehensive Course*, New York: Dover.

Peressini, A. L., and D. R. Sherbert. 1971. *Topics in Modern Mathematics for Teachers*, New York: Holt.

Posamentier, A. S. 2002. *Advanced Euclidean Geometry*, Emeryville, Calif.: Key.

Proclus. 1992. *A Commentary on the First Book of Euclid's Elements*, reprint ed., tr. G. R. Morrow, with notes and introduction, Princeton, N.J.: Princeton University Press.

Reynolds, W. 1992. "Hyperbolic Geometry on a Hyperboloid," *American Mathematical Monthly*, **100**(5):442–455.

Saccheri, G. 1986. *Girolamo Saccheri's Euclides Vindicatus*, 2nd ed., tr. G. B. Halsted, New York: Chelsea.

Stahl, S. 1993. *The Poincaré Half-Plane: A Gateway to Modern Geometry*, Boston: Jones and Bartlett.

Stewart, Ian. 2007. *Why Beauty Is Truth: A History of Symmetry*, New York: Perseus.

Sved, M. 1991. *Journey into Geometries*, Washington, D.C.: Mathematical Association of America.

Taylor, E. F., and J. A. Wheeler. 1992. *Spacetime Physics*, 2nd ed., New York: W. H. Freeman and Company.

Trudeau, R. J. 1987. *The Non-Euclidean Revolution*, Boston: Birkhäuser.

Weeks, J. R. 2002. *The Shape of Space*, 2nd ed., New York: Marcel Dekker.

Weyl, H. 1952. *Symmetry*, Princeton, N.J.: Princeton University Press.

Yaglom, I. M. 1979. *A Simple Non-Euclidean Geometry and Its Physical Basis*, tr. A. Shenitzer, New York: Springer-Verlag.

Mathematically Advanced

Artin, E. 1957. *Geometric Algebra*, New York: Wiley.

Bachmann, F. 1973. *Aufbau der Geometrie aus dem Spiegelungsbegriff*, New York: Springer-Verlag.

Bachmann, F., H. Behnke, K. Fladt, and H. Kunle, eds. 1974. *Geometry*, tr. S. H. Gould, vol. 2 of *Fundamentals of Mathematics*, Cambridge, Mass.: MIT Press.

Beardon, A. F. 1983. *The Geometry of Discrete Groups*, New York: Springer.

Benedetti, R., and C. Petronio. 1992. *Lectures on Hyperbolic Geometry*, New York: Springer.

Blumenthal, L. M., and K. Menger. 1970. *Studies in Geometry*, New York: W. H. Freeman and Company.

Borsuk, K., and W. Szmielew. 1960. *Foundations of Geometry*, Amsterdam: North-Holland.

Casson, A., and S. Bleiler. 1988. *Automorphisms of Surfaces after Nielsen and Thurston*, New York: Cambridge University Press.

Coxeter, H. S. M. 1998. *Non-Euclidean Geometry*, 6th ed., Washington, D.C.: Mathematical Association of America.

Diamond, F., and J. Shurman. 2006. *A First Course in Modular Forms*, New York: Springer.

Do Carmo, M. 1976. *Differential Geometry of Curves and Surfaces*, Englewood Cliffs, N.J.: Prentice-Hall.

——. 1992. *Riemannian Geometry*, tr. F. Flaherty, Boston: Birkhäuser.

Einstein, A. 2006. *Relativity: The Special and the General Theory*, Masterpiece Science (paperback), with supplements by R. Penrose, R. Geroch, and D. Cassidy, New York: Plume.

Ewald, G. 1971. *Geometry: An Introduction*, Belmont, Calif.: Wadsworth.

Eymard, P., and J.-P. Lafon. 2004. *The Number* π, tr. S. S. Wilson, Providence, R.I.: American Mathematical Society.

Fenchel, W. 1989. *Elementary Geometry in Hyperbolic Space*, New York: de Gruyter.

Fenchel, W., J. Nielsen, and A. L. Schmidt. 2002. *Discontinuous Groups of Isometries in the Hyperbolic Plane*, New York: de Gruyter.

Gallo, D. M., and R. M. Porter, eds. 1987. *Analytical and Geometric Aspects of Hyperbolic Space*, New York: Cambridge University Press.

Gray, A., E. Abbena, and S. Salamon. 2006. *Modern Differential Geometry of Curves and Surfaces with Mathematica*, 3rd ed., Boca Raton, Fla.: Chapman and Hall/CRC.

Greenberg, M. J. 1979. "On J. Bolyai's Parallel Construction," *Journal of Geometry*, **12(1)**:45–64.

——. 1988. "Aristotle's Axiom in the Foundations of Hyperbolic Geometry," *Journal of Geometry* **33**:53–57.

Greenberg, M. J., and J. Harper. 1982. *Algebraic Topology: A First Course*, New York: Perseus.

Hartshorne, R. 2000. *Geometry: Euclid and Beyond*, New York: Springer.

Hartshorne, R., and R. Van Luijk. 2006. "Non-Euclidean Pythagorean Triples, A Problem of Euler and Rational Points on K3 Surfaces," *arXiv:math* AG/0606700, **1**:1–11.

Helgason, S. 2001. *Differential Geometry, Lie Groups, and Symmetric Spaces*, Providence, R.I.: American Mathematical Society.

Henkin, L., P. Suppes, and A. Tarski, eds. 1959. *The Axiomatic Method*, Amsterdam: North-Holland.

Hilbert, D. 1988. *Foundations of Geometry*, 2nd English ed., tr. L. Unger from 10th German ed., revised and enlarged by P. Bernays, Peru, Ill.: Open Court division of Carus.

Hubbard, J. H. 2006. *Teichmüller Theory and Applications to Geometry, Topology, and Dynamics*, vol. I, Ithaca, N.Y.: Matrix Editions.

Klein, F. 1968. *Vorlesungen über Nicht-Euklidische Geometrie,* Berlin: Springer-Verlag.

Lenz, H. 1967. *Nichteuklidische Geometrie,* Mannheim: Bibliographisches Institut.

McCleary, J. 1994. *Geometry from a Differentiable Viewpoint,* New York: Cambridge University Press.

Milnor, J. 1982. "Hyperbolic Geometry—The First 150 Years," *Bulletin of the American Mathematical Society* **6**:9–24.

Moise, E. E. 1990. *Elementary Geometry from an Advanced Standpoint,* 3rd ed., Reading, Mass.: Addison-Wesley.

O'Neill, B. 2006. *Elementary Differential Geometry,* revised 2nd ed., New York: Academic Press.

Pambuccian, V. 2001. "Fragments of Euclidean and Hyperbolic Geometry," *Scientiae Mathematicae Japonicae* **53**:361–400.

———. 2005. "Euclidean Geometry Problems Rephrased in Terms of Midpoints and Point Reflections," *Elemente der Mathematik,* **60(1)**:19–24.

Pejas, W. 1961. "Die Modelle des Hilbertschen Axiomensystems der absoluten Geometrie," *Mathematische Annalen,* **143**:212–235.

Prékopa, A., and E. Molnár, eds. 2005. *Non-Euclidean Geometries: János Bolyai Memorial Volume,* New York: Springer. See in particular V. Pambuccian, "Axiomatizations of Hyperbolic and Absolute Geometries," pp. 119–153; J. Gray, "Gauss and Non-Euclidean Geometry," pp. 61–80; S. Helgason, "Non-Euclidean Analysis," pp. 367–384.

Ramsay, A., and R. D. Richtmyer. 1995. *Introduction to Hyperbolic Geometry,* New York: Springer.

Ratcliffe, J. G. 2006. *Foundations of Hyperbolic Manifolds,* 2nd ed., New York: Springer.

Schwerdtfeger, H. 1980. *Geometry of Complex Numbers.* New York: Dover.

Sharpe, R. W. 1997. *Differential Geometry: Cartan's Generalization of Klein's Erlangen Program,* New York: Springer.

Sommerville, D. M. Y. 1958. *Elements of Non-Euclidean Geometry,* New York: Dover.

Spivak, M. 1999. *A Comprehensive Introduction to Differential Geometry,* 5 vols., Waltham, Mass.: Publish or Perish Press.

Stillwell, J. 1992. *Geometry of Surfaces,* New York: Springer.

Thurston, W. P. 1997. *Three-Dimensional Geometry and Topology,* Princeton, N.J.: Princeton University Press.

Toth, L. F. 1964. *Regular Figures,* New York: Macmillan.

Wolfe, H. E. 1945. *Introduction to Non-Euclidean Geometry,* New York: Holt, Rinehart and Winston.

Yaglom, I. M. 1962. *Geometric Transformations,* 3 vols., Washington, D.C.: Mathematical Association of America.

Symbols

P ɪ l	P is incident with l (p. 12)
$\overline{AB}$	real number length of segment AB in an Archimedean Hilbert plane (p. 170)
$\{l\}$	the set of points lying on line l (p. 14)
AB	segment with endpoints A and B (p. 16)
$\overleftrightarrow{PQ}$	line through P and Q (p. 16)
$\cong$	congruent (p. 16)
K	the field of constructible real numbers (p. 141)
$\overrightarrow{AB}$	ray emanating from A through B (p. 18)
$\sphericalangle A$	angle with vertex A (p. 19)
$l \parallel m$	line l parallel to line m (p. 20)
$l \perp m$	line l perpendicular to line m (p. 23)
$\mathbb{Q}$	the field of rational numbers (p. 86)
$\mathbb{R}$	the field of real numbers (p. 86)
$\mathbb{C}$	the field of complex numbers (p. 86)
$\triangle ABC$	triangle with vertices A, B, C (p. 25)
$\square ABCD$	quadrilateral with successive vertices A, B, C, D (p. 44)
$S \cup T$	union of S and T (p. 45)
$S \cap T$	intersection of S and T (p. 45)
$H \Rightarrow C$	statement H implies statement C (p. 55)
RAA	reductio ad absurdum method of proof (p. 58)
$\mathbb{N}$	the infinite set of natural numbers (p. xxx)
$\sim S$	negation of statement S (p. 60)
$\sim$	equivalence relation (p. 83)
$\mathscr{A}^*$	projective completion of $\mathscr{A}$ (p. 83)
$A * B * C$	point B is between points A and C (p. 108)

$AB < CD$	segment AB is smaller than segment CD (p. 124)
$\sphericalangle ABC < \sphericalangle DEF$	angle ABC is smaller than angle DEF (p. 128)
$n \cdot CD$	segment CD laid off n times (p. 132)
R_m	reflection across line m (pp. 153, 337)
$\lvert AB \rvert$	length of segment AB (p. 155)
$(\sphericalangle A)^\circ$	number of degrees in angle A (p. 169)
$\triangle DEF \sim \triangle ABC$	triangle ABC is similar to triangle DEF (p. 215)
δABC	defect of triangle ABC (p. 252)
$r \vert s$	ray r is limiting parallel to ray s (p. 257)
$\Pi(PQ)^\circ$	number of degrees in angle of parallelism associated to PQ (p. 260)
$[ABCD$	biangle ABCD (p. 276)
$A)(B$	open chord with endpoints A and B (p. 298)
$P(l)$	pole of the chord l (p. 308)
(AB, CD)	cross-ratio of ordered tetrad ABCD (p. 319)
$d(AB)$	Poincaré length of Poincaré segment AB (p. 320)
$d'(AB)$	Klein length of segment AB (p. 343)
$p(A)$	polar of the point A (p. 360)
I	the identity transformation (p. 399)
H_A	half-turn about A (p. 414)
$\mathscr{P}^1(K)$	projective line over K (p. 439)
$PGL(2, K)$	projective group over K (p. 440)
$PSL(2, \mathbb{R})$	the real projective special linear group (p. 441)
C_n	cyclic group of order n (p. 449)
D_n	dihedral group of order $2n$ (p. 451)
$(\sphericalangle A'AB')^r$	radian measure of $\sphericalangle A'AB'$ (p. 486)
$\sinh x$	hyperbolic sine of x (p. 488)
$\cosh x$	hyperbolic cosine of x (p. 488)
$\tanh x$	hyperbolic tangent of x (p. 488)
$k = 1$	normalization of the distance scale to be 1 (p. 489)
$\Pi(x)$	angle of parallelism in radians (p. 491)
x^*	complementary length to x (p. 501)
$(A, B \vert C, D)$	A and C separate B and D in an elliptic plane (p. 542)
ds	infinitesimal change in distance (p. 555)
K	Gaussian curvature of surface S (p. 567)
$\chi(S)$	Euler characteristic of surface S (p. 567)
g	genus of a surface (p. 567)
$\mu(AB)$	multiplicative length of a hyperbolic segment (p. 577)
$\tan(\theta/2)$	tangent half angle in hyperbolic plane (p. 578)
k	"curvature" constant of an H-plane (p. 592)

Name Index

Abel, N. H., 297, 400
Ackerman, Wilhelm, 68
Alembert, J. L. R. d', 225
Alexander the Great, 7
Apollonius, 7, 313, 333
Archimedes, 9, 33, 39
Archytas, 6
Aristophanes, 520
Aristotle, 5, 12, 16, 23, 67, 78
Artin, Emil, 52, 391, 591
Augustine, Saint, 275

Bachmann, F., 447, 588, 591
Banchoff, Tom, xxiii, 563
Baudhayana, 2
Bell, Eric Temple, 6n, 247, 394
Beltrami, Eugenio, xx, 247, 248, 292,
    293–297, 346, 484, 487
Bernays, Paul, 15, 68
Bessel, F. W., 244, 372
Birkhoff, G., 15
Bishop, Errett, xvii, 68, 377
Bolyai, Farkas, xxiii, 104, 239, 476,
    478
  attempted proof of parallel
    postulate, 229, 240
Bolyai, János
  angle of parallelism, 480
  constructions in hyperbolic plane,
    xxviii, 520–528
  discovery of non-Euclidean
    geometry, xxvii, 239–242, 247,
    290

letter to his father, 226
life and works, xx–xxiii, 239–242
limiting parallel ray construction,
    xiv, 258–259
squaring the circle, 520
Bonnet, O., 562, 566–570
Boole, George, 67
Boy, Werner, 563, 564
Brianchon, C.-L., 89, 100, 351
Brouwer, L. E. J., 68, 377, 381

Cantor, George, 16, 67, 77, 376, 377,
    378, 595n
Cardan, J., 277
Carroll, Lewis. *See* Dodgson, Charles
Cartan, E., 249
Cayley, Arthur, 85–86, 100, 247,
    248, 292, 370, 401–402, 487
Chern, S.-S., 559
Church, Alonzo, 68
Clairaut, Alexis-Claude, 219–221
Clavius, 177, 213, 254
Clifford, W. K., 374
Cohen, Paul J., 378n, 382
Commandino, 8
Courant, R., 376
Coxeter, H. S. M., xxv, 442n, 547

De Morgan, Augustus, 67
Dedekind, Richard, xx, 16, 67,
    134–138
Dehn, M., xix, 266, 573, 592
Desargues, Girard, 89, 98

613

Subject Index

617